T0306123

Whale-watching
Sustainable Tourism and Ecological Management

Within little more than a generation, whale-watching has been subject to global industrial development. It has been portrayed by destinations and business operators, and advocated by environmental groups, as a sustainable activity and an alternative to whaling. However, in recent years the sustainability of these activities has increasingly been questioned, as research shows that repeated disturbance by boat traffic can severely disrupt critical behaviours of cetaceans in the wild.

Bringing together contributions by international experts, this volume addresses complex issues associated with commercial whale-watching, sustainable development and conservation of the global marine environment. It highlights widely expressed concerns for the failure of policy, planning and management, and pinpoints both long-standing and emerging barriers to sustainable practice. Featuring numerous case studies, the book provides critical insights into the diverse socio-cultural, political, economic and ecological contexts of this global industry, highlighting the challenges and opportunities that arise along the pathways to sustainability.

James Higham is Professor of Tourism at the University of Otago, New Zealand, and Visiting Professor of Sustainable Tourism at the University of Stavanger, Norway. His research interests focus on various aspects of tourism and environmental change.

Lars Bejder is an Associate Professor at Murdoch University, Australia, and an Adjunct Associate Professor at Duke University, USA. His research interests include analysis and development of quantitative methods to evaluate complex animal social structures, evaluation of the impacts of human activity on cetaceans, and fundamental biology and ecology.

Rob Williams is a Canadian marine conservation biologist and a Marie Curie Senior Research Fellow with the Sea Mammal Research Unit at the University of St Andrews, Scotland. His research focuses on estimating wildlife abundance and distribution, and assessing impacts of human activities on behaviour and energetics of marine mammals.

Whale-watching

Sustainable Tourism and Ecological Management

Edited by

James Higham
University of Otago, New Zealand

Lars Bejder
Murdoch University, Western Australia

Rob Williams
University of St Andrews, Scotland, UK

CAMBRIDGE
UNIVERSITY PRESS

Shaftesbury Road, Cambridge CB2 8EA, United Kingdom

One Liberty Plaza, 20th Floor, New York, NY 10006, USA

477 Williamstown Road, Port Melbourne, VIC 3207, Australia

314–321, 3rd Floor, Plot 3, Splendor Forum, Jasola District Centre, New Delhi – 110025, India

103 Penang Road, #05–06/07, Visioncrest Commercial, Singapore 238467

Cambridge University Press is part of Cambridge University Press & Assessment, a department of the University of Cambridge.

We share the University's mission to contribute to society through the pursuit of education, learning and research at the highest international levels of excellence.

www.cambridge.org
Information on this title: www.cambridge.org/9780521195973

First published 2014
3rd printing 2016

A catalogue record for this publication is available from the British Library

Library of Congress Cataloging-in-Publication data
Whale-watching : sustainable tourism and ecological management / edited by James Higham, University of Otago, New Zealand, Lars Bejder, Murdoch University, Western Australia, Rob Williams, University of St Andrews, Scotland, UK.
   pages   cm
Includes bibliographical references and index.
ISBN 978-0-521-19597-3 (hardback)
1. Whale watching – Environmental aspects.   2. Dolphin watching – Environmental aspects.   3. Cetacea – Conservation.   I. Higham, James E. S.   II. Williams, Rob, 1972–
QL737.C4W435   2014
599.5072′3 – dc23   2013034910

ISBN   978-0-521-19597-3   Hardback

Additional resources for this publication at www.cambridge.org/9780521195973

To our partners – Linda, Michelle and Erin

And to Amy Samuels (1950–2008)

Contents

Colour plates will fall between pages 204 and 205.

Acknowledgements

As is commonly the case, we have incurred various debts of favour during the course of this book project. First and foremost, we have been most grateful for the strong support of Megan Waddington (Editor – Life Sciences) and Martin Griffith (Commissioning Editor – Life Sciences) at Cambridge University Press (United Kingdom). The enthusiasm and support of Megan and Martin were important to the completion of this book. The planning of this book was initiated by conversations that took place between James and Lars at the Australian Wildlife Tourism conference (2007) in Western Australia for which Prof. Ross Dowling (Edith Cowan University) and his conference organizing team are acknowledged. There followed significant input from Dr David Lusseau (University of Aberdeen), which considerably and positively influenced the scope and direction of this book. We acknowledge the investment of significant time and effort on the part of all chapter authors, who have provided their original and insightful empirical research contributions to this volume. We are also most grateful to our five invited experts who freely and willingly contributed their expertise to inform Chapter 2: Dr Randall Reeves (Chairman of the Cetacean Specialist Group of the International Union for the Conservation of Nature), Dr Barbara Taylor (Marine Mammal Genetics Group at NOAA's Southwest Fisheries Science Center), Associate Professor Liz Slooten (University of Otago), Dr Christine Erbe (Director of the Centre for Marine Science & Technology, Curtin University) and Dr Brandon Southall (Senior

Scientist for Southall Environmental Associates). We gratefully acknowledge the research assistance provided by Debbie Hopkins and Wiebke Finkler (University of Otago) and the administrative support provided by Jo O'Brien throughout the course of this project. Their help and support greatly eased progress to completion of this volume.

This book was a particular focus of James' attention during a period of study leave in 2012. During this time, much of which was spent in the United Kingdom, he received the great support of colleagues, family and friends including Polly and Charles Higham, Tom Higham and Katerina Douka, Emma Holt, Caroline Orchiston, Richard and Jane Higham, John and Jo Higham, Rob Higham and Alpa Shah, Tim and Joan Higham, Ted Higham, Judy Bennett, Alex Armitage and Deepa Shah, Rob and Jacqui Askew, Dave Duffus, Phil Dearden, Tom and Lorraine Hinch, Scott Cohen, Stefan Gössling, Paul Peeters, Bernard Lane, Jan Vidar Haukeland, Odd Inge Vistad, Kreg Lindberg, Truls Engstrøm and Åsa Grahn, Christina Cavaliere, Debbie Hopkins, Wiebke Finkler, Janet Hoek, Dorian Owen, Liz and Fergus Moynihan, Tim and Claire Moynihan and Martin and Sarah Talks.

Lars expresses his gratitude to the people who have particularly inspired and influenced the course of his academic career: Steve Dawson, Dave Johnston, Neil Loneragan, David Lusseau, Peter Madsen, Ken Pollock, the late Amy Samuels and Hal Whitehead. Steve Dawson initially took Lars under his wing, introducing him to the world of marine science. This was despite his limited experience in this field. He is grateful for the leap of faith and early guidance extended to him. During his PhD years, Lars worked under the guidance of Hal Whitehead and Amy Samuels. While Hal has an unsurpassed capacity for envisioning 'the big picture', Amy had an unprecedented eye for the finer details. It was a privilege obtaining feedback from these two ends of the spectrum and an exciting and fruitful challenge balancing these approaches during his dissertation. Lars also thanks David Lusseau for endless and fruitful discussions on developing quantitative techniques for evaluating impacts of human activities on cetaceans – and for keeping him somewhat sane during the annual Scientific Committee meetings of the International Whaling Commission. Lars acknowledges Peter Madsen, his fellow Dane, for being a source of inspiration that pushes him to be a better scientist. Thanks to David Johnston and Ken Pollock for always being a great sounding board and source of encouragement. Lars is grateful to Neil Loneragan for his mentorship and friendship over the past six years. In some way, each of these people has had a role in the development and creation of this book. Above all, Lars acknowledges his wife, Michelle, his daughter, Melia, and the rest of his family, for their constant support and encouragement.

Rob attributes his career in marine mammal science to a lucky break that allowed him to conduct some of the first control-exposure experiments on wild whales, to measure behavioural responses of killer whales to whale-watching traffic. He expresses his gratitude to Andrew Trites, David Bain and John Ford for giving him that opportunity. He sustained a career in marine mammal science thanks to some excellent advice from his good friend and colleague, Alexandra Morton, who promised him that if he built his life in the path of the whales, the rest would fall into place. It has. Since then, Rob's research in this discipline has benefited from generous input from Erin Ashe, Lynne Barre, Ian Boyd, Kim Crosbie, Volker Deecke, Philip Hammond, David Lusseau, Dawn Noren, Randall Reeves, Meike Scheidat, Jodi Smith, Len Thomas and Terrie Williams. Rob's interest in anthropogenic impacts on cetaceans led naturally to research on acoustics, and he is grateful to Chris Clark, Christine Erbe, Mark Johnson, Alex MacGillivray, Dimitri Ponirakis, Peter Tyack and Vincent Janik for their guidance. During the development of this book, Rob had the good fortune of spending two years with the Sea Mammal Research Unit at the University of St Andrews on a Marie Curie International Incoming Fellowship, under the supervision of Phil Hammond and Len Thomas. Rob thanks his colleagues at SMRU for many interesting discussions about how to quantify the subtle impacts of repeated disturbance to cetaceans, and more importantly, to propose solutions. He is grateful that Len and Phil saw this book as a complement to the themes of his research fellowship. Mostly, he thanks Erin Ashe, who shares his philosophy of building a life around work that gives meaning.

Last, but by no means least, this book was made possible by the great support of our immediate families; Linda Buxton, Ali, Katie and George Higham; Michelle and Melia; and Erin Ashe.

James Higham
Dunedin, New Zealand

Lars Bejder
Perth, Australia

Rob Williams
St Andrews, Scotland

Contributors

Simon J. Allen

Research Fellow, Murdoch University Cetacean Research Unit, School of Veterinary and Life Sciences, Murdoch University, Western Australia, Australia

Tommy D. Andersson

School of Business, Economics and Law, University of Gothenburg, Sweden

David E. Bain

Friday Harbor Laboratories, University of Washington, WA, USA

Isabel Beasley

School of Earth and Environmental Studies, James Cook University, Townsville, Australia

Lars Bejder

Cetacean Research Unit, Centre for Fish, Fisheries and Aquatic Ecosystems Research, School of Veterinary and Life Sciences, Murdoch University, Western Australia, Australia

Guy Cantin

Fisheries and Oceans Canada, Oceans Management Division, Mont-Joli (Québec), Canada

Carole Carlson

Provincetown Center for Coastal Studies; Dolphin Fleet of Provincetown, Provincetown, MA, USA

Clément Chion
Department of Geography, Université de Montréal, Complex Systems Lab, Montréal (Québec), Canada

Fredrik Christiansen
Centre for Integrative Ecology, School of Life and Environmental Sciences, Deakin University, Warrnambool, Victoria, Australia

Rochelle Constantine
School of Biological Sciences, University of Auckland, Auckland, New Zealand

Peter Corkeron
Bioacoustics Research Program, Cornell Lab of Ornithology, Ithaca, NY, USA

Suzan Dionne
Parks Canada, Natural Resource Conservation, Haute-Ville, Québec, Canada

Wiebke Finkler
Department of Tourism, University of Otago, Dunedin, New Zealand

Paul Forestell
Long Island University, New York, NY, USA

Susanna E. Gothall
University of Gothenburg, Sweden

James Higham
Department of Tourism, University of Otago, Dunedin, New Zealand

Erich Hoyt
Whale and Dolphin Conservation Society, North Berwick, Scotland, UK

Genevieve Johnson
Zoos Victoria, Werribee South, Victoria, Australia

David W. Johnston
Duke University Marine Laboratory, Division of Marine Science and Conservation, Nicholas School of the Environment, Duke University, Beaufort, NC, USA

Hidehiro Kato
Cetacean & Marine Mammals, Laboratory of Cetacean Biology, Faculty of Marine Science, Tokyo University of Marine Science and Technology (TUMSAT), Tokyo, Japan

John N. Kittinger
Center for Ocean Solutions, Woods Institute for the Environment, Stanford University, Monterey, CA, USA

Philippe Lamontagne
National Research Council Canada, Research & Development, Ottawa, Ontario, Canada

Jacques-André Landry
l'École de Technologie Supérieure, University of Québec, Montreal, Canada

Neil Loneragan
Centre for Fish, Fisheries and Aquatic Ecosystems Research, School of Veterinary and Life Sciences, Murdoch University, Western Australia, Australia

David Lundquist
Department of Anatomy, University of Otago, Dunedin, New Zealand

David Lusseau
University of Aberdeen, School of Biological Sciences, Aberdeen, UK

Cynde McInnis
Cape Ann Whale Watch, Gloucester, MA, USA

Kepā Maly
Kumu Pono Associates LLC, Hawai'i, USA

Danielle Marceau
Department of Geomatics Engineering, University of Calgary, Canada

Helene Marsh
School of Earth and Environmental Studies, James Cook University, Townsville, Australia

Cristiane C.A. Martins
Department of Geography, Université de Montréal,
Complex Systems Lab, Montréal (Québec), Canada

Naomi McIntosh
US Office of National Marine Sanctuaries, Honolulu,
Hawai'i, USA

Nadia Ménard
Parks Canada, Saguenay St-Lawrence Marine Park,
Tadoussac (Québec), Canada

Robert Michaud
Group for Research and Education on Marine
Mammals (GREMM), Québec, Canada

Sue Muloin
Adult Learner Social Inclusion Project, Griffith
University, Australia

Mark Orams
New Zealand Tourism Research Institute, AUT
University, Auckland, New Zealand

Lael Parrott
Department of Geography, Université de Montréal,
Complex Systems Laboratory, Montréal (Québec),
Canada

E.C.M. Parsons
Department of Environmental Science and Policy,
George Mason University, Fairfax, USA and University
Marine Biological Station Millport, University of
London, UK

Marianne Rasmussen
The University of Iceland's Research Center in Húsavik,
Húsavík, Iceland

Naomi A. Rose
Humane Society International

David G. Simmons
Faculty of Environment Society and Design, Lincoln
University, Christchurch, New Zealand

Jonathon Spring
New Zealand Tourism Research Institute, AUT
University, Auckland, New Zealand

Andrew W. Trites
Marine Mammal Research Unit, Fisheries Centre,
University of British Columbia, Vancouver, Canada

Samuel Turgeon
Department of Geography, Université de Montréal,
Complex Systems Laboratory, Montréal (Québec),
Canada

Julian Tyne
Centre for Fish, Fisheries and Aquatic Ecosystems
Research, School of Veterinary and Life Sciences,
Murdoch University, Western Australia, Australia

Beatrice D. Wende
University of Gothenburg, Sweden

Rob Williams
Sea Mammal Research Unit, School of Biology,
University of St Andrews, Fife, Scotland, UK

Heather Zeppel
Australian Centre for Sustainable Business &
Development, University of Southern Queensland,
Australia

Abbreviations

ABM	agent-based model
AKWIC	Antarctic Killer Whale International Catalogue
CAWW	Cape Anne Whale Watch
CITES	Convention on International Trade in Endangered Species of Wild Fauna and Flora
CRRU	Cetacean Research and Rescue Unit
DOC	(New Zealand) Department of Conservation
DMM	decision-making model
DSP	Dolphin Space Program
EBFM	Ecosystem-Based Fishery Management
EBM	ecosystem-based management
GIS	geographic information system
HWDT	Hebridean Whale and Dolphin Trust
IBM	individual-based model
ICES	International Council for the Exploration of the Sea
IFAW	International Fund for Animal Welfare
ITIA	Icelandic Travel Industry Association
IUCN	International Union for Conservation of Nature
IWC	International Whaling Commission
LAC	Limits of Acceptable Change
LaWE	Large-scale Whale Watching Experiment
LEK	local ecological knowledge
MBL	mesopelagic boundary layer
MMC	Marine Mammal Commission
MMNB	minimization of the mean normalized bias
MMPA	Marine Mammal Protection Act
MMPR	Marine Mammals Protection Regulations

MNPL	Maximum Net Productivity Level	SBNMS	Stellwagen Bank National Marine Sanctuary
MPA	marine protected area		
MRI	Marine Research Institute	SMMRD	Santaurio de Mamiferos Marinos de la República Dominicana
NAMMCO	North Atlantic Marine Mammal Commission	SWG-WW	Standing Working Group on Whalewatching
NASS	North Atlantic Sighting Surveys		
NGO	non-government organizations	TEK	traditional ecological knowledge
NOAA	National Oceanic and Atmospheric Administration	UNWTO	United Nations World Tourism Organisation
ONMS	Office of National Marine Sanctuaries	WDCS	Whale and Dolphin Conservation Society
PBR	potential biological removal		
PCAD/PCoD	Population Consequences of Acoustic Disturbance	WWF	World Wildlife Fund
		KHR	kernel home-range
SAC	Special Area of Conservation	DLW	doubly labelled water

Tourism, cetaceans and sustainable development

Moving beyond simple binaries and intuitive assumptions

James Higham, Lars Bejder and Rob Williams

The majestic aspects of whales – their size; the apparent intelligence of some whales; the songs of others – led to rediscovery of the old iconography – whales as magnificent in their own right

(Corkeron, 2006: 161)

Introduction

Of the few iconic experiences available in the natural world, little compares to killer whales (*Orcinus orca*) outwitting and overpowering their prey, a cooperative group of feeding humpback whales (*Megaptera novaeangliae*), the flukes of a diving sperm whale (*Physeter macrocephalus*) or the spectacular aerial displays of socializing bottlenose dolphins (*Tursiops* sp.). These stunning megafauna experiences explain the widespread rejection of whale hunting and the phenomenal growth of whale-watching in recent decades (Hoyt, 2001; O'Connor *et al.*, 2009; Cisneros-Montemayor *et al.*, 2010). Under the circumstances, it is remarkable that these animals, and indeed all species of cetaceans (whales, dolphins and porpoises), vary so widely in the legal and management protection they receive in jurisdictions around the world.

Cetaceans trigger sentiments of awe, inspiration and excitement. 'Few creatures carry more emotion...than whales; and few issues arouse as much passion as whaling' (Hammond, 2006: 54).

These emotions and passions give rise to deeply entrenched and, at times, bitterly conflicting views on whales in terms of utility, identity, nationhood and sovereignty. The sinking of the *Ady Gil*, flagship of the Sea Shepherd Conservation Society, while protesting against the Japanese whaling fleet in the Southern Ocean whale sanctuary[1] in 2010 highlights these entrenched views. Few wildlife species are contested as intensely as whales.

It is not very long since whales were the focus of industrial-scale exploitation (Hammond, 2006), a practice that brought many populations of great whales dangerously close to extinction. Indeed, the closure of whale-processing factories and the discontinuation of commercial whaling has been due almost entirely to over-efficiency, resulting in 'economic extinction' (the depletion of stocks to the point of commercial non-viability; Hammond, 2006). It was this status of 'near obliteration' that gave momentum to the 'Save the Whales' global environmental movement in the 1970s; since then, whales have become the 'standard bearers of marine environmental issues' (Corkeron, 2006: 161). The shift from whale-hunting to whale-watching

---

[1] The Southern Ocean Whale Sanctuary was established by the IWC in 1994 to ban commercial whaling in the oceans surrounding Antarctica. Representing an area of approximately 50 million km^2, it generally exists to the south of latitude 40°S (in the Indian Ocean it adjoins the Indian Ocean Whale Sanctuary at 55°S; adjacent to the South American continent it exists from 60°S).

Whale-watching: Sustainable Tourism and Ecological Management, eds J. Higham, L. Bejder and R. Williams.
Published by Cambridge University Press. © Cambridge University Press 2014.

has not been universal. While the International Whaling Commission (IWC) adopted the moratorium on commercial whaling in 1982 (International Fund for Animal Welfare, 1995), and in 1993 'formally recognised whale-watching as a legitimate tourism industry which provided for the sustainable use of these animals' (Orams, 2000: 561), whale-hunting practices continue (see Chapter 6).

There are inherent tensions between different world views associated with whales. The efforts of environmental non-government organizations (NGOs) such as the International Fund for Animal Welfare (IFAW), Greenpeace and the World Wildlife Fund (WWF) have for many years steadfastly promoted whale-watching as an economically viable alternative to whaling. In 1987, for example, dedicated efforts in the Azores targeted attitudes towards the practice of sperm whale hunting, which continued in the Atlantic at that time. There followed a transition from a whale-hunting to a whale-watching economy in the Azores (Neves, 2010). The following year, IFAW documented and disseminated the feasibility of whale-watching in Iceland to establish an economically viable counter to the proposed resumption of whaling (O'Connor et al., 2009). Initial success and international support for whale-watching in Iceland has, however, been complicated by the resumption of whaling (Higham & Lusseau, 2008; see Chapters 7 and 8). Some find the processing of whale parts sufficient to justify a tourist boycott of Iceland (Williams, 2006). Others continue to support the development of a lucrative Icelandic whale-watching industry in the hope that this growing economic pressure will eventually serve as a countervailing force to the whaling industry.

The transition from whale-hunting to whale-watching seems logical when considered in terms of sustainable utility and marine conservation. However, the wheels of change have turned slowly. Three decades passed between the first whale-watching trips in California, USA in 1952 and the IWC whaling moratorium in 1982. Since that time, the public appetite for viewing cetaceans in the wild has become insatiable, growing itself to industrial scale. The transformation that was witnessed in the Azores

has become a global trend (Neves, 2010). Maritime communities have realized the considerable socio-economic benefits available through the development of whale-watching (Garrod & Fennell, 2004). Indeed, commercial whale-watching has become the economic lifeblood of many peripheral coastal regions. The case of Kaikoura (New Zealand), for example, is one of transformation from a depressed and decaying rural community to a thriving regional tourism economy built principally upon whale-watching. The Kaikoura story is also one of cultural renaissance (see Chapter 22). Indeed, whale-watching has become the economic mainstay of many coastal communities in both the developed (Hoyt, 2001; Hoyt & Hvenegaard, 2002) and developing worlds (Mustika et al., 2012a).

The socioeconomic benefits of whale-watching have sustained remarkable growth; 12% per annum growth in global whale-watch numbers throughout the 1990s coupled with increases in tourist expenditure of 18.6% per annum (Garrod & Fennell, 2004). From approximately 2 million whale-watchers in 1990, the industry grew to 9 million participants in 1999 (Hoyt, 2001). IFAW estimates that whale-watching is an industry that now exceeds $US 2.1 billion per annum, 13 million whale-watchers, generating 13,000 jobs (O'Connor et al., 2009). Such figures point to a rapidly changing socio-political and economic context, both globally and regionally, as communities and nations recognize and seek to pursue the economic benefits of whale-watching. With appropriate business models (Neves, 2010), these benefits are generally more equitably available at the community level than extractive practices (Herrera & Hoagland 2006; Parsons & Draheim, 2009).

Growth in whale-watching has matured in the developed world, but shows few signs of slowing in developing world contexts (Lusseau et al., 2013). Indeed, Cisneros-Montemayor et al. (2010) indicate that an additional $US 413 million and 5700 jobs could exist in the global whale-watching system, with much of this capacity available to developing world nations (Mustika et al., 2012b). This existing and latent capacity equates to a whale-watch industry of over $US 2.5 billion, supporting 19,000 jobs globally (Cisneros-Montemayor *et al.*, 2010: 1275). However, Kuo *et al.* (2012) demonstrate that achieving the latent capacity of the global whale-watch industry requires the discontinuation of commercial whale-hunting practices, observing that whaling reduces the capacity of the global whale-watch system, most particularly in those countries that continue to engage in whale-hunting (Higham & Lusseau, 2007). This is a line of debate that has been ignored in countries such as Iceland and Norway which seek to prove that hunting and watching whales are practices that can coexist.

This edited volume addresses the phenomenon of whale-watching, which we define as commercial tourist ventures including opportunities for people to observe, swim with, touch or feed wild cetaceans from shore, sea or air. The term whale-watching is used 'to denote a wide range of activities involving human interaction with various species of whales, dolphins, and porpoises, collectively known as cetaceans' (Garrod & Fennell, 2004: 335). From humble origins whale-watching has grown largely without restraint to industrial scale – bringing with it a host of planning and management challenges. Over 25 years of accumulated science demonstrates that human interactions with cetaceans can affect animal behaviour (Baker & Herman, 1989). However, without acceptance that altered behaviours could have broader biological and ecological consequences (Corkeron, 2006; Neves, 2010) whale-watching has continued to grow in the almost complete absence of regulatory and management frameworks (Higham *et al.*, 2009). Given the economic importance of whale-watching, it is remarkable that such disregard for sustainable management has so widely prevailed.

The disservice of binary debates and assumptions

There is little doubt that the search for sustainability has been hindered by the simplification of complex issues (see Neves, 2010). Despite the efforts of the scientific community, a range of binary debates have dominated whale-watching discourses. These binaries, we argue, have sustained a number of unhelpful assumptions that need to be more critically interrogated. In this chapter we outline five binaries, and question the misplaced assumptions that they may have perpetuated.

1. Whale-hunting is bad so whale-watching must be good

The groundswell of abhorrence towards the practice of whale-hunting perhaps dates to 1922 when, according to Corkeron (2006: 161), 'Sir Sidney Harmer of the British Museum described Norwegian whaling in British sub-Antarctic waters as "*insensate slaughter arousing feelings of horror and disgust*"' (emphases added). Since then, there has been an upwelling of general opinion that treating whales as extractive resources has 'cheapened humanity' (Corkeron, 2006), culminating in the 1970s in one of the first truly global socio-environmental movements (McCormick, 1989). Subsequently, whale-watching has come to be viewed as an extension of the great victories of the environmental movement of the later twentieth century, such as the moratorium on commercial whaling (1982) and the creation of the Southern Ocean Whale Sanctuary (1994). Experiencing whales in the wild, according to some, apparently equates to conservation of the marine environment (Neves, 2010).

However widespread these sentiments may be, they are not universal. Commercial whaling practices (based in some cases on claims of science and sustainable harvest) continue despite the uncertainty of current whale population estimates and the impossibility of achieving a high degree of scientific certainty in those estimates, in large part because the quota-setting algorithms are designed to take that uncertainty into account (Hammond, 2006). It could be argued that the emotional stakes of whale-hunting have led to management frameworks to account for uncertainty that have improved management procedures for fish stocks

(Hammond, 2006). Japan, Norway and Iceland have continued 'scientific' whale-hunting claiming the need for research to understand how whale populations 'interact' with 'other components of marine ecosystems, notably commercially important fish species' (Corkeron, 2006: 162). The sale of whale meat for commercial markets is a by-product of this science (in much the same way that some countries pay for fishery stock assessments by chartering fishing boats to do the surveys, and using fish catches to meet the costs of data collection). Coastal communities in parts of Scandinavia strongly resist the external imposition of views on whale-hunting, claiming autonomy and identity arising from the tradition of seasonal whale-hunting practices (Ris, 1993; Smested, 1997). Indigenous whale-hunting in the Arctic regions of North America arise as representations of cultural identity, self-governance and indigenous rights (Hinch, 1998).

The diversity of whale-hunting practices is mirrored by whale-watching. If the act of whale-hunting in any form is considered barbaric, it is intuitively appealing to consider whale-watching as (comparatively) benign. This may have been an acceptable position in the 1970s when whale-watching emerged as a viable alternative to extractive whale industries. It may also have been excusable in the 1980s when phenomenal growth in whale-watching, and other forms of 'ecotourism development' so conclusively advocated by the United Nations World Tourism Organisation (UNWTO) (Hall, 1994), profited from the 'green tourism' discourses of the day. However, Knight (2009: 180) observes that we now live 'in an age when our visual appetite for wildlife has never been greater'. Wildlife viewing, once the domain of dedicated enthusiasts, or 'specialists' (Duffus & Dearden, 1990), has moved into the mainstream of commercial tourism (Knight, 2009). With this has come a proliferation and diversification of opportunities to encounter wildlife (Higham *et al.*, 2008). Under the circumstances, uncritical treatment of 'whale-watching as good' does a disservice to the pursuit of sustainability.

This binary assumption conceals an inescapable tension. Knight (2009: 167) identifies a fundamental contradiction in wildlife viewing in that 'wild animals are generally human-averse; they avoid humans and respond to human encounters by fleeing and retreating to cover'. This tension has perhaps been overlooked on socioeconomic grounds, in much the same way that decisions are made about fishery by-catch or ship strikes (see Chapter 2). The regional politics of whale-watching has been driven by the economic development agenda, to the extent that efforts to adequately protect whales have at best been neglected and, at worst, resisted. Meanwhile, at the global level, the international politcs of whale-watching has been deliberately and consistently located in relation to the practices of whale-hunting (Neves, 2010). While the case for continued whale-hunting has been perpetuated on 'scientific' and socio-cultural grounds, the case for whale-watching has been stated in unitary terms, as an alternative to all forms of whale-hunting. Neves (2010: 719) critiques the 'monolithic presentation of whale-watching as the antithesis of whale hunting', arguing that the 'homogenized portrayal of whale-watching in mainstream environmental discourse as diametrically opposite to whale hunting…obscures the existence of bad whale-watching conduct'.

2. Industrial mass tourism is high in impact, so ecotourism must be relatively low impact

The view that ecotourism is a 'green' economic activity that is synonymous with wildlife conservation is longstanding. Since the 1960s, the term 'ecotourism' has been used to describe a benign form of tourism that offers the potential of mutual interests in economic development and nature conservation (Hetzer, 1965). By the 1980s, following the rise of global environmental issues in the 1970s, the term 'ecotourism' had become firmly established. In 1987, Ceballos-Lascurain defined ecotourism as '*that tourism that involves travelling to relatively undisturbed or uncontaminated natural areas with the specific object of studying, admiring and enjoying the scenery and its wild plants and animals'* (Ceballos-Lascurain 1991: 31, emphases added). Taken on face value, ecotourism is a passive and appealing form of tourism (Dowling & Sharpe, 1996).

A plethora of terminology can be applied to the whale-watch phenomenon. Wildlife tourism, including tourism that focuses on free-ranging cetaceans, is generally perceived to be inherently beneficial, and so tends to be considered 'ecotourism'. Cetacean-focused tourism and ecotourism are both subsets of nature-based tourism, which encompasses a variety of ways for people to enjoy nature (Bejder & Samuels, 2003). Ecotourism by definition requires *contributions to the conservation of species or habitats* (Higham, 2007). It is not enough merely to mitigate harm; ecotourism should provide a net benefit to conservation. Strictly defined, ecotourism is environmentally responsible travel that contributes to conservation of biodiversity, sustains the well-being of local people, is inclusive at the local community level, offers learning experiences for tourists, involves responsible action on the part of tourists and the tourism industry, and requires the lowest possible consumption of nonrenewable resources (e.g. UNEP, 2002).

Indeed, it is true that at the local scale of analysis, ecotourism may contribute to the protection of natural environments and conservation of endangered species (Higham, 2007). It may also foster economic transition, regional development, community empowerment and the creation of employment opportunities in peripheral areas and for indigenous communities (Butler & Hinch, 1996; Hall & Boyd, 2003). Advocates also point to the argument that ecotourism businesses may contribute to the communication of conservation messages to the general public (Beaumont, 2001). The potential for well-developed education programmes to contribute to this end has been explored, highlighting the importance of business philosophy (Higham & Carr, 2002; Neves, 2010), education programme design (Orams, 1997) and the critical role of the guide/interpreter (Weiler & Ham, 2001; and see Chapters 9–11). In best practice cases it is evident that interests in environmental conservation and conservation advocacy can be advanced through this form of tourism (see Higham & Carr, 2002). These sentiments allow ecotourism to be seen as a 'caring partner for the environment' (Becken & Schellhorn, 2007: 87).

There is a counterview that does not deny these potential benefits, but calls for a more balanced and critical appraisal of ecotourism. The counterview questions the assumption that alternative forms of tourism development such as ecotourism are desirable simply because they are preferable to popular forms of mass tourism (Butler, 1990). Wheeler (1991) offered an early articulation of this view, describing ecotourism as an 'elaborate ruse' and effective marketing tool for building further demand for tourism. Indeed, Orams (1995) observes that many definitions of ecotourism are so broad as to make ecotourism indistinguishable from any other form of tourism. Studies of the social values of 'ecotourists' have demonstrated that they are no more likely to have 'green' social values than others (Blamey & Braithwaite, 1997). Wheeler (1991: 95) is more direct. 'Veracious wolf in lamb's clothing, the sensitive traveller is the real perpetrator of the global spread of tourism and in this capacity must take responsibility for some of tourism's adverse impact'.

Indeed, a range of specific environmental management challenges have emerged in association with ecotourism. They arise from the fact that ecotourism operations should take place in unmodified (Valentine 1993), natural (Orams 1995; Blamey & Braithwaite, 1997; Fennell, 1998) or pristine (Ceballos-Lascurain, 1991) areas. With this emerges a raft of challenges associated with the management of visitor activities in environments that are fragile, finite and valued primarily for conservation. This operational environment is difficult to reconcile with the further definitional requirement that ecotourism visitor operations and activities should be low in impact (Lindberg & McKercher, 1997; Orams, 1995; Wight, 1993). Furthermore, the pace of ecotourism development has been the cause of policy paralysis as management agencies are forced to act reactively to rapid shifts in demand (Higham *et al.*, 2009). Variations of these challenges arise in the developing-world context due to factors that include 'shortages in the endowments of human, financial and social capital within the community, lack of mechanisms for a fair distribution of the economic benefits of ecotourism, and (resource) insecurity' (Coria & Calfucura, 2012: 47; see also Chapter 21, this volume).

The sheer weight of demand for ecotourism experiences has resulted in the swift transition from 'alternative tourism' to the commercial mainstream. Indeed, the evolution of ecotourism in the 1980s can be seen quite simply as one form of periodic transformation in the global capitalist economy (Neves, 2010). Wheeler (1991: 96) highlights the capitalist underpinnings of ecotourism, observing that 'by clothing itself in a green mantle, the industry is being provided with a shield with which it can both deflect valid criticism and improve its own image while, in reality, continuing its familiar short term tourism commercial march'. Indeed, from the outset, Hall (1994) described ecotourism as a new form of ecological imperialism and western economic domination, a line of debate that has extended to the politics of the IWC (Bailey, 2012).

Lately, the global scale of analysis has been applied to the environmental outputs of ecotourism. Much science has attended to locally contextualized, site-specific impacts of whale-watching (Higham & Lusseau, 2004), with temporal scale (short-term–long-term impacts) being a critical consideration (Bejder *et al.*, 2006). The effects of local exhaust fumes on resident killer whales has been addressed by Lachmuth *et al.* (2011). However, in addressing the broader spatial scale of analysis, it is necessary to also accept that tourists – and most particularly 'ecotourists' (Hall, 2007) – contribute significantly to global environmental change, perhaps most notably climate change by way of CO_2 emissions from international air travel. Little scholarly attention has been paid to the likely consequences of global climate change for

whale-watching (Lambert *et al.*, 2010) although Neves (2010) does briefly contemplate the ecological footprint of 10 million ecotourists. Becken and Schellhorn (2007) call for an 'open-system' approach to understanding ecotourism, arguing that local/regional studies are incomplete and flawed. They state that the 'open-systems approach and the link to global issues clearly challenge the widely accepted conceptual link between ecotourism and nature conservation' (Becken & Schellhorn, 2007: 99). These issues also challenge the mainstream ecotourism rhetoric in ways that can no longer be conveniently ignored.

3. Whale-watching is a non-consumptive alternative to extractive (consumptive) whale-hunting practices

Like all forms of wildlife tourism, whale-watching has been perceived uncritically as a non-consumptive activity (Knight, 2009) which underpins the false assumption of ecological sustainability (Lusseau *et al.*, 2013). Wildlife viewing has been described as non-consumptive in contrast to the immediate and lethal outcomes of hunting (Duffus & Dearden, 1990). This terminology dates to the early 1980s when the IWC co-sponsored the first whale-watching conference, *Whales Alive* (1983) in Boston, MA. This conference recommended that new forms of 'non-consumptive' utility should be specifically considered by the IWC in managing global whale stocks (O'Connor *et al.*, 2009). As Knight (2009: 168) observes, 'on the face of it, the two activities – viewing and hunting – could not appear more different'. There are, in fact, inherent contradictions in such unitary terminology (Tremblay, 2001). It has been suggested that the 'consumptive versus non-consumptive dichotomy that is often associated with numerous definitions of wildlife tourism may be somewhat misleading' (Lemelin, 2006: 516).

A number of scholars have critiqued the consumptive/non-consumptive dichotomy at a number of different levels. One line of writing has it that hunting, while lethal at the level of individual animals, is not necessary deletive at the population level. Hunting, therefore, may not be inconsistent with species conservation (Tremblay, 2001). Conversely wildlife viewing can be positioned as a form of 'ocular consumption' (Lemelin, 2006), which may then frame questions of impact upon focal animals (and wider animal populations) being subject to the tourist gaze (Urry, 2002) and, therefore, consumption by the human eye. 'While the gaze itself may be virtually harmless, this form of leisure is still dependent on the transformation of landscapes…and tourism infrastructures (transportation, accommodations, services, etc.)', which may or may not be sustainable (Lemelin, 2006: 518). Such avenues of development, for supposedly benign 'non-consumptive' activities are, in fact, implicated in the sustainable capacity of whale-watch systems (Higham & Lusseau, 2007).

Knight (2009) takes this critique further, stating the case that hunting and viewing are fundamentally similar. Tourists who seek to observe wild animals at close quarters must (be assisted to) locate and approach focal animals. This requires precisely the same techniques as the hunter. Both engage in systematically locating, identifying and pursuing target animals, which are generally 'wary of human presence and reluctant to expose themselves to human eyes' (Knight, 2009: 169). Neves' (2010) Marxist critique also 'reveals significant continuities between whale hunting and whale-watching, especially the fetishized commoditization of cetaceans and the creation of a metabolic rift in human–cetacean relations. In both contexts nature is produced first and foremost according to capitalist principles, which problematizes the pervasive assumption that whale-watching correlates primarily and directly with conservation' (Neves, 2010: 719).

The directed, intensive and sustained tourist gaze offers further parallels with hunting (and predation more generally), which trigger alarm and anti-predatory responses to avoid detection and minimize close and/or prolonged interaction (Tremblay, 2001; Knight, 2009). The importance of managing 'human–wildlife viewing interactions' (including pursuit, intensive gaze and proximal interaction) receives further emphasis given that tourist satisfaction is commonly associated with close-up, unconstrained and prolonged interactions with wild animals (Orams, 2000), the experience of critical behaviours (e.g. hunting, feeding, socializing and courtship) and, in some cases, immediate proximity extending to touch (e.g. Muloin, 1998). In addressing this issue, Bejder *et al.* (2009) apply aspects of evolutionary theory for decision-making under the risk of predation to make predictions about how individual animals respond to non-lethal forms of human disturbance. This approach assumes that animals use analogous decision processes to evaluate responses to the risks presented by natural predators and those presented by anthropogenic agents of disturbance. If so, individual animals will take the same ecological considerations into account when they experience human disturbance as they do when they perceive the risk of predation (Lima & Dill, 1990; Beale & Monaghan, 2004b). In fact, experimental studies have shown that, when approached by whale-watching boats, killer whales adopt evasive tactics that look surprisingly like those used by moths to evade bats (Williams *et al.*, 2002). The net effect of repeated disturbance is a reduction of time spent feeding (Williams *et al.*, 2006), and this energetic cost is a concern for populations of at-risk, food-limited whales. The body of whale-watching science would strongly suggest discontinuation of the term 'non-consumptive' and the unhelpful assumptions that this terminology supports. If it is consumptive, to quote Meletis and Campbell (2007), 'call it consumption!'

4. Whale-watching equates to conservation of the marine environment

If one accepts the problematization of 'non-consumptive' wildlife tourism, then the assumption that whale-watching is akin to conservation of the marine environment is also drawn into question. It is intuitively appealing to assume that whale-watching, as an alternative to whale-hunting, is a form of stewardship that contributes in some way to species conservation. Neves (2010) observes that this assumption has been perpetuated through the public communication strategies of environmental NGOs, and the marketing practices of most whale-watch companies. She contests 'the reductionism that is entailed in taking for granted that the relationship between (whale-watching), economic development/growth, and conservation is essentially and universally benign' (Neves, 2010: 721). Such practices have contributed to a deep-set predicament insofar that they 'undermine the possibility of distinguishing between different types of whale-watching and the degree to which they effectively live up to conservationist goals' (Neves, 2010: 721).

The problem is that successful commercial wildlife viewing requires that visitors are concentrated in well-defined locations where interactions with wild animals are predictable (Whittaker & Knight, 1997). Viewing wildlife naturally takes place where sightings are consistent, focal animals can be viewed in abundance or where spectacular behaviours may be predictably observed (see Chapters 17 and 19). For resident killer whales, the species aggregates in critical habitats, and for reasons of efficiency, whale-watching traffic is concentrated in precisely the habitats that are most critical to the whales' survival (Williams et al., 2009). The critical nature of these locations, in terms of site ecology and wildlife behaviours, raises two important points. First, the behavioural state of wild animals varies significantly over time (e.g. over both stages of the breeding cycle and life course); and second, animal responses to external stimuli (e.g. including the presence of tourists) are likely to vary over time, as influenced by these temporal determinants (Williams et al., 2006). Knight (2009) questions how wildlife intolerance of humans and industrial-scale tourist interactions with wild animals can be reconciled. He argues that 'wild animals are only viewed on this scale because they have been made viewable through human intervention' (Knight, 2009: 167). Such interventions include attraction (e.g. food provisioning), capture

and confinement (e.g. for display in aquaria) and habituation (i.e. a waning in flight response to repeated stimuli) (Knight, 2009). All are considered to produce diminished behaviours in 'wild' animals and reduce population fitness (Knight, 2009; Bejder et al., 2009; Higham & Shelton, 2011).

The dangers of assuming that tourist interactions with cetaceans in the wild are benign are clearly evident. Extensive field-based behavioural studies have been peer-reviewed and published over the last 25 years. Most studies have focused on behavioural changes depending on the presence and density of boats. They find that groups of animals tend to tighten when boats are present (e.g. Blane & Jaakson, 1995; Novacek et al., 2001; Bejder et al., 2006). Some species show signs of active avoidance. Responses range from changes in movement patterns (Edds & MacFarlane, 1987; Salvado et al., 1992; Campagna et al., 1995; Bejder et al., 1999; Novacek et al., 2001), increases in dive intervals (Baker et al., 1988; Baker & Herman, 1989; Janik & Thompson, 1996; Bejder et al., 2006) and increases in swimming speed (Blane & Jaakson, 1995; Williams et al., 2002). These signs of avoidance can be a result of not only the presence of boats, but also the manoeuvring of boats including sudden changes in vessel speed or rapid approaches (Gordon et al., 1992; Constantine, 2001). While many of these papers make management recommendations, few have been acted upon.

These studies illustrate the folly of equating animal observation to species conservation. They also perhaps point to the global politics of whale conservation. Neves (2010) argues that 'the efforts of some of the world's most prominent environmental NGOs to save whales from being hunted to extinction have produced and propagated whale-watching as a quintessentially and uniformly benign activity'. Efforts to counter this deeply embedded assumption have been slow to gain traction, although a milestone was achieved in May 2006 with the IWC's St Kitts Declaration on dolphin and whale-watching tourism. This declaration recognizes that the rapid development of cetacean viewing activities has been largely unchecked, accepts that cetacean populations can be significantly affected by these activities and states the importance of moving towards sustainable practices. In a marked change in rhetoric the St Kitt's declaration (2006: np) also observes that 'cetacean watching operations should be confined to those populations best able to sustain *exploitation*' (emphasis added). Meanwhile, the portrayal of whale-watching as nature conservation (Neves, 2010) has contributed to perpetuating unsustainable practices (Lusseau *et al.*, 2013).

5. Whale-hunting and whale-watching are mutually exclusive

A fifth binary assumption is that whale-hunting and whale-watching are mutually exclusive. This reductionism has been central to the widely held view that whale-watching will ultimately displace global hunting practices (Corkeron, 2006). The fact that whale-hunting has continued uninterrupted in some regions, and has been resumed in recent years in others, clearly contradicts this assumption (Higham & Lusseau, 2007). In fact, there is widespread evidence that whale-hunting and whale-watching are not mutually exclusive (see Chapters 4, 7 and 8). Rather, whales represent a site of political contestation. Whale-watching is seen by some as a form of cultural imperialism that is imposed globally by urban liberals in the global north. Bailey (2012: 490) documents one view of the IWC's hunting moratorium as 'an imperialistic infringement of sovereignty by industrialized, urbanized countries, and English-speaking countries'. However, she also observes the growing influence of whale-watching interests in the political dynamics of the IWC with specific reference to the interests of the global south. 'At the 62nd Annual Meeting of the IWC in Agadir in 2010, a group of Latin American and Caribbean states known as the Buenos Aires Group (BAG), acted in concert to support the conservationist position' (Bailey, 2012: 490), an action that is seen to counter claims that the whaling moratorium is an act of neo-imperialism.

In fact, Corkeron (2006: 165) notes that 'if there is one clear message, it is that the relationship between whaling and whale-watching is not simply the case of one replacing the other'. He observes that whaling nations (where whale-watching activities also take place without exception) see whales as commodities for both culinary and ocular consumption. Different species of whale may be differentiated in terms of conservation status and 'sustainable harvest'. Perhaps most critically, whaling nations have in the past and continue to see 'the act of killing whales as an expression of national identity' (Corkeron, 2006: 165). Given these sentiments, the view that whales can be utilized in more than one commercial manner, and serve a range of economic and non-economic outcomes, prevails in some regions of the world.

It is now the case that whale-watching revenues exceed hunting revenues in all nations that continue to practise whale-hunting (Kuo *et al.*, 2012). Reporting figures dating to the late 1990s, Kuo *et al.* (2012) demonstrate that the revenues generating from whale-watching in both Japan and Norway had at that time already exceeded whale-hunting revenues. There is little doubt that the potential capacity of the whale-watch systems in these countries has been significantly constrained by the continued practice of whale-hunting (Higham & Lusseau, 2007; Kuo *et al.*, 2012). Indeed, several studies claim that tourists who seek to achieve interactions with cetaceans in the wild hold strong pro-environmental values (Parsons *et al.*, 2003; Rawles & Parsons, 2004), which strongly dictate various aspects of their tourist decision-making and behaviour (Parsons & Draheim, 2009). Parsons and Rawles (2003) demonstrate that whale-watchers would not only boycott whale-watching, but in many cases will abandon altogether intended visits to countries that continue to practise commercial whaling, with broader-ranging implications for the tourism/hospitality and service sector industries in those countries. Similarly, Björgvinsson (2003, cited by Kuo *et al.*, 2012) reports that the emergent whale-watching sector in Iceland was compromised by the resumption of commercial whaling, with wider implications for tourism-related sectors (e.g. transport, hospitality and retail) (Kuo *et al.*, 2012; see also Chapters 7 and 8 of this volume). Virtually nothing is known about tourists' views on aboriginal subsistence whaling. Clearly the relationship between whale-hunting and whale-watching is complex, and the view that whale-watching will inevitably prevail over whale-hunting due to mutual exclusivity must be more critically questioned.

Challenging the reductionist binaries: the search for sustainable whale-watching

These binary debates have given rise to a range of unhelpful assumptions. A more complex and critical treatment of the whale-watching phenomenon may usefully be informed by consideration of whale-watching in terms of the theory of the tourism system (Figure 1.1). Systems theory recognizes that 'global tourism is a highly complex system (consisting) of a multitude of actors who interact at crosscutting levels to produce certain outcomes' (Cornelissen, 2005: 4). The system is an abstract representation of geographic/human, biological/ecological and industrial elements, that are linked in complex relationships and treated as a whole or set of elements (Hall, 2004). Therefore, Figure 1.1 conceptualizes whale-watching as an open system. Whale-watching practices can be seen to interact with a wide range of dynamic external forces, both directly and indirectly related to tourism, in a manner that is dynamic (Weaver & Lawton, 2009). It affords recognition of the fact that whale-watching affects and is affected by the broader local–global, socio-cultural, economic, ecological and political environments within which it exists. Whale-watching does not exist in isolation of these wider contexts, but rather is subject to evolutionary dynamics and stochastic events that play out at the local, regional and global levels of the open tourism system (Higham & Lusseau, 2007; Higham *et al.*, 2009).

In recognizing the importance of scale, this book begins by addressing the global context (Part I).

It contemplates whales as a global common pool resource (Moore & Rodger, 2010), which are subject to a range of global environmental threats. Such an approach highlights the uncertainty of local tourism impacts when set within broader global threats to cetacean morbidity and mortality (see Chapter 2, this volume), such as the development of transport networks (e.g. vessel strikes), fisheries by-catch and habitat degradation (e.g. noise pollution and changing global climatic systems). The urgency of such insights is highlighted by recent warnings from the IWC's Scientific Committee about the 'grave state' of two critically endangered species of cetaceans: the vaquita porpoise (*Phocoena sinus*) of Mexico, and New Zealand's Maui's dolphin (*Cephalorhynchus hectori maui*), both of which are being pushed towards extinction due to by-catch entanglement in fishing nets (World Wildlife Fund, 2012). By necessity, treatment of the global context reaches across the environmental, political and socio-cultural domains. These chapters consider global issues of relevance to cetaceans, and how tourism contributes to those issues.

The resolution of analysis then shifts from the global to the local/regional to address the human and ecological dimensions of whale-watching. The former attends to the interplay of whale-hunting and whale-watching, visitor experiences, and the potential for indigenous/traditional ecological knowledge to contribute to important aspects of sustainability (Part II). It also considers the potential for whale-watching to contribute to cultural renaissance. It critically addresses the effectiveness of environmental education programmes, widely considered so important to raising awareness of marine conservation among both visitors and host communities (Garrod & Fennell, 2004). The extent to which visitor education programmes influence the attitudes and behaviours of visitors in an enduring manner upon return to the tourist generating region (see Figure 1.1) is critical to these discussions. The latter explores the ecological effects of whale-watching on cetaceans (Part III), in order to address the behavioural ecology and

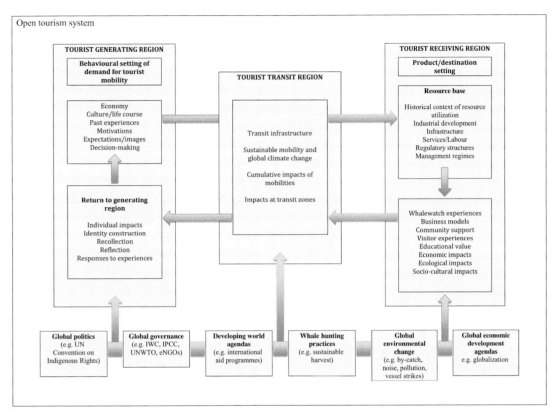

Figure 1.1 The whale-watching open tourism system (adapted from Hall, 2004).

ecological constraints of cetaceans. Here, most critically, close attention is paid to the temporal links between short-term disturbance and long-term effects, and to the potential for disturbances of individual animals to result in population-wide consequences. Consideration in Part III extends to the management responses that are required to begin to address the ecological impacts of industrial whale-watching on the marine resources base of the destination region (Figure 1.1).

These chapters highlight the urgency of responses on the part of governments, NGOs, policy-makers, resource managers, local communities and tourists. They highlight the 'growing chorus of voices' calling for appropriate regulations and enforcement (Kessler & Harcourt, 2010), which is so critical to local sustainable practices. These matters are addressed in Part IV, which adopts a case-study approach to exploring the planning and management actions required to foster interests in sustainability. Here, questions are asked about widespread failures to adequately protect animals that are the focus of the billion dollar global whale-watch industry, general lack of government/regulatory responses and the ineffectiveness of management interventions (Bejder *et al.*, 2006). These chapters respond to highly variable management regimes, approaches and outcomes, which have been described as 'haphazard' and 'highly fragmented' (Gjerdalen & Williams, 2000; Garrod & Fennell, 2004; Allen *et al.*, 2007). In recognition of the global context of whale-watching in the developed world, Part IV also addresses the exponential growth of whale-watching that is currently taking place in the developing world context, and the unique sustainability challenges that this raises.

Conclusion

Global ecotourism and the industrial development of whale-watching represent important transformations in the global capitalist economy (Neves, 2010). A consequence has been increasing sustainability concerns (Baker & Herman, 1989; Butler, 1990; Wheeler, 1991). Responses to the challenge of sustainability, within the policy, planning, marketing and tourism management domains, have failed to arrest those concerns. The global scale of whale-watching, the biodiversity crisis and the transformation of local/regional economies underscores the urgency of achieving steps towards sustainability. This lends considerable weight to the case to protect cetaceans from unsustainable practices, whether those practices take the form of commercial, scientific or indigenous hunting, fisheries by-catch, live capture for public display or sublethal impacts of whale-watching that add up, cumulatively, to unsustainable 'consumption'. In addressing the binary debates highlighted in this chapter, this volume pursues the case for nuanced and fine-grained analyses of whale-watching, set within the context of the open tourism system. The chapters that follow address a myriad active discourses and debates that contribute to our shifting understandings of whale-watching from simple binaries to more advanced understandings of human phenomena.

REFERENCES

Allen, S., Smith, H., Waples, K. & Harcourt, R. (2007). The voluntary code of conduct for dolphin watching in Port Stephens, Australia: Is self regulation an effective management tool? *Journal of Cetacean Resource Management* 9(2), 159–166.

Bailey, J.L. (2012). Whale-watching, the Buenos Aires Group and the politics of the International Whaling Commission. *Marine Policy* 36, 489–494.

Baker, C.S. & Herman, L.M. (1989). Behavioural responses of summering humpback whales to vessel traffic: Experimental and opportunistic observations. Final report to the National Park Service, Alaska Regional Office, Anchorage, Alaska.

Baker, C.S., Perry, A. & Vequist, G. (1988). Humpback whales of Glacier Bay, Alaska. *Whalewatcher* Fall, 13–17.

Beale, C.M. & Monaghan, P. (2004a). Behavioural responses to human disturbance: A matter of choice. *Animal Behavior* 68, 1065–1069.

Beale, C.M. & Monaghan, P. (2004b). Human disturbance: People as predation-free predators? *Journal of Applied Ecology* 41, 335–343.

Beaumont, N. (2001). Ecotourism and the conservation ethic: Recruiting the uninitiated or preaching to the converted? *Journal of Sustainable Tourism* 9(4), 317–341.

Becken, S. & Schellhorn, M. (2007). Ecotourism, energy use and the global climate: Widening the local perspective. In J.E.S. Higham (Ed.), *Critical Issues in Ecotourism: Understanding a complex tourism phenomenon*. Oxford: Elsevier, pp. 85–101.

Bejder, L. & Samuels, A. (2003). Evaluating impacts of nature based tourism on cetaceans. In N. Gales, M. Hindell & R. Kirkwood (Eds), *Marine Mammals: Fisheries, tourism and management issues*. Collingwood: CSIRO, pp. 229–256.

Bejder, L., Dawson, S.M. & Harraway, J.A. (1999). Responses by Hector's dolphins to boats and swimmers in Porpoise Bay, New Zealand. *Marine Mammal Science* 15(3), 738–750.

Bejder, L., Samuels, A., Whitehead, H., *et al.* (2006). Decline in relative abundance of bottlenose dolphins exposed to long-term disturbance. *Conservation Biology* 20, 1791–1798.

Bejder, L., Samuels, A., Whitehead, H., Finn, H. & Allen, S. (2009). Impact assessment research: Use and misuse of habituation, sensitisation and tolerance in describing wildlife responses to anthropogenic stimuli. *Marine Ecology Progress Series* 395, 177–185.

Björgvinsson, A. (2003). Iceland, whale v/s whale-watching. Available from http://www.global500.org/news_83.html

Blamey, R. & Braithwaite, V. (1997). A social values segmentation of the potential ecotourism market. *Journal of Sustainable Tourism* 5(1), 29–45.

Blane, J.M. & Jaakson, R. (1995). The impact of ecotourism boats on the Saint Lawrence beluga whales. *Environmental Conservation* 21(3), 267–269.

Butler, R. & Hinch, T. (Eds) (1996). *Tourism and Indigenous Peoples*. London: International Thomson Business Press.

Butler, R. W. (1990). Alternative tourism: Pious hope or trojan horse? *Journal of Travel Research* 28(3), 40–45.

Campagna, C., Rivarola, M.M., Greene, D. & Tagliorette, A. (1995). Watching southern right whales in Patagonia. Unpublished report to UNEP, Nairobi.

Ceballos-Lascurain, H. (1991). Tourism, ecotourism and protected areas. *Parks* 2(3), 31–35.

Cisneros-Montemayor, A.M., Sumaila, U.R., Kaschner, K. & Pauly, D. (2010). The global potential for whale-watching. *Marine Policy* 34, 1273–1278.

Constantine, R. (2001). Increased avoidance of swimmers by wild bottlenose dolphins (*Tursiops truncatus*) due to long-term exposure to swim-with-dolphin tourism. *Marine Mammal Science* 17(4), 689–702.

Coria, J. & Calfucura, E. (2012). Ecotourism and the development of indigenous communities: The good, the bad, and the ugly. *Ecological Economics* 73, 47–55.

Corkeron, P.J. (2006). How shall we watch whales? In D.M. Lavigne (Ed.), *Gaining Ground: In pursuit of ecological sustainability*. Guelph, ON: International Fund for Animal Welfare, pp. 161 170.

Cornelissen, S. (2005). *The Global Tourism System*. Ashgate: Aldershot.

Dowling, R.K. and Sharp, J. (1996). Conservation-tourism partnerships in Western Australia. *Tourism Recreation Research* 22(1), 55–60.

Duffus, D.A., & Dearden, P. (1990). Nonconsumptive wildlife-oriented recreation: A conceptual framework. *Biological Conservation* 53(3), 213–231.

Edds, P.L. & MacFarlane, J.A.F. (1987). Occurrence and general behaviour of balaenopterid cetaceans summering in the Saint Lawrence Estuary, Canada. *Canadian Journal of Zoology* 65, 1363–1376.

Fennell, D.A. (1998). Ecotourism in Canada. *Annals of Tourism Research* 25(1), 231–234.

Garrod, B. & Fennell, D.A. (2004). An analysis of whale-watching codes of conduct. *Annals of Tourism Research* 31(2), 334–352.

Gjerdalen G. & Williams PW. (2000). An evaluation of the utility of a whale watching code of conduct. *Tourism Recreation Research* 25(2), 27–37.

Gordon, J., Leaper, R., Hartley, F.G. & Chappell, O. (1992). *Effects of whale watching on the surface and underwater acoustic behaviour of sperm whales off Kaikoura, New Zealand*. Department of Conservation Science and Research Series. No. 52.

Hall, C.M. (1994). Ecotourism in Australia, New Zealand and the South Pacific: Appropriate tourism or a new form of ecological imperialism? In E. Cater & G.L. Lowman (Eds), *Ecotourism: A sustainable option?* Chichester, U.K.: John Wiley and Sons, pp. 137–158.

Hall, C.M. (2004). *Tourism: Rethinking the social science of mobility*. Pearson: Harlow, UK.

Hall, C.M. (2007). Scaling ecotourism: The role of scale in understanding the impacts of ecotourism. In J.E.S. Higham (Ed.), *Critical Issues in Ecotourism: Understanding a complex tourism phenomenon*. Oxford: Elsevier, pp. 243–255.

Hall, C.M. & Boyd, S. (2003). *Ecotourism in Peripheral Areas*. Clevedon: Channel View Publications.

Hammond, P. (2006). Whale science – and how (not) to use it. *Significance* June 2006, 54–58.

Herrera, G.E., & Hoagland, P. (2006). Commercial whaling, tourism, and boycotts: an economic perspective. *Marine Policy* 30, 261–269.

Hetzer, D. (1965). Environment, tourism, culture. *Links* 1, np.

Higham, J.E.S. (Ed). (2007). *Critical Issues in Ecotourism: Understanding a complex tourism phenomenon*. Oxford: Elsevier Butterworth-Heinemann.

Higham, J.E.S. & Carr, A. (2002). Profiling visitors to Ecotourism Operations. *Annals of Tourism Research* 29(4), 1168–1171.

Higham, J.E.S. & Lusseau, D. (2004). Ecological impacts and management of tourist engagements with Cetaceans. In R. Buckley (Ed.), *Environmental Impacts of Ecotourism*. Wallingford: CAB International, Wallingford.

Higham, J.E.S. & Lusseau, D. (2007). Urgent need for empirical research into whaling and whale-watching. *Conservation Biology* 21(2), 554–558.

Higham, J.E.S., & Lusseau, D. (2008). Slaughtering the goose that lays the golden egg: Are whaling and whale-watching mutually exclusive? *Current Issues in Tourism* 11(1), 63–74.

Higham, J.E.S. & Shelton, E. (2011). Tourism and wildlife habituation: Reduced population fitness or cessation of impact? *Tourism Management* 32(4), 1290–1298.

Higham, J.E.S., Lusseau, D. & Hendry, W. (2008). The viewing platforms from which animals are observed in the wild: A discussion of emerging research directions. *Journal of Ecotourism* 7(2/3), 132–141.

Higham, J.E.S., Bejder, L. & Lusseau, D. (2009). An integrated and adaptive management model to address the long-term sustainability of tourist interactions with cetaceans. *Environmental Conservation* 35, 294–302.

Hinch, T.D. (1998). Ecotourists and indigenous hosts: Diverging views on their relationship with nature. *Current Issues in Tourism* 1(1), 120–124.

Hoyt, E. (2001). *Whale-watching 2000: Worldwide tourism numbers, expenditures, and expanding socioeconomic benefits.* Crowborough, East Sussex, UK: International Fund for Animal Welfare.

Hoyt, E., & Hvenegaard, G.T. (2002). A review of whale-watching and whaling with applications for the Caribbean. *Coastal Management* 30(4), 381–399.

International Fund for Animal Welfare (IFAW) and Tethys European Conservation (1995). *Report of the Workshop on the Scientific Aspects of Managing Whale-watching.* International Fund for Animal Welfare, Tethys European Conservation, Montecastello di Vibio, Italy.

Janik, V.M. & Thompson, P.M. (1996). Changes in surfacing patterns of bottlenose dolphins in response to boat traffic. *Marine Mammal Science* 12, 597–602.

Kessler, M. & Harcourt, R. (2010). Aligning tourist, industry and government expectations: A case study from the swim with whales industry in Tonga. *Marine Policy* 34, 1350–1356.

Knight, J. (2009). Making wildlife viewable: Habituation and attraction. *Society and Animals* 17, 167–184.

Kuo, H.-I., Chen, C.-C. & McALeer, M. (2012). Estimating the impact of whaling on global whale-watching. *Tourism Management* 33(6), 1321–1328.

Lachmuth, C.L., Barrett-Lennard, L.G., Steyn, D.Q. & Milsom, W.K. (2011). Estimation of southern resident killer whale exposure to exhaust emissions from whale-watching vessels and potential adverse health effects and toxicity thresholds. *Marine Pollution Bulletin* 62(4), 792–805.

Lambert, E., Hunter, C., Pierce, G.J., & MacLeod, D. (2010). Sustainable whale-watching tourism and climate change: Towards a framework of resilience. *Journal of Sustainable Tourism* (Special Issue: Tourism: Adapting to Climate Change and Climate Policy) 18(3), 409–427.

Lemelin, R.H. (2006). The gawk, the glance, and the gaze: Ocular consumption and polar bear tourism in Churchill, Manitoba, Canada. *Current Issues in Tourism* 9(6), 516–534.

Lima, S.L. & Dill, L.M (1990). Behavioral decisions made under the risk of predation: A review and prospectus. *Canadian Journal of Zoology* 68, 619–640.

Lindberg, K. and McKercher, B. (1997). Ecotourism: A critical overview. *Pacific Tourism Review* 1, 65–79.

Lusseau, D., Bejder, L., Corkeron, P., Allen, S. & Higham, J.E.S. (2013). Learning from past mistakes: A new paradigm for managing whale-watching. *Conservation Letters* (under review).

McCormick, J. (1989). *The Global Environmental Movement.* London: Belhaven Press.

Meletis, Z.A. & Campbell, L.M. (2007). Call it consumption! Re-conceptualizing ecotourism as consumption and consumptive. *Geography Compass* 1, 850–870.

Moore, S.A. & Rodger, K. (2010). Wildlife tourism as a common pool resource issue: Enabling conditions for sustainability governance. *Journal of Sustainable Tourism* 18(7), 831–844.

Muloin, S. (1998). Wildlife tourism: The psychological benefits of whale-watching. *Pacific Tourism Review* 2, 199–213.

Mustika, P.L.K., Birtles, A., Welters, R. & Marsh, H. (2012a). The economic influence of community-based dolphin watching on a local economy in a developing country: Implications for conservation. *Ecological Economics* 79, 11–20.

Mustika, P.L.K., Birtles, A., Everingham, Y. & Marsh, H. (2012b). The human dimensions of wildlife tourism in a developing country: Watching spinner dolphins at Lovina, Bali, Indonesia. *Journal of Sustainable Tourism* 20(2), 1–23.

Neves, K. (2010). Cashing in on Cetourism: A critical ecological engagement with dominant E-NGO discourses on whaling, cetacean conservation, and whale-watching. *Antipode* 42(3), 719–741.

Novacek, S.M., Wells, R.S. & Solow, A.R. (2001). Short-term effects of boat traffic on bottlenose dolphins, *Tursiops truncatus*, in Sarasota Bay, Florida. *Marine Mammal Science* 17(4), 673–688.

O'Connor, S., Campbell, R., Cortez, H. & Knowles, T. (2010). *Whale-watching Worldwide: Tourism numbers, expenditures and expanding economic benefits.* Special report for the International Fund for Animal Welfare prepared by Economists at Large. Yarmouth, MA: International Fund for Animal Welfare.

Orams, M.B. (1995). Towards a more desirable form of ecotourism. *Tourism Management* 16(1), 3–8.

Orams, M.B. (1997). The effectiveness of environmental education: Can we turn tourists into 'greenies'? *Progress in Tourism and Hospitality Research* 3, 295–306.

Orams, M.B. (2000). Tourists getting close to whales, is it what whale-watching all about? *Tourism Management* 21, 561–569.

Parsons, E.C.M. (2003). Seal management in Scotland: Tourist perceptions and the possible impacts on the Scottish tourism industry. *Current Issues in Tourism* 6(6), 540–546.

Parsons, E.C.M., & Draheim, M. (2009). A reason not to support whaling – A tourism impact case study from the Dominican Republic. *Current Issues in Tourism* 12(4), 397–403.

Parsons, E.C.M. & Rawles, C. (2003). The resumption of whaling by Iceland and the potential negative impact in the Icelandic whale-watching market. *Current Issues in Tourism* 6(5), 444–448.

Parsons, E.C.M., Warburton, C.A., Woods-Ballard, A., Hughes, A. & Johnston, P. (2003). The value of conserving whales: The impacts of cetacean-related tourism on the economy of rural West Scotland. *Aquatic Conservation* 13(5), 397–415.

Rawles, C.J.G. & Parsons, E.C.M. (2004). Environmental motivation of whale-watching tourists in Scotland. *Tourism in Marine Environments* 1(2), 129–132.

Ris, M. (1993). Conflicting cultural values: Whale tourism in Northern Norway. *Arctic* 46(2), 156–163.

Salvado, C.A.M., Kleiber, P. & Dizon, A.E. (1992). Optimal course by dolphins for detection avoidance. *Fishery Bulletin* 90, 417–420.

Smestad, T.H. (1997). *Images of Whales, Whaling and Whalers: A rhetorical study of the controversies over Norwegian Minke whaling*. M.A. thesis. University of Oslo, University of Maastricht.

Tremblay, P. (2001). Wildlife tourism consumption: Consumptive or non-consumptive? *International Journal of Tourism Research* 3, 81–86.

UNEP. (2002). Tour operators launch new performance indicators to promote environment friendly tourism. Press release 13 November. London/Nairobi. http://www.grida.no/newsroom.cfm?pressReleaseItemID=30

Urry, J. (2002). *The Tourist Gaze* (2nd edn). London: Sage.

Valentine, P. S. (1993). Ecotourism and nature conservation: A definition with some recent developments in Micronesia. *Tourism Management* 14(2), 107–115.

Weaver, D. & Lawton, L. (2009). *Tourism Management* (4th edn). Milton, Australia: Wiley.

Weiler, B. & Ham, S. (2001). Tour guides and interpretation. In D. Weaver (Ed.), *The Encyclopaedia of Ecotourism*. Wallingford, UK: CABI Publishing, pp. 549–564.

Wheeler, B. (1991). Tourism's troubled times: Responsible tourism is not the answer. *Tourism Management* 12(2), 91–96.

Whittaker, D. & Knight, R.L. (1998). Understanding wildlife responses to humans. *Wildlife Society Bulletin* 26, 312–317.

Wight, P. (1993). Ecotourism: Ethics or eco-sell? *Journal of Travel Research* 31(3), 3–9.

Williams, N. (2006). Iceland shunned over whale hunting. *Current Biology*, 16(23), R975–R976.

Williams, R., Trites, A.W. &, Bain, D.E. (2002). Behavioural responses of killer whales (*Orcinus orca*) to whale-watching boats: Opportunistic observations and experimental approaches. *Journal of Zoology* 256, 255–270.

Williams, R., Lusseau, D. & Hammond, P. (2006). Estimating relative energetic costs of human disturbance to killer whales (*Orcinus orca*). *Biological Conservation* 133(3), 301–311.

Williams, R., Lusseau, D. & Hammond, P. (2009). The role of social aggregations and protected areas in killer whale conservation: The mixed blessing of critical habitat. *Biological Conservation* 142, 709–719.

World Wildlife Fund (2012). *Progress for whales at IWC meeting*. Retrieved from http://www.wwf.org.uk/news_feed.cfm?unewsid=6112 (9 July 2012).

Part I

The historical and contemporary contexts

Threats facing cetacean populations

The global context

Rob Williams

Introduction

This chapter provides a brief introduction to the current state of knowledge of threats facing cetacean populations globally, through interviews with some of the leading researchers in the field. Understanding the global status of cetacean populations, and the threats that they face, is critically important to their conservation. This chapter considers the global status of cetacean populations, the information needs required to manage those populations, anthropogenic stressors (including whale-watching), and approaches to understanding the effects of sublethal stressors on cetacean populations. Informed by one-on-one discussions with a diverse group of cetacean experts around the world, this chapter highlights the challenges inherent in seeking to accurately and confidently assess the status of cetacean populations. Yet it is critically important within the context of species conservation to have some appreciation of the status of cetacean populations, and the myriad threats that they face. A great deal of research in recent decades has gone into developing quantitative methods to estimate the size of cetacean populations and to evaluate the sustainability of human-caused mortality through direct harvest or indirect removals, such as by-catch in fishing gear or ship strikes. In recent years, much of the work of marine mammal conservation has been focused on

understanding the ways that non-lethal threats, such as ocean noise or reduction of prey species through overfishing, could cause population-level effects. This chapter provides a brief introduction to those broad themes, but really, each topic deserves a chapter (at least) in its own right. Readers are strongly encouraged to explore the publications of the following experts for a more thorough treatment of these disciplines.

This chapter draws upon five expert interviews with participants selected based on their specific expertise, and long-standing contributions to their fields of research. Specifically, this chapter presents the views of Dr Randall Reeves (Chairman of the Cetacean Specialist Group of the International Union for the Conservation of Nature), Dr Barbara Taylor (Marine Mammal Genetics Group at NOAA's Southwest Fisheries Science Center), Dr Liz Slooten (University of Otago), Dr Christine Erbe (Director of the Centre for Marine Science & Technology, Curtin University) and Dr Brandon Southall (Senior Scientist for Southall Environmental Associates). The style of writing adopted in this chapter is conversational so as to present the views of the chapter participants verbatim. The overarching aim of the chapter is to provide state-of-the-art insights through informal interviews conducted with research leaders in specific areas of importance. In doing so, it provides some context to frame the discussions that are presented in the chapters

Whale-watching: Sustainable Tourism and Ecological Management, eds J. Higham, L. Bejder and R. Williams.
Published by Cambridge University Press. © Cambridge University Press 2014.

that follow. The biographies of the interview participants are provided at the end of the chapter.

Global overview of status of cetacean populations: Dr Randall Reeves

RW: Didn't we save the whales in the 1980s? How are cetacean populations doing, globally?

RRR: There's no simple answer. Some are in serious trouble, some in not-so-serious trouble, some doing well by all appearances, and of course in many cases, it would be presumptuous to even hazard a judgement.

RW: Can we make some generalizations? Are populations that were depleted by whaling recovering after the moratorium on commercial whaling? Are small cetaceans doing better than large whales? Are small cetaceans primarily affected by by-catch, rather than directed hunts?

RRR: Let's deal with 'the moratorium' first. Yes many, probably most, of the populations that were depleted by whaling are recovering. But this is not in any way related to 'the moratorium'. By the time the so-called moratorium on commercial whaling came into force, virtually all of the species and populations that had been seriously depleted by whaling were already protected from commercial whaling under the IWC's management system in place at the time (early 1980s).

By-catch is 'direct mortality' and this is an important point to keep in mind, especially in the present context where I think the distinction between lethal and 'sublethal' issues is fundamental. The lethal threat of by-catch is a major problem not just for 'small cetaceans' but also for large ones, notably North Atlantic right whales. In fact I would argue that the lethal threats of by-catch in fishing gear and ship strikes are, together, a significant problem for most if not all of those species and populations of formerly commercially important large whales that are not recovering, or not recovering as fast as we'd like them to be doing, in the absence

of commercial whaling. So it's important not to relegate the 'primarily affected by by-catch' phrase to small cetaceans only. Having said that, by-catch in fishing gear continues to be the most pressing and acute threat to species and populations of small cetaceans generally, in my view. It is particularly discouraging that this was made abundantly clear more than 20 years ago (Perrin *et al.*, 1994) and since then, we have lost the baiji and are coming very close to losing the vaquita (as we speak) and I suspect other losses have either happened without being documented (at least in the case of local populations) or are imminent. Of course in most cases, perhaps all of them, multiple factors are involved and the populations are being affected by other 'stressors' and not just by-catch, but by-catch is the biggie. And although considerable investment has been made in developing fixes like pingers, stiff nets, alternative gear, etc., the sad truth is that it still looks like the only real answer to the by-catch problem is to separate the animals from the gear in time or space (see also Chapter 17). Recent work by researchers in New Zealand has shown that time/area closures can be effective at stopping and even reversing the damage caused by by-catch. There seems little doubt that *pingers* – acoustic alarms to alert cetaceans to the presence of fishing gear – 'work' for some cetacean species, i.e. they keep the animals from entangling in gillnets under some circumstances. And although they are not a perfect solution, pingers have, at the very least, bought us some time in those areas where compliance and monitoring has proven feasible (e.g. Gulf of Maine, California offshore).

RW: Can't you make one generalization about the global state of cetacean populations? Can we say that river dolphins are the most threatened as a family?

RRR: This seems like a fair generalization. We've lost one truly ancient family of river dolphins (Turvey *et al.*, 2007), very recently (Lipotidae). Following currently accepted taxonomy, four families of river cetaceans [broadly construed, for argument's sake, to include species that occur only in freshwater systems (Platanistidae and Iniidae) as well as species or genera represented by both freshwater and marine populations (Delphinidae and Phocoenidae)] are still out there. The franciscana is, of course, not a river dolphin ecologically but has some features that link it to some species that are, so we can regard its family (Pontoporiidae) a fifth, as another in the group. I think the families Platanistidae, Iniidae, and Pontoporiidae must always be viewed as among the highest priorities for cetacean conservation, and all three are mostly threatened by either by-catch (franciscana) or a combination of by-catch, intentional killing, and serious habitat degradation and loss (*Inia* spp. and the Ganges and Indus dolphins; see also Chapter 21).

RW: If you had to pick one species, which is the most threatened?

RRR: Vaquita. Based on sheer numbers, there may be fewer Maui's dolphins alive today, but they are a subspecies and therefore I think it is fair to insist that the loss of the vaquita, a well-defined, long-accepted species, would be more grave. But I'd rather not get into this game of ranking species (or other units) against each other as though there is competition for which is 'most threatened'. We need to concentrate on saving everything, without prejudice, and if that means more skilled, committed people must be provided with the tools and resources to work on the task, let's offer that strategy as an alternative to triage.

RW: What about the most threatened large whale? North Pacific right whale?

RRR: There may be fewer North Atlantic right whales than North Pacific right whales alive today, and there is plenty of evidence that the former are threatened by an array of factors that come with living cheek-by-jowl with so many of us. The 'eastern' population of North Pacific right whales is critically endangered and needs to be scrupulously protected. However, again, much like with Maui's dolphins, if this population were to disappear, there would still be right whales in the western North Pacific. Similar situations still exist in the eastern North Atlantic, where right whales can probably be regarded as functionally extinct, and in the eastern South Pacific off Chile and Peru where only a small remnant of southern right whales persists.

Information needs to manage populations: Dr Barbara Taylor

RW: When Dr Reeves notes above that the status of some populations is simply more dire than that of other populations, how do we know that? What information goes into a 'status assessment'?

BT: Status is generally the state of a population relative to its 'normal' or 'historical' level. Of course, we can infer historical population size for very few populations or species. Sometimes historical records of numbers killed are used to back-calculate to estimated numbers. Most estimates for large whales use this method. In other cases, the historical state is inferred by contraction of the range. Most status estimates for river dolphins use this type of information. For some populations, the status is known to be dire simply because numbers are small and still decreasing. Vaquita are an example of such a case. For the vast majority of species and populations there is a great deal of uncertainty. For example, tens of thousands of marine mammals were killed in high-seas driftnet fisheries. There is no doubt that such levels of kills changed the status of the affected populations and species, but we have no information on historical or current abundance. Typically a status assessment needs a defined unit to conserve (sometimes a population and at other times a subspecies or species), an estimate of abundance for that unit together with the precision of that estimate, some knowledge about how much the population could grow and an estimate of human-caused mortality. Although this minimum information is not sufficient to infer where a population is relative to historical numbers, it is sufficient in most cases to infer whether the population can sustain the human-caused mortality.

RW: What is the state of information on cetacean populations, generally? Is it fair to say that we're focused on hunted populations of whales (whether commercial, scientific or aboriginal subsistence whaling)? Or populations for which we know that human-caused mortality is a problem? Are some species completely under the radar? I know that the Committee on the Status of Endangered Wildlife In Canada and the US Marine Mammal Protection Act mandate periodic status updates, but how well are we doing at keeping up with the demand for assessments?

BT: There is no doubt that some species have very good information while others have almost none. Of course, there is good reason that the status of whales that were hunted to near-extinction have received a good deal of research focus. But even there, knowledge is very poor for some species. Generally, our knowledge is better for species that are relatively easy to study and that have high public interest. Species that are easy to study have good markings that make it easy to identify individuals (humpback and killer whales) or species that have coastal distributions that make study possible either from land or from small boats (bottlenose dolphins, bowhead and grey whales). Species that have some or all of their distributions far from land are generally poorly understood (fin and sei whales, many dolphins and all beaked whales). The porpoises are often neglected globally, perhaps because of their small size and for many relatively shy behaviour.

RW: What about our information on the status of large whales versus small cetaceans?

BT: The IUCN evaluated all cetacean species, although many were listed as data-deficient. There are fewer large whales listed as data-deficient than among the species of small cetaceans, especially if you include beaked whales and other medium-sized oceanic whales (like false killer whales and pilot whales) in that category. Many scientists, myself included, felt that evaluating status at the species level was not particularly useful for cetaceans. Consider, for example, blue and fin whales that have several subspecies and many independent populations. In such a case, risk can really only be evaluated at the population/subspecies level. Once you get down to this level, even our understanding of the status of large whales is surprisingly poor. A group of us asked the question of stocks assessed in US waters: 'For what percent of populations could we detect a precipitous decline (defined as losing 50% of the population in 15 years)?' I expected to have a poor ability using typical standards for detecting declines for beaked whales and small cetaceans and was not surprised. What was surprising was that small cetaceans and large whales were very similar: we couldn't detect a precipitous decline for about 70% of the populations. To 'improve' at flagging potential declines we either need to increase how often and how intensely we survey or we need to change our standards of what evidence triggers concern.

RW: If you were giving advice (to an early-career cetacean researcher or manager), what would you want them to know about statistical power, based on the lessons you've learned about monitoring cetacean populations?

BT: Statistical power is a term that describes the probability of detecting an effect when the effect is happening: for example, the probability of detecting a precipitous decline when the decline has happened. The context is that in hypothesis testing, which is commonly used by scientists, the hypothesis that the data are used to question is that the population is stable. There are several schools of statistical thought about how to best interpret data and none are particularly intuitive and are therefore not easy to explain. Perhaps the best advice, therefore, is for managers and scientists to work closely together to carefully lay out what the conservation goals are. For example, if the goal is to maintain healthy populations of all marine mammals, then the management must take into account that some species are difficult (and expensive) to get data for. Therefore, to meet the above stated goal the uncertainties will need to be explicitly accounted for. Because many cetaceans are difficult to monitor, it is important not to be over-confident about the state of our knowledge and to continually think about potential new threats and how we can tell whether and how much they affect cetaceans.

RW: What do we do in light of this low statistical power?

BT: Poor precision in abundance estimates can be accounted for by using a minimum abundance estimate rather than the mean estimate. Imagine two estimates: one with a mean of 1000 and poor precision such that 95% of the time the true abundance was between 100 and 1900 and one with good precision with a mean of 1000 and a 95% range between 900 and 1100. Basing management, say an allowable kill in a fishery, on the mean would result in the same management for both cases even though we have little confidence in the poor precision case. If instead, we based our management on a number we felt, let's say, 80% sure that there were at least that number we would base management on a number around 300 for the poor precision case and more like 950 for the precise case. A management scheme that incorporates uncertainty in this way is carried out in the US Marine Mammal Protection Act. Such a treatment is called for given the objectives to maintain *all* marine mammals as functioning elements of their ecosystem. If poor-data species were treated without accounting for the level of our ignorance, then we would be more likely to fail to meet management objectives for such species.

We can also find innovative technological solutions to improve our ability to detect trends.

For example, passive acoustic monitoring[1] allows locations to be monitored for long periods of time. Further progress is required to be able to use such methods for all species, but we're pretty close to having the technical abilities to monitor some species that we've done very poorly with using visual methods, like sperm whales.

RW: How well equipped are we, analytically and at the science–policy interface, at identifying and mitigating human activities that cause sublethal takes?

BT: Sublethal takes are currently poorly handled in management. Consider as an example ocean noise. Let's imagine that animals are displaced from the best habitat by ocean noise. Displaced animals have less food, which reduces their reproduction and in lean years could contribute to some unobserved deaths. For cetaceans, such population-level effects are extremely difficult to quantify. Instead, we generally use indirect evidence. For example, if we compare the trends in abundance between a stock that lives in a noisy area to one that lives in a quiet area we might be able to infer what the population-level effect of noise is. However, that requires us to estimate trends in abundance, which in turn depends on monitoring the population. We need to develop more precise methods for monitoring trends for species that are difficult to survey with

[1] Passive acoustic monitoring is the use of hydrophones to listen to underwater sounds from animals, human operations and the natural physical environment (wind, waves, precipitation). For short-term monitoring, this can be as simple as a hydrophone lowered over the side of a vessel, recording onto a laptop onboard. For long-term monitoring, this usually involves the deployment of autonomous recorders that are anchored to the seafloor and record and store sound for many months. Arrays of multiple hydrophones further allow the localization and tracking of sound sources underwater. Passive acoustic monitoring is extremely efficient and useful in remote locations (e.g. polar regions, under ice), for long-term monitoring (e.g. to determine trends), and for short-term mitigation monitoring of vocal animals around industrial operation (specifically in poor visibility where visual surveys are impossible).

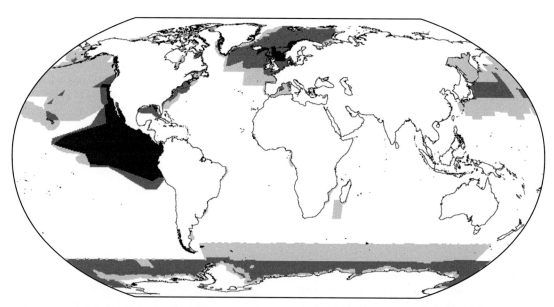

Figure 2.1 A global view of where published cetacean line-transect surveys have taken place (from Kaschner *et al.*, 2012). While this global analysis misses out on many fine-scale cetacean studies, especially those that use photo-identification or direct census methods rather than line-transect surveys, it still makes the general point that our picture of cetacean status is heavily influenced by studies that have taken place in waters under the jurisdiction of wealthy, industrialized countries, especially those in the northern hemisphere. Image © Kaschner *et al.*

traditional methods like mark–recapture photo-identification or visual line-transect surveys. Passive acoustic monitoring is a promising new area that is being used for vaquita and being developed for beaked whales.

RW: As an aside, Kaschner et al. (2012) recently published a paper that shows, visually, how well we are doing at providing the information needed to manage cetacean populations. Figure 2.1 illustrates global coverage of cetacean survey data, to put the statements above in context.

BT: Global understanding of the distribution and abundance of cetaceans is generally pretty poor and definitely patchy. Surveys with big ships have been driven in large part by known management issues like commercial whaling, the tuna fishery that targets dolphin schools to capture fish and Navy operations. Large parts of the world's oceans, especially in the Southern Hemisphere,

remain unsurveyed. However, many of the most remote areas of the oceans also have few threats to cetaceans. Areas with high human populations adjacent to cetacean populations are more worrisome from a conservation perspective, particularly where gillnet fishing has been promoted to feed those human populations. Coastal and island waters in the tropics are areas of special concern for the poorly known cetacean populations.

Cetaceans and anthropogenic stressors (including whale-watching): Dr Liz Slooten

RW: Imagine a marine mammal population that appears to be in trouble. How do you begin to identify and rank factors, both natural and anthropogenic, that are contributing to the population's decline so that we are identifying management actions that are most likely to promote recovery?

LS: If there isn't already good information on distribution and abundance, a population survey is usually the first step. This is what Steve Dawson and I did when we first started our research on New Zealand dolphins (aka Hector's dolphin) in 1984. We didn't know that there was any conservation problem – our intention was to study the behaviour and acoustics of the New Zealand dolphin. For the population survey, we used a small inflatable boat, surveying around most of New Zealand and counted dolphins. We kept finding dead dolphins on beaches, and talking with fishermen revealed that many of the gillnetters fishing in dolphin habitat were catching on the order of 10–20 New Zealand dolphins per year. Just adding up the numbers that they told us they had caught made it clear that there was a problem.

We then re-directed our research from what was meant to be a study of behaviour and acoustics of New Zealand dolphins to a study focused on the information needed to determine whether the level of by-catch was sustainable. This meant using photo-ID to estimate survival and reproductive rates, doing population viability analyses, and so on. Of course, the division between pure and applied science is not black and white. For many field studies, all of the research that goes into assessing the conservation status of a population ends up teaching us a lot about the basic biology of a species, and if you know a species' biology, you can usually identify a short list of plausible factors that could be influencing the dynamics of that population.

For most marine mammals, fishing is by far the most serious conservation threat. In fact, fishing is considered to be the most serious human impact on the marine environment, full stop. This includes direct impacts like by-catch, which is estimated to cause some 300,000 deaths per year (Read *et al.*, 2006). By-catch in fishing gear is a global problem affecting most species of small cetaceans, which is why it is named in endangered species listing and recovery documents for small cetaceans around the world. As the field of marine mammal science has matured and new

information is obtained on other anthropogenic threats, our list of candidate threats has grown to include: climate change; the indirect effects of fishing (e.g. less prey available for marine mammals, seabirds and other marine life); chemical pollution; ocean noise; harmful algal blooms; oil spills; entanglement in or ingestion of marine plastics; and ship strikes. These individual threats are almost always substantially different in terms of the likely magnitude of impact; quantity and quality of data available; potential for managing or mitigating the threat; and the political willingness to manage the threat.

RW: Can you walk us through an example?

LS: I supervised David Lusseau's PhD project (see also Chapters 13 and 16), at the University of Otago (New Zealand), on the impact of tourism on bottlenose dolphins in Fiordland. There are several other potential threats to this population, including a hydro-electric power station that dumps very cold, fresh water into Doubtful Sound. In addition, past (and to a lesser extent current) fishing activity may have further altered the dolphins' environment. David made excellent progress in determining the impact of tourism (Lusseau, 2003). Marta Guerra-Bobo, for her MSc project, is gathering more detailed data on the tourism impact. Studying the other threats, including environmental changes caused by fishing and the power station has proven much more difficult. This does not, of course, mean that these other threats are less serious than tourism, just that it's more difficult to demonstrate the 'smoking gun'.

Another challenge is that, in most countries, different threats are the responsibility of different agencies. In New Zealand, for example, tourism is managed by the Department of Conservation, fisheries mortality is managed by the Ministry of Fisheries, pollution is managed by the Ministry for the Environment and local government agencies. What this boils down to is that it seems to be nobody's responsibility to determine the total impact on a population, or its habitat, or to ensure that the cumulative impact is sustainable. Each agency only wants to deal with 'their' threat. And each agency often wants to manage their particular threat at the maximum sustainable level. You can see that's not going to go well.

Political willingness to manage threats depends on who causes the threat, the level of funding and lobbying skill of the industry or interest group causing the threat, and the willingness of politicians and other decision-makers to ensure the activity is sustainable. Imagine, for example, a species that lives in an area where fishing, mining and tourism are the main human activities. If the fishing and mining industry are well-funded, well-organized and highly skilled at political lobbying, it will be tempting for managers to direct most of their attention to regulating tourism rather than the other two activities, even if this is not the activity that causes the most serious impact. Regulators often go for the 'low hanging fruit' or 'path of least resistance' rather than, necessarily, the most serious threat. If there are 'other threats', this often reduces the enthusiasm of any one of the relevant agencies to reduce the threat they are responsible for. It also confuses the public debate. For example, right now the fishing industry are arguing that their impact on the New Zealand dolphin is sustainable, and it is the 'other' threats that must be causing the declines.

It's important to focus your research on what is likely to be the most serious threat, and to keep your eyes open for other, additional threats. If it is clear that fishing is the most serious threat, but there are additional threats, including pollution, mining and tidal energy generation, then by-catch needs to be reduced to well below the maximum number of deaths that the dolphins could cope with. The population needs to have something to come and go on – in case some of the other threats turn out to be worse than you thought, or in case something else comes along (like a series of bad years for predation or prey availability).

RW: *I take it that the cumulative impact on the population is more than just the sum of the individual impacts?*

LS: No doubt. Measuring any one of these threats is already a serious challenge. Estimating the total impact on a particular population or species is almost impossible. Cumulative impacts are a huge challenge. This is one of the big questions facing marine mammal scientists around the world right now. For many species, multiple stressors are the norm rather than the exception, and sometimes the threats can be synergistic (e.g. the effect of climate change might be stronger, for a population already compromised by pollution) rather than just additive.

RW: *What makes whale-watching different from the other human impacts we have mentioned?*

LS: For all of the other human impacts, there is a benefit for humans and a cost for cetaceans. For cetacean-centred tourism, there is also a potential benefit for cetaceans. The challenge is making better use of that potential. We would be so much further ahead if all whale- and dolphin-watching companies provided excellent educational material – not just about the biology of the species that their business relies on, but also about the conservation threats to that species. Most people, when they have just been on a whale- or dolphin-watching trip, are just smiling from ear to ear and highly motivated to do something to protect that species. Often that potential is wasted, which is such a shame.

Anthropogenic ocean noise: Dr Christine Erbe

RW: *How do we know that cetaceans are disturbed by noise from whale-watching boats or other human activities in the marine environment?*

CE: To 'disturb' implies some knowledge of the animals' emotional or mental state. I'd rather say we know that noise can have certain types of impact on marine mammals. If you picture a sound source underwater, e.g. a fisheries sonar, as you go further and further away from the source in any direction, the received level decreases. At the longest ranges, a sound source might just be audible above ambient noise. For animals to respond behaviourally, you would expect the received level to be substantially higher. However, in areas that are still mostly free from anthropogenic noise, or where animals have been sensitized to certain types of noise, behavioural responses have been documented at very long ranges, modelled to be at the limit of audibility (e.g. beluga whales responding to icebreakers in the Arctic).

We also know that underwater noise can mask animal communication. Given that ocean water conducts light very poorly yet sound very well, marine mammals rely primarily on their acoustic sense for communication, navigation and sensing their environment. From behavioural and electrophysiological studies, we know that noise can mask these sounds to the point where they are no longer recognizable or detectable. There's an increasing amount of data documenting hearing loss in marine mammals after temporary, intense noise exposure. Several studies are investigating noise-induced stress.

We're piecing together a picture of bioacoustic impacts from controlled experiments with individuals of a few 'accessible' species (e.g. small cetaceans in captivity, or wild cetaceans in coastal areas). The biological significance of these – mostly temporary and/or individual effects – remains largely unknown. When does noise put an entire population at risk? How much noise of what type is too much? One common problem with behavioural studies in the wild is the lack of knowledge of context. If an animal moves away from a whale-watching boat, is this escape behaviour or would the animal have changed its path in the absence of the boat as well, or were there other factors that influenced the behaviour (e.g. did the animal detect conspecifics or prey in the other direction)?

Context is hardly ever measured in behavioural studies, yet we have the tools to do so quite efficiently these days. We can very easily map the current physical environment using physical oceanographic sensors or remote sensors. We can measure the biological environment via *in situ* sampling, acoustic or visual imaging and remote sensing. We can quantify the ambient soundscape using passive acoustic loggers. I think we need to do many more correlations, primarily of physical, acoustical and biological oceanographic measures. We also need to collect enough data for the results to have statistical power. While there remains a lot we don't understand about noise impacts, we now have recordings spanning several decades in a few selected regions, and these show a consistent trend of increase in underwater noise levels of about 3 dB/decade (that's a doubling in acoustic energy every decade).

Experimental and modelling approaches to understanding effects of sublethal stressors on cetaceans: Dr Brandon Southall

RW: Can you tell us what you have learned about conducting experiments on whales in the wild and predicting consequences of sublethal stressors, such as anthropogenic noise, on whale populations?

BLS: As you can imagine, it is difficult to conduct experiments on individual high-resolution whale behavioural responses to potential stressors in the wild. Control-exposure experiments offer a powerful tool to identify cause-and-effect relationships between the exposure to a known stressor and response, but they produce the most powerful results when experimenters can maintain various contextual things constant while modifying only one experimental treatment. For whales in their natural habitat, all other things are rarely constant and logistical constraints often result in relatively low sample sizes or limit our ability to account for confounding effects, which may be extremely important to mediating responses, such as the social and ecological context in which exposures occur.

In our recent work on beaked whales (Tyack *et al.*, 2011), we took a multi-scale approach to studying whale responses to naval sonar. We used a compound approach to this study with approaches on very different spatial and data resolution scales. This included the use of high-resolution acoustic and movement sensors to obtain fine-scale information about the direct responses of individual whales to sonar; satellite tags to provide a longer temporal and larger spatial perspective on individual response, but with an associated loss of data resolution; and passive acoustic monitoring to indicate population-level observations on broad scales that could be correlated with sonar activity but that lack any individual resolution. Taken as a whole, these studies gave us more information than any one approach could on its own. This involves some subjective trade-offs among data resolution, sample size and the level of proof that the study design offers. In my experience, researchers interested in sublethal effects of human activities on marine mammals will benefit from designing studies that complement experimental with observational studies, and that explore the problem from the perspective of both individuals and populations. It is particularly important that we scientifically approach these issues from both the individual and population level since each approach has relevance in different ways in terms of how we manage protected species and critical habitats.

The question of 'burden of proof' really speaks to the importance of good communication between scientists and managers. Over the years, I have seen a shift in terms of where we place the burden of proof with respect to consequences of noise to marine mammals. At the beginning of my career, policy-makers were looking for evidence that noise could have serious impacts on the behaviour or fitness of individual marine mammals, largely because that was the level at which management decisions of impact (or 'take' in the USA) were assessed and corresponding mitigation decisions made. Now, there seems to be an increasing realization that there are significant and important considerations beyond just lethal or injurious impacts on individuals in close proximity to intense anthropogenic events. There is more realization that chronic activities

and sublethal impacts at the population level can be particularly important. There are important and very logical consequences of this realization, including more comprehensive assessments of cumulative impacts and consideration of broad-scale sublethal effects such as communication-masking.

However, an interesting and potentially troubling direction this has the potential to take seems to be demanding evidence of unsustainable population-level consequences before managers act. In the last decade, a lot of effort has gone into parameterizing the Population Consequences of Acoustic Disturbance (PCAD/PCoD) framework with real data. It is exciting to see statisticians and ecological modellers working with biologists to put observed behavioural responses into a population-level framework. It is an important research direction, because most of the relevant policies apply to populations, not individuals. But I worry that if taken to the extreme in a management scheme that fails to identify effects other than demonstrable population-level impacts and trends, this change in direction will have unanticipated negative consequences. The PCAD framework may well provide a very useful way for regulatory agencies to assess the sum total of cumulative effects of many different activities in a region during programmatic reviews and identify particularly sensitive types of impacts or life stages, for instance. However, it was never intended nor will it be likely to be appropriate to evaluate whether a particular development or activity should or should not proceed. We would not want a population-level approach to become the only metric by which we assess sustainability of an individual industrial activity. From a conservation perspective, there are so many sources of uncertainty and bias in our studies that if we waited for definitive evidence at the population level before invoking management action, we run the risk of exceeding limits long before identifying what those limits are.

In my view, this is where the lessons of marine mammal science teach us that, as we broaden our view of the appropriate ways of balancing individual/local and regional/population levels of impact we should take a precautionary and adaptive approach to management (e.g. Taylor *et al.*, 2000). Given the difficulties of effectively detecting population-level effects of sublethal impacts of human activities on marine mammals, we need to pay particularly close attention to the few cases where that has been done, and we need to manage our expectations of science. In my view, developing an increasingly sophisticated scientific understanding of impacts on individuals will remain an essential part of our overall scientific and management approach. This is true both because these questions are themselves still fundamentally important (particularly in the context of endangered species as many marine mammals are), and also to inform the broader population-level assessments that should form the basis of ecosystem-level cumulative impact assessments. The trick is managing a complementary scientific and regulatory approach that works on both levels, deals responsibly with what may be high levels of uncertainty, and ensures that individual impacts that are deemed acceptable do not result in irreversible population-level trends before we have the ability to see them coming.

About the contributors

Dr Christine Erbe [CE] is Director of the Centre for Marine Science & Technology (CMST) at Curtin University in Perth, Western Australia. Her expertise lies in underwater acoustics, marine soundscapes, noise generation and propagation, and noise impacts on marine life. CMST is part of the Australian Sustainable Development Institute, and much of CMST's research is centred around the sustainable development of offshore industries (oil & gas, construction, fishing, tourism) by monitoring, mitigating and managing impacts on the marine ecosystem.

Dr Randall R. Reeves [RRR] is the chairman of the Cetacean Specialist Group of the International Union for the Conservation of Nature, co-chair of the Marine Mammal Species Specialist Subcommittee of the Committee on the Status of Endangered Wildlife in Canada and a long-standing member of the Scientific Committee of the International Whaling Commission.

Dr Liz Slooten [LS] works at the University of Otago (New Zealand). Her research, and that of her graduate students, is focused on marine mammal conservation – including the impacts of fishing, tourism and other human activities. Liz is a member of the Cetacean Specialist Group of the IUCN, and has represented New Zealand at the Scientific Committee of the International Whaling Commission for many years. She chairs the organizing committee for the 20th Biennial Conference of the Society for Marine Mammalogy, held in New Zealand in 2013.

Dr Brandon Southall [BLS] is President and Senior Scientist for Southall Environmental Associates (SEA), Inc., a Research Associate with the University of California, Santa Cruz (UCSC), and an Adjunct Assistant Professor at Duke University. Brandon has extensive technical expertise in leading laboratory and field research programmes and applying science in national and international policies through his role in leading the Ocean Acoustics Program for the US National Oceanic and Atmospheric Administration. He also serves as a technical advisor to international corporations and environmental organizations regarding offshore energy development and commercial shipping and has published nearly 50 peer-reviewed scientific papers and technical reports, and has given hundreds of presentations to scientific, regulatory, Congressional, and general audiences around the world.

Dr Barbara Taylor [BT] leads the Marine Mammal Genetics Group at NOAA's Southwest Fisheries Science Center, actively participates in the IUCN Cetacean Specialist Group, and chairs the Marine Mammal Society's Conservation Committee. Her first 10 years in marine mammal research were spent studying harbour porpoise, harbour seals, bowhead whales and humpback whales, mostly in Alaska. Since receiving her PhD at the University of California, San Diego, her research has shifted from a field orientation to a quantitative approach. Her research interests include genetics focusing on identifying units to conserve; population dynamics of small populations; conservation biology; demography; population viability analysis and decision analysis. Since witnessing the extinction of baiji, vaquita have become her first priority in conservation research where she serves on the steering committee for acoustic monitoring, on the international recovery team and on a special committee to advise the President of Mexico on the efficacy of recovery actions for this critically endangered porpoise.

REFERENCES

Kaschner, K., Quick, N.J., Jewell, R., Williams, R. & Harris, C.M. (2012). Global coverage of cetacean line-transect surveys: Status quo, data gaps and future challenges. *PLoS ONE* 7(9), e44075. doi:10.1371/journal.pone.0044075

Lusseau, D. (2003). Effects of tour boats on the behavior of bottlenose dolphins: Using Markov chains to model anthropogenic impacts. *Conservation Biology* 17(6), 1785–1793.

Perrin, W.F., Donovan, G.P. & Barlow, J. (Eds) (1994). Gillnets and cetaceans. *Report of the International Whaling Commission* (Special Issue) 15, 629 pp.

Read, A.J., Drinker, P. & Northridge, S. (2006). Bycatch of marine mammals in U.S. and global fisheries. *Conservation Biology* 20, 163–169.

Taylor, B.L., Wade, P.R., De Master, D.P. & Barlow, J. (2000). Incorporating uncertainty into management models for marine mammals. *Conservation Biology* 14, 1243–1252.

Turvey, S.T., Pitman, R.L., Taylor, B.L., *et al.* (2007). First human-caused extinction of a cetacean species? *Biology Letters* 3, 537–540.

Tyack, P.L., Zimmer, W.M.X., Moretti, D., Southall, B.L. & Claridge D.E. (2011). Beaked whales respond to simulated and actual Navy sonar. *PLoS ONE* 6(3), e17009. doi:10.1371/journal.pone.0017009

3

From exploitation to adoration

The historical and contemporary contexts of human–cetacean interactions

Simon J. Allen

Introduction

Cetaceans have long held great significance in the lives of humans. This significance has been both practical and spiritual and is reflected in artefacts dating back some 10,000 years, in spoken myths and legends, in the writings of Greek and Roman philosophers and many more recent chroniclers of the encounters between humans and wildlife (Alpers, 1963; Cousteau & Diolé, 1972; Constantine, 2002; Neimi, 2010). Some cultures have feared cetaceans, others have been indifferent, while still others revered them, erecting statues and totems in their honour and even deifying them in some cases (Constantine, 2002). Broadly speaking, the historical view of cetaceans was divided along loose phyletic lines: dolphins being seen as sleek, graceful and a good omen for sailors (or simply as pests to fishers), while the great whales were monstrous, mysterious and fearful.

With increasingly thorough exploration and exploitation of riverine and oceanic habitats over the course of the last few centuries, human–cetacean interactions have moved far beyond opportunistic encounters and come to span the broadest possible spectrum (Whitehead *et al.*, 2000). Humans and cetaceans have engaged in commensal and mutually beneficial foraging relationships; humans have hunted cetaceans for food, bait, oil, baleen and other resources, culled cetaceans over fears of competition for fish resources, and captured them for display in aquaria; and cetaceans have, in turn, exploited various human activities for assisted locomotion, play and the procurement of food (e.g. Busnel, 1973; Shane *et al.*, 1986; van Waerebeek & Reyes, 1994; Orams, 1997; Samuels & Tyack, 2000; Read, 2008). This spectrum of interactions has resulted in negative, neutral and positive outcomes for both humans and cetaceans involved.

Despite this diverse suite of interactions and the reverence in which some cultures have viewed cetaceans, the relationship through the nineteenth and much of the twentieth century could only be characterized as grossly exploitative on the part of humans and negative in outcomes for cetaceans. Hundreds of thousands of dolphins and porpoises have been taken in either directed hunts or incidentally caught in fisheries targeting other wildlife (fish and invertebrates) and many of the great whales were hunted to the brink of extinction, these practices more recently being carried out against national and international legal mandates (e.g. Bannister, 1986; Yablokov, 1994; Perry *et al.*, 1999; Gosliner, 1999; Northridge & Hofman, 1999; Mangel *et al.*, 2010).

Coincident with the expansion of aquaria displaying dolphins and the use of great whales as an emblem for the green movement, there was a sweeping change in attitudes in at least some western cultures in the 1970s (Lavigne *et al.*, 1999;

Whale-watching: Sustainable Tourism and Ecological Management, eds J. Higham, L. Bejder and R. Williams.
Published by Cambridge University Press. © Cambridge University Press 2014.

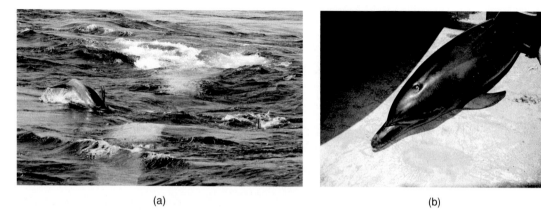

(a)   (b)

Figure 3.1 (a) Exemplifying the diversity of human–cetacean interactions within a time and place, common bottlenose dolphins (*Tursiops truncatus*) forage on fish escaping from trawl nets during winch-up off northwestern Australia (photo: S. Allen). (b) A dolphin caught in a trawl net during the same fishing trip (photo: S. Allen). These dolphins follow trawlers around for days and weeks at a time, benefitting from the concentrated food source, but independent observer reports suggest that around 50 dolphins are caught per annum (Allen & Loneragan, 2010). Underwater video footage taken inside actively fishing trawl nets suggests that by-catch is under-estimated, as some dolphins fall out of the net before being landed (Jaiteh *et al.*, 2013). Dolphins, like all marine mammals, are protected in Australian waters. (See colour plate section.)

Samuels & Tyack, 2000; Chapter 4, this volume). Between the 1970s and 1990s, some nations passed laws prohibiting the harm of particular individuals, populations or indeed the order Cetacea in its entirety (e.g. Baur *et al.*, 1999). These regulations have generally been introduced in recognition of the significant roles cetaceans play in ecosystems, and also in recognition of their cognitive abilities and the complex social societies in which they live. Nevertheless, some cultures still hold a primarily utilitarian view of cetaceans and the broad suite of human–cetacean interactions of prior centuries continues across the globe to this day, even within the very regions where cetaceans fall under full protection from disturbance, harassment and takes as a result of human activities (Northridge & Hofman, 1999; Whitehead *et al.*, 2000; Shaughnessy *et al.*, 2003; Mangel *et al.*, 2010; Chapters 2, 4, 6 and 18, this volume; Figures 3.1a, b).

As has been the case with many a faunal icon, human perceptions of cetaceans and, thus, the ways in which humans interact with cetaceans, have evolved with a greater understanding of their biology and habits. Whether this acquisition of

knowledge and the resultant increased awareness was designed to achieve more efficient harvests, to serve empirical science, or to implement improved conservation measures, cetaceans being subject to various forms of scientific research is yet another in the suite of human–cetacean interactions. Research on cetaceans, through relatively recent history and now contemporaneously, has also ranged from being lethal (e.g. associated with whaling), through varying levels of invasion (e.g. capture and handling) to being benign (e.g. shore-based observations) in its outcomes for cetaceans (see detailed reviews in Samuels & Tyack, 2000; Boyd *et al.*, 2010).

After humble beginnings in the 1950s, recent decades have also seen the rise and spread of the nature-based tourism industry that revolves around free-ranging cetacean populations – 'whale-watching' hereafter (e.g. Connor & Smolker, 1985; Duffus & Dearden, 1993; Samuels *et al.*, 2000; Chapter 5, this volume). In the late 1990s, conservative estimates tallied a global whale-watching industry involving over 2000 licensed operators taking tourists to observe, or indeed swim with or feed, cetacean communities in the waters of over 80 different countries and territories, yielding around US$1 billion per annum (Hoyt, 2001). Just a decade on, updated figures indicated over 3300 licensed operators taking tourists to interact with cetaceans in around 120 countries and territories, and more than a doubling in global revenue to over US$2 billion per annum (O'Connor *et al.*, 2009). The number of unlicensed tour operators is difficult to quantify, but would no doubt inflate these figures substantially. This rapid growth and diversification of whale-watching, combined with increasing evidence of the impacts it can have on targeted cetacean populations (e.g. Corkeron, 2004; Lusseau, 2004; Lusseau *et al.*, 2006; Bejder *et al.*, 2006; Williams *et al.*, 2006), present ongoing challenges to the management of this broad category of human–cetacean interactions (Chapter 25, this volume).

In this chapter, the historical and contemporary contexts of human–cetacean interactions are overviewed. The body of the chapter is broken into three 'ages' in which particular human–cetacean interactions are used as loosely representative of those times: early chance encounters and opportunism; through the industrialization of fishing and whaling and the beginnings of scientific exploration into cetacean behaviour and biology; followed by the rise of nature-based tourism based on cetaceans. The chapter culminates in the present time, in which these interactions still range from exploitation to adoration. This concluding section touches on what the twenty-first century has already brought in human–cetacean interactions.

The 10,000-year age of chance encounters and opportunism

Records of human–cetacean interactions date back as far as the Stone Age. There are carvings of killer whales (*Orcinus orca*) in rocks of northern Norway that are estimated to be some 9000 years old. From the same area and era, archaeological excavations have revealed single-edge slate knives,

likely used for flensing, fashioned in the shape of diving cetaceans (Constantine, 2002; Neimi, 2010; Figure 3.2a). Coast-dwelling peoples have long used stranded whale carcasses on an opportunistic basis, while and hand harpoons and porpoise skulls found in burial mounds in Norway and Japan suggest that the active hunting of small cetaceans probably dates from several thousand years BC.

A whale on the beach is wealth for seven villages. (Japanese proverb)

Bowhead whale (*Balaena mysticetus*) DNA extracted from artefacts collected from excavations in Greenland indicate use by Eskimo and Inuit cultures of over 4000 years ago (Sinding *et al.*, 2012). A recent review of Turkish cetacean fisheries suggests beginnings involving small cetacean hunting as early as several centuries BC (Tonay & Öztürk, 2012). Numerous Roman and Greek myths and legends reflect these people's great affinity for the sea and for dolphins. Indeed, while some accounts of the interactions between humans and dolphins have far-fetched elements, many are supported by the more factual writings of Aristotle and the circumstances of some contemporary human–cetacean interactions (Alpers, 1963). One Roman philosopher, for example, told the story of a Mediterranean peasant boy who befriended a dolphin he called 'Simo'. The boy supposedly fed the dolphin and was rewarded by being able to ride on the dolphin's back. Many Roman artefacts portray a close relationship between humans and dolphins (e.g. Figure 3.2b). There are multiple instances of lone dolphins seeking out human contact and social interaction in more recent times, many of the characteristics of which bear striking resemblance to these fables (Lockyer, 1990; Orams, 1997; Constantine, 2002).

The Ancient Greeks held dolphins in such high regard that injuring or killing one was viewed in the same light as killing a human, and was punishable by death. While they held dolphins in great reverence, nearby people, such as the Mossynoics and Thracians, were harpooning dolphins for their meat and blubber (Alpers, 1963; Tonay & Öztürk, 2012).

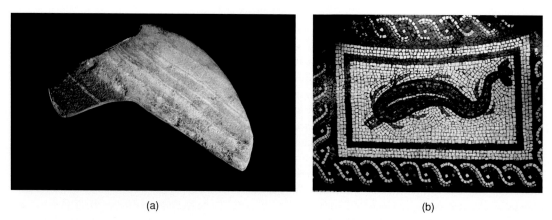

(a)   (b)

Figure 3.2 (a) Shaped to evoke association with diving whales and dolphins, a single-edged slate knife that was likely to have been used for flensing fish and marine mammals by communities of northern Norway in the Stone Age (photo: A. Icagic/Tromsø University Museum). (b) A mosaic floor section depicting a dolphin from a third-century AD Roman villa in the south of England (photo: D. Allen, Wolf Design). (See colour plate section.)

There is evidence of mutually beneficial foraging relationships between various coastal tribes and groups of dolphins from many different locations (Busnel, 1973; Orams, 1997). Those involving coastal fishers and Atlantic humpback dolphins (*Sousa teuszii*) in Mauritania, Africa, and common bottlenose dolphins in Brazil, South America, still occur today (Orams, 1997; Daura-Jorge *et al.*, 2012). Another interesting example is that from Moreton Bay, southeast Queensland, where Australian Aborigines were reported to work with dolphins (likely Indo-Pacific bottlenose dolphins, *T. aduncus*) to herd and capture migrating mullet (*Mugil* spp.) (Fairholme, 1856). In 1799, English explorer Matthew Flinders described the catching of dugongs (*Dugong dugon*) by Aborigines in this same locale (Mackaness, 1979); so related marine mammals were clearly viewed in differing lights. The tribal elders with whom Fairholme (1856) spoke suggested that the fishing association with dolphins had occurred for as long as they could remember. Given the likelihood that Aborigines have inhabited the island continent for over 50,000 years (Veth & O'Connor, 2013), this may represent an oral history link to the most ancient of human–cetacean interactions with a positive outcome for both. In other cultures, however, both dolphins and whales have been viewed throughout history as more akin to fish, and,

therefore, purely for harvest. It was the lack of physical and technological prowess that meant many of the human–cetacean interactions with negative outcomes for cetaceans stayed largely opportunistic (for humans, at least) and they persisted through some thousands of years.

The 150-year age of no-chance encounters, exploration and exploitation

Aboriginal Australians engaging in mutually beneficial foraging relationships with delphinids provides a link between the age of opportunism and chance encounters to the age of no-chance encounters in, perhaps, one of the more remarkable human–cetacean interactions documented. In Twofold Bay, southern New South Wales, commercial whalers established an apparently cooperative hunting bond with killer whales in the pursuit of humpback whales (*Megaptera novaeangliae*) in the late 1800s (Wellings, 1944). The killer whales reportedly alerted the shore-based whalers, including both colonial and Aboriginal participants, to the presence of migrating humpbacks by swimming into the bay and performing surface-acstive behaviours, such as breaching, tail-slapping and lob-tailing. The killers would then lead the whaleboats to the humpbacks and assist in their capture (Figure 3.3a). Once the humpbacks were killed by a combination of harassment by killer whales and harpooning by whalers, the whalers would attach buoys to the carcasses, leaving them for the killers to feast on the lips, tongue and genitals, then returning later to tow the baleen whales to shore for processing. The seasonal onset of the humpback migration would see the arrival of the killers, each of which became known to the whalers by their distinctive dorsal fin. This apparent mutualism, although clearly to the detriment of baleen whales, persisted for many decades, until the death of the perceived leader of the killer whale pod, 'Old Tom', in 1930. With a diminishing humpback population and without the killer whales for assistance, shore-based commercial whaling was no longer financially viable in Twofold Bay and ceased in the same year (Wellings, 1944; Mead, 1961). Like the interaction between Aborigines and marine mammals in Moreton Bay, this situation aptly reflects how humans viewed closely related animals in different ways according to the time, location and cultural perspective (see also Figure 3.3b).

There were significant improvements in technology in the late nineteenth century and into the twentieth century that resulted in a change in human–cetacean interactions. This shift was not in the diversity of these interactions, which were already quite broad, but in the intensity of these interactions. The advent of the harpoon cannon, the industrialization of fishing fleets and the use of factory ships allowing whaling crews to stay at sea for months on end and process their catch 'in real time' meant an exponential increase in the extent of human exploration of the seas and exploitation of its inhabitants. This is perhaps aptly illustrated by the following quotes from the French oceanic explorer, entrepreneur and filmmaker, Jacques Cousteau:

Until the nineteenth century…man's career as a killer of whales was limited by the relatively primitive means of pursuit and slaughter at his command – ships propelled by sails or oars, hand winches, flimsy ropes, and hand-thrown harpoons. At this time, size was a decided advantage to

a whale, for a man could attack and kill only the smaller specimens. With the invention of the harpoon gun in 1864, however, the balance of power shifted. (Cousteau & Diolé, 1972)

The harpoon cannon not only destroyed myth, legend, poetry, and Moby Dick, the white whale; it also brought into question the very survival of the largest beings on the face of the earth. (Cousteau & Diolé, 1972)

The spread of whaling technology occurred rapidly and the exploitation of the great whales became one of history's 'best' examples of unsustainable, profit-driven natural resource abuse. The lack of foresight in managing whaling led to the collapse of many populations of great whales and, accordingly, the demise of many large commercial whaling companies in the twentieth century (see also Chapters 2, 4 and 18, this volume). Dawbin (1986), for example, reports on the decimation of Southern right whale (*Eubalaena australis*) stocks around New Zealand and southeastern Australia in the 1800s. Similarly, right whale stocks off southwestern Australia that were depleted by the mid-1860s did not recover until half a century after their protection in 1935, because of illegally continued whaling by the former Soviet Union from 1950 to 1970 (Yablokov, 1994; Bannister, 2001). In an example of an anthropogenically induced (fisheries) collapse of a small cetacean population, Fontaine *et al.* (2012) recently used genetic data to infer a harbour porpoise (*Phocoena phocoena*) population reduction of about 90% in the Black Sea within the past five decades. Harder to quantify, but no less detrimental to cetacean populations, were both the direct and indirect impacts of industrialized commercial fisheries targeting fish and invertebrates. Hundreds of thousands of cetaceans have been injured or killed through interacting with fisheries operations (e.g. Read *et al.*, 2006), and many more have been impacted by the indirect effects of depleting fish stocks and the changes that occur to ecosystems when just one functional group is depleted, or, indeed, entirely removed (e.g. Myers & Worm, 2003).

The first in-depth research into cetacean biology occurred as a direct result of our commercial

(a)

(b)

(c)

(d)

Figure 3.3 (a) The remarkable human–cetacean interaction involving killer whales and shore-based whalers, working together in hunting humpback whales off southeastern Australia. Here, a whaleboat and a killer whale, 'Old Tom', flank a humpback whale (photo: C. Wellings/courtesy National Library of Australia). (b) The image of a whale carved into a boab (*Adansonia gregorii*) on the Kimberley coast of northwestern Australia. Humpback whales were, and still are, revered here, while dugongs were, and still are, hunted for consumption by Aboriginal people. Given the boab's longevity, this carving could be decades, hundreds or even over 1000 years old (photo: S. Allen/permission by D. Woolagoodja, senior custodian of the Dambimangari/Worwoorra People). (c) Cetacean researchers conducting shore-based observations of whales from the Big Island of Hawai'i (photo: S. Allen). (d) Researchers conducting boat-based photographic identification of dolphins in southwestern Australia (photo: S. Allen). (See colour plate section.)

interest in the great whales over the last 150 years and generally resulted in the fatality of the research subject. This research was driven by a quest for improved knowledge of whale population biology, with the end goal of increased harvests or better stock management as the resource declined from over-exploitation (Samuels & Tyack, 2000). Research methods generally involved the analysis of carcasses and historical whaling statistics, but also the use of detailed descriptions of whale behaviour in response to approaching vessels and harpooning attempts. Whalers with an intimate knowledge of their quarry's behaviour could not only better exploit the resource, but also reduce the serious risks to people associated with whaling. Smaller cetaceans were generally harvested seasonally and from shore-based stations, so that behavioural data obtained by spending extended periods at sea and in close proximity to the animals were lacking (Samuels & Tyack, 2000).

In the last 70 years or so, however, the proliferation of aquaria displaying cetaceans also led to research on captive delphinids being undertaken, shedding light on many aspects of their behaviour, physiology and their sensory and cognitive abilities (e.g. McBride & Hebb, 1948; Tavolga & Essapian, 1957; Bateson, 1974; Herman, 1980). While critical to raising awareness and appreciation of cetaceans, at least in some western cultures, there were obvious limitations to the applicability of research in the captive setting to free-ranging cetacean behavioural ecology. For example: relatively few cetacean species adapt to captivity, essentially a few delphinids, and only after considerable 'trial and error'; conditions in the captive setting are often far removed from unconstrained natural conditions; and, furthermore, there are growing concerns over the welfare and rights of cetaceans in captivity (Whitehead *et al.*, 2000; Stewart & Marino, 2009).

In seeking to raise the public's awareness of the plight of cetaceans in the seas and oceans of the world, pioneers in the realm of marine research employed techniques such as: harpooning or branding instead of photographing for individual identification; temporarily restraining animals by lasso or net for behavioural observation; and capturing them for the purposes of attaching bulky video cameras or transmitters. These techniques would be considered unacceptable by today's standards. For example,

Having succeeded in grabbing and holding on to a whale calf this morning, he conceived of the idea of – stunt-riding on the back of a whale…From that moment on, the most vital question aboard *Polaris* has become: who will be the first to be towed by a bucking whale-bronco? (Cousteau & Diole, 1972)

Less-invasive, scientific outcome-oriented endeavours crucial to the understanding and conservation of free-ranging cetaceans have more recently complemented the early research on captive delphinids (Figure 3.3c,d). Ever more detailed insights into many facets of cetacean social structure, behavioural ecology and human impacts thereon have been gained from the long-term study of individually recognizable animals within wild populations (e.g. Clapham, 2000; Connor *et al.*, 2000; Chapters 14–17, this volume). In research through whaling (lethal), the captive setting and that conducted on free-ranging cetaceans, there is at least a loose scale of 'most through least' detrimental to cetaceans. Loose only, as some research on free-ranging cetaceans still involves the use of invasive techniques that can result in considerable stress to individuals or groups of cetaceans. Indeed, some research/monitoring techniques may have contributed to the death of already-compromised individuals (e.g. Bearzi, 2000; Moore *et al.*, 2012). The ethics surrounding scientific research on cetaceans has developed into an area that requires due consideration by all practitioners (detailed in Gales *et al.*, 2010). Research that is ethically rigorous in both its design and in its conduct has become critical to furthering our understanding of human–cetacean interactions.

The 30-year age of continued whaling and the rise of whale-watching

Member states of the International Whaling Commission (IWC) adopted a moratorium on commercial whaling of all species almost three decades ago. Whaling is, thus, considered largely defunct as a commercial enterprise (Cisneros-Montemayor *et al.*, 2010). Nevertheless, human–cetacean interactions that include whaling in numerous guises, even if not always for commercial gain, continued through the 1980s and onward. For example: Aboriginal communities in various regions (e.g. Canada, Greenland and the US) continue to hunt cetaceans in traditional or subsistence whaling endeavours; the Japanese continue to hunt whales for 'scientific' purposes, while various dolphin species are driven into coastal regions and killed for food, killed as part of a cull (to reduce perceived competition for fish resources), or sold for display in aquaria; the nations of Norway and Iceland continue to hunt whales on a commercial basis; the people of the Danish province of the Faroe Islands conduct annual or more frequent 'grinds' resulting in the deaths of hundreds of pilot whales (*Globicephala melas*) and several other odontocete species; and various African, Asian, South and central American nations, as well as West Indian and South Pacific Islanders hunted cetaceans throughout this period (Whitehead *et al.*, 2000).

During this same period, tourism that focuses on free-ranging cetaceans grew enormously in the number of participating communities and the diversity of targeted cetacean species (Hoyt, 2001; O'Connor et al., 2009). Many countries that once exploited cetaceans for food and other consumables, and some that still do, replaced whaling with whale-watching as a less-invasive use of a natural resource (e.g. Romero & Hayford, 2000; Parsons & Rawles, 2003; Kogi *et al.*, 2004). Whale-watching brought, and continues to bring, measurable benefits to local communities including: economic gains through an inward flow of visitor expenditures; employment opportunities; expanded infrastructure; exposure of visitors to unfamiliar wildlife communities and natural areas; logistic and financial support for scientific endeavours; marine environmental education; and the promise of a 'more sustainable' use of cetaceans than directed harvests (Orams, 1997, 2001; Hoyt, 2001; Bejder & Samuels, 2003; Garrod & Fennell, 2004). Benefits to cetaceans through tourism include: non-lethal encounters with humans instead of directed harvests; and the indirect conservation value to populations, species and ecosystems through increased public awareness and activism.

Human–cetacean interactions through whale-watching include dolphin provisioning programmes, swimming with dolphins and whales, and viewing numerous cetacean species. In the early 1990s, whale-watching was supported by the IWC (and various influential NGOs) as a sustainable, non-consumptive use of animals, with member countries encouraged to manage industry development. In the mid- to late 1990s and early twenty-first century, however, the wisdom of the largely uncontrolled and inconsistently managed growth of whale-watching without parallel application of research and monitoring was questioned (e.g. Corkeron, 2004, 2006). Doubt was also cast over whether the industry deserved the label of ecotourism without further scrutiny (Samuels & Bejder, 2004). These challenges followed and led to broader debate as to whether ecotourism amounts to anything more than nature-based tourism with a green label applied to increase its appeal, and also, whether whale-watching could actually be considered a sustainable endeavour (e.g. Orams, 1999; Bejder & Samuels, 2003; Lusseau *et al.*, 2006).

Provisioning cetaceans has provided thousands of humans with 'up close and personal' encounters with dolphins, but short-term behavioural impacts similar to those reported for provisioned terrestrial mammals have been reported. These included overt aggression between animals and toward people (e.g. Connor *et al.*, 1992; Orams *et al.*, 1996), and risks to dolphins have included injuries associated with fishing gear, decreased maternal behaviour, dependency of juveniles on provisioning, and physical abuse by humans (Mann & Kemps, 2003; Samuels *et al.*, 2003). While these interactions should be carefully managed, they are unlikely to impact upon the viability of populations, species or ecosystems.

The demand for encounters that involve swimming with free-ranging cetaceans experienced rapid growth in the 1990s (Würsig, 1996). A decade ago there were already over 20 species of cetaceans being targeted in swim programmes around the world (Samuels *et al.*, 2000). Swimming with cetaceans usually involves the use of snorkelling equipment combined with either: free swimming; being towed on a rope beside or behind a vessel; being towed on ropes behind a motorized underwater scooter; or hanging onto a rope near a stationary vessel. Most interactions involve dolphins, but operations involving swimming with whales include dwarf minke whales (*Balaena acutorostrata*) in the Great Barrier Reef Marine Park, Australia, and humpback whales in the waters of the South Pacific island nation of Tonga (Birtles *et al.*, 2002; Orams, 2002).

Swimming with cetaceans has proven contentious among scientists, managers and tourism

(a)   (b)

Figure 3.4 (a) A dolphin calf, habituated to close interactions with humans through provisioning and a swim-with-dolphins programme, with fishing line entanglement. This entanglement prevented suckling, resulting in slow starvation (photo: Bunbury Dolphin Discovery Centre). (b) Recreational boaters breaching minimum approach distance regulations to Indo-Pacific bottlenose dolphins, Rottnest Island, Western Australia. After repeated breaches, some individuals started tail slapping and then the resting group moved away into an area where boats could not approach (photo: S. Allen). (See colour plate section.)

operators. Proponents of swim programmes suggest that cetaceans can choose to maintain or terminate interactions if disturbed (e.g. Dudzinski, 1998; Birtles *et al.*, 2002), while others caution that animals exposed to, and tolerant of, regular contact with humans may be more prone to vessel strike, entanglement or vandalism (e.g. Spradlin *et al.*, 1998; Figure 3.4a). Impacts recorded when swimming with cetaceans have included aggression toward humans; avoidance of swimmers in the short and long term; physical abuse of animals; disruption of natural behaviour; and risk of injury to both humans and animals (Shane *et al.*, 1993; Bejder *et al.*, 1999; Constantine, 2001; Samuels *et al.*, 2003). Accordingly, swimming with cetaceans is banned in the waters of nations including Argentina, Brazil, Mexico, South Africa and Spain (Marsh *et al.*, 2003) and is not supported by some NGOs.

Watching cetaceans occurs from land-, aircraft- and boat-based platforms, with the latter being by far the most prevalent. Again, many thousands of humans have enjoyed first-hand encounters with charismatic animals and been provided with opportunities to learn about megafauna and the environments they inhabit. Cetaceans often approach vessels and extended interactions that are benign in outcome occur. However, when industries intensify and competition for the tourist dollar mounts, problems have arisen. Documented responses to vessel approaches by dolphins include: short-term changes to respiration rates; behavioural state; acoustic behaviour; movements and habitat use (e.g. Janik & Thompson, 1996; Scarpaci *et al.*, 2000; Hastie *et al.*, 2003; Constantine *et al.*, 2004; Figure 3.4b). Similar findings have been reported for at least four species of great whales (Baker & Herman, 1989; Gordon *et al.*, 1992; Richardson *et al.*, 1995; Heckel *et al.*, 2001).

Injuries or fatality from vessel strike through whale-watching activities are less common, but nonetheless traumatic for the individuals involved and warrant concerted mitigation efforts as human usage of estuaries and coastal zones continues to increase. As is the case with swim programmes, the issue of some communities of cetaceans becoming habituated to regular contact with vessels may render them more prone to vessel strike and/or entanglement (Wells & Scott, 1997; Stone & Yoshinaga, 2000; Laist *et al.*, 2001). With regard to deleterious impacts from exposure to whale-watching activities over the longer term, changes in: daily behavioural budgets; patterns of residency; relative abundance; and even energetics have been detected in several delphinid communities (Constantine, 2001; Lusseau, 2003, 2004, 2005; Bejder *et al.*, 2006; Williams *et al.*, 2006). The results of these exemplary studies challenge claims that whale-watching is sustainable (e.g. Lusseau *et al.*, 2006). Indeed, the IWC's consensus has shifted in the last decade from the promotion of whale-watching as a sustainable, non-consumptive use of animals to:

There is new compelling evidence that the fitness of individual odontocetes repeatedly exposed to whalewatching vessel traffic can be compromised and that this can lead to population-level effects. (IWC, 2006)

Most whale-watching operations do not yet involve collaborations between operators, research and management and are, thus, unable to identify, let alone ameliorate, negative impacts or claim to be managed for sustainability. Whale-watching's environmentally educational value, sustainable management and contributions to the conservation of animals and ecosystems (or at least, minimal environmental degradation) are 'best practice' goals toward which the industry should aspire. Under past and current circumstances, whale-watching is generally established and growing before any form of research or impact assessment is conducted. Management regimes have typically been implemented on a reactionary basis as a result of perceived or measured impacts and the variability evident in these regimes has been described as 'haphazard' (Garrod & Fennell, 2004). Where assessed, operator compliance to management guidelines and regulations has not proven adequate in fulfilling the goals of reducing exposure of cetaceans to tour vessel activity (e.g. Whitt & Read, 2006; Allen *et al.*, 2007). Furthermore, the onus to prove or disprove impacts has rested largely upon researchers, rather than on those that are statutorily obliged to manage human activities or those that profit from whale-watching. While whale-watching has contributed significantly to the adoration of cetaceans in some western cultures, it has proven

deleterious impacts on cetaceans and needs to be managed accordingly.

Human–cetacean interactions into the twenty-first century: the age of extinctions?

So it remains that human–cetacean interactions vary between the cultural perspectives and motivations of the humans involved, between locations and between cetacean species, covering just as broad a spectrum today as they have done in prior centuries, albeit in differing ways. Humans in disparate locations around the globe: still engage in mutually beneficial foraging relationships with various cetaceans; humans still hunt cetaceans; humans still display cetaceans in aquaria; and free-ranging cetaceans still exploit human activities in a variety of ways (e.g. Figure 3.5a). The dissemination of insightful and innovative research and the continuing spread of whale-watching have, in at least some circumstances, contributed to a broader awareness of the conservation needs of cetaceans and marine environs as a whole. There are now numerous NGOs and influential bodies lobbying on behalf of cetacean conservation. However, despite more detailed knowledge of cetacean biology and anthropogenic impacts thereon, increased public awareness and far more widespread empathy toward the welfare and rights of cetaceans in the twenty-first century, the outcomes of many human–cetacean interactions remain predominantly detrimental to cetaceans individually and at population levels.

Human activities and influence continue to encroach further, deeper and more pervasively into cetacean habitats. While many might perceive there to be fewer cetaceans suffering the effects of directed harvests than did in the nineteenth and twentieth centuries, for example, most species of cetacean are still intentionally killed in at least some of their respective distributions (Whitehead *et al.*, 2000; Mangel *et al.*, 2010). Some species that are returning from the brink of extinction at the hands of commercial whaling now face ever-increasing obstacles to successful migration and breeding in

(a)

(b)

Figure 3.5 (a) Indo-Pacific bottlenose dolphins leaping in the pressure wave off the bow of a ship leaving the Swan River, Western Australia (photo: S. Allen) (b) Further upstream on the very same day, an emaciated resident dolphin with gross skin lesions and fishing line entanglement perishes adjacent to the city of Perth (photo: S. Allen). (See colour plate section.)

the form of coastal development, including the proliferation of marine aquaculture, port development and shipping activity (Allen & Bejder, 2003; Smith *et al.*, 2012). Furthermore, the two most severe threats to cetacean diversity and persistence through the twenty-first century remain of anthropogenic origin: the direct and indirect impacts of interactions with industrialized commercial fisheries; and the effects of climate change (Halpern *et al.*, 2007; Read, 2008). Indeed, while they may have been interpreted as pessimistic at the time of publication, some of Anderson's (2001) and DeMaster *et al.*'s (2001) objective and dispassionate predictions are already proving correct only a decade into this century. For example:

In the late 20th century new threats arose through contamination of the marine environment and these will be exacerbated and added to as human numbers double and economies grow over the next century. (Anderson, 2001)

High contaminant concentrations have been found in northeast Pacific killer whales; there is evidence of increasing uptake of brominated flame retardants in striped dolphins (*Stenella coeruleoalba*) around Japan; Atlantic bottlenose dolphin populations in southeastern USA carry a suite of organic chemicals at or above the level where adverse effects have been reported in both wildlife and humans; numerous bottlenose dolphin carcasses exhibiting gross

skin lesions have washed ashore around cities and urban centres of Queensland, Victoria and Western Australia in recent years; the concentrations of pollutants (including DDE and PCBs) detected in the tissues of dolphins that died in the Swan River adjacent to Perth, Western Australia, in the winter of 2009 exceeded published thresholds for effects on immune function and levels of the banned pesticide dieldrin were among the highest reported globally (Ross *et al.*, 2000; Isobe *et al.*, 2009; Fair *et al.*, 2010; Holyoake *et al.*, 2010; Figure 3.5b). What, then, is happening in locations where the facilities or; indeed; the interest in detecting such anthropogenic influences are lacking?

And from DeMaster *et al.* (2001):

Because of this [commercial fisheries adversely affecting marine mammals], the number of extant populations and species richness of marine mammals will be reduced by the end of the 21st century, and coastal populations will be affected more negatively than will noncoastal species. (DeMaster *et al.*, 2001)

The baiji (or Yangtze River dolphin, *Lipotes vexillifer*) is now extinct as a result of entanglement in fishing gear and habitat modification, representing the first loss of a cetacean species directly attributable to human influences (Turvey *et al.*, 2007). This will undoubtedly precede others: the Yangtze finless porpoise (*Neophocaena phocaenoides*) of the same estuary; the vaquita (*Phocoena sinus*) of the Sea of Cortez; Maui's dolphin (*Cephalorhynchus hectori maui*) of New Zealand's North Island; and the North Atlantic right whale (*Eubalaena borealis*) face similar fates should human usage of the estuarine and coastal regions they inhabit not change in the very near future (Dawson *et al.*, 2001; Kraus *et al.*, 2005; Slooten *et al.*, 2006; Jaramillo-Legorreta *et al.*, 2007; Turvey *et al.*, 2007). The 'common' dolphin (*Delphinus delphis*) is rapidly approaching extirpation in various parts of the Mediterranean Sea due to widespread overfishing; fishery-impacted stocks of spinner and spotted dolphins (*S. longirostris* and *S. attenuata*, respectively) of the Eastern Tropical Pacific are not recovering despite considerable reductions in by-catch in recent decades; and the ongoing by-catch of dolphins and other protected and critically endangered marine fauna in northwestern Australia is invariably deemed 'acceptable' by the Department of Fisheries, despite there being no assessments of the ability of these wildlife populations to absorb this human-caused mortality (Bearzi *et al.*, 2003; Cramer *et al.*, 2008; Allen & Loneragan, 2010; Department of Fisheries, 2011).

It is against such an ominous backdrop that whale-watching, the central theme of this volume, has recently been viewed in relative terms, branded 'sustainable' and promoted as 'ecotourism'. For example:

Responsible whale watching is the most sustainable, environmentally friendly and economically beneficial 'use' of whales in the 21st century. At a time when the global economy, our planet's great whales and international whale conservation measures are all under threat, it is encouraging to see coastal communities the world over continuing to reap increasing benefits from this rapidly developing form of ecotourism. Animals and people both do better when whales are seen and not hurt. (Ramage, 2009)

However, 'sustainable, environmentally friendly' whale-watching usually involves burning fossil fuels, adding noise and other pollutants to cetacean habitats and has proven detrimental to some targeted communities. In at least some instances, it

would seem that cetaceans are, in fact, 'hurt' as a result of being seen. Of course, human–cetacean interactions in the form of whale-watching incur less-obvious impacts than whaling and commercial fisheries interactions, but less impact does not mean no impact. It seems a prudent time for the whale-watching industry to be recognized as consumptive in nature and managed accordingly. With those first (not insignificant) steps taken, the whale-watching industry could then be at the forefront of championing cetacean conservation: leading by example; disseminating information about the current reality of human–cetacean around the globe; spreading environmentally sound messages; and encouraging tourists to change their behaviour to reduce the negative aspects of these interactions (Corkeron, 2006; Chapters 4 and 25, this volume).

Cetaceans commonly rank highly in the animal popularity stakes. For example, both dolphins and whales featured in the world's 'top ten most popular animals' in a survey of over 50,000 members of the public from 73 countries (BBC, 2004). If broader society is not comfortable with the notion of the twenty-first century being the age of extinctions for some of the most popular animals on the planet, as it has inauspiciously begun, the large proportion of human–cetacean interactions that result in cetaceans being disturbed, injured or killed will need to be curbed, while those that result in benign or positive outcomes for cetaceans will need to spread as quickly as whale-watching did in the late twentieth century.

The greatness of a nation and its moral progress can be judged by the way its animals are treated. (M. Gandhi)

REFERENCES

Allen, S. & Bejder, L. (2003). Southern Right Whale *Eubalaena australis* sightings on the Australian coast and the increasing potential for entanglement. *Pacific Conservation Biology* 9, 228–233.

Allen, S.J. & Loneragan, N.R. (2010). *Reducing dolphin bycatch in the Pilbara finfish trawl fishery*. Final Report to the Fisheries Research and Development Corporation. Perth, Australia: Murdoch University, p. 59. ISBN: 978-0-86905-926-5.

Allen, S., Smith, H., Waples, K. & Harcourt, R. (2007). The voluntary code of conduct for dolphin watching in Port Stephens, Australia: Is self-regulation an effective management tool? *Journal of Cetacean Research & Management* 9, 159–166.

Alpers, A. (1963). *The Dolphins*. London: John Murray.

Anderson, P.K. (2001). Marine mammals in the next one hundred years: Twilight for a Pleistocene megafauna? *Journal of Mammalogy* 82, 623–629.

Baker, C.S. & Herman, L.M. (1989). *Behavioral responses of summering humpback whales to vessel traffic: experimental and opportunistic observations*. Report from the University of Hawaii to the US National Park Service, Anchorage, Alaska, p. 50.

Bannister, J.L. (1986). Southern Right Whales: Status off Australia from the twentieth-century 'incidental' sightings and aerial survey. In *Report to the International Whaling Commission (Special Issue 10)*. Cambridge: IWC, pp. 153–158.

Bannister, J.L. (2001). Status of Southern Right Whales (*Eubalaena australis*) of southern Australia. *Journal of Cetacean Research and Management (Special Issue)* 2, 103–110.

Bateson, G. (1974). Observations of a cetacean community. In J. McIntyre (Ed.), *Mind in the Waters*. New York, NY: Scribner's, pp. 146–165.

Baur, D.C., Bean, M.J. & Gosliner, M.L. (1999). The laws governing marine mammal conservation in the United States. In J.R. Twiss & R.R. Reeves (Eds), *Conservation and Management of Marine Mammals*. Washington, DC: Smithsonian Institute Press, pp. 48–86.

Bearzi, G. (2000). First report of a common dolphin (*Delphinus delphis*) death following penetration of a biopsy dart. *Journal of Cetacean Research and Management* 2, 217–221.

Bearzi, G., Reeves, R.R., Notarbartolo di Sciara, G., *et al.* (2003). Ecology, status and conservation of short-beaked common dolphins *Delphinus delphis* in the Mediterranean Sea. *Mammal Review* 33, 224–252.

Bejder, L. & Samuels, A. (2003). Evaluating the effects of nature-based tourism on cetaceans. In N. Gales, M. Hindell & R. Kirkwood (Eds), *Marine Mammals: Fisheries, tourism and management issues*. Collingwood: CSIRO, pp. 229–256.

Bejder, L., Dawson, S.M. & Harraway, J.A. (1999). Responses by Hector's dolphins to boats and swimmers in Porpoise Bay, New Zealand. *Marine Mammal Science* 15, 738–750.

Bejder, L., Samuels, A., Whitehead, H., *et al.* (2006). Decline in relative abundance of bottlenose dolphins (*Tursiops* sp.) exposed to long-term disturbance. *Conservation Biology* 20, 1791–1798.

Birtles, R.A., Arnold, P.W. & Dunstan, A. (2002). Commercial swim programs with dwarf minke whales on the northern Great Barrier Reef, Australia: Some characteristics of the encounters with management implications. *Australian Mammalogy* 24, 23–38.

Boyd, I.L., Bowen, W.D. & Iverson, S.J. (2010). *Marine Mammal Ecology and Conservation: A handbook of techniques*. Oxford: Oxford University Press.

British Broadcasting Corporation (2004). *BBC Wildlife Magazine* 22.

Busnel, R.G. (1973). Symbiotic relationship between man & dolphins. *Transactions of the New York Academy of Sciences* 35, 112–135.

Cisneros-Montemayor, A.M., Sumaila, U.R., Kaschner, K. & Pauly, D. (2010). The global potential for whale watching. *Marine Policy* 34, 1273–1278.

Clapham, P. (2000). The humpback whale: Seasonal feeding and breeding in a baleen whale. In J. Mann, R.C. Connor, P.L. Tyack & H. Whitehead (Eds), *Cetacean Societies: Field studies of dolphins and whales*. Chicago, IL: University of Chicago Press, pp. 173–196.

Connor, R.C. & Smolker, R.A. (1985). Habituated dolphins (*Tursiops* sp.) in Western Australia. *Journal of Mammalogy* 66, 398–400.

Connor, R.C., Smolker, R.A. & Richards, A.F. (1992). Dolphin alliances and coalitions. In A.H. Harcourt & F.B.M. de Waal (Eds), *Coalitions and Alliances in Humans and other Animals*. Oxford: Oxford University Press, pp. 415–443.

Connor, R.C., Wells, R.S., Mann, J. & Read, A.J. (2000). The bottlenose dolphin: Social relationships in a fission-fusion society. In J. Mann, R.C. Connor, P.L. Tyack & H. Whitehead (Eds), *Cetacean Societies: Field studies of dolphins and whales*. Chicago, IL: University of Chicago Press, pp. 173–196.

Constantine, R. (2001). Increased avoidance of swimmers by wild bottlenose dolphins (*Tursiops truncatus*) due to long-term exposure to swim-with-dolphin tourism. *Marine Mammal Science* 17, 689–702.

Constantine, R. (2002). Folklore and legends. In W.F. Perrin, B. Wursig and J.G.M. Thewissen (Eds), *The Encyclopedia of Marine Mammals*. San Diego, CA: Academic Press, pp. 448–450.

Constantine, R., Brunton, D.H. & Dennis, T. (2004). Dolphin-watching tour boats change bottlenose dolphin (*Tursiops truncatus*) behaviour. *Biological Conservation* 117, 299–307.

Corkeron, P.J. (2004). Whale watching, iconography, and marine conservation. *Conservation Biology* 18, 847–849.

Corkeron, P.J. (2006). How shall we watch whales? In D. Lavigne (Ed.), *Gaining Ground: In pursuit of ecological sustainability*. Guelph, ON: International Fund for Animal Welfare, pp. 161–170.

Cousteau, J.-Y. & Diolé, P. (1972). *The Whale: Mighty monarch of the sea*. New York, NY: Arrowood Press.

Cramer, K.L., Perryman, W.L. & Gerodette, T. (2008). Declines in reproductive output in two dolphin populations depleted by the yellowfin purse-seine fishery. *Marine Ecology Progress Series* 369, 273–285.

Daura-Jorge, F.G., Cantor, M., Ingram, S.N., Lusseau, D. & Simões-Lopes, P.C. (2012). The structure of a bottlenose dolphin society is coupled to unique foraging cooperation with artisanal fishermen. *Biology Letters* 8, 702–705.

Dawbin, W.H. (1986). Right whales caught in waters around south eastern Australia and New Zealand during the nineteenth and early twentieth centuries. In *Report to the International Whaling Commission (Special Issue 10)*, Cambridge: IWC, pp. 261–267.

Dawson, S., Pichler, F., Slooten, E., Russell, K. & Baker, C.S. (2001). The North Island Hector's dolphin is vulnerable to extinction. *Marine Mammal Science* 17, 366–371.

DeMaster, D.P., Fowler, C.W., Perry, S.L. & Richlen, M.F. (2001). Predation and competition: The impact of fisheries on marine-mammal populations over the next one hundred years. *Journal of Mammalogy* 82, 641–651.

Department of Fisheries. (2011). *State of the Fisheries and Aquatic Resources. Report 2010/11* (W.J. Fletcher & K. Santoro, Eds), p. 359. Department of Fisheries, Western Australia. Available at http://www.fish.wa.gov.au/About-Us/Publications/Pages/Annual-Report.aspx.

Dudzinski, K.M. (1998). The best-kept secret in dolphin swim programs is in Japan. *Whalewatcher* 31, 14–17.

Duffus, D.A. & Dearden, P. (1993). Recreational use, valuation, and management, of killer whales (*Orcinus orca*) on Canada's Pacific Coast. *Environmental Conservation* 20, 149–156.

Fair, P.A., Adams, J., Mitchum, G., et al. (2010). Contaminant blubber burdens in Atlantic bottlenose dolphins (*Tursiops truncatus*) from two southeastern US estuarine areas: Concentrations and patterns of PCBs, pesticides, PBDEs, PFCs, and PAHs. *Science of the Total Environment* 408, 1577–1597.

Fairholme, J.K.E. (1856). The Blacks of Moreton Bay and the porpoises. *Zoological Society of London, Proceedings* 24, 353–354.

Fontaine, M.C., Snirc, A., Frantzis, A., et al. (2012). History of expansion and anthropogenic collapse in a top marine predator of the Black Sea estimated from genetic data. *Proceedings of the National Academy of Sciences* 109, E2569–E2576.

Gales, N.J., Johnston, D., Littnan, C. & Boyd, I.L. (2010). Ethics in marine mammal science. In I.L. Boyd, W.D., Bowen & S.J. Iverson (Eds), *Marine Mammal Ecology and Conservation: A handbook of techniques*. Oxford: Oxford University Press, pp. 1–15.

Garrod, B. & Fennell, D.A. (2004). An analysis of whale-watching codes of conduct. *Annals of Tourism Research* 31, 334–352.

Gordon, J., Leaper, R., Hartley, F.G. & Chappell, O. (1992). *Effect of whale-watching vessels on the surface and underwater acoustic behaviour of sperm whales off Kaikoura, New Zealand*. Science and Research Series No. 52. Wellington, New Zealand: Department of Conservation, p. 64.

Gosliner, M.L. (1999). The tuna–dolphin controversy. In J.R. Twiss & R.R. Reeves (Eds), *Conservation and Management of Marine Mammals*. Washington, DC: Smithsonian Institute Press, pp. 48–86.

Halpern, B.S., Selkoe, K.A., Micheli, F. & Kappel, C.V. (2007). Evaluating and ranking the vulnerability of global marine ecosystems to anthropogenic threats. *Conservation Biology* 21, 1301–1315.

Hastie, G.D., Wilson, B., Tufft, L.H. & Thompson, P.M. (2003). Bottlenose dolphins increase breathing synchrony in response to boat traffic. *Marine Mammal Science* 19, 74–84.

Heckel, G., Reilly, S.B., Sumich, J.L. & Espejel, I. (2001). The influence of whalewatching on the behaviour of migrating gray whales (*Eschrichtus robustus*) in Todos Santos Bay and surrounding waters, Baja California, Mexico. *Journal of Cetacean Research and Management* 3, 227–237.

Herman, L.M. (Ed.) (1980). *Cetacean Behaviour: Mechanisms and functions*. New York, NY: Wiley-Interscience.

Holyoake, C., Finn, H., Stephens, N., et al. (2010). *Technical Report on the Bottlenose Dolphin* (Tursiops aduncus) *Unusual Mortality Event within the Swan Canning Riverpark, June-October 2009*. Report to the Swan River Trust, Perth, Western Australia.

Hoyt, E. (2001). *Whale Watching 2001: Worldwide tourism numbers, expenditures and expanding socioeconomic benefits.* Yarmouth Port, UK: International Fund for Animal Welfare, p. 158.

International Whaling Commission. (2006). Report of the scientific committee. *Journal of Cetacean Research and Management (Suppl.)* 8, 1–65.

Isobe, T., Ochi, Y., Ramu, K., *et al.* (2009). Organohalogen contaminants in striped dolphins (*Stenella coeruleoalba*) from Japan: Present contamination status, body distribution and temporal trends (1978–2003). *Marine Pollution Bulletin* 58, 396–401.

Jaiteh, V.F., Allen, S.J., Meeuwig, J.J. & Loneragan, N.R. (2013). Subsurface behaviour of bottlenose dolphins (*Tursiops truncatus*) interacting with fish trawl nets in north-western Australia. *Marine Mammal Science* 29, E266–E281.

Janik, V.M. & Thompson, P.M. (1996). Changes in surfacing patterns of bottlenose dolphins in response to boat traffic. *Marine Mammal Science* 12, 597–602.

Jaramillo-Legorreta, A., Rojas-Bracho, L., Brownell Jr., R.L., *et al.* (2007). Saving the Vaquita: Immediate action, not more data. *Conservation Biology* 21, 1653–1655.

Kogi, K., Hishii, T., Imamura, A., Iwatani, T. & Dudzinski, K. (2004). Demographic parameters of Indo-Pacific bottlenose dolphins (*Tursiops aduncus*) around Mikura Island, Japan. *Marine Mammal Science* 20, 510–526.

Kraus, S.D., Brown, M.W., Caswell, H., *et al.* (2005). North Atlantic right whales in crisis. *Science* 309, 561–562.

Laist, D.W., Knowlton, A.R., Mead, J.G., Collet, A.S. & Podesta, M. (2001). Collisions between ships and whales. *Marine Mammal Science* 17, 35–75.

Lavigne, D.M., Scheffer, V.B. & Kellert, S.R. (1999). The evolution of North American attitudes toward marine mammals. In J.R. Twiss & R.R. Reeves (Eds), *Conservation and Management of Marine Mammals.* Washington, DC: Smithsonian Institute Press, pp. 10–47.

Lockyer, C. (1990). Review of incidents involving wild, sociable dolphins, worldwide. In S. Leatherwood & R.R. Reeves (Eds), *The Bottlenose Dolphin.* San Diego, CA: Academic Press, pp. 337–353.

Lusseau, D. (2003). The effects of tour boats on the behavior of bottlenose dolphins: Using Markov chains to model anthropogenic impacts. *Conservation Biology* 17, 1785–1793.

Lusseau, D. (2004). The hidden cost of tourism: Detecting long-term effects of tourism using behavioural information. *Ecology and Society* 9, 2. Available at: http://www.ecologyandsociety.org/vol9/iss1/at2

Lusseau, D. (2005). Residency pattern of bottlenose dolphins (*Tursiops* spp.) in Milford Sound, New Zealand, is related to boat traffic. *Marine Ecology Progress Series* 295, 265–272.

Lusseau, D., Slooten, E. & Currey, R.J.C. (2006). Unsustainable dolphin-watching tourism in fiordland, New Zealand. *Tourism in Marine Environments* 3, 173–178.

Mackaness, G. (1979). *The Discovery and Exploration of Moreton Bay and the Brisbane River.* Dubbo: Review Publications.

Mangel, J.C., Alfaro-Shigueto, J., Van Waerebeek, K., *et al.* (2010). Small cetacean captures in Peruvian artisanal fisheries: High despite protective legislation. *Biological Conservation* 143, 136–143.

Mann, J. & Kemps, C. (2003). The effects of provisioning on maternal care in wild bottlenose dolphins, Shark Bay, Australia. In N. Gales, M. Hindell & R. Kirkwood (Eds), *Marine Mammals: Fisheries, tourism and management issues.* Collingwood: CSIRO Publishing, pp. 304–317.

Marsh, H., Arnold, P., Freeman, M., *et al.* (2003). Strategies for conserving marine mammals. In N. Gales, M. Hindell & R. Kirkwood (Eds), *Marine Mammals: Fisheries, tourism and management issues.* Collingwood: CSIRO Publishing, pp. 1–19.

McBride, A.F. & Hebb, D.O. (1948). Behaviour of the captive bottlenose dolphin *Tursiops truncatus. Journal of Comparative Physiology and Psychology* 41, 111–123.

Mead, T. (1961). *The Killers of Eden.* Sydney: Angus and Robertson Publishers.

Moore, M., Andrews, R., Austin, T., *et al.* (2012). Rope trauma, sedation, disentanglement, and monitoring-tag associated lesions in a terminally entangled North Atlantic right whale (*Eubalaena glacialis*). *Marine Mammal Science* DOI: 10.1111/j.1748–7692.2012.00591.x.

Myers, R.A. & Worm, B. (2003). Rapid worldwide depletion of predatory fish communities. *Nature* 423, 280–283.

Neimi, A.R. (2010). Life by the shore. Maritime dimensions of the late Stone Age, Arctic Norway. In W. Østreng (Ed.), *Transference. Interdisciplinary Communications.* Available at http://www.cas.uio.no/publications_/transference.php, ISBN: 978–82–996367–7–3.

Northridge, S.N. & Hofman, R.J. (1999). Marine mammal interactions with fisheries. In J.R. Twiss & R.R. Reeves (Eds), *Conservation and Management of Marine Mammals.* Washington: Smithsonian Institute Press, pp. 99–119.

O'Connor, S., Campbell, R., Cortez, H. & Knowles, T. (Eds) (2009). *Whale Watching Worldwide: Tourism numbers, expenditures and expanding economic benefits.* Report to International Fund for Animal Welfare. Melbourne: Economists at Large.

Orams, M.B. (1997). Historical accounts of human–dolphin interaction and recent developments in wild dolphin based tourism in Australasia. *Tourism Management* 18, 317–326.

Orams, M.B. (1999). *Marine Tourism: Development, impacts and management.* London: Routledge.

Orams, M.B. (2001). From whale hunting to whale watching in Tonga: A sustainable future? *Journal of Sustainable Tourism* 9, 128–136.

Orams, M.B. (2002). Humpback whales in Tonga: An economic resource for tourism. *Coastal Management* 30, 361–380.

Orams, M.B., Hill, G.J.E. & Baglioni, Jr, A.J. (1996). 'Pushy' behaviour in a wild dolphin feeding program at Tangalooma, Australia. *Marine Mammal Science* 12, 107–117.

Parsons, E.C.M. & Rawles, C. (2003). The resumption of whaling in Iceland and the potential negative impact in the Icelandic whale-watching market. *Current Issues in Tourism* 6, 444–448.

Perry, S.L., DeMaster, D.P. & Silber, G.K. (1999). The great whales: History and status of six species listed as endangered under the U.S. Endangered Species Act of 1973. *Marine Fisheries Review* 61, 1–74.

Ramage, P.R. (2009). Preface. In S. O'Connor, R. Campbell, H. Cortez & T. Knowles (Eds), *Whale Watching Worldwide: Tourism numbers, expenditures and expanding economic benefits.* Report to International Fund for Animal Welfare. Melbourne: Economists at Large.

Read, A.J. (2008). The looming crisis: Interactions between marine mammals and fisheries. *Journal of Mammalogy* 89, 541–548.

Read, A.J., Drinker, P. & Northridge, S. (2006). Bycatch of marine mammals in U.S. and global fisheries. *Conservation Biology* 20, 163–169.

Richardson, W.J., Greene, C.R., Jr., Malme, C.I. & Thomson, D.H. (1995). *Marine Mammals and Noise.* San Diego, CA: Academic Press.

Romero, A. & Hayford, K. (2000). Past and present utilisation of marine mammals in Grenada, West Indies. *Journal of Cetacean Research and Management* 2, 223–226.

Ross, P.S., Ellis, G.M., Ikonomou, M.G., Barrett-Lennard, L.G. & Addison, R.F. (2000). High PCB concentrations in free-ranging Pacific killer whales (*Orcinus orca*): Effects of age, sex and dietary preference. *Marine Pollution Bulletin* 40, 504–515.

Samuels, A. & Bejder, L. (2004). Chronic interaction between humans and free-ranging bottlenose dolphins near Panama City Beach, Florida, USA. *Journal of Cetacean Research and Management* 6, 69–77.

Samuels, A. & Tyack, P. (2000). Flukeprints: A history of studying cetacean societies. In J. Mann, R.C. Connor, P.L. Tyack & H. Whitehead (Eds), *Cetacean Societies: Field studies of dolphins and whales.* Chicago, IL: University of Chicago Press, pp. 9–44.

Samuels, A., Bejder, L. & Heinrich, S. (2000). *A Review of the Literature Pertaining to Swimming with Wild Dolphins.* Report to the U.S. Marine Mammal Commission, Maryland.

Samuels, A., Bejder, L., Constantine, R. & Heinrich, S. (2003). Swimming with wild cetaceans, with a special focus on the Southern Hemisphere. In N. Gales, M. Hindell & R. Kirkwood (Eds), *Marine Mammals: Fisheries, tourism and management issues.* Collingwood: CSIRO Publishing, pp. 277–303.

Scarpaci, C., Bigger, S.W., Corkeron, P.J. & Nugegoda, D. (2000). Bottlenose dolphins (*Tursiops truncatus*) increase whistling in the presence of 'swim-with-dolphin' tour operations. *Journal of Cetacean Research and Management* 2, 183–185.

Shane, S.H., Wells, R.S. & Wursig, B. (1986). Ecology, behaviour and social organization of the bottlenose dolphin: A review. *Marine Mammal Science* 2, 34–63.

Shane, S.H., Tepley, L. & Costello, L. (1993). Life-threatening contact between a woman and a pilot whale captured on film. *Marine Mammal Science* 9, 331–336.

Shaughnessy, P., Kirkwood, R., Cawthorn, M., Kemper, C. & Pemberton, D. (2003). Pinnipeds, cetaceans and fisheries in Australia: A review of operational interactions. In N. Gales, M. Hindell & R. Kirkwood (Eds), *Marine Mammals: Fisheries, tourism and management issues.* Collingwood: CSIRO Publishing, pp. 136–152.

Sinding, M.-H.S., Gilbert, M.T.P., Grønnow, B., *et al.* (2012). Minimally destructive DNA extraction from archaeological artefacts made from whale baleen. *Journal of Archaeological Science* 39, 3750–3753.

Slooten, E., Dawson, S.M., Rayment, W.J. & Childerhouse, S.J. (2006). A new abundance estimate for Maui's dolphin: What does it mean for managing this critically endangered species? *Biological Conservation* 128, 576–581.

Smith, J.N., Grantham, H.S., Gales, N., *et al.* (2012). Identification of humpback whale breeding and calving habitat in the Great Barrier Reef. *Marine Ecology Progress Series* 447, 259–272.

Spradlin, T.R., Terbush, A.D. & Smullen, W.S. (1998). National Marine Fisheries Service update on human/dolphin interactions in the wild. *Soundings* 23, 25–27.

Stewart, K.L. & Marino, L. (2009). *Dolphin–Human Interaction Programs: Policies, problems and alternatives.* Michigan: Animals and Society Institute.

Stone, G.S. & Yoshinaga, A. (2000). Hector's dolphin *Cephalorhynchus hectori* calf mortalities may indicate new risks from boat traffic and habituation. *Pacific Conservation Biology* 6, 162–170.

Tavolga, M.C. & Essapian, F.S. (1957). The behavior of the bottlenose dolphin (*Tursiops truncatus*): Mating, pregnancy, parturition and mother–infant behavior. *Zoologica* 42, 11–31.

Tonay, A.M. & Öztürk, A.A. (2012). Historical records of cetacean fishery in the Turkish seas. *Journal of the Black Sea/Mediterranean Environment* 18, 388–399.

Turvey, S.T., Pitman, R.L., Taylor, B.L., *et al.* (2007). First human-caused extinction of a cetacean species? *Biology Letters* 3, 537–540.

Van Waerebeek, K. & Reyes, J. (1994). Post-ban small cetacean takes off Peru: A review. *Report of the International Whaling Commission* (*Special Issue*) 15, 503–519.

Veth, P. & O'Connor, S. (2013). Australia: The last 50,000 years. In A. Bashford & S. Macintyre (Eds), *Cambridge History of Australia, Volume 1 Colonial Australia.* Cambridge: Cambridge University Press, pp. 17–42.

Wellings, C.E. (1944). The killer whales of Twofold Bay, NSW, Australia *Grampus orca. Australian Zoologist* 10, 291–294.

Wells, R.S. & Scott, M.D. (1997). Seasonal incidence of boat strikes on bottlenose dolphins near Sarasota, Florida. *Marine Mammal Science* 13, 475–480.

Whitehead, H., Reeves, R.R. & Tyack, P. (2000). Science and the conservation, protection, and management of wild cetaceans. In J. Mann, R.C. Connor, P.L. Tyack & H. Whitehead (Eds), *Cetacean Societies: Field Studies of Dolphins and Whales.* Chicago, IL: University of Chicago Press, pp. 308–332.

Whitt, A. & Read, A. (2006). Assessing compliance to guidelines by dolphin-watching operators in Clearwater, Florida, USA. *Tourism in Marine Environments* 3, 117–130.

Williams, R., Lusseau, D. & Hammond, P. S. (2006). Estimating relative energetic costs of human disturbance to killer whales (*Orcinus orca*). *Biological Conservation* 133, 301–311.

Würsig, B. (1996). Swim-with-dolphin activities in nature: Weighing the pros and cons. *Whalewatcher* 30, 11–15.

Yablokov, A.V. (1994). Validity of whaling data. *Nature* 367, 108.

4

Human attitudes and values

Tradition and transformation and zombies

Peter Corkeron

Introduction

Tradition is commonly cited in arguments justifying the continuation of commercial whaling (e.g. Morishita, 2006). Rather than explore the meaning and context of whaling traditions, I shall focus on the more general question of how we perceive animals, and what those perceptions mean for how people relate to animals. Then I shall describe the recent transformations in those who support whaling, and those who oppose it, focusing on whale-watching. Finally, I question whether the non-government organization (NGO) support for whale-watching as a replacement for whaling has become, itself, a tradition that has morphed into a zombie.

Zombies may seem an odd addition to the discussion of the role of nature-based tourism in conservation, but here, I discuss a zombie idea, rather than the shambling undead. Recently, the construct of 'zombie' ideas has been introduced (Quiggin, 2010) in macroeconomics. A zombie idea is one that should have been killed off long ago, given available evidence that it is wrong, and yet it just will not die. Although the impetus for zombie concepts came from macroeconomics (Quiggin, 2010), the construct has been taken up by authors in other areas: for example, military history (Stockings, 2010), environmental economics (Reyes, 2011) and ecology (Fox, 2011).

The *Oxford English Dictionary* defines *tradition* as 'the transmission of customs or beliefs from generation to generation', and *transformation* as 'a thorough or dramatic change in form or appearance'. Although tradition is used as a justification for whaling by whalers, I note in passing that there are several aspects of tradition that are glossed over in the whaling debate. A history of whaling is not a sufficient condition for maintaining whaling – many towns along Australia's southern coast began as whaling settlements, and whale products were the first exports from the British penal colonies that were to become Australia (Dakin, 1934). Australia remained a whaling nation until 1978 (Suter, 1982), yet Australian society is now staunchly anti-whaling (e.g. Hamazaki & Tanno, 2001).

Japan has a tradition of recalcitrance in international negotiations over whaling, dating back to League of Nations treaties prior to World War II (Tønnessen & Johnsen, 1982) – that is, ever since multinational negotiations to regulate whaling started. The continuing revelations of illegal commerce in whale products (e.g. Baker *et al.*, 2010) show that the tradition of those engaged in whaling to flout international agreements aimed at managing the industry (Tønnessen & Johnsen, 1982) still flourishes. A tradition of this behaviour does not justify its existence. Appeals to traditions per se without other justification for a particular behaviour

Whale-watching: Sustainable Tourism and Ecological Management, eds J. Higham, L. Bejder and R. Williams.
Published by Cambridge University Press. © Cambridge University Press 2014.

suggests that the other justifications, if they exist, are probably weak.

In this chapter, I discuss how NGOs' capacity to oppose whaling has been transformed by their traditional support for whale-watching. This is because whale-watching and whaling can and do co-occur. All major whaling nations – Japan, Iceland and Norway – have had whale-watching industries for decades now (O'Connor *et al.*, 2009). I argue that supporting whale-watching in whaling nations restricts NGOs' range of action against whaling in those nations, given NGOs' history (and, by now, tradition) of supporting whale-watching in these countries. Finally, comparing the relative monetary value of whaling and whale-watching (another NGO tradition) has been rendered meaningless. But first, I briefly review some background on how we value animals.

How we value animals

Elsewhere (Corkeron, 2006a), I have briefly sketched out ideas on how the manner in which people perceive animals affects the way in which they manage their interactions with them, derived from Caughley (1985). Put simply, animals can be divided into four categories (Caughley, 1985: 129):

- *Nasties* – 'we fear or loathe and which we would like to thin out';
- *Lovelies* – 'we like, revere or honour and which we wish therefore to conserve';
- *Commodities* – 'we use as domesticants or harvest as wild animals'; and
- *Irrelevancies* – 'that seldom impinge on us and towards which we direct no strong feelings'.

There are different ways in which we manage our dealings with animal populations. We tend to control Nasties, cherish Lovelies, use Commodities, and ignore Irrelevancies. It has been demonstrated that, in many countries, whales have gone from being Commodities to Lovelies (Corkeron, 2006a). However, another transformation has also occurred. Members of the fisheries establishments of whaling nations now portray whales as competitors for

fish (e.g Corkeron, 2006b; Morishita, 2008). That is, whales have transformed from Commodities to Nasties. This is as profound a change as that to Lovelies in the non-whaling nations.

Transformations among supporters of whaling

In 2006, the International Whaling Commission (IWC), the multinational institution responsible for overseeing international management of large whales, passed the St Kitts & Nevis Declaration (Anonymous, 2006, np), which includes the following statement: 'ACCEPTING that scientific research has shown that whales consume huge quantities of fish making the issue a matter of food security for coastal nations and requiring that the issue of management of whale stocks must be considered in a broader context of ecosystem management since ecosystem management has now become an international standard'.

Most generally accepted conceptions of 'Ecosystem-Based Fishery Management', or an 'Ecosystem Approach to Fisheries' (hereafter, EBFM) do not include marine mammal culls as an integral component (e.g. Pikitch *et al.*, 2004). A definition that encapsulates the general view of EBFM succinctly (inasmuch as a 'general view' exists) is that given by Pikitch and coauthors in *Science* (2004: 1892): 'EBFM reverses the order of management priorities so that the objective of sustaining ecosystem structure and function supersedes the objective of maximizing fisheries yields'.

However, spokespeople for nations engaged in whaling and other commercial marine mammal hunts have been claiming that marine mammal populations negatively affect the abundance of commercial fish (for details, see Corkeron, 2006b, 2009b). They also argue that culls of marine mammals, ostensibly to benefit fisheries, are an integral part of EBFM. Apart from the 'St Kitts & Nevis Declaration' above, other examples include published work by the Japanese government's chief negotiator at the IWC (e.g. Morishita, 2008), and the 'Safety Net' initiative (a new model for how the IWC could operate, with an 'ecosystem approach' that included culling), developed by members of the Japanese delegation at the IWC through 2006–2008 (e.g. Iliff, 2010).

The idea that marine mammals must be culled to protect fisheries is not new. For example, when the Norwegian spring-spawning herring (*Clupea harengus*) population collapsed completely in the late 1960s the Norwegian government organized a hunt of killer whales (*Orcinus orca*) that were known to eat herring. Over 700 killer whales were killed between 1969 and 1980 (Øien, 1988). What is new is the notion that culling marine mammals is a *primary* component of effective fisheries management that takes account of ecosystem interactions. Understandably, there is a reluctance among some in the fisheries science establishment to accept that EBFM is being construed to including predator culling as a primary component (e.g. Harwood, 2007; Murawski, 2007).

Transformations in the application of science

One result of this change to seeing whales as Nasties has been the co-option of some aspects of government-supported marine science in whaling nations. That 'scientific whaling' is a misuse of the methods of marine science has been argued for some time now (e.g. Clapham *et al.*, 2003; Gales *et al.*, 2005; Corkeron, 2009a). The notion of 'scientific whaling' is diametrically opposed to the growing tradition of researcher impact mitigation (e.g. Anonymous, 2012). What is less clear is the manner in which aspects of marine ecological research in whaling nations have been subverted by the desire to demonstrate that marine mammals need to be culled. Elsewhere I have demonstrated that some Norwegian marine science has been overly focused on demonstrating the deleterious effects of marine mammals on fisheries, while ignoring evidence that their effects are relatively trivial (Corkeron, 2009b).

Spokespeople for the fisheries establishments of whaling nations assert that they are interested in

using scientific approaches to manage whale (and seal) hunting (e.g. Ludvigsen, 2005), the implication being that those who oppose lethal use of marine mammals are taking an unscientific stance. For example, the announcement for a recent (cancelled) conference on 'Marine Ecosystem Management: How to make it sustainable' included the following statement: 'There are different opinions on how to interpret the concept of sustainable ecosystem management. The perspective is vastly different if one compares the views shared by communities dependent on harvesting from marine ecosystems, with the views of the public far away from such realities, typically living in the cities of the large industrialized countries' (NAMMCO, 2010: np). Implicit in this statement is the idea that city-dwellers do not understand the realities of what EBFM should be. However, those 'realities' (NAMMCO, 2010) are distorted by spokespeople for fisheries establishments, as demonstrated in the following paragraph.

Illustration of this distortion, using the manner in which the results of surveys in 2007, conducted by Icelandic marine researchers, have been portrayed in Icelandic media has been previously reported (Corkeron, 2007). In reporting preliminary data from the survey, when sightings of some species had declined (e.g. minke whales and white-beaked dolphins), the media reports (correctly) included a caveat that data were preliminary, and should not be interpreted as demonstrating an actual decline. Yet raw data from the same survey included more fin whale sightings than in previous surveys. Interpretations of these data (in the popular press) as indicating an increase in fin whale numbers did not include a qualifier as to their preliminary nature. On the contrary, they were used by Iceland's Fisheries Minister at the time, Mr Gudfinsson, to argue that whale stocks were large and increasing, providing an ecosystem-based argument for whaling (for details, see Corkeron, 2007).

A disappointing aspect of the politicization of some marine research in Scandinavian nations is that Scandinavian governments, particularly the Norwegian government, have a long tradition of supporting excellent marine research. Some of the very early research that provided the intellectual framework for modern fisheries science was conducted in Norway (e.g. Hjort, 1914), as was some excellent early work on the biology of baleen whales (e.g. Ingebrigtsen, 1929). Professor Birger Bergersen, a Norwegian anatomist and outspoken critic of unsustainable whaling prior to World War II, was the first chair of the IWC (Tønnessen & Johnsen, 1982). Some Norwegian institutions maintain this tradition of excellence in marine biology (e.g. Jørgensen *et al.*, 2007). The desire to portray marine mammals as Nasties has influenced Norwegian marine mammal research to the point where it has become an embarrassment to some Norwegian ecologists (e.g. Lislevand, 2009).

Transformations among opponents of whaling: commoditizing 'Lovelies'

How have the whale conservation movement, and countries that have become opposed to commercial whaling, responded to these developments, and what role has whale-watching played in these responses? In the early days of modern industrial whaling, most attitudes towards whales were utilitarian ('Commodities'), although there were exceptions. For example, in 1922, Sir Sidney Harmer, a British scientist, described Norwegian whaling in the subAntarctic as 'insensate slaughter arousing feelings of horror and disgust' (Tønnessen & Johnsen, 1982: 342). That whaling drove many populations of large whales close to extinction (Tønnessen & Johnsen, 1982) provided the initial impetus to the 'save the whales' movement. The majestic aspects of whales, particularly their size and the apparent intelligence of some, led to the idea that whales' intrinsic value was sufficient justification for not hunting them. The move to supporting whale-watching by members of the conservation NGO community is, in one sense, a rational development of this change.

Initially, whale-watching offered a commercial alternative to whaling, allowing members of the NGO community to argue that communities could benefit materially from stopping whaling (e.g.

Hoyt, 2001). Whale-watching grew internationally through the latter years of the last century (Hoyt, 2001), and as whaling scaled down, it seemed that whale-watching could replace whaling as the primary commercial use of whales. Unfortunately for the proponents of this idea, in the first decade or so of this century, the number of whales being killed by whaling – both in terms of individuals and species – is increasing (IWC, 2011a, 2011b), and all major whaling nations have whale-watching industries.

A result of the IWC's decision to set commercial quotas at zero until a workable management regime is enacted (the so-called 'moratorium' on whaling) is that whaling post-'moratorium' has continued at a relatively low level. Ongoing whaling, whether by commercial quota under objection to the 'moratorium' (currently in Norway and Iceland) or through the abuse of Article VIII of the ICWR (so-called 'scientific whaling') carried out at different times through national programmes by Norway, Iceland and Japan, still occurs. However, it provides substantially less whale meat for major markets (i.e. in Japan) than in the days before the 'moratorium'. One result of this has been that the conservation NGO community has been able to argue that whale-watching is more economically valuable than is whaling (e.g. IFAW, 2009).

Green Marxist critique of the role of the multinational NGO community in promoting ecotourism addresses some of the larger philosophical issues raised by the manner in which members of the NGO community have adopted the neoliberal economic world view (e.g. Brockington & Duffy 2010; Carrier, 2010; Igoe *et al.*, 2010). Neves (2010) developed an interesting Green Marxist critique of the capitalist logic behind NGO support for whale-watching, and the problems this generates.

My intent in the rest of this chapter is more prosaic. I'll review some basic information on the recent history of whaling and of whale-watching in whaling nations. I use this to argue that the strategy of supporting whale-watching as a replacement for whaling in whaling nations has not, and – from the evidence to date – will not, work.

Whale-watching and whaling can and do co-occur

A problem with supporting whale-watching as an alternative to whaling is that whale-watching and whaling occur in the same places at the same time. Whale-watching industries exist in Norway, Iceland and Japan, the three major whaling nations. Although in most cases, the animals watched and the animals hunted are different species, this is not necessarily the case. In Iceland, and now in western Greenland, whale-watching and whaling industries target the same species (minke and humpback whales, respectively) in the same place at the same time. Off western Greenland, it is possible that whaling will target individual animals known to be resident in the area where whale-watching occurs (Boye et al., 2010).

The hope that whale-watching in whaling nations would lead to a change in societal attitudes to viewing whales as Lovelies (Ris, 1993) appears by now to be misplaced. In the three major whaling nations, whale-watching has been ongoing for over two decades (Hoyt, 2001; O'Connor et al., 2009). During this time, the notion that marine mammal culls are a primary component of EBFM has also taken hold (see above), so if anything, the general perception of whales has worsened, rather than improved, over time.

The extent to which whaling affects whale-watching in whaling nations remains unclear – do potential tourists not visit whaling nations because whaling occurs there (Higham & Lusseau, 2007)? One effect of tourism – whether going whale-watching or not – in one whaling nation is that it appears to be keeping the local whaling industry running. Recently, the Whale and Dolphin Conservation Society (WDCS, 2011) estimated that up to 40% of minke whale meat consumed in Iceland is eaten by tourists. This has led to NGOs that oppose whaling and support whale-watching in Iceland recently running campaigns requesting tourists not to eat whale meat (e.g. the International Fund for Animal Welfare (IFAW, 2011; WDCS, 2011). Given this, it is something of a stretch to equate tourism,

even whale-watch tourism, with a successful anti-whaling campaign in Iceland.

Supporting whale-watching in whaling nations restricts NGOs' range of action against whaling

This raises another reason to question the strategy of supporting whale-watching in whaling nations. This support creates a hidden cost that has not, to my knowledge, been discussed – it limits the scope for action by conservation NGOs. Whale-watching enterprises in Iceland and Norway have received financial and moral support from conservation NGOs (e.g. Ris, 1993). Having provided this support, NGOs are not in a position to call for tourism boycotts of these countries, as to do so would damage the whale-watching companies that NGOs support.

Recently, commercial whaling quotas in Iceland have increased substantially. Whether a tourism boycott of Iceland could be an effective tool in the campaign to stop whaling is not the issue here. The point is that the support that the NGO community has given to Icelandic whale-watching companies constrains options for establishing a boycott. Given the substantial growth in commercial whaling in Iceland over recent years (IWC, 2011a), it is difficult to see that the current strategy of the anti-whaling NGOs is working well (or at all), yet the strategy of supporting whale-watching constrains their possibilities for further action.

Comparing the relative monetary value of whaling and whale-watching has become meaningless

Another argument made over the past years for supporting whale-watching in whaling nations has been that whale-watching is more valuable than whaling (e.g. IFAW, 2009). Although this may be true at the moment, it has become irrelevant. At the start of whale-watching in the early 1980s, there was reason to believe that whale-watching could displace whaling as a commercial use for whales. That whale-watching could lead to the end of all commercial whaling as a rationale now for the antiwhaling movement's support of whale-watching is based on a set of hopes that have not been fulfilled.

Whales remain international common-pool resources. The economic value of whale-watching in a non-whaling country is irrelevant to people in a whaling nation who wish to kill some of a population of whales that is being watched elsewhere. The only way it becomes relevant is when geopolitical concerns impinge in a manner that makes whaling a component of broader negotiations. The value of whale-watching might matter in countries that may be influenced to start whaling (Corkeron, 2006a), but it is of little relevance where whaling is well established.

Within whaling nations, direct comparisons of values of the two forms of utilization (whaling and whale-watching) are, demonstrably, irrelevant to those responsible for managing both industries. The argument from whaling nations is that it is possible to run both industries simultaneously (e.g. Icelandic Ministry of Fisheries and Agriculture, 2009). As both have coexisted in Norway and Iceland for about two decades (Hoyt, 2001), this argument can be seen as valid, even if a nation's whaling record may affect the likelihood that people will holiday there (Higham & Lusseau, 2007).

Other objections exist to the utility of making comparisons of the relative monetary value of whale-watching and whaling. If large-scale commercial whaling for meat and other, more value-added products (e.g. WDCS, 2010) recommences, the monetary value of whaling, overall, may prove more valuable than whale-watching. Further, the belief in whaling nations that culling marine mammals will lead to increased fisheries yields (see above) includes a belief that the value of whaling is enhanced by inclusion of extra fisheries yields made available by whaling (Anonymous, 2010). The abstract of a recent report from the University of Iceland puts this added value, if 150 fin and 150 minke whales were hunted each year in Icelandic waters, at approximately two orders of magnitude

more than the total earnings of Icelandic whale-watching (Anonymous, 2010). This has never been addressed in NGO comparisons of the value of whale-watching over whaling.

Finally, a market-based approach to resolving the whaling debate has been suggested (Costello *et al.*, 2012). The suggestion put forward is to create an international market for whale quotas that could be bought either by those wishing to hunt, or not to hunt, whales. Yet implicit in the long-standing argument for whale-watching to replace whaling is market-based thinking – that is, that the two uses of whales are inherently fungible. It has been hoped for over three decades that the market value of whale-watching will outweigh the market value of the products from whaling to the point that whaling will disappear with no adverse economic or social impacts to communities that currently earn a living from whaling. This thinking has failed – the proponents of the new market-based approach do not even identify it when discussing the history of market-based approaches to solving the whaling issue (Costello *et al.*, 2012).

Conclusion: when a tradition transforms into a zombie?

The number – of both species and individuals – of whales being killed by whaling in recent years has increased (IWC, 2011a, 2011b). Given this, why is the NGO community still supporting work that demonstrates the value of whale-watching, without caveats as to the success of whale-watching as a replacement for whaling (IFAW, 2009)? Support for whale-watching seems to have become an end in itself.

Most discussions of attitudes and values regarding whales, and the role that whale-watching has in influencing those values, seem trapped in a 1980s time warp. Yet, over the past three decades, we have seen the promotion of whales-as-Nasties by supporters of whaling, and the commoditization of whales-as-Lovelies by those opposed to whaling. Despite the failure of promoting whale-watching as a replacement for whaling, encouraging the 'whale-watching and whaling are fungible' idea remains a tradition among anti-whaling NGOs. Further, this focus on whale-watching has limited NGOs' responses to the transformations that have happened over recent decades in whaling nations.

Market liberalism (*sensu* Quiggin, 2010) based approaches to nature conservation can be criticized on several grounds (e.g. Brockington & Duffy 2010; Carrier, 2010; Igoe *et al.*, 2010). In the case of supporting whale-watching as a replacement for whaling, the criticism is simple. It hasn't worked, and there is no evidence that it will work in the foreseeable future. That NGOs supporting whale-watching need to run campaigns asking tourists not to eat whale meat illustrates the depth of this failure.

Whale-watching can offer a variety of social benefits (e.g. Wilson & Tisdell, 2003), but it has not worked as a replacement for whaling in whaling nations. There are, no doubt, other approaches that could be taken to oppose continued commercial and 'scientific' whaling, but the idea that whale-watching will replace whaling has become a zombie concept. For those engaged in campaigns to stop whaling, it's time to lay the idea to rest.

Acknowledgements

My thanks to J. Higham, J. Gatzke and M. Simpkins for comments that improved a previous version of this manuscript.

REFERENCES

Anonymous (2006). 2006 Resolutions The resolutions made at the 58th IWC annual meeting in St. Kitts & Nevis, 2006. Available at: http://iwcoffice.org/meetings/resolutions/resolution2006.htm

Anonymous (2010). *Nr. 21/2010 – Þjóðhagslegt gildi hvalveiða*. Available at: http://www.sjavarutvegsraduneyti.is/frettir/frettatilkynningar/nr/10007

Anonymous (2012). Guidelines for the treatment of animals in behavioural research and teaching. *Animal Behaviour*, 83, 301–309.

Baker, C.S., Steel, D., Choi, Y., *et al.* (2010). Genetic evidence of illegal trade in protected whales links Japan with the US and South Korea. *Biology Letters* 6, 647–650.

Boye, T.K., Simon, M. & Madsen P.T. (2010). Habitat use of humpback whales in Godthaabsfjord, West Greenland, with implications for commercial exploitation. *Journal of the Marine Biological Association of the United Kingdom* 90, 1529–1538.

Brockington, D. & Duffy, R. (2010). Capitalism and conservation: The production and reproduction of biodiversity conservation. *Antipode* 42, 469–484.

Carrier, J.G. (2010). Protecting the environment the natural way: Ethical consumption and commodity fetishism. *Antipode* 42, 672–689.

Caughley, G. (1985). Problems in wildlife management. In H. Messel (Ed.), *The Study of Populations*. Sydney: Pergamon Press, pp. 129–135.

Clapham, P.J., Berggren, P., Childerhouse, S., *et al.* (2003). Whaling as science. *Bioscience* 53, 210–212.

Corkeron, P.J. (2006a). How shall we watch whales? In D.M. Lavigne (Ed.), *Gaining Ground: In pursuit of ecological sustainability. Proceedings of an international forum*. Guelph, ON: The International Fund for Animal Welfare and Limerick, Ireland: University of Limerick, pp. 161–170.

Corkeron, P.J. (2006b). Opposing views of the 'ecosystem approach' to fisheries management. *Conservation Biology* 20, 617–619.

Corkeron, P.J. (2007). *Iceland, whaling and ecosystem-based fishery management*. Available at: http://www.wdcs-de.org/docs/Iceland_Corkeron_Report.pdf

Corkeron, P.J. (2009a). Reconsidering the science of scientific whaling. *Marine Ecology Progress Series* 375, 305–309.

Corkeron, P.J. (2009b). Marine mammals' influence on ecosystem processes affecting fisheries in the Barents Sea is trivial. *Biology Letters* 5, 204–206.

Costello, C., Gerber, L. & Gaines, S. (2012). A market approach to saving whales. *Nature* 481, 139–140.

Dakin, W.J. (1934). *Whalemen Adventurers: The story of whaling in Australian waters and other southern seas related thereto, from the days of sail to modern times*. Sydney: Angus and Robertson.

Fox, J. (2011). *Zombie ideas in ecology.* Available from: http://oikosjournal.wordpress.com/2011/06/17/zombie-ideas-in-ecology/

Gales, N.J., Kasuya, T., Clapham, P.J. & Brownell, R.L. (2005). Japan's whaling plan under scrutiny. *Nature* 435, 883–884.

Hamazaki, T. & Tanno, D. (2001). Approval of whaling and whaling-related beliefs: Public opinion in whaling and nonwhaling countries. *Human Dimensions of Wildlife* 6, 131–144.

Harwood, J. (2007). Is there a role for ecologists in an ecosystem approach to the management of marine resources? *Aquatic Conservation: Marine and Freshwater Ecosystems* 17, 1–4.

Higham, J.E.S. & Lusseau, D. (2007). Urgent need for empirical research into whaling and whale watching. *Conservation Biology* 21, 554–558.

Hjort, J. (1914). Fluctuations in the great fisheries of northern Europe viewed in the light of biological research. *Rapports, Conceil Permanent International pour l'Exploration de la Mer* 20, 1–228.

Hoyt, E. (2001). *Whale Watching 2001: Worldwide tourism numbers, expenditures, and expanding socioeconomic benefits.* Yarmouth Port, MA: International Fund for Animal Welfare, pp. i–vi, 1–158.

Icelandic Ministry of Fisheries and Agriculture (2009). *Questions and Answers about sustainable whaling.* Available at: http://eng.sjavarutvegsraduneyti.is/news-and-articles//nr/9604

IFAW (2009). Press release. 'New Report Documents Massive Growth in Whale Watching'. Available at: http://www.ifaw.org/ifaw_united_states/media_center/press_releases/6_23_2009_55364.php

IFAW (2011). Press release. 'New whale protection campaign urges tourists visiting Iceland to avoid eating whale meat'. Available at: http://www.ifaw.org/in/news/new-whale-protection-campaign-urges-tourists-visiting-iceland-avoid-eating-whale-meat

Igoe, J., Neves, K. & Brockington, D. (2010). A spectacular eco-tour around the historic bloc: Theorising the convergence of biodiversity conservation and capitalist expansion. *Antipode* 42, 486–512.

Iliff, M. (2010). Contemporary initiatives on the future of the International Whaling Commission. *Marine Policy* 34, 461–467.

Ingebrigtsen, A. (1929). Whales caught in the North Atlantic and other seas. *Rapports, Conseil Permanent International pour l'Exploration de la Mer* 56, 123–135.

IWC (2011a). *Catches taken: Under objection or Under Reservation. Catches under Objection since 1985.* Available at: http://iwcoffice.org/conservation/table_objection.htm

IWC (2011b). *Special Permit Catches since 1985.* Available at: http://iwcoffice.org/conservation/table_permit.htm

Jørgensen, C., Enberg, K. & Dunlop, E.S., *et al.* (2007). Managing evolving fish stocks *Science* 318, 1247–1248.

Lislevand, T. (2009). *Nødvendig med norsk hval- og selfangst?* Available at http://www.forskning.no/artikler/2009/mai/220385

Ludvigsen, S. (2005). The Norwegian position on culling. *Science* 305, 497–498.

Morishita, J. (2006). Multiple analysis of the whaling issue: Understanding the dispute by a matrix. *Marine Policy* 30, 802–808.

Morishita, J. (2008). What is the ecosystem approach for fisheries management? *Marine Policy* 32, 19–26.

Murawski, S. (2007). Ten myths concerning ecosystem approaches to marine resource management. *Marine Policy* 31, 681–690.

NAMMCO (2010). Marine Ecosystem Management Conference is now canceled. Available at: http://www.nammco.no/Nammco/Mainpage/DocumentsAndInformation/marine_ecosystem_management_conference_is_now_cancelled.html

Neves, K. (2010). Cashing in on cetourism: A critical ecological engagement with dominant E-NGO discourses on whaling, cetacean conservation and whale-watching. *Antipode* 42, 719–741.

O'Connor, S., Campbell, R., Cortez, H. & Knowles, T. (2009). *Whale Watching Worldwide: Tourism numbers, expenditures and expanding economic benefits.* A special report from the International Fund for Animal Welfare. Yarmouth, MA: Economists at Large.

Pikitch, E.K., Santora, C., Babcock, E.A., *et al.* (2004). Fishery management and culling: Response. *Science* 306, 1892.

Quiggin, J. (2010). *Zombie economics: How dead ideas still walk among us.* Princeton, NJ: Princeton University Press. 216 pp.

Reyes, O. (2011). Zombie carbon and sectoral market mechanisms. *Capitalism Nature Socialism* 22, 117–135.

Ris, M. (1993). Conflicting cultural values: Whale tourism in northern Norway. *Arctic* 46, 156–163.

Stockings, C. (Ed.) (2010). *Zombie Myths of Australian Military History.* Sydney: University of New South Wales Press. 279 pp.

Suter, K.D. (1982). Australia's new whaling policy: Formulation and implementation. *Marine Policy* 6, 287–302.

Tønnessen, J.N. & Johnsen, A.O. (1982). *The History of Modern Whaling* (transl. by R.I. Christophersen). Berkeley, CA: University of California Press.

WDCS (2010). Reinventing the Whale. The whaling industry's development of new applications for whale oil and other products in pharmaceuticals, health supplements and animal feed. 12 pp. Available at: http://www.wdcs.org/publications.php

WDCS (2011). Press release. 'WDCS calls on visitors to Iceland not to eat whale meat' Available at: http://www.wdcs.org/stop/killing_trade/tourists.php

Wilson, C. & Tisdell, C. (2003). Conservation and economic benefits of wildlife-based tourism: Sea turtles and whales as case studies. *Human Dimensions of Wildlife* 8, 49–58.

Øien, N. (1988). The distribution of killer whales (*Orcinus orca*) in the North Atlantic based on Norwegian catches. *Rit Fiskideildar* 11, 65–78.

5

The whale-watching industry

Historical development

Erich Hoyt and E.C.M. Parsons

Introduction

The story of commercial whale-watching spans half a century from the first $1 USD earned on the back of a grey whale (*Eschrichtius robustus*) in 1955 to the $2.1 billion USD industry of today (Hoyt, 2009a; O'Connor *et al.*, 2009). As with many new entertainment 'industries' of our era ranging from surfing to the iPods, iPhones and iPads of Apple Computer, Inc., it all started in California. The surprise perhaps lies in how popular and pervasive worldwide the whale-watching industry has become – to the extent that we now focus on the implications from 'too much success', try to limit the industry in some areas, and ask if true sustainability is achievable.

The origins and early historical development of the whale-watching industry have shaped the industry to this day. Considering the developmental stages of tourism and rate of growth experienced by the whale-watch industry helps us understand its impact. Whale-watching has been profoundly influenced by a number of factors: (1) the platform used, including the types of boats, as well as the background and location of the whale-watching owner-operators; (2) the species being watched and the peculiar geography of the ecosystem where they are found; and (3) the typology of the visitors or tourists taking the trips and their expectations. Looking at all these aspects helps us grasp what the industry has become today and the implications for the future.

Early history

The first 'official' whale-watching trip was conducted by a fisherman from San Diego, California, named Chuck Chamberlin who put out a sign saying 'See the whales: $1' (Hoyt, 1984, 2009b). These were winter-to-spring boat trips to see grey whales as they migrated back and forth between the lagoons of Baja California, México and Alaska. Chamberlin was certainly influenced by the land-based student whale-watching 'counts' that had occurred since the late 1940s as part of a University of California (La Jolla) research and government monitoring project by the pioneer whale researcher Carl Hubbs. In 1950, Cabrillo National Monument was converted from an old US Army gun station into the first public whale-watching lookout, hosting some 10,000 people that first year (Hoyt, 1984, 2001). The other factor was that Chamberlin had a boat and was not doing much fishing in the winter (Hoyt, 1984; Gilmore, pers. comm., 1983).

Grey whales – at the time recovering from very low numbers after nineteenth-century whaling – gained huge popularity from whale-watching. Following the success of whale-watching at Cabrillo, lookouts, formal and informal, some with naturalist guides, sprang up all along the California coast to witness the grey whale migration spectacle (Jones & Swartz, 2009). Californians adopted the grey whale as their state marine mammal, and the whale became the

Whale-watching: Sustainable Tourism and Ecological Management, eds J. Higham, L. Bejder and R. Williams.
Published by Cambridge University Press. © Cambridge University Press 2014.

symbol of the US conservation movement in the late 1960s (Forestell, 2009). It was a stirring thought that this animal, heavily hunted and thought to be nearly extinct, was returning to coastal waters in ever-increasing numbers year after year.

Until the late 1970s, land-based whale-watching was far more popular than boat-based whale-watching, mainly driven by the numerous lookouts all along the California, Oregon and Washington coasts, as well as by California whale festivals (Hoyt, 2001). Thus, even though whale-watching was becoming ever more popular, *commercial* whale-watching was still embryonic.

During this period, commercial whale-watching spread first to the Mexican lagoons, with long-range, naturalist-led trips out of San Diego, and then up the California coast with boat trips by fishermen and organized school trips on larger tourist boats out of the Los Angeles and San Francisco areas. A local Hawaiian whale club, the Wailupe Whale-watchers, sponsored informal trips to see humpback whales (*Megaptera novaeangliae*). In 1973, the Montreal Zoological Society began offering whale-watching trips down the St Lawrence River in Canada (Hoyt, 2009b) to see various baleen whales. In 1975, whale-watching opened up to see the humpback whales feeding in New England waters on the east coast of the US. This was to prove a turning point.

Whale-watching becomes big business

A successful partnership between science, education and commercial whale-watching began in Provincetown, Massachusetts, in 1975, when fisherman Al Avellar, of what would become the Dolphin Fleet, asked Charles 'Stormy' Mayo to be his naturalist. Mayo soon saw the possibilities for using the boat as a platform for studying whales. He set up the Center for Coastal Studies as a research and educational institution, and the close ties with commercial whale-watching here have been maintained ever since (Hoyt, 1995, 2009b).

The arrangement works as follows. The Center provides naturalist guides for the Dolphin Fleet.

They are paid to help direct the boat to the whales, presenting an informal educational lecture, and answering questions. The Center sells T-shirts and other merchandise on board. Most important, Center researchers can conduct their own photo-ID research and collect other data. This key partnership between science, education and commerce proved to be successful. Proximity to large human populations (Boston and New York) gave a huge potential audience. The Center for Coastal Studies–Dolphin Fleet partnership set the bar high for New England whale-watching. By the early 1990s, 18 of the 21 whale-watching operators had naturalists guiding boats and lecturing whale-watchers, while 10 operations were taking and contributing ID photos (Hoyt, 1995).

The New England model of successful whale-watching, education and research, like Yankee whaling from an earlier century, spread its influence far and wide. New England has had an impact on the development of whale-watching in locales as diverse as the Gulf of St Lawrence in Quebec; northern Norway; and Dominica in the eastern Caribbean. Operators from these areas worked with, were visited by or in some cases actively studied the New England whale-watch model. Unfortunately, over time, the model lost its impact as whale-watching was transplanted further and further afield, where operators sometimes with little or no knowledge of whales using different kinds of boats encountered different economic challenges, with different customers.

Measuring the value of whale-watching

The first attempts to measure the commercial impact of whale-watching came at the 'Whales Alive' Conference in Boston in 1983, where several researchers presented the first papers looking at the industry from the perspective of economics and social science. They found that whale-watchers were generally well-educated, among other things, and numbered at the time in the low hundreds of thousands in what was then a North American

Table 5.1 Estimated growth of whale-watching worldwide.

Year	No of countries and territories with commercial whale-watching	Number of whale-watchers	Direct expenditure in millions of US $	Total expenditure in millions of US $	Average annual increase % from previous period	Sources
1981	3	400,000	$4.1	$14	–	Kaza, 1982; Kelly, 1983; Sergeant, pers. comm., 1984
1988	–	1,500,000	$11–16	$38.5–56	20.8%	Kraus, 1989
1991	31	4,046,957	$77	$317.9	39.2%	Hoyt, 1992
1994	65	5,425,506	$122.4	$504.3	10.3%	Hoyt, 1995
1998	87	9,020,196	$299.5	$1,049	13.6%	Hoyt, 2001
2008	119	12,977,218	$872.7	$2,113.1	3.7%	O'Connor et al., 2009; Hoyt & Iñíguez, 2008

industry (Kaza 1982; Kelly 1983; Tilt 1985a, 1985b). The first estimate of the value of whale-watching covered the year 1981 and amounted to $14 million in total tourist expenditure (Kaza 1982; Kelly, 1983; Sergeant, pers. comm.; see Table 5.1). This included the cost of tickets plus food, accommodation, travel and souvenirs associated with the whale-watching trip. This simplified economic methodology has been used ever since to chart the growth of whale-watching, although other approaches such as contingent valuation method and rate of return have given valuable insights into the commercial success of whale-watching.

Whale-watching goes global

In 1984, when *The Whale Watcher's Handbook* was published, the first world guide to whale-watching, commercial whale-watching was in just four countries (Hoyt, 1984). By the mid to late 1980s, whale-watching began to expand rapidly to areas such as Argentina, Canary Islands (Tenerife), New Zealand, the UK and Ireland. A world survey of commercial whale-watching in 1991 turned up an estimate of 4 million whale-watchers spending $317.9 million USD. The number of whale-watchers grew to 5.4 million in 1994, and 9 million in 1998 (Hoyt 1995,

2001). In 1991, only 31 countries were involved in whale-watching. By 1998, some 87 countries and overseas territories had commercial whale-watch tours with some 9 million people a year going whale-watching (Hoyt, 2001). The growth rate from 1991 to 1998, based on numbers of whale-watchers, was 12.1% average annual increase per year: 3–4 times the growth rate of all tourism arrivals during the period (Hoyt, 2001). This was the greatest period of growth experienced by whale-watching, and this unchecked growth set up the problems with whale-watching and the challenges to sustainability that we face today.

By the early 2000s, whale-watching growth was slowing down. The 9/11 travel blip, SARS, and the threat of Bird Flu depressed tourism in general, although only for a few years; mainly, it has shown positive growth from 2000 to 2010. By 2008, the whale-watch numbers had climbed to nearly 13 million people spending $2.1 billion USD on whale-watch tours in 119 countries (O'Connor et al., 2009). However, the average increase from the period 1998 to 2008 was a much more reasonable 3.7% per year, compared to global tourism growth of 4.2% per year. Yet, in many areas of the world, the damage from such fast, unchecked growth in the late 1980s and through the 1990s has created enduring problems (Hoyt, 2009b). Also, even though overall the levels have evened out, this is largely a factor of the maturity of the North American industry which represents nearly half of all whale-watching tourists (O'Connor *et al.*, 2009). Fast growth during the past decade continues in young whale-watching countries such as mainland China (107% growth per annum since 1998), Maldives (86%), Cambodia and Laos together (79%), St Lucia (74%), Madeira (73%), Venezuela (58%), Costa Rica and Nicaragua (both 56%) and Panama (53%) (O'Connor *et al.*, 2009). Efforts to create effective regulatory frameworks in these countries are essential if they are not to repeat the mistakes of other countries in the past. Most countries have allowed whale-watching to start up with few, if any, controls, opting for regulations only after problems begin to develop. New Zealand is an example of a country that, right from the beginning, regulated whale-watching with a permit system and set of rules based on scientific studies and precautionary management. In communities and countries where whale-watching is recent or new, there is the opportunity to learn from both this good example and the experiences of other countries where problems became severe due to lack of regulation. It may be easier for managers to take an early proactive and precautionary approach than to have to try to exercise control after whale-watching is well established. The concern over the fast growth of unregulated whale-watching and the severe impacts that could occur to cetacean populations, particularly in developing countries, led to the whale-watching subcommittee of the International Whaling Commission adopting this issue as an annual item on their agenda (IWC, 2012)

Although too-fast growth has led to considerable problems, current problems with whale-watching can be understood by looking at the platforms used, the species being watched and the peculiar geography where they are found, and the typology of the visitors or tourists taking the trips and their expectations. In terms of platform, land-based whale-watching and whale-watching on large ships has accommodated growth better than fleets of smaller boats or low-flying aircraft. Some species – for example, southern right whales (*Eubalaena*

australis), humpback whales and grey whales – can be watched from land on migration or on mating grounds; other species require boats. When whale or dolphin species are found in confined geographical areas such as narrow straits or bays, multiple boats become a problem. In terms of the visitors or tourists, mass tourists, in particular, can have expectations for large hotels and other amenities that can put a strain on the infrastructure of small coastal communites. These concepts are discussed in greater depth below.

Categories of commercial and other whale-watching

As whale-watching has evolved, it has come to encompass many different types of activity. For example, whale-watching can be land-based, boat-based or even aerial (whale-watching from helicopters, planes or balloons; Parsons *et al.*, 2006). Moreover, the activity might be 'commercial whale-watching', where tourists pay to go out on a trip, versus 'recreational whale-watching', where cetaceans are viewed from personal vessels. Parsons *et al.* (2006) note that the distinction is important as recreational whale-watching trips may not be regulated and managed, and the organizers may be unaware of their potential impacts on the target species. Whale-watching is also separated into 'directed whale-watching' where cetaceans are being specifically sought out versus 'opportunistic whale-watching', when a trip is not specifically focused on seeing whales (e.g. a scenic marine tour), but cetaceans are often seen on such trips and are likely mentioned in the company's marketing materials (Parsons *et al.*, 2006). Boat-based whale-watching may also occur as 'trips' which may be a few hours to a day long, or multi-day 'tours'. The vessel used might also be 'powered' or 'unpowered' (e.g. kayaks, or sailing boats), and again the distinction is important as powered vessels may produce more noise, and hence greater impacts. Further details about the diversity of whale-watching platforms are listed in Table 5.2.

Table 5.2 The diversity of whale-watching platforms (and the impact on the character and size of the industry, including the capital, staff and infrastructure needed, and the visitor types).

Platform or vessel type	Origin	Visitor types	Capital, staff, and infrastructure needed	Days	Visitor capacity
Cruise ships (large)	N, I	GP	H, H, L	1–7*	1000+
Cruise ships (medium)	N, I	GP	H, H, L-M	1–15	25–150
Purpose-built large w-w boats	L, N	All	H, M-H, M-H	1	150–300
Research vessels	L, N, I	HE, GP	L-H, M-H, M	1–15	10+
Sailboats	L, N, I	All	M, L-M, L-M	1–15	5+
Dinghy, panga	L	HE, GP	L, L, L	1	2–5
Zodiac/other inflatable	L, N	HE, GP	L, L, L	1	2–15
Kayak/canoe	L, N, I	HE, GP	L, L, L	1–15	1–2
Fishing/whaling boat	L, N	All	M, M, M-H	1	5–25
Ferry	L, N, I	All	L-M, L-M, M-H	1	50–300
Aircraft	L, N	All	H, M-H, H	1	1–4
Land-based	L	All	L, L, L	1–7	1–1000

KEY

Origin: L, Local; N, National; I, International.

Visitor types: GP, General public; HE, Higher education (13+); SC, School children (8–12).

Capital, Staff, & Infrastructure Needed:

• capital needed for platform

• staff needed (in relation to visitor numbers)

• local infrastructure needed

L, Low; M, Medium; H, High.

Days: *Days can range from 1 to 30, but actual whale-watching days are usually 1 but sometimes up to 7.

Visitor capacity: Number of whale-watchers on a single trip.

Source: Table adapted from IFAW (1999) and Hoyt (2005), with permission.

Although perhaps not recognized as being commercial whale-watching per se, cetacean researchers often subsidize the costs of their research trips by taking on board paying tourists, who may also be involved in collecting data for the researchers. As income is derived from taking paying tourists (sometimes referred to as 'volunteers') to watch whales, however, even though the income may be ploughed back into research and not for profit, this type of activity should technically be considered to be commercial whale-watching. The data gathered by these 'whale research trips/tours' can prove valuable for conservation and the management of the cetaceans being observed. Parsons *et al.* (2006) highlight that for more effective data-gathering, and to ensure that bona fide scientific research is indeed taking place, ideally such trips or tours should ensure that:

(1) Scientists are involved in providing advice on the design and analysis of the survey/expedition.

(2) The data collected on the survey/expedition are analysed and published (ideally as peer-reviewed journal articles or reports).

(3) The results of the survey/expedition are disseminated to the paying participants and appropriate authorities and stakeholders.

(4) The paying participants actively assist with the collection of data and/or logistical aspects of the survey/expedition.

(5) Appropriate and detailed training is given to the paying participant (Parsons *et al.*, 2006: 250).

The above, however, is different from 'whale-watching-aided research' when, for example, whale-watching operations allow scientists to use the vessel as a 'platform of opportunity', or when sightings and environmental data are collected in a methodical manner and subsequently submitted to scientists for analysis. Whale-watching aided research has been extremely valuable in providing data for cetacean conservation and management (IWC, 2003, 2004, 2005, 2006, 2007, 2008, 2009, 2010).

Some whale-watching trips involve placing tourists in the water with cetaceans. These 'swim-with-cetacean' activities most commonly occur when dolphins are a target species, and the number of countries where this occurs is increasing. Some locations, such as Tonga, the Dominican Republic and Australia, also have operations that allow swimming with large whales, typically humpback or northern minke whales (*Balaenoptera acutorostrata*). In some of these operations, tourists are tethered to lines and animals control the encounter (IWC, 2003), whereas in other operations some pursuit is involved, or tourists are placed in the path of oncoming cetaceans. Parsons *et al.* (2006) define the former as 'passive' swim-with-cetacean trips, and the latter 'active'. There is some concern over the fast growth and potential impacts of swim-with-cetacean tourism, particularly swim-with-whale tourism, for the animals as well as the potential risks to humans from such large animals (IWC, 2003), although the 'passive' type is less of a problem in terms of impacts on the target population.

With respect to potential impacts on cetaceans, Parsons *et al.* (2006) define two types of 'intrusive' whale-watching, which have appeared in recent years in some locations. These types of whale-watching may be particularly harmful to cetaceans:

(1) **Physically intrusive** operations involve the tourists or tour operators/crew physically touching the cetaceans in any way except for accidental or incidental contact; and

(2) **Ecologically intrusive** operations include activities that alter the natural history and behavioural ecology of

cetaceans, such as introducing food to cetaceans (Parsons *et al.*, 2006, p. 250).

Is commercial whale-watching ecotourism?

Whale-watching is often described as 'ecotourism', but ecotourism has a specific definition. Ceballos-Lascurain (1991) was one of the first to define this type of tourism activity as:

Tourism that involves traveling to relatively undisturbed or uncontaminated areas with the specific objective of studying, admiring, and enjoying the scenery and its wild plants and animals, as well as any existing cultural manifestations (both past and present) found in these areas. (p. 25)

In 1996, the IUCN officially adopted a similar definition of ecotourism:

Ecotourism is environmentally responsible travel and visitation to relatively undisturbed natural areas, in order to enjoy and appreciate nature (and accompanying cultural features, both past and present) that promotes conservation, and provides for beneficially active socioeconomic involvement of local populations. (Ceballos-Lascurain, 1996)

Honey (1999) provides a similar definition that has become widely adopted and emphasizes education, contributions to conservation and benefits to local host communities:

Ecotourism is travel to fragile, pristine, and usually protected areas that strive to be low impact and (usually) small scale. It helps educate the traveler; provides funds for conservation; directly benefits the economic development and political empowerment of local communities; and fosters respect for different cultures and for human rights. (p. 25)

A profusion of descriptions of ecotourism led Fennell (1999) to review 15 definitions of ecotourism and ultimately come up with his own composite definition:

Ecotourism is a sustainable form of natural resource-based tourism that focuses primarily on experiencing and learning about nature, and which is ethically managed to be low-impact, non-consumptive, and locally oriented (control, benefits, and scale). It typically occurs in natural areas, and should contribute to the conservation or preservation of such areas. (p. 43)

Therefore, by definition, ecotourism is a tourism activity that is specifically designed to reduce its impacts, has a component of local host population involvement and participation, and promotes conservation. As such, probably few whale-watching operations could be classified as genuine 'ecotourism'. The International Whaling Commission has gone further and their whale-watching sub-committee specifically defined 'whale ecotourism' as a commercial whale-watching operation that attempts to:

(a) Actively assist with the conservation of their resource (cetaceans), such as co-operating with research groups and other scientists and with research projects or allowing vessels to be used by scientists/research groups as platforms of opportunity;

(b) Provide appropriate, accurate and detailed interpretative/educational materials or activities for their clientele about the cetaceans viewed and their associated habitat;

(c) Minimize their environmental impact (such as reducing emissions or disposing of refuse appropriately);

(d) Adhere to whalewatching regulations or an appropriate set of guidelines, if no specific regulations are available for the area; and

(e) Provide some benefits to the local host community within which the company operates. Such benefits could include a policy of preferential employment of local people, selling local handicrafts, or supporting (either financially or in kind) local community-based conservation, education, cultural or social projects or activities (for example financially or in kind support for a voluntary marine rescue service or providing non-profit trips for local schools). (Parsons *et al.*, 2006: 250–251)

Certain operations in locations such as North America, Australia and Europe could qualify as whale ecotourism using the above definition, but at present these would be a small minority of operations. To reduce the impacts of whale-watching on cetaceans, and to enhance the economic benefits of whale-watching at a local level, encouraging whale-watching operations to become whale ecotourism would ultimately promote the sustainability of the industry. In addition to the boat-based trips that comprise most of the whale-watching industry, it should be noted that the above definition could easily be applied to land-based whale-watching operations, or cetacean-themed museums and visitor centres.

Problems related to whale-watching growth

Whale-watching started in California as a passion and only later became a business. The fundamental aspects that helped create the business were: (1) that whales, initially the grey whales, were close enough to land to be seen, initially by land-based spectators and later by customers in small boats, (2) the availability of customers who were keen, and (3) the availability of boats to take them to get a closer look. The whales and customers were essential, of course, but equally important was the availability of boats that were not being used during the periods that whales were present. These were fishing boats, mainly, but also local tourist sightseeing boats, cruisers, diving skiffs, and in some cases ferries.

However, as the industry developed, in both California and elsewhere in the world, there was a proliferation of dedicated whale-watching tours and vessels. As such, the density of vessels around groups of cetaceans in some locations began to escalate. Even in areas where there are regulated limits on the number of vessels allowed in close proximity to whales, once one vessel leaves, another can approach almost immediately, leaving the animals almost constantly exposed to disturbances.

With the success of whale-watching in New England, the boats changed. Bigger boats were custom-designed for whale-watching (with flat decks, seats, toilets, and an ability to sell concessions on board including seasick tablets, sunscreen and refreshments). This allowed New England whale-watching to move more gracefully into a mature phase, attracting wider groups of tourists, and larger numbers without necessarily increasing the number of vessels. Larger vessels, however, may have greater impacts, being potentially noisier and less manoeuvrable. The facility to take out larger numbers of tourists may also lead to a broadening of the tourist types taking trips, from more specialist tourists to more general or mass tourists, which may change tourism expectations (Duffus & Dearden, 1990).

In other locations, however, the problem of vessel crowding has become particularly serious. In the Canary Islands, for example, the proliferation of boats and lack of regulation in the late 1980s and early 1990s led to a crisis situation with close to 100 boats on the water. Similar situations have been reported off southern Vancouver Island with vessels watching killer whales (*Orcinus orca*) from both US and Canadian ports converging on a single endangered population of killer whales – the so-called 'southern residents'. Scientists in this region are concerned about the impact of such crowding, which is not currently regulated (Williams *et al.*, 2009).

There is also an increasing trend for faster and faster vessels to get to whale-watching locations more rapidly, and thus more tourists can be taken out over the course of a day. The potential impacts of high-speed whale-watching vessels, defined as those travelling at more than 13 knots (IWC, 2005), have led to some international concerns (e.g. IWC, 2003, 2004; see Box 5.1 for details about the IWC). These faster vessels are more likely to kill cetaceans if collisions occur; the risk of collision is potentially increased, due to less time for whales or vessels to manoeuvre out of the way; and these faster vessels often produce more noise pollution that can disturb animals.

In some locations the number of companies are limited (via licensing systems), and there are controls on the number and duration of trips. In Kaikoura, New Zealand, for example, there is only one permit for sperm whale watching with a restricted number of boats operated by one

Box 5.1 The International Whaling Commission and whale-watching

The International Whaling Commission (IWC) was established in 1946 to address the declines in many large whale stocks due to commercial whaling, and to manage these stocks. Through the IWC, whaling bans were successively introduced for a variety of depleted whale species, and in 1982 a moratorium was imposed on all commercial whaling (which came into effect in 1986). In 1993, the IWC passed a resolution (resolution 1993–9), noting that whale-watching was a sustainable use of whales as a resource, and further noted the growing economic value of the industry. The following year (via resolution 1994–14) the IWC Scientific Committee was asked to provide scientific advice on whale-watching issues, including guidelines for whale-watching operations. A third resolution in 1996 (resolution 1996–2) emphasized that because the IWC was tasked to consider the use, management and conservation of whales as resources, and as whale-watching was a major economic use of cetaceans, it had a role in overseeing whale-watching and its management internationally.

To provide the scientific advice, the IWC Scientific Committee established a whale-watching subcommittee. Every year, the subcommittee discusses scientific studies that provide information on the impacts of whale-watching activities on cetaceans, and methods to collect scientific data from whale-watching vessels to aid management of whale stocks. The whale-watching subcommittee also developed a set of international whale-watching guidelines (http://www.iwcoffice.org/conservation/wwguidelines.htm), and compiled a compendium of whale-watching laws, regulations, and codes of conduct from around the world (IWC, 2006). Moreover, in 2004, it organized and held a workshop on the scientific aspects of sustainable whale-watching (IWC, 2005; available from http://www.iwcoffice.org/_documents/sci_com/WW_Workshop.pdf). The subcommittee also discusses arising and sometimes controversial aspects of the whale-watching industry, such as dolphin feeding stations, swim-with-whale tourism, and the impacts of high-speed whale-watching vessels.

Since 2009, the Conservation Committee of the IWC has also been developing an action plan to deal with whale-watching management issues, again recognizing the importance of whale-watching as an economic use of cetaceans, and the potential impacts that the activity poses. The Conservation Committee may become even more involved in international oversight of the industry.

company. Additional permits in this area were awarded to dolphin watching, dolphin swimming, and watching from helicopter and fixed-wing aircraft. Each company has become a specialist and the renewal of permits depends on good practice. In Puerto Pirámides, Argentina, only six companies are allowed to operate southern right whale tours. For some years, South Africa did not allow whale-watching boats, preferring to encourage a land-based industry, and when they did finally allow boat-based whale-watching, restricted it to two permits per area. In general, in such areas where the regulations are in place and are strictly enforced, the industry is arguably more sustainable because there is less pressure on the animals, and the existing businesses have less competition and may be more willing to invest in their own company. However, there are many areas where there are licensing systems and regulations on paper, but compliance with these may be low (e.g. Scarpaci *et al.*, 2003, 2004), unless appropriate monitoring occurs and the regulations are enforced.

With an increase in the number and availability of boat-based whale watching trips, and what may arguably be a desire by tourists to get closer to cetaceans, land-based whale-watching, which was initially an important component of whale tourism, has become less important as a sector of the whale-watching industry. This is unfortunate, as land-based whale-watching has no direct impact on the cetaceans being viewed. Some threatened species of cetaceans, such as southern right whales, can be watched benignly from land platforms.

The propensity of some species of whales and dolphins to be curious and to approach boats has not only determined the species that are most likely to be watched, but has shaped the kind of whale-watching that occurs. Some species of large whale such as grey whales, humpback whales, and, to an extent, northern minke whales (*Balaenoptera acutorostrata*), frequently approach boats. Many dolphin species approach vessels to bow-ride on fast-moving vessels. This tendency for animals to approach vessels can lead to management problems, especially when local regulations require vessels to maintain some distance from cetaceans. In contrast to birds or other land-based wildlife where it is accepted that best practice is for the observer to be unobtrusive or invisible to the wildlife while watching from a blind, whales and dolphins are approached by often noisy boats and the animals may be aware of the whale-watchers.

However, not all cetaceans are curious and readily approach vessels. Some species may be 'shy', attempting to avoid whale-watching boats, such as harbour porpoises (*Phocoena phocoena*). There may be certain age classes of individual species, or certain times of the year (e.g. breeding seasons), or even times of day (e.g. feeding or resting with spinner dolphins, *Stenella longirostris*, or killer whales) when cetaceans are susceptible to disturbance. Some species are vulnerable or endangered, and these species perhaps should not be watched at all, unless watching is from discrete land-based platforms (e.g. Irrawaddy dolphins *Orcaella brevirostris*; see Beasley *et al.*, 2010). A 'one size fits all' attitude to whale-watching operations and management, and an inability to recognize that different species and populations, under a complex set of variables, behave differently around boats, may lead to more sensitive species being stressed, or guidelines and regulations being put in place that are inappropriate or ineffective (IWC, 2012).

Partially because of the propensity of some species to be 'curious' towards whale-watching vessels, the whales' natural curiosity has in effect shaped wildlife experiences into 'encounters', such that this becomes the expectation for many if not most whale-watching trips (Hoyt, 2003). Moreover, whale-watchers can develop unrealistic expectations from marketing materials of whale-watching companies and conservation groups (NGOs) showing pictures or videos of close encounters with cetaceans, and footage of spectacular cetacean behaviours or from seeing hand-feeding of dolphins and circus-style tricks in captive cetacean facilities.

As a result, operators may feel pressured to get closer to animals or manoeuvre their vessels to solicit behaviour such as bow-riding or breaches to satisfy their clients (Orams, 2000). This encounter-driven whale-watching, largely conducted from large boats, has serious implications in terms of developing sustainable, ecological whale-watching management.

As whale-watching industries have developed and as the capacity to accommodate ever more tourists has increased, there has been a shift in the kind of tourist from a specialist type, such as the dedicated whale enthusiast who might be knowledgeable about cetaceans and concerned about cetacean conservation and environmental issues, the so-called 'pioneer ecotourist' (e.g. Rawles & Parsons, 2004), to a more general or 'mass' tourist (Duffus & Dearden, 1990). These general tourists have different expectations about whale-watching trips, may be less environmentally motivated, have less understanding of cetacean behaviour, may demand quicker, more action-packed and encounter-driven trips, and thus may also impose fewer demands towards the development of an ecologically responsible industry.

The geography of whale-watching locations is another crucial factor as to whether problems may develop as whale-watching grows. 'Geographically restricted', or somewhat confined, areas such as Stellwagen Bank National Marine Sanctuary, southern Tenerife and southern Vancouver Island are notable for having concentrated whale populations in one area that are served by numerous whale-watching operators from various ports all converging on the same area. This can also happen with cetacean populations confined in rivers, fjords, estuaries or bays. In contrast, whales that are spread out over extended coastlines (e.g. California, South Africa, parts of Australia) or around large islands (e.g. Iceland) may have only 1–2 operators from each port going to different populations or different portions of a population and not putting stress on the same individuals repeatedly.

It is important to note that significant impacts can occur on cetaceans (Parsons, 2012), sometimes with only a small amount of whale-watch activity or growth in the industry. For example, Bejder et al. (2006a, 2006b) documented impacts on a dolphin population in Australian coastal waters, including a decline in the population (Bejder et al., 2006b), with just two dolphin-watching vessels operating in the area.

Whale-watching development in whaling countries

Part of the growth of whale-watching has seen its spread into the three main whaling countries of Japan and Norway (from 1988), and Iceland (1991). Norway hunts northern minke whales (*Balaenoptera acutorostrata*), approximately 500–700 a year, under a reservation to the IWC commercial whaling moratorium (see Box 5.1). Japan also hunts whales – northern (*B. acutorostrata*) and Antarctic (*B. bonaerensis*) minke whales, Bryde's (*B. edeni*), sei (*B. borealis*), fin (*B. physalus*), and sperm whales (*Physeter macrocephalus*) – via an IWC provision that allows whales to be killed for scientific research. After scientific samples are taken, the whale meat is processed and sold in food markets. This scientific whaling programme is somewhat controversial and has been heavily criticised for the poor quality of its 'science' (e.g. Clapham et al., 2003, 2007). Iceland has conducted scientific whaling in recent years, but then switched to commercial whaling, and currently takes northern minke whales and fin whales.

Whale-watching in Japan started first in the remote tropical islands of Ogasawara of far southern Japan and took a few years to move to the mainland. At one time more than 20 communities had whale-watching trips all over Japan and today nearly 200,000 people a year go whale-watching. This is widely seen as part of a new wave of interest by young people who may have watched whales in Guam, Hawai'i, British Columbia or Mexico, and have no interest in eating whales (Morikawa & Hoyt, 2011). For the most part, the Japanese whale-watching locations are well removed from whaling areas, and attempts to develop whale-watching near locations where cetaceans are hunted (e.g. Taiji) have failed due to 'discord between drive and hand-harpoon fishermen, and tourists, resulting from different opinions over animal welfare' (Endo & Yamao, 2007: 180).

In Norway, whale-watching has grown much more slowly, perhaps because Norway's tourism audience is largely from European countries where whaling is considered a thing of the past. Whale-watching in Norway is largely targeted towards killer whales and sperm whales, and not species that are hunted. Also, whaling and whale-watching occur in different locations, although there have been instances of whales being killed by whaling operations in the view of passengers of whale-watching vessels, leading to consternation among whale-watchers, especially young passengers (Berglund, 2006).

In Iceland, whale-watching became well established and took off dramatically in the years before Iceland resumed first so-called scientific whaling (in 2003) and then commercial whaling (in 2006). For a number of years in the late 1990s and early 2000s, Iceland commanded one of the highest growth rates for whale-watching in the world (see Chapters 7 and 8). Proximity to whaling, particularly in Faxafloi off Reykjavik, has led to friction between whale-watching operators and whalers, including whale-watching companies trying to blockade whaling boats (Iceland Review Online, 2008). Whale-watching companies in Iceland have also protested that sightings of northern minke whales – one of the main target species for Icelandic whale-watching – have decreased due to removals of animals by whaling operations (Iceland Review Online, 2007). The whale-watching operator association in Iceland has argued for a cessation of whaling, releasing several official statements to this effect. It has been debated whether whale-watching businesses have suffered in Iceland due to whaling. However, some foreign companies in the UK and Germany certainly cancelled trips after whaling re-started in Iceland (Williams, 2006), but

others renewed their efforts to get customers to support the whale-watching operators who offer a non-lethal alternative economic use for whales. Some argue that countries that attract wildlife and nature tourists and yet continue to engage in whaling risk diminishing their tourism image and their potential audience and that it can take decades to recreate an effective tourism brand (Hoyt & Hvenegaard, 2002; Hoyt, 2007). This view has been backed with research data, however. In a survey in Scotland, more than 90% of whale-watchers stated that they would not go on a whale-watching trip in a country that hunted whales and nearly 80% would boycott taking a holiday in a whaling country altogether (Parsons & Rawles, 2003).

It is important to note that several conservation NGOs have actively supported and advocated whale-watching as an alternative to whaling in these countries. There may be some difficult compromises when the calls are made to boycott travel to whaling countries, but the same NGOs have emphasized that whale-watch operators in these countries should in fact be supported as they are often the strongest voice within the country opposing the whaling decisions made by fisheries, if not from a conservation perspective then from an arguably more tangible economic perspective.

Whale-watching is a substantive industry in the Caribbean (Hoyt & Hvenegaard, 2002), in particular in countries such as the Dominican Republic, Dominica and Belize (Hoyt, 2005; O'Connor et al., 2009). Several Caribbean countries have taken an active position supporting commercial whaling bodies such as the IWC, and whale-watching researchers have highlighted the potential economic impacts that whaling could have on tourism (Hoyt & Hvenegaard, 2002). A recent study has tried to quantify this, and found that more than three-quarters of tourists surveyed at a Caribbean tourism destination stated that if a Caribbean country supported the hunting or capture of whales or dolphins they would be less likely to visit it on holiday (Draheim et al., 2003 2010). Due to the enormous role that tourism plays in the economy of Caribbean islands, data such as this should at least make governments think twice before being a vocal proponent of whaling, or engaging in commercial whaling themselves. Although more data are needed on the compatibility between whaling and whale-watching in the same location (Higham & Lusseau, 2007, 2008), data do seem to suggest that whaling is detrimental to the growth and development of a whale-watching industry.

Conclusions

Whale-watching started from humble beginnings, and rapidly grew to become a substantial international industry. A recent analysis of whale-watching potential based upon known data on whale and dolphin distributions suggested that there are still areas of high cetacean abundance where whale-watching could expand, and that if the opportunities in these places were exploited, whale-watching could generate an additional $413 million USD (2009) in annual revenue which when added to the current $2.1 billion USD estimates for the economic value for whale-watching could bring the annual value of the international whale-watching industry to more than $2.5 billion USD (Cisneros-Montemayor *et al.*, 2010). However, many of these potential locations for growth are in developing countries, and care must be taken to ensure that the industry develops sensibly, whales as a resource are not impacted, and sensitive species and populations, in particular, are not being stressed. Following the intensive growth in the 1990s, it is clear that a 'one size fits all' approach to whale-watching operations and management does not work. The nature of the whale-watching activities, the location, and species being watched are all factors that must be considered when developing management plans to reduce the impacts of whale-watching; for example, how large and fast are vessels? Educating and communicating with tourists is also important – the tourists should have realistic expectations about what they are likely to see, and also be aware of the need to reduce the impacts of their whale-watching activities on the animals they wish

to encounter. A key requirement of whale-watching management is that compliance with guidelines and regulations be monitored, and regulations enforced.

As commercial whale-watching moves through the second decade of the twenty-first century and its seventh decade since inception, the challenge now is to develop management approaches that are site- and species-specific. Whale-watching needs to examine past problems and mistakes, and the scientific research quantifying the dimensions of the problems, as well as to take on board management advice. In this way, it can become the sustainable and ecologically responsible industry that many of its proponents so clearly want it to be.

REFERENCES

Beasley, I., Bejder, L. & Marsh, H. (2010). Dolphin-watching tourism in the Mekong River, Cambodia: A case study of economic interests influencing conservation. Paper presented to the Scientific Committee at the 62nd Meeting of the International Whaling Commission, 30 May–11 June 2010, Agadir, Morocco. SC62/WW4.

Bejder, L., Samuels, A., Whitehead, H. & Gales, N. (2006a). Interpreting short-term behavioural responses to disturbance within a longitudinal perspective. *Animal Behaviour* 72, 1149–1158.

Bejder, L., Samuels, A., Whitehead, H., *et al.* (2006b). Decline in relative abundance of bottlenose dolphins exposed to long-term disturbance. *Conservation Biology* 20, 1791–1798.

Berglund, N. (2006). Whale shot in front of tourists. *Aftenposten*, 4 July. Retrieved from http://www.aftenposten.no/english/local/article1376980.ece/ (10 August 2008).

Ceballos-Lascurain, H. (1991). Tourism, eco-tourism and protected areas. *Parks* 2(3), 31–35.

Ceballos-Lascurain, H. (1996). *Tourism, Ecotourism, and Protected Areas.* Gland, Switzerland: IUCN. 315 pages.

Cisneros-Montemayor, A.M., Sumaila, U.R., Kaschner, K. & Pauly, D. (2010). The global potential for whale watching. *Marine Policy* 34, 1273–1278.

Clapham, P.J., Berggren, P., Childerhouse, S., *et al.* (2003). Whaling as science. *BioScience* 53, 210–212.

Clapham, P.J., Childerhouse, S., Gales, N.J., Rojas-Bracho, L., Tillman, M.F. & Brownell, R.L. (2007). The whaling issue: Conservation, confusion, and casuistry. *Marine Policy* 31, 314–319.

Draheim, M., Bonnelly, I., Bloom, T., Rose, N. & Parsons, E.C.M. (2010). Tourist attitudes towards marine mammal tourism: An example from the Dominican Republic. *Tourism in Marine Environments* 6, 175–183.

Duffus, D.A. & Dearden, P. (1990). Non-consumptive wildlife-oriented recreation: A conceptual framework. *Biological Conservation* 53, 213–231.

Endo, A. & Yamao, M. (2007). Policies governing the distribution of by-products from scientific and small-scale coastal whaling in Japan. *Marine Policy* 31, 169–181.

Fennell, D.A. (1999). *Ecotourism: An introduction.* New York, NY: Routledge.

Forestell, P.H. (2009). Popular culture and literature. In W.F. Perrin, B. Würsig & J.G.M. Thewissen (Eds), *Encyclopedia of Marine Mammals* (2nd edn). San Diego, CA: Academic Press, pp. 899–913.

Higham, J.E.S. & Lusseau, D. (2007). Urgent need for empirical research into whaling and whalc watching. *Conservation Biology* 21, 554–558.

Higham, J.E.S. & Lusseau, D. (2008). Slaughtering the goose that lays the golden egg: Are whaling and whale-watching mutually exclusive? *Current Issues in Tourism* 11, 63–74.

Honey, M. (1999). *Ecotourism and Sustainable Development: Who owns paradise?* Washington, DC: Island Press.

Hoyt, E. (1984). *The Whale Watcher's Handbook.* Garden City, NY: Doubleday.

Hoyt, E. (1992). Whale watching around the world: A report on its value, extent and prospects. *International Whale Bulletin* 7, 1–8.

Hoyt, E. (1995). Whalewatching takes off. *Whalewatcher* 29(2), 3–7.

Hoyt, E. (2001). *Whale Watching 2001: Worldwide tourism numbers, expenditures, and expanding socioeconomic benefits.* Yarmouth Port, MA: International Fund for Animal Welfare, pp. 1–157.

Hoyt, E. (2003). Toward a new ethic for watching dolphins. In T. Frohoff & B. Peterson (Eds), *Between Species: Celebrating the dolphin–human bond.* San Francisco, CA: Sierra Club Books, pp. 168–177.

Hoyt, E. (2005). Sustainable ecotourism on Atlantic islands, with special reference to whale watching, marine protected areas and sanctuaries for cetaceans. *Biology and Environment: Proceedings of the Royal Irish Academy,* 105B(3), 141–154.

Hoyt, E. (2007). *A Blueprint for Dolphin and Whale Watching Development.* Washington, DC: Humane Society International (HSI), pp. i–iii, 1–28.

Hoyt, E. (2009a). Afterword by Erich Hoyt. In S. O'Connor, R. Campbell, H. Cortez & T. Knowles (Eds), *Whale Watching Worldwide: Tourism numbers, expenditures and economic benefits.* Yarmouth Port, MA: IFAW, pp. 283–287.

Hoyt, E. (2009b). Whale watching. In W.F. Perrin, B. Würsig & J.G.M. Thewissen (Eds), *Encyclopedia of Marine Mammals* (2nd edn). San Diego, CA: Academic Press, pp. 1219–1223.

Hoyt, E. & Hvenegaard, G. (2002). A review of whale watching and whaling with applications for the Caribbean. *Coastal Management* 30(4), 381–399.

Hoyt, E. & Iñíguez, M. (2008). *The State of Whale Watching in Latin America.* Chippenham, UK: WDCS; Yarmouth Port, MA: IFAW; and London: Global Ocean.

Iceland Review Online. (2007). Fewer minke whales in whale watching areas. Iceland Review Online. 2 September. Retrieved from http://icelandreview.com/icelandreview/search/news/Default.asp?ew_0_a_id=288600 (5 August 2010).

Iceland Review Online. (2008). Whale watchers at war with whale hunters. Iceland Review Online. 10 July. Retrieved from http://www.icelandreview.com/icelandreview/daily_news/?cat_id=16567&ew_0_a_id=308980 (5 August 2010).

IFAW. (1999). *Report of the Workshop on the Socioeconomic Aspects of Whale Watching.* Yarmouth Port, MA: International Fund for Animal Welfare.

IWC. (2003). Report of the Sub-Committee on Whalewatching. *Journal of Cetacean Research and Management* 5(Suppl), 382–391.

IWC. (2004). Report of the Sub-Committee on Whalewatching. *Journal of Cetacean Research and Management* 6(Suppl), 335–346.

IWC. (2005). Report of the Sub-Committee on Whalewatching. *Journal of Cetacean Research and Management* 7(Suppl), 327–337.

IWC. (2006). Report of the Sub-Committee on Whalewatching. *Journal of Cetacean Research and Management* 8(Suppl), 241–251.

IWC. (2007). Report of the Sub-Committee on Whalewatching. *Journal of Cetacean Research and Management* 9(Suppl), 326–340.

IWC. (2008). Report of the Sub-Committee on Whalewatching. *Journal of Cetacean Research and Management* 10(Suppl), 322–335.

IWC. (2009). Report of the Sub-Committee on Whale-watching. *Journal of Cetacean Research and Management* 11(Suppl), 334–343.

IWC. (2010). Report of the Sub-Committee on Whale-watching. *Journal of Cetacean Research and Management* 11(Suppl 2), 332–345.

IWC. (2012). *Report of the Sub-Committee on Whale-watching.* Cambridge: International Whaling Commission.

Jones, M.L. & Swartz, S.L. (2009). Gray whale, *Eschrichtius robustus*. In W.F. Perrin, B. Würsig & J.G.M. Thewissen (Eds), *Encyclopedia of Marine Mammals* (2nd edn). San Diego, CA: Academic Press, pp. 563–611.

Kaza, S. (1982). Recreational whale-watching in California: A profile. *Whalewatcher* 16(1), 6–8.

Kelly, J.E. (1983). The value of whalewatching. Paper presented at Whales Alive Conference, Boston, 7–11 June 1983, pp. 1–5, i–vi.

Kraus, S.D. (1989). Whales for profit. *Whalewatcher* 23(2):18–19.

Morikawa, J. & Hoyt, E. (2011). Of whales, whaling and whale watching in Japan: A conversation. In P. Brakes & M.P. Simmonds (Eds), *Whales and Dolphins: Cognition, culture, conservation and human perceptions.* London: Earthscan/Routledge and Taylor & Francis, pp. 89–99.

O'Connor, S., Campbell, R., Cortez, H. & Knowles, T. (2009). *Whale Watching Worldwide: tourism numbers, expenditures and expanding economic benefits.* Yarmouth Port, MA: International Fund for Animal Welfare, 2695 pp.

Orams, M.B. (2000). Tourists getting close to whales, is it what whale-watching is all about? *Tourism Management* 21(6), 561–569.

Parsons, E.C.M. (2012). The negative impacts of whale-watching. *Journal of Marine Biology,* doi:10.1155/2012/807294.

Parsons, E.C.M. & Rawles, C. (2003). The resumption of whaling by Iceland and the potential negative impact in the Icelandic whale-watching market. *Current Issues in Tourism* 6, 444–448.

Parsons, E.C.M., Fortuna, C.M., Ritter, F., *et al.* (2006). Glossary of whalewatching terms. *Journal of Cetacean Research and Management* 8(Suppl.), 249–251.

Rawles, C.J.G. & Parsons, E.C.M. (2004). Environmental motivation of whale-watching tourists in Scotland. *Tourism in Marine Environments* 1, 129–132.

Scarpaci, C., Nugegoda, D. & Corkeron, P.J. (2003). Compliance with regulations by 'swim-with-dolphins' operations in Port Philip Bay, Victoria, Australia. *Environmental Management* 31, 342–347.

Scarpaci, C., Nugegoda, D. & Corkeron, P.J. (2004). No detectable improvement in compliance to regulations by 'swim-with-dolphin' operators in Port Philip Bay, Victoria, Australia. *Tourism in Marine Environments* 1, 41–48.

Tilt, W. (1985a). *Whalewatching in California: Survey of knowledge and attitudes.* New Haven, CT: Yale School of Forestry and Environmental Studies.

Tilt, W. (1985b). *Whales and Whalewatching in North America with Special Emphasis on Whale Harassment.* New Haven, CT: Yale School of Forestry and Environmental Studies.

Williams, N. (2006). Iceland shunned over whale hunting. *Current Biology* 16(23), R975–R976.

Williams, R., Bain, D., Smith, J.C. & Lusseau, D. (2009). Effects of vessels on behavior patterns of individual southern resident killer whales *Orcinus orca. Endangered Species Research* 6, 199–209.

6

The International Whaling Commission (IWC) and whale-watching

Carole Carlson, Naomi Rose, Hidehiro Kato and Rob Williams

Introduction

The International Whaling Commission (IWC) is the global body charged with ensuring the conservation and management of whales. It performs a central role in the discussion of all aspects of whaling and, since 1975, whale-watching. In terms of whale-watching, the brief of the IWC extends to the scientific, legal, socioeconomic and educational aspects of the activity. A critical function of the IWC for more than three and a half decades has been to collate, analyse and disseminate leading edge research and information on whale-watching to member (and non-member) governments. It also provides a forum for the discussion and dissemination of scientific studies addressing all aspects of whale-watching in a timely manner. As such, the IWC contributes at the global level to the pursuit of sustainable whale-watching, while addressing the educational, socio-cultural and economic development opportunities that are widely associated with it.

This chapter provides a brief review of the activities and outcomes achieved by the IWC in the 37 years since whale-watching became incorporated into the IWC agenda. Central to this historical review is the critical role that the IWC has played in the transition from extractive to non-extractive practices. It also seeks to underline the role of the IWC in facilitating the development of sustainable whale-watching; a future in which education, research, conservation, and management are seamlessly engaged and mutually informing. This goal arises from the philosophy that commercial whale-watching and rigorous science can take place simultaneously, particularly with whale-watching vessels used as 'platforms of opportunity' to collect data. In addition, commercial whale-watching passengers can be actively engaged in the research and conservation efforts of the scientific community and management agencies, such that all stakeholders play an active part in learning about whale conservation and informing sustainable tourism management. Lastly, the chapter explores the case of Antarctic whale-watching, illustrating the interplay of commercial whale-watching, science and conservation.

Whale-watching and the International Whaling Commission

The International Whaling Commission's interest in whale-watching dates to 1975, when concerns were raised by the IWC Scientific Committee (SC) regarding the interactions of tourist boats and breeding whales and the potential negative impacts that might arise from those interactions. Of specific concern at that time were boat excursions into the grey

Whale-watching: Sustainable Tourism and Ecological Management, eds J. Higham, L. Bejder and R. Williams.
Published by Cambridge University Press. © Cambridge University Press 2014.

whale breeding lagoons of Mexico, and the possibility of detrimental impacts upon grey whale mothers and calves.[1] There followed, in 1976, a request submitted by the IWC SC to the governments of the USA and Mexico, to establish regulations to manage tourist interactions with, and afford adequate protection to, grey whales in their breeding lagoons.[2] In that year, the IWC adopted a Resolution, proposed by Denmark, that grey whales be protected and that governments responsible for the protection of grey whales in their waters establish regulations to ensure such protection as an urgent priority. Thus, the initial involvement of the IWC in matters relating to whale-watching was centred on protection of whales in response to a recognized concern for the welfare of breeding animals.

Following the initial interest of the IWC into the regulation of the nascent whale-watching industry in the mid-1970s, a concerted focus on whale-watching emerged in the early years of the following decade. At the 1982 IWC meeting the USA proposed, and the IWC agreed, to co-sponsor a special meeting to allow dedicated and detailed discussion of the 'non-consumptive' utilization of cetaceans, considering research, recreation, education and culture.[3] That forum took place in 1983, with the hosting of the inaugural whale-watching conference 'Whales Alive' in Boston, Massachusetts. The title of the conference underlines the philosophy underpinning the IWC's engagement with whale-watching and whale conservation. The Secretary of the IWC attended the conference as an observer. Outcomes of the conference were reported to the IWC in 1984, highlighting the significant transition to non-extractive practices and the need for the IWC to ensure careful oversight of the emerging 'non-consumptive' use paradigm.

With the rapid increase of whale-watching globally in the 1990s, the IWC adopted its first whale-watching Resolution in 1993. The Resolution allowed for the creation of a Working Group on Whalewatching to be convened at the 1994 IWC meeting to draw together and review the global whale-watching industry.[4] The Working Group received a review document prepared by the Secretariat, drawing upon information provided by 15 member governments on the extent and economic and scientific values of whale-watching activities.[5] In 1994, the Commission adopted a Resolution on whale-watching that (1) requested the submission of information by all contracting parties on whale-watching, (2) sought the advice of the SC on setting guidelines for the conduct of whale-watching, and (3) requested that the IWC maintain all aspects of whale-watching under review.[6]

With the advent of the Working Group on Whalewatching in 1995, the focused efforts of the IWC fell upon a range of issues and debates. In 1996, an IWC Resolution underlined its future role in monitoring and advising on whale-watching activities and committed the Commission to discuss educational, economic and social aspects of whale-watching at its Annual Meeting in 1997.[7] The first SC Whalewatching Working Group met in 1996 and determined that the overall objective in developing guidelines for the management of whale-watching was to 'ensure that the development of whalewatching is ecologically sustainable and meets, to the extent possible, the requirements of the industry and expectations of the wider community'.[8]

The following objectives were adopted by the Commission as the basis for further consideration of issues related to the management of whale-watching.

1. Ensuring that whale-watching does not significantly increase the risk to the survival or ecological functioning of local populations or species or their environment; and therefore, in the short

---

[1] *Rep. Int. Whal. Commn* 27, 209–11, 1976.
[2] *Rep. Int. Whal. Commn* 28, 68, 1977.
[3] *Rep. Int. Whal. Commn* 33, 31–32, 1983.
[4] *Rep. Int. Whal. Commn* 44, 33–34, 1994.
[5] Reviews were received from the governments of Argentina, Australia, Brazil, Chile, France, Ireland, Japan, Mexico, New Zealand, Norway, Oman, Spain, Sweden, UK (including British Virgin Islands, Turks & Caicos Islands and other British territories), and USA.
[6] *Rep. Int. Whal. Commn* 45, 33, 1995.
[7] *Rep. Int. Whal. Commn* 47, 20–21, 1997.
[8] *Rep. Int. Whal. Commn* 47, 250–251, 1997.

term, that whale-watching does not result in significant adverse change in population dynamics such as birth or mortality rates, or impede normal patterns of habitat use or activity, including feeding, resting and reproduction.

2. The development and maintenance of viable and responsible whale-watching activities.

The following priority topics were identified for future work.

1. A more detailed review of the approach distances, effort and activity limitations in place in existing operations for a range of species and information on the basis for such controls.

2. An assessment of current studies of the effects of different approach distances and platforms.

3. A review of the qualitative methods used to assess the short-term reactions of cetaceans and the basis of judgements of adverse effects.

4. Comparative studies on different approaches/ distances and other controls which may be required on areas important for feeding, resting and reproduction.[9]

Since the initial Resolution, the IWC SC has reviewed and addressed a wide spectrum of issues arising from commercial whale-watching activities (Table 6.1). The penultimate point in Table 6.1 underlines the IWC's dedication to the search for a symbiosis between commercial whale-watching and scientific research interests. Following the 1996 Resolution, a subcommittee on Whalewatching within the SC was created in 1998.

In 1997 the USA submitted a report on the potential educational opportunities afforded by whale-watching, with guidelines on how best to capture those opportunities in full measure. In 1998 the focus of dedicated attention fell upon the socioeconomic elements of whale-watching, highlighting development opportunities available to coastal communities through engagement in whale-watching, economic benefits arising from, and the potential for whale-watching to facilitate opportunities for, non-lethal research including educational benefits and the development of

[9] *Rep. Int. Whal. Commn* 47, 20, 1997.

Table 6.1 Selected scientific issues addressed by the Working Group and Sub-committee on Whalewatching since 1995.

Issues addressed

- Identifying and assessing the possible effects of whale-watching operations on cetaceans/whales
- Examining current status of methods of assessment of impacts, including assessment of behavioural change
- Providing advice on the management of future whale-watching based on assessment of impacts
- Reviewing information on noise production from vessels and aircraft and its effects on cetaceans
- To draw up a set of general principles to assist coastal states in the management of whale-watching, based on the experience of member countries
- Considering the assessment of possible short- and long-term effects of whale-watching and some special situations such as 'swim-with' programmes and dolphin provisioning programmes
- Utilizing the opportunities for scientific research conducted from whale-watching boats
- Research on the effectiveness of, and compliance with, management measures.

empirical research methods. As the millennium drew to a close, the IWC specifically addressed the legal framework for whale-watching, part of which included a comprehensive review of legislation and guidelines existing in different regions of the world.

In 2000 the attention of the IWC centred on the increasing value of whale-watching for small island states in the developing world; communities that had previously advocated whaling as a means of survival for economically marginalized societies. In that year, the IWC also dedicated a two-day workshop preceding the SC to the long-term impacts and potential consequences of whale-watching for cetaceans. The focus in 2000 and the year following was on whale-watching as a non-consumptive and sustainable use of whales, acknowledging the fact that at that time the concept of 'non-consumptive' use was coming under increasingly close academic scrutiny. In 2001 the New Zealand government Commissioner, citing a report produced by IFAW

(2001), noted that the global whale-watch industry was at that time worth more than $US 1 billion. Such was the apparent growth of global whale-watching that in 2002 the SC raised concerns on the effects of noise on whales, reviewing the effectiveness of, and compliance with, national whale-watching guidelines and regulations.

The pressures of growing demand for tourist interactions with cetaceans in the wild led the IWC in 2006 to host a workshop to address the strategic planning of large-scale whale-watching. This workshop sought to concentrate the focus of research efforts on understanding the interplay between whale-watching impacts on cetaceans with other anthropogenic factors (cumulative and synergistic impacts) and wider ecological dynamics. The workshop highlighted a range of issues and challenges arising from uncertainly surrounding global whale stocks (see Chapter 2). First and foremost, it highlighted the need to address existing data sources that might provide insights into population dynamics at the species level (e.g. local exploitation histories, documented local whale-watching activities; baseline, longitudinal and ecological data sources). It also promoted the need for appropriate and systematic research design to address life histories and exposure to changing environmental factors at the species/generic level, and the use of experimental studies with emerging technologies, analytical tools and modelling techniques (e.g. see Chapter 20).

In 2007, the IWC adopted a Resolution proposed by Argentina and co-sponsored by 15 countries on the non-lethal use of cetaceans.[10] During April of the following year, the SC dedicated a workshop to the strategic planning of large-scale whale-watching research. This workshop represented a commitment to improving the long-term efforts to study the impacts of whale-watching on cetaceans. The focus on impact management continued in 2008, when Brazil requested that the management of whale-watching become a subject regularly and routinely addressed by the Conservation Committee

[10] *Rep. Int. Whal. Commn* 57, 3, 2007.

(established in 2003 through a Resolution by the Commission). Brazil's request acknowledged the increasing acceptance that large-scale commercial whale-watching, while non-extractive, does bring with it specific management challenges and potential impacts. While the scientific aspects of whale-watching had been addressed by the SC since 1975, that committee was not well placed to provide recommendations related to whale-watching management, whereas the Conservation Committee (CC) was ideally placed to do so. This proposal was welcomed by the SC.

The focus of the last four years (2009–2012) has been directed towards the priority development of a five-year strategic plan. In 2008 an inter-sessional correspondence group was established to look at all aspects of whale-watching and make recommendations for future workshops. The group identified three key areas of interest to the IWC and its members:

1. research and assessment;
2. management and capacity building; and
3. development.

These key areas served as objectives that the Commission would seek to promote as part of an integrated body of work over time. The focus of the objectives would be to develop tools to assess and understand the opportunities and risks for whale-watching, support effective management of sustainable whale-watching activities, based on science; and realize the social and economic potential of whale-watching for the global community. The inter-sessional correspondence group also recommended that the Standing Working Group on Whalewatching (SWG-WW) of the CC be established to prepare, in consultation with the SC, a five-year strategic plan for consideration at IWC/62 (2010). That recommendation was endorsed by the CC and an inter-sessional workshop to initiate the strategic plan was held in Argentina in late 2010, with financial support from Australia and the USA.

The 2010 workshop in Argentina produced a range of recommendations and outputs. It highlighted that a five-year strategic plan centred on responsible whale-watching might be well served

Table 6.2 Selected issues outlined and addressed in the IWC's web-based Handbook on Whale Watching (2011) (*Source*: www.iwc.int).

While management responsibility lies with national governments, the IWC must continue to find ways to serve an effective advisory role in the development of responsible whale-watching.

Local issues require local solutions. Whale-watching is a diverse phenomenon and there is no universal best practice. The SWG-WW handbook must provide avenues of development that take clear account of local circumstances and context.

In order to ensure full coverage of fields of expertise (some of which lies beyond the IWC), the bringing in of additional expertise on national delegations, and improving/establishing cooperation with other relevant international, national and regional bodies and organizations (e.g. The Buenos Aires group, the UN World Tourism Organization and whale-watching industry associations) are required. This could include those working with the socioeconomic aspects of tourism, regional bodies considering whale-watching regulations.

Due to incomplete scientific, socioeconomic and other advice on several key aspects related to sustainable whale-watching, the development of guidance should be seen as an iterative process. A commitment to continual updating is required as new scientific outputs and other insights arise.

Consideration should be given by the Commission to developing formal 'conservation' and 'user' objectives for whale-watching, against which to monitor the success or otherwise of measures taken. This point recognizes that in the case of other anthropogenic activities (e.g. whaling), the Commission has assigned highest priority to conservation objectives when establishing the balance between conservation and user objectives.

by the development of a web-based handbook on whale-watching to provide concise and focused information and advice to a range of target stakeholders on all aspects of responsible whale-watching. The handbook recognizes and is based upon a range of important issues (Table 6.2). In

acknowledging these issues, the handbook provides information and guidance on a range of factors that should be accommodated when both establishing new, and improving existing, whale-watching operations. It also provides assessments of the relative merits, strengths and weaknesses of various management approaches under different development scenarios. The handbook places specific emphasis on monitoring interactions, to ensure the effectiveness of management measures. The web-based format allows the handbook to be regularly revised and updated, to assist relevant authorities to develop best practice approaches to achieve sustainable, responsible whale-watching. In 2011 the CC received and considered a report on the work of its SWG-WW. The report included an updated Strategic Plan for Whalewatching, to be implemented in association with the web-based Handbook on Whalewatching that was adopted in 2012.

IWC model of whale-watching and conservation management

The continuing work of the IWC demonstrates the foremost commitment to the conservation of whale species in relation to all anthropogenic uses of cetaceans. Evidence of the potential for symbiosis between scientists and tourism operators in pursuit of mutual conservation management outcomes is becoming increasingly apparent. Such evidence, in the Antarctic tourism context, is provided in some detail by Williams and Crosbie (2007). They note that both 'careful management and dedicated research are needed to ensure that the growing Antarctic marine tourism industry does not inadvertently harm (whale) populations, which are recovering from heavy exploitation in the early part of the 20th century' (Williams & Crosbie, 2007, p. 195). Given the logistical challenges and costs constraints of conducting seaborne Antarctic research, and the interest of the IWC in monitoring and understanding whale population recovery, it is evident that responsible boat-based tourism is well placed to contribute to Antarctic whale conservation and research through collaboration. Williams and Crosbie (2007) document various examples of collaboration between the Antarctic tourist industry operators and whale researchers.

Foremost among these is the potential for Antarctic tour operations to serve as a platform of opportunity for research. It has been noted elsewhere that in contrast to the northern hemisphere, our understanding of southern whale stocks is particularly poor due to the far less comprehensive coverage of the vast southern ocean environments (see Chapter 2). This gap is being addressed with the support of Antarctic tour operators. Productive partnerships have hitherto included transporting researchers to and from remote scientific bases, and providing ship time for researchers working on well-defined cetacean research projects in the Southern Ocean (Pitman & Ensor, 2003; Williams *et al.*, 2006). Tour operators are also making direct contributions to the Antarctic Humpback Whale Catalogue and the Antarctic Killer Whale International Catalogue (AKWIC) (Allen *et al.*, 2006; AKWIC, 2006; Williams & Crosbie, 2007). Identification photographs collected from whale-watching vessels have been crucial in recent findings documenting humpback whale movements between breeding grounds in different ocean basins (Stevick *et al.*, 2010, in press), and identifying seasonal migratory destinations (Stevick *et al.*, 2004, 2006; Dalla Rosa *et al.*, 2012), including the movement of whales between the northern and southern hemispheres (Rasmussen *et al.*, 2007). Passengers are also encouraged to contribute photographs of individual whales to well-defined research projects to inform survey coverage of cetaceans in the Southern Ocean (Pitman & Ensor, 2003; Williams *et al.*, 2006b). 'In all, this healthy cooperation between industry and science is estimated by the authors to be worth approximately US$1 million of in-kind support' (Williams & Crosbie, 2007, p. 197).

Williams and Crosbie (2007) also identify areas of possible future partnerships. Such collaborations might include contributions to understanding stock boundaries in southern hemisphere baleen whale populations; identifying the timing of peak migration (e.g. of humpback whales); understanding ecological relationships between ice cover and whale distribution; understanding abundance, distribution and movement patterns of southern whale species; and the opportunistic collection of identification photographs of individual blue whales in the southern hemisphere. This symbiosis between whale-watching operators and science is directly acknowledged in the IWC's commitment to draw on the specialist expertise that is available through organizations such as whale-watching industry associations (e.g. the International Association of Antarctic Tour Operators (see Table 6.2)) and the tour operators themselves. It also underlines the commitment of the IWC to exploring and achieving important conservation outcomes through whale-watching.

Conclusions

This chapter reviews the contributions of the IWC to the pursuit of sustainable, responsible whale-watching since initial concerns regarding boat interactions with whales in breeding grounds were voiced four decades ago. In 2013, one agenda item of the IWC Subcommittee on Whalewatching centres on scientific aspects of the Five-year Strategic Plan for Whalewatching. In doing so, the Subcommittee will continue to focus on the impacts of whale-watching on cetaceans and review whale-watching activities in different regions of the world. In terms of research, implementation of the Large-scale Whale Watching Experiment (LaWE) is a continuing priority, as is the development of an online database for the global tracking of commercial whale-watching and associated data collection. Subjects of continued review include swim-with-whale/dolphin operations and in-water interactions, dolphin provisioning programmes, whale-watching guidelines and regulations and collision risks to cetaceans from whale-watching vessels. Through these avenues of endeavour, the IWC will continue to serve an advisory role in relation to commercial whale-watching that facilitates the conservation of all species of cetaceans.

REFERENCES

Allen, J., Carlson, C. & Stevick, P. (2006). *Interim Report: IWC Research Contract 16, Antarctic Humpback Whale Catalogue SC/58/SH19*. Retrieved from http:// www.iwcoffice. org.

Antarctic Humpback Whale Catalogue. (2006). Retrieved from http://199.33.141.23:591/alliedwhale/login.html (21 May 2006).

Antarctic Killer Whale Identification Catalogue. (2006). Retrieved from http:// www.orcaresearch.org/akwic.htm (21 May 2006).

Dalla Rosa, L., Félix, F., Stevick, P.T. *et al.* (2012). Feeding grounds of the eastern South Pacific humpback whale population include the South Orkney Islands. *Polar Research* 31, 17331–17324, DOI: 17310.13402/polar.v 17331i17320.17324.

Pitman, R.L. & Ensor, P. (2003). Three different forms of killer whales in Antarctic waters. *Journal of Cetacean Research and Management* 5(2), 131–139.

Rasmussen, K., Palacios, D.M. & Calambokidis, J., *et al.* (2007). Southern Hemisphere humpback whales wintering off Central America: Insights from water temperature into the longest mammalian migration. *Biology Letters* 3, 302–305. doi:310.1098/rsbl.2007.0067.

Stevick, P.T., Aguayo, A., Allen, J., *et al.* (2004). Migrations of individually identified humpback whales between the Antarctic Peninsula and South America. *Journal of Cetacean Research and Management* 6, 109–113.

Stevick, P.T., Pacheco de Godoy, L., McOsker, M., Engel, M.H. & Allen, J. (2006). A note on the movement of a humpback whale from Abrolhos Bank, Brazil to South Georgia. *Journal of Cetacean Research and Management* 8, 297–300.

Stevick, P.T., Neves, M.C., Johansen, F., *et al.* (2010). A quarter of a world away: female humpback whale moves 10 000 km between breeding areas. *Biology Letters* 7, 299–302.

Stevick, P.T., Allen, J.M., Engel, M.H., Félix, F., Haas, B. & Neves, M.C. (in press). Inter-oceanic movement of an adult female humpback whale between Pacific and Atlantic breeding grounds off South America. *Journal of Cetacean Research and Management.*

Williams, R. & Crosbie, K. (2007). Antarctic whales and antarctic tourism. *Tourism in Marine Environments* 4(2), 195–202.

Williams, R., Hedley, S.L. & Hammond, P.S. (2006). Modelling distribution and abundance of Antarctic baleen whales using ships of opportunity. *Ecology and Society* 11(1). http://www.ecologyandsociety.org/vol11/iss1/ art1/

Part II

Human dimensions of whale-watching

7

The whaling versus whale-watching debate

The resumption of Icelandic whaling

Marianne Rasmussen

Introduction

Iceland is a unique example of a country being a former whaler nation transforming to a whale-watching country, where whale-watching today is a very important industry for the Icelandic economy. The aim of this chapter is to describe in a neutral way the historical aspect of Icelandic whaling history and how the whale-watching industry has evolved during the last 20 years. The whaling history includes old whaling, the time period before modern whaling, which started around 1870 with the Industrial Revolution and the use of a new kind of whaling vessel that increased catch numbers. Iceland is now one of the few places in the world where both whaling and whale-watching of the same species (Minke whales) in the same bay occurs. Of course, this causes many conflicts and a lot of political discussions both nationally and internationally. The changes from an old whaling nation to a whale-watching nation also reflect general changes in society from being a country mainly dependent on the fishing industry to being a country where tourism is a very important industry.

Whaling history in Iceland

The utilization of whale resources has been a traditional part of Iceland's history, providing an important dietary component throughout the ages. Lindquist (1997) described peasant fishermen whaling in the Northeast Atlantic area ca. 900–1900 AD. From 1 to 400 AD, the Norse considered fishes including cetaceans to be in the domain of the god Njörðr and when collected or caught they were Njörðr's gift. From the eleventh century, Njörðr's gift translated into the Christian God's gift (Lindquist, 1997). The Icelandic word 'Hvalraki' means beached whale and it also means the lucky one, referring to the old days where finding a beached whale could be a matter of life or death when food sources were often in limited supply, and therefore the person finding a beached whale would be the lucky one. The king's mirror (in Icelandic, *Konungs skuggjá*) was written in the thirteenth century, probably for the sons of Hakon, the old King of Norway. In there it is written: 'In the oceans off Iceland, little seems to me worth remembering or discussing except the whales there in the waters which are quite various' (from Ægisson *et al.*, 1997). Lindquist (1997) mentions an example from 1367, where Staður church in Steingrímsfjörður, Strandýsla, Iceland owned: 'the full half of whale comings' and also in the year 1385 a blue whale came ashore in Norse Greenland with a shot mark recognized as belonging to a certain person in Ísafjarpðardjúp, northwest Iceland.

Whaling was first recorded by the Norse Vikings of Scandinavia between 800 and 1000 AD. North

Whale-watching: Sustainable Tourism and Ecological Management, eds J. Higham, L. Bejder and R. Williams.
Published by Cambridge University Press. © Cambridge University Press 2014.

European whaling history has been divided into the Old commercial whaling trade (seventeenth to nineteenth centuries) and the Modern whaling industry (since 1870). During the early Middle Ages the Basques developed their unique hand harpoon tow whaling technique for the taking of black right whales, *Eubalaena glacialis*. The black right whale and the bowhead whale, *Balaena mysticetus*, are both relatively slow swimmers and have thicker blubber compared with the rorquals (blue whales, *Balaenoptera musculus*, fin whales, *Balaenoptera physalus*, sei whales *Balaenoptera borealis* and minke whales, *Balaenoptera acutorostrata*) and were therefore easier to catch and recover because they would usually float when dead; moreover, they offered far more blubber (oil) and longer baleen plates. Basque vessels are first documented whaling inshore in Icelandic waters in the year 1608 (Lindquist, 1997). Jón Guðmundsson the Learned wrote *On Iceland's Diverse Nature* in which he mentioned in 1640–1644 that the black right whales existed in great numbers around Iceland, but the foreign whalers pursued only this species, which was rapidly depleted (Guðmundsson, 1924). Foreign Old whaling continued inshore at the northwest and west Iceland until the eighteenth century. Remains from the old foreign whaling station from the seventeenth century have been found at Strákatangi in Strandasýsla in the northwestern part of Iceland (Rafnsson & Edvardsson, 2009). From the late seventeenth to the early eighteenth century peasant fishermen of Eyjafjörður in north Iceland took bottlenose whales, *Hyperoodon ampullatus*, in hand harpoon tow whaling.

The Danish–Norwegian government in 1750 worked for the creation of a small Icelandic Old whaling industry, but the whaling project was unsuccesful and during 1778–1787, the Danish–Norwegian government tried to establish hand harpoon tow whaling in Iceland. In 1788 the Icelandic Althing (government) offered grants for the purchase of whaling gear at the most suitable fjord, notably in northwest and west Iceland. A final attempt to establish hand harpoon tow whaling Basque-style in Iceland was made in 1829–1832. In March 1829 merchant Friðrik Svendsen received considerable financial support from the government (at this time the Danish government) in order to establish a 'whaling works at Ísafjarðardjúp' ('Hvalfangerie-Anlæg ved Isefjords Dyb') but he was not succesful, and in 1833 the whaling boats and other equipment were auctioned off.

Cetaceans were speared from small boats in Iceland until the mid 1890s. The Norwegian Svend Foyn then developed a new method of hunting. It was to harpoon the whales using a gun; after the whale had been killed by an explosive in the point of the harpoon, air was pumped into the carcass to keep it afloat. The carcasses of whales killed in this way were towed ashore and processed in large whaling stations. In 1883 Norway was granted permission by Icelandic authorities to set up whaling operations in Iceland. At first eight stations were established in the west fjords, but five were added in the east coast of Iceland subsequently. A total of 1305 whales were processed at these stations in 1902. Whaling off the east fjords lasted little more than a decade and was discontinued shortly after 1913. By 1915, the Norwegians had killed about 17,000 whales when the Icelandic parliament passed a law on the protection of whales in Icelandic waters (Björgvinsson & Lugmayr, 2002).

Icelanders began whaling in the modern sense in 1935 on the basis of new legislation on whaling in Icelandic waters. The new legislation granted Icelanders the exclusive right to hunt whales in Icelandic territorial waters, and futhermore decreed that all whales killed should be fully utilized. The new law provided the foundation for the first modern Icelandic whaling station, which was established in Tálknafjörður and operated in the years 1935–1939. A whaling station in Hvalfjörður (Whale fjord) was established in 1948. During the following four decades that the station was operational, about 300–400 whales were processed every year, representing a total of some 15,000 whales (Björgvinsson & Lugmayr, 2002).

International Whaling Commission (IWC) and Iceland

The International Whaling Commision (IWC) was set up under the International Convention for the Regualtion of Whaling which was signed in Washington DC on 2nd December 1946. The purpose of the Convention is to provide for the proper conservation of whale stocks and thus make possible the orderly development of the whaling industry. The main duty of the IWC is to keep under review and revise as necessary the measures laid down in the Schedule to the Convention which govern the conduct of whaling throughout the world. These measures, among other things, provide for the complete protection of certain species; designate specified areas as whale sanctuaries; set limits on the numbers and size of whales which may be taken; prescribe open and closed seasons and areas for whaling; and prohibit the capture of suckling calves and female whales accompanied by calves. The compilation of catch reports and other statistical and biological records is also required (www.iwcoffice.org). Today the IWC is represented by 88 member countries. (International Whaling Commission, 1946)

Iceland has been a member of IWC since 1947. It was in 1982 that the IWC agreed to stop all commercial whaling by the 1986 season. Iceland did not make an objection to the decision, as did other whaling countries such as Japan and Norway, and these countries therefore continued to hunt whales for scientific purposes. After the moratorium took effect, Iceland continued a small scientific whaling programme, and killed some 60 whales a year until 1989, and withdrew from the IWC in 1992. Iceland rejoined the IWC in 2002 with a legally disputed reservation against the moratorium.

A major new development is the IWC's involvement in whale-watching as a sustainable use of cetacean resources. In 1993, the IWC invited Contracting Governments to undertake a preliminary assessment of the extent, and economic and scientific value, of whale-watching activities. In 1996 it adopted a Resolution that underlined the IWC's future role in monitoring and advising on the subject. The Scientific Committee has agreed the following general guidelines for whale-watching and produced a compilation of whale-watching regulations from around the world. The following objectives were adopted by the Commission as the basis for further consideration of issues relating to the management of whale-watching.

1. Ensuring that whale-watching does not significantly increase the risk to the survival or ecological functioning of local populations or species or their environment; and therefore, in the short-term, that whale-watching does not result in significant adverse change in population dynamics such as birth or mortality rates, or impede normal patterns of habitat use or activity, including feeding, resting and reproduction.
2. The development and maintenance of viable and responsible whale-watching activities.

(International Whaling Commission, 1996)

The Icelandic Marine Research Institute

The Icelandic Marine Research Institute (MRI; www.hafro.is) was established in 1965 and is a government institute under the auspices of the Ministry of Fisheries. One of its aims is to provide advice to the government on catch levels and conservation measures. North Atlantic Sighting Surveys (NASS) have been conducted by MRI since 1986 and in 2001 the population of minke whales (*Balaenoptera acutorostrata*) from aerial surveys was estimated to be 43,633 (95% CI: 30.148–63.149; Borchers *et al.*, 2009). Iceland resumed scientific whaling in 2003 and in a five-year research programme killed a total of 200 minke whales from 2003 to 2007. The main objectives of the research were to collect basic information on the feeding ecology of minke whale in Icelandic waters by analysing the stomach contents and include research on the energetic, food requirements and seasonal and spatial variations in whale abundance.

A multi-species model was applied to the management scheme and included six animal species: three whale species (minke whales, humpback whales, *Megaptera novaeangliae*, and fin whales, *Balaenoptera physalus*) and one shrimp species, capelin (*Mallotus villosus*) and cod (*Gadus morhua*). The multi-species model is based on the paper by Stefánsson *et al.* (1997). The secondary objectives of the research were to investigate the stock structure of the minke whale in the North Atlantic by genetic methodology and satellite telemetry, to investigate parasites and diseases in the minke whale in Icelandic waters, to collect information on age and reproduction of minke whales in Icelandic waters, and to investigate the concentration of organ chlorines and trace elements in various organs and tissue types. A report of the preliminary results from the research was published by Vikingsson *et al.* (2006) and in Icelandic on the marine research webpage (Vikingsson, 2008). The results shows the different prey species consumed by minke whales such as sandeel (*Tobianus* sp.), herring (*Clupea harengus*), capelin, cod, redfish (*Sebastes mentella*) and krill (*Meganychtiphanes norwegica* and *Thysanoessa raschi*). The average level of mercury in Icelandic minke whales was 125 μg/g, lower than that found in Norwegian minke whales (Kleivane & Börsum, 2003). Dioxin-like PCBs in Icelandic minke whales had an average level of 13 ng WHO-TEQ/kg, w.w., lower than that from minke whales in the North Sea (Vikingsson, 2008).

Iceland resumed commercial whaling under its reservation to the IWC moratorium in 2006, killing 7 out of 9 fin whales and 7 out of 30 minke whales in the 2006/2007 commercial quota issued by the Icelandic Ministry of Fisheries. The Icelandic Fishery Ministry issued a quota for 40 minke whales in 2008, of which it killed 38 whales. No fin whales were killed commercially in 2007/2008. In January 2009 the out-going Icelandic Fisheries Minister took the decision to increase the commercial hunting quotas for both minke and fin whales. In 2009 Iceland caught 125 fin whales and 81 minke whales. In March 2010, the new Fisheries Minister went further still, setting quotas for 2010 at 200 fin and 200 minke whales, with a possible carry-over of 20% of any unused quota from 2009. In 2010, 148 fin whales were caught and no fin whaling was conducted

in 2011 and 2012. In 2010, 60 minke whales were caught and in 2011, 58 minke whales were caught in Icelandic waters. To date, four minke whales have been caught in 2012.

North Atlantic Marine Mammal Commission (NAMMCO) and Iceland

The North Atlantic Marine Mammal Commission (NAMMCO) is an international body for cooperation on the conservation, management and study of marine mammals in the North Atlantic (www. nammco.no). The NAMMCO agreement was signed in Nuuk in 1992 by Norway, Iceland, Greenland and the Faroe Islands. NAMMCO provides a mechanism for cooperation on conservation and management for all species of cetaceans (whales and dolphins) and pinnipeds (seals and walruses). The Council, the decision-making body of the Commission, reviews advice requested from the Scientific Committee, coordinates recommendations for further scientific research and reviews hunting methods for marine mammals in member countries. Each member state proposes a quota for hunting based on current population estimates of the species and the Scientific Committee will accept the quota if it is sustainable for the population.

Non-governmental organizations (NGOs) and anti-whaling campaigns

The first anti-whaling voyage of the *Rainbow Warrior* (Greenpeace) to Iceland was in 1978, where direct efforts were engaged to interfere with whale-hunting. When the *Rainbow Warrior* returned to Iceland in 1979, Hvalur (the whaling ships) fired harpoons over the protesters. Iceland began sending naval escorts with the whalers and twice seized the *Rainbow Warrior* with gunboats. A second incident occurred in international waters and Greenpeace zodiacs were taken (Day, 1987). These protests were followed by millions of people throughout the world and triggered considerable public pressure on the

IWC to declare the moratorium in 1982. In November 1986 activists linked to the Sea Shepherd Conservation Society vandalized a whaling station at Hvalfjörd by damaging machinery and computers. They also opened the seacocks on two of Iceland's four whaling ships and sank the vessels, which were still anchored in Reykjavik harbour (Darby, 2007).

The International Fund for Animal Welfare (IFAW) sponsored the first feasibility study of whale-watching in Iceland (Lindquist & Tryggvadóttir, 1990) and also sent their research vessel, *Song of the Whale*, on two research expeditions to Iceland in 2004 and 2006 (IFAW, 2004, 2006a). In 2008 they sponsored an international whale-watching conference in association with Icewhale (the Icelandic association of whale-watching companies in Iceland) held in Selfoss in the southern part of Iceland with participants from Australia, the US, Brazil and Norway. During the conference delegates from 11 countries issued a joint statement declaring that commercial whaling poses a threat to the success of whale-watching and ecotourism. The Whale and Dolphin Conservation Society (WDCS) has also published several reports about the whaling situation in Iceland (see Cockeron, 2007).

Whale-watching history in Iceland

IFAW has actively supported and facilitated the development of whale-watching in Iceland. It sponsored a feasibility study in Iceland and Lindquist and Tryggvadóttir concluded in 1990 that whale-watching in Iceland would be feasible. Whale-watching started in Höfn (Figure 7.1) in the southeastern part of Iceland in 1991 (Björgvinsson, pers. comm.) and about 100 people went whale-watching in the first year (Fischer, 1997). Whale-watching boats were operated from Höfn in the summers of 1992 and 1993, but because of frequent bad weather and no protection along the south coast of Iceland, whale-watching from Höfn was stopped.

In 1994, whale-watching started from Keflavik (Figure 7.1) in the southwestern part of Iceland by the company 'Dolphin and Whale Spotting' (www.

dolphin.is), and in 1995 the company North Sailing (www.northsailing.is) began commercial operators in Húsavík (Figure 7.1). In 2001, another whale-watching company 'Gentle Giants' started in Húsavík (www.gentlegiants.is). 'North Sailing' and 'Gentle Giants' are currently the only two whale-watching companies operating in Skjálfandi Bay from Húsavík. The whale-watching season in Húsavík is from April until October and a normal whale-watching tour is three hours. Most tourists visits Húsavík during peak season, June–August, and in 2010 approximately 50,000 tourists visited Húsavík to go whale-watching. In March 2010, Húsavík was nominated as one of the 10 best places in the world to go whale-watching (Homewood, 2010). The transformation of Húsavík from a fishing village to a whale-watching town has been described in Einarsson (2009). Húsavík has about 2300 inhabitants today; comparable fishing villages of similar size in Iceland do not have the same amount of restaurants, hotels or souvenir shops, and every year new hotels, shops and restaurants open.

A small-scale whale-watching company also now operates in Eyjafjörður (www.bataferdir.is) from Dalvik (Figure 7.1). Since whale-watching started in Iceland it has increased every year, and today many different boats and companies conduct commerical whale-watching operations in Iceland. The Icewhale keeps updating the number of whale-watching companies in Iceland (www.icewhale.is). The largest increase has been in the southwestern part of Iceland. The company 'Dolphin and Whale Spotting' is a good example of the increase and how the whale-watching market has been developing in Iceland. The company began operating from Keflavik using a small fishing vessel with room for about 10 tourists, and in 1999 the company rented a bigger vessel, which had space for 30 passagengers.

In 2000, an even bigger vessel which could carry 45 passengers was rented, and finally in 2001 the company bought the boat *Moby Dick*, which has a capacity of 70 tourists, changing its company name at that time to 'Moby Dick Whale Watching'. The *Moby Dick* was used from Keflavik from

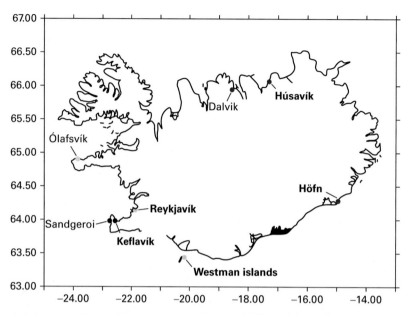

Figure 7.1 Map of Iceland with locations of different whale-watching areas marked.

2001 to 2007, when the company ceased operating, most likely because of increased competition from whale-watching operators from Reykjavik. Whale-watching in Faxaflói Bay can depart from either Reykjavik or Keflavik, but most tourists stay in Reyk-javik and it is easier for them to go whale-watching from there. There is currently no whale-watching operation from Keflavik.

The company 'Elding' (www.elding.is) started operating from Sandgerði (Figure 7.1) in the south-western part of Iceland in 2000. They moved their boat to Hafnarfjördur (close to Reykjavik) in 2001 and then to Reykjavik in 2002, where they have been operating from since. 'Elding' started to offer winter tours in weekends from 2009 and now offer whale-watching all year round, with daily tours from April until October. In 2012 about 10,000 tourists went whale-watching during the first four months. Today, four different companies operate from Reyk-javik (Hvalalif, www.hvalalif.is; Sjósigling, www.sjosigling.is; and Special Tours, www.specialtours.is). Some of the boats have a capacity of up to 145 passengers. 'Big Whale Watching' was started

in Ólafsvík (Figure 7.1) at Snæfellsness Peninsula in the western part of Iceland in the year 2000 by the company SeaTours (www.seatours.is) and until summer 2004 it was possible to watch blue whales (*Balaenoptera musculus*) off the western part of Ice-land, but often the trips were long, and could be up to 14 hours in rough seas (pers. obs.). After 2004 blue whales have often been observed from the whale-watching boats from Húsavík (Iversen *et al.*, 2009), and today whale-watching no longer operates from Ólafsvík. Boat tours are conducted around the West-man Islands (Figure 7.1) in the summer by the com-pany 'Viking Tours' (www.vikingtours.is), which is not a specialist whale-watching tour, but a scenic and wildlife tour around the islands. In July it is possible to see killer whales (*Orcinus orca*) and the boat will follow the killer whales for some time (pers. obs.).

IFAW (2009) published a summary report about whale-watching worldwide. At that point, the majority of tourists went whale-watching in Iceland from Reykjavik (51%) followed by Húsavík (36%). The most common species sighted in Faxaflói Bay

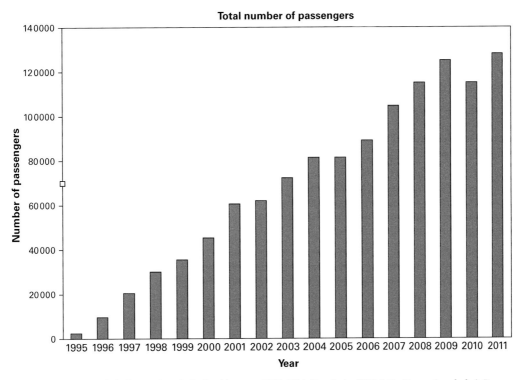

Figure 7.2 Number of whale-watchers in Iceland by year, 1995–2011 (Icewhale, 2012: http://www.icewhale.is/).

and in Skjálfandi Bay were white-beaked dolphins, minke whales, humpback whales and harbour porpoises (Salo, 2004; Cecchetti, 2006; Magnúsdóttir, 2007; Bertuilli, 2010). The newest data available from 2009 from Icewhale (Figure 7.2) indicate that 125,000 tourists went whale-watching in Iceland in 2009 and approximately 128,000 in 2011 (Sigursteinn Másson, IFAW, pers. comm.).

IFAW also sponsored a report on possible whale-watching in the west fjords (Cecchetti & Rasmussen, 2008) and it was suggested that it would be feasible to conduct commercial whale-watching from the west fjords. One location suggested was Drangsness in Steingrimsfjörður, where boat-based nature watching was already being conducted. However, this business has not increased over the last few years. Instead, some whale-watching tours were established in Isafjörður in summer 2012.

Code of conducts for whale-watching in Iceland

Currently there is no official code of conduct for whale-watching in Iceland. The 'Wild North Project' was started in 2008 (www.thewildnorth.org). The aim of this project was to find ways to contribute to the sustainable development of wildlife tourism in the Northern Periphery, and thereby ensure the sustainable foundations of commercial operations, the livelihood of its workers and the long-term integrity of natural resources. Within this project were members from Greenland, Faroe Islands, Iceland and Norway. Partners were a mixture of scientists, tourist operators and governmental institutions. The aim was to have a maximum number of tourists watching the wildlife with a minimal impact on focal wildlife species. Species include whales (in Greenland and Iceland), birds (in Faroe Islands,

Norway and Iceland), seals (in Iceland) and foxes (in Iceland). Codes of conduct were then developed based on best practice and research conducted at each location.

Guidelines from Greenland can be found at the webpage for the Greenlandic Institute of Natural Resources (Boye *et al.*, 2011). Many studies throughout the world have shown that increased boat traffic and whale-watching have an impact on whales. For example, humpback whales have been shown to avoid whale-watching boats (Scheidat *et al.*, 2004; Schaffer *et al.*, 2009) or to change their dive pattern (Boye *et al.*, 2010). Increased whale-watching has also been shown to affect the fitness and survival of local populations of dolphins (Bejder *et al.*, 2006). Whale-watching traffic is increasing in Iceland with, for example, 16 trips per day from Húsavík in Skálfandi Bay, where up to four or five different boats can be watching the same whale at the same time.

Four different whale-watching boats operating from Reykjavik also watch the same whale simultaneously at times. Icewhale is aware of international codes of conduct used by whale-watching companies – for example, the review of guidelines around the world by Carole Carlson (2008). Both 'Elding' and 'North Sailing' whale-watching have codes of conducts on their webpages, but currently the guidelines are all implemented on a voluntary basis. Some of these guidelines include approach distance: the whale-watching boats should keep a distance of at least 200 m from the whale. It is also important what engine is used (and how engines are used) to regulate noise levels. For example, it has been shown that killer whales change the frequency in their calls off the coast of Canada when many whale-watching boats are present (Foote *et al.*, 2004). As most cetaceans depend primarily on acoustics as the major sense, heavy boat traffic can be a major cause of disturbance in feeding and reproduction. Grossman (2010) also stated that noise reduces ocean habitat for whales. Some species in Iceland might be at more risk than others. For example, white-beaked dolphins are found primarily in coastal Icelandic waters (Pike *et al.*,

2009; Rasmussen & Miller, 2002) and they are one of the main target species for whale-watching in Iceland. White-beaked dolphins have been observed mating at the end of August in Faxaflói Bay (Rasmussen, 2004) and in general the mating season has been described as July–September, with calves born in May–August (Vikingsson & Ólafsdóttir, 2004; Reeves *et al.*, 1999). On 25 August 2010 the birth of a calf was observed from a whale-watching boat (Schmidt, pers. obs.) and a few weeks later a dead white-beaked dolphin calf was found on the beach close by. Of course, we do not know if it was the same calf, but it is important to point out that it might be of some risk for white-beaked dolphins to have whale-watching vessels close by during their calving season.

In order to be able to implement operator guidelines or codes of conduct within the 'Wild North Project', it has been important that these guidelines are developed based on knowledge from local sites. In Húsavík, research has been conducted from the local lighthouse in order to obtain site-specific information about the whale-watching boats and the behaviour of the whales (e.g. swimming speed and direction) when whale-watching boats are present and when not. A pilot study was conducted in the summer of 2008 (Guðrúnardóttir, 2008), showing that the project was feasible using the lighthouse in Húsavík. The project was continued in the summers of 2009 and 2010 using the lighthouse in Húsavík and a digital theodolite (Geodimeter 600) was connected to a laptop computer and the software program Cyclops (Newcastle University) was used to perform analyses. In Húsavík, studies were conducted using the lighthouse as a research platform for land-based observations and a change in the swimming speed of minke whales around boats was observed (Mágnúsdóttir *et al.*, 2011), but no changes in humpback whale behaviour around boats were observed (Revery, 2011). The speed and direction of whales around boats were also observed (Martin, 2012). In Faxaflói Bay, Christiansen *et al.* (2011) showed that minke whales were interrupted in their feeding behaviour and the diving pattern and the directness index from the whale-watching vessels were different from a land-based control site at Gardskagi lighthouse.

The whaling versus whale-watching debate: the resumption of Icelandic whaling

The international media has followed the debate about whaling versus whale-watching in Iceland. Can Iceland both watch and kill whales? Other whaling countries usually watch other whale species that they are killing – for example in Norway, where sperm whale-watching is popular and minke whaling is conducted. In Iceland, minke whales are one of the main target species for whale-watching and are also a target for the whaling boats. The Icelandic Travel Industry Association (ITIA) opposes whaling due to concerns about reputational damage that may threaten the whale-watching business. The ITIA identifies strong opposition to whaling in Iceland's main tourist markets. Tourism accounted for 13% of Iceland's export markets in 2003 (Walton, 2003). ITIA stated at that time that Iceland's image would be damaged by a return to whaling (BBC News, 6 August 2003). It was also stated in the French Daily *Le Monde* at this time that 'You don't touch one of the most mythical animals in the world with impunity', in reference to the decision by Iceland to resume what it called a scientific whaling programme after a 14-year moratorium (BBC News, 19 August 2003). *Le Monde* wrote that for many Icelanders whaling is about their small country's prestige and pride. Local surveys showed that 70–80% of the nation supported whaling when asked, 'should Iceland resume whaling?' The majority of Icelandic people believe that whales are a problem for the fisheries and should therefore be culled (Björnvinsson, 2003).

Malmquist (2003) argued against the position of the Ministry of Fisheries that whaling in Icelandic waters is necessary in order to manage fish stocks. The international Cetacean Society also protested against Iceland resuming whaling and encouraged people to contact the Icelandic ambassador or ministers to express their views (O'Connell, 2003).

However, most worries came from those organizing whale-watching trips, who were concerned for the harm that bad publicity would cause. Siglausson (2005) published a report estimating the economic value of whale-watching in Iceland. In the report he estimated that the direct value of whale-watching businesses in Iceland was close to US$ 8,550,000 and the total revenue for the Iceland economy was estimated to be US$ 14,000,000 in 2003. Parsons and Rawles (2003) published a report about the potential negative impact on the Icelandic whale-watching market, stating that 79% of whale-watchers would boycott a country that conducted hunting for cetaceans, and a further 12.4% stated that although they would visit a country conducting whaling operations they would not go whale-watching in that country.

On 17 October 2006, Iceland's fisheries minister, Mr Einar K. Gudfinsson, announced that he had decided to allow the resumption of commercial whaling. Iceland was in the media spotlight once again (e.g. Black, 2006). The *New York Times* wrote that resumption of whaling would hurt Iceland's tourism (Conlin, 2006). In *Iceland Review Online* (2006), Heimir Harðarson from 'North Sailing' in Húsavík said that tourists had already started to cancel their trips to Iceland after this announcement. Björgvinsson (2007) stated that whaling undermines whale-watching in Iceland. A poll conducted by IFAW in 2006 indicated that only 1.1% of Icelandic households eat whale meat weekly (IFAW, 2006b).

When resuming commercial whaling in May 2008, the Icelandic fisheries minister commented that whale-hunting is part of the culture and that minke sashimi is a popular dish. In May 2008, for the first time since the early 1990s, both Iceland and Norway exported whale meat to Japan (e.g. Black, 2008). In June 2008, Iceland exported 80 tons of fin whale meat to Japan despite the trade in whale products being restricted under CITES (the Convention on International Trade in Endangered Species of Wild Fauna and Flora).

In October 2008 the Icelandic currency collapsed and Iceland was again in the world's media because of its financial crisis. In January 2009 the Icelandic

Ministry of Fisheries posted a continuation of sustainable whaling. It stated that 'the total allowable take of fin and minke whales for the next five years will be according to scientific recommendations of the Marine Research Institute'. In January 2009 the Icelandic newspaper *morgunblaðid* published an advertisement for the whaling companies, where it is stated that whales around Iceland eat 2 million tonnes of fish per year.

Several public meetings were held in Reykjavik in 2009 with representatives from both the whale-watching and whaling industry. One meeting was also held at the University of Iceland, where Dr Hilmar Malmquist argued against scientific whaling. In *Iceland Review Online* (2009), Rannveig Grettarsdóttir, the director of 'Elding' whale-watching, expressed her concern that some tourists had already cancelled their tours and that some travel agencies had announced that they would remove 'Elding' from their list if Iceland pursued commercial whaling. It was suggested in the Icelandic media in 2009 that whaling would help to rebuild the Icelandic economy. According to a poll in 2009 over 67% of Icelanders were either 'very' or 'rather supportive' of commercial whaling, with only 19.7% opposed to it. In 2009 the MRI suggested some areas of protection in Iceland where whaling should not be permitted (Figure 7.3). These included the main whale-watching areas at the time. The projected areas include the part of Faxafloi Bay, where most whale-watching occurs, Eyjafjördur and Skjálfandi Bay and Steinsgrimsfjördur, where whale-watching was suggested as a possibility in Cecchetti and Rasmussen (2008).

In the summer of 2009 a survey was conducted onboard the whale-watching boat, 'Elding' from Reykjavik and the results were presented by Bertuilli *et al.* (2010). Tourists were asked if they would come to Iceland or not and if they would try whale meat. Most people answered that they would try whale meat, but only once. The vast majority of those surveyed expressed opposition to hunting whales (76%). Contrary to Parsons and Rawles (2003), 81% of the whale-watchers with no prior knowledge of the whaling situation in Iceland before planning

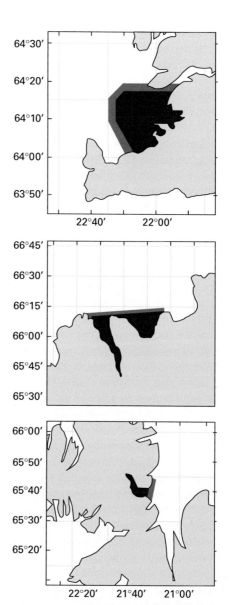

Figure 7.3 Whale-watching areas in Iceland, protected areas where whaling is not allowed (from the top: Faxaflói Bay, Eyjafjörður and Skjálfandi Bay and Steingrimsfjörður).

their trip mentioned that whaling was not a good reason to avoid visiting Iceland. In October 2009, 26 countries issued a joint statement putting diplomatic pressure on Iceland to abandon whaling. The joint demarche, a formal diplomatic protest signed by the UK, the US, Australia and 12 European Union countries including Germany, France, Portugal and Spain, states that the 26 governments are 'deeply disappointed' with the former Icelandic government's decision to authorize commercial whaling (Environmental News Service, 2009).

In spring 2010 a new financial report was published from the Economics Department at the University of Iceland to the Icelandic Fishery Ministry. It estimated that the salaries for people working in the fin whaling industry would be 750 Mill ISK (US$ 5.925 Mill), for minke whaling 21.7 Mill ISK (US$ 171,430 Mill) and the salary from the whale-watching companies was estimated to be 200–300 Mill ISK (US$ 1.58–2.37 Mill). It was also stated that, 'If 150 fin whales and 150 minke whales were hunted every year, it would mean that you would be able to catch 2,200 tons more of cod, 4,900 tons more of saithe and 13,800 tons more of capelin'. This prompted new discussions between the whalers and the whale-watching companies. Malmquist from the Icelandic Natural History museum argued against the simplicity in the models used by the economists. Emeritus Professor Arnor Garðarson also wrote critically about the report in Visir in April 2010. Both Malmquist and Garðarson, argued against the simplified model as it did not accommodate other animals (such as sea birds, bigger fish and seals) eating cod, saithe and capelin as well. Veal (2010) wrote that Greenpeace protestors in Rotterdam intercepted seven containers with 140 kg of fin whale meat from Iceland, destined for Japan in 2009. They claimed that the import of whale meat to the Netherlands is illegal, but Dutch authorities turn a blind eye to consignments destined elsewhere. Very little fin whale meat was sold in 2009, however. Statistics in Iceland state that a whole 3 kg of whale meat was exported, worth ISK 5442 (US$ 43) in 2009. The main benefit of whaling, however, would come from the fish that could be caught if not taken by whales. On a long-term basis, the populations of cod, capelin and haddock would increase and allow extra quotas for the fishing industry of 2200 tonnes of cod, 4900 tonnes of haddock and 13,800 tonnes

of capelin, say the study authors. These figures have been disputed, however. Hilmar Malmquist, curator of the Natural History Museum of Kopavogur, has long been critical of the science used by pro-whaling interests. 'The report is badly written and lacks scientific credibility. It is highly pro-whaling and biased towards interests in the fishing and whaling sector, such as the Federation of Icelandic Fishing Vessel Owners', he says. While estimates of the value of whale-watching are always subject to debate, the IFAW put the global value in 2008 at over $2 billion (O'Connor et al., 2009). Lambert et al. (2010) give a figure of $1 billion for cetacean (whale, dolphin and porpoise) tourism. Locally in Húsavík, the value of whale-watching was estimated to be 300 Mill ISK (Guðmundsdóttir & Ívarsson, AV 2008), corresponding to $US 2.7 million today.

In August 2010, the minke whaling company Hrefnuveidimenn started a new business called 'whale-watching with whale-hunters' (www.hrefna. is). They offer a four-hour whale-watching tour, where the tourists will get to try to shoot with the harpoon and they will also be offered various parts of whale meat to try on the tour, but so far this has not been done in practice. However, the whale-hunters did not get the correct permissions and these tours have never started. Whale meat is, however, sold in many restaurants in Reykjavik, and as 700,000 tourists visited Iceland in 2011 (Icelandic statistics) and if all the tourists try the whale meat just once, it is enough to maintain the market. Most tourists only visit Iceland once. The population of Iceland was only about 320,000 in 2011. In October 2010 a brief report was published by Iceland Nature Conservation Association and The International Fund for Animal Welfare which estimated the value of Icelandic fin whale meat. In 2009, 1500 tonnes of fin whale meat were caught and in 2010, 1776 tonnes. In 2010, 501 tonnes were exported to Japan, and 0 tonnes in 2009; 55 tonnes were imported to Japan in 2010 and 90% were still in Icelandic customs. Only 1.7% of the total catch was exported to Japan. The Icelandic Ministry of Fisheries estimated in 2009 the annual value of whale meat to be

$US 41 million. The actual annual value of exported meat was $US 3.5 million and the actual value of meat imported into Japan in 2010 was $US 0.8 million. In 2011 a preliminary report was presented to the IWC addressing the change in behaviour of minke whales around whale-watching boats (Christiansen *et al.*, 2011). This was published in the Icelandic paper, *Morgunblaðid* and translated from English in *Iceland Review Online* without any of the authors being contacted or interviewed (July 2011). The study was used in the whaling debate to ask if whale-watching had a negative effect on the minke whales, then maybe whaling was not so bad. *Iceland Review Online* published an article about man versus whale in 2012, where both Konráð Eggertsson as a whaler and Hördur Sigurbjarnarson, manager of the whale-watching company 'North-Sailing', comment on their opinion on the topic.

Conclusion

Both whale-watching and the whaling of minke whales continue in Faxaflói. The whale-watching companies want Faxaflói Bay to be a sanctuary for whale-watching. The government decided to make Faxaflói Bay a sanctuary in the spring of 2013; however, the new government in spring 2013 changed it again and at the moment minke whaling is still allowed inside Faxaflói Bay. IceWhale is planning on a meeting for all whale-watching companies in Iceland and hopefully they will agree to follow best-practice guidelines on a voluntary basis.

REFERENCES

BBC News. (6 August 2003). Retrieved from http://news.bbc.co.uk/2/hi/science/nature/3128863.stm (accessed 4 July 2012).

BBC News. (19 August 2003). Retrieved from http://news.bbc.co.uk/2/hi/europe/3162281.stm (accessed 4 July 2012).

Bejder, L., Samuels, A., Whitehad, H., & Gales, N. (2006). Interpreting short time behaviour response to disturbance within longitudinal perspective. *Animal Behaviour* 72, 1149–1158.

Bertulli, C.G. (2010). Minke whale (*Balaenoptera acutorostrata*) and white-beaked dolphin (*Lagenorhynchus albirostris*) feeding behaviour in Faxaflói Bay, south-west Iceland. Masters Thesis, Institute of Biology, University of Iceland, Reykjavik, Iceland.

Bertulli, C.G., Barreau, T. & Matassa, S.D. (2010). Whale-watching vs whaling in Iceland a survey of whale-watching tourists attitudes towards conservation issues. Poster presentation. European Cetacean Society conference, Stralsund, Germany, 22–24 March.

Björgvinsson, Á (2003). *Iceland. Whaling v/s Whale watching*. Retrieved from http://www.global500.org/news_83.html (accessed 4 July 2012).

Björgvinsson A. (2007). Whaling undermines whale watching in Iceland. *WWF Arctic Bulletin* 1.07, 18–19.

Björgvinsson, Á. & Lugmayr, H. (2002). *Whale Watching in Iceland*. Reykjavik: JPV Publishers.

Black, R. (2006). *Iceland begins commercial whaling*. Retrieved from http://news.bbc.co.uk/2/hi/science/nature/6059564.stm (accessed 4 July 2012).

Black, R (2008). *Japan sells Icelandic whale meat*. Retrieved from http://news.bbc.co.uk/2/hi/science/nature/7767716.stm (accessed 4 July 2012).

Borchers, D.L., Pike, D.G., Gunnalugsson, T. & Víkingsson, G.A. (2009). Minke whale abundance estimation from the NASS 1987 and 2001 aerial cue-counting surveys taking appropriate account of distance estimation errors. *NAMMCO Scientific Publications* 7, 95–110.

Boye, T., Simon, M. & Madsen, P. (2010). Habitat use of humpback whales in Godthaabsfjord, West Greenland with implications for commercial exploitation. *Journal of the Marine Biological Association of the United Kingdom* 90, 1529–1538.

Boye, T., Simon, M. & Ugarte, F. (2011). *A note on guidelines for sustainable whale watching in Greenland. With special focus on humpback whales*. 26 pp. Retrieved from http://www.natur.gl/fileadmin/user_files/Dokumenter/Raadgivning/A_note_on_whale_watching_guidelines_2011.pdf.

Carlson, C. (2008). *A review of whale watch guidelines and regulations around the world version 2008*. Retrieved from http://www.wdcs.org/submissions_bin/ww_regulations_2008.pdf

Cecchetti A. (2006). The spatial and temporal distribution of cetaceans within Skálfandi Bay, North East Iceland. MSc Thesis, University of Wales, Bangor, UK.

Cecchetti, A. & Rasmussen, M. (2008). Whale Watching Feasibility Study in the West fjords, NW Iceland. Unpublished report for IFAW, 25 pp.

Christiansen, F., Rasmussen, M. & Lusseau, D. (2011). Whalewatching boats disrupt the foraging activities of Minke whales in Faxaflói bay, Iceland. SC/63/WW2.

Cockeron, P. (2007). Iceland, whaling and ecosystem-based fishery management. Report for WDCS, 33 pp.

Conlin, J. (2006). The Resumption of Whaling Hurts Iceland Tourism. *New York Times*. Retrieved from http://travel.nytimes.com/2006/11/12/travel/12transwhale.html (accessed 4 July 2012).

Darby, A. (2007). *Harpoon into The Heart of Whaling*. Sydney: Allen and Unwin.

Day, D. (1987). *The Whale War*. London: Routledge.

Einarsson, N. (2009). From good to eat to good to watch: Whale watching, adaptation and change in Icelandic fishing communities. *Polar Research* 28, 129–138.

Environmental News Service. (2009). Retrieved from http://www.ens-newswire.com/ens/oct2009/2009-10-05-01.html (accessed 4 July 2012).

Fischer, S. (1997). Whale Watching in Iceland. SC/50 WW7. Bath: WDCS.

Foote, A., Osborne, R.W. & Hoelzel, A.R. (2004). Environment: Whale-call response to masking boat noise. *Nature* 428, 910.

Garðarsson, A. (2010). Athugasemdir við skýrslu Hagfræðistofnunar. Retrieved from http://www.visir.is/article/2010285734004 (in Icelandic) (accessed 4 July 2012).

Grossman, E. (2010). Noise reduces ocean habitat for whales. *Scientific American*. Retrieved from http://www.scientificamerican.com/article.cfm?id=noise-reduces-ocean-habitat-for-whales (accessed 4 July 2012).

Gudfinsson, E.K. (2006). Retrieved from http://eng.sjavarutvegsraduneyti.is/news-and-articles//nr/1300 (accessed 4 July 2012).

Guðmundsson, J. (1924) [1640–1644]. Ein stutt undireitting um Íslands adskilianlegar nátturur'. *Hermansson* 1924: 1–26, incl. 9 plates.

Guðmundsdóttir, R. & Ívarsson, A.V. (2008). *Efnahagsleg áhrif ferðaþjónustu á Húsavík-tilkoma hvalaskoðunar.* Iceland: Húsavík Academic Center. 65 pp.

Guðrúnardóttir, H. (2008). Do the whale watching boats have an effect on the distribution of whales in Skjálfandi Bay? Unpublished report (in Icelandic), 22 pp.

Homewood, V. (2010). Retrieved from http://www.telegraph.co.uk/travel/activityandadventure/7397122/Whale-watching-10-of-the-best.html (accessed 4 July 2012).

Icelandic Ministry of Fisheries. (2009). Retrieved from http://eng.sjavarutvegsraduneyti.is/news-and-articles/nr/9554 (accessed 4 July 2012).

Iceland Nature Conservation Association and The International Fund for Animal Welfare. (2010). Iceland and Trade in Whale Products.

Iceland Review Online. (July 2011). Retrieved from http://www.icelandreview.com/icelandreview/search/news/Default.asp?ew_0_a_id=379731 (accessed 4 July 2012).

Iceland Review Online. (January 2012). Retrieved from http://www.icelandreview.com/icelandreview/search/news/Default.asp?ew_0_a_id=385879 (accessed 4 July 2012).

International Fund for Animal Welfare. (2004). Report on scientific research in Icelandic waters from the International Fund for Animal Welfare's vessel 'Song of the Whales', July–August, 2004, 19 pp.

International Fund for Animal Welfare. (2006a). Report on scientific research in Icelandic waters from the International Fund for Animal Welfare's vessel 'Song of the Whales', July–August, 2006, 26 pp.

International Fund for Animal Welfare. (2006b). Capacent Gallup Poll for Iceland.

International Fund for Animal Welfare. (2009). Whale Watching Worldwide Tourism numbers, expenditures and expanding economic benefits. 295 pp.

International Whaling Commision. (1946). Convention. 13 pp.

International Whaling Commision. (1996). Retrieved from http://iwcoffice.org/conservation/whalewatching.htm (accessed 4 July 2012).

Iversen, M., Rasmussen, M.H., Cecchetti, A., Sigurðardóttur, V., Wald, E. & Vikingson, G.A. (2009). Seasonal occurrence and potential movement of feeding area of blue whales around Iceland – Preliminary studies. ECS conference, Istanbul, Turkey.

Kleivane, L. & Börsum, J. (2003). Undersökelse av kvikksölvnivaaer i muskel fra vågehval 2002. Veterinærinstituttet, Februar 2003.

Lambert, E., Hunter, C., Pierce, G.J., MacLeod, C.D., Scott, D. & Becken, S. (2010). Sustainable whale-watching tourism and climate change: Towards a framework of resilience. *Journal of Sustainable Tourism*, 18, 409–427.

Lindquist, O. & Tryggvadóttir, M.H. (1990). Whale watching in Iceland. A feasibility study. / Frumkönnun á möguleikum á hvalaskoðun við Ísland'. Unpublished IFAW report, Akureyri, Iceland. 30 pp.

Lindquist, O.L. (1997). *Peasant Fisherman Whaling in the Northeast Atlantic Area ca 900–1900 AD*. Iceland: University of Akureyri. 69 pp.

Magnúsdóttir, E.E. (2007). Year-round distribution and abundance of white-beaked dolphins (*Lagenorhynchus albirostris*) off the southwest coast of Iceland. Master Paed., Institute of Biology, University of Iceland, Reykjavik, Iceland.

Magnúsdóttir, E.E., Cummings, J. & Rasmussen, M.H. (2011). Effects of boat traffic on the minke whale's (*Balaenoptera acutorostrata*) swimming behaviour in Skjálfandi bay, NE-Iceland. ECS conference, Cadiz, Spain.

Malmquist, H. (2003). Vafasöm hvalveiðistefna og varasöm. Morgunblaðið, laugardagur 21. September 2003. Skoðun bls. 31 (in Icelandic).

Martin, S. (2012). Whale watching in Iceland: An assessment of whale watching activities on Skjálfandi Bay. Masters Thesis, University of Akureyri, 87 pp.

O'Connell, K. (2003). Iceland resumes whaling. *Whales Alive!* XII, No. 4.

O'Connor, S., Cambell, R., Cortez, H. & Knowles, T. (2009). *Whale Watching World Wide: Tourism numbers, expenditures and expanding economics benefits.* A special report from the International Fund for Animal Welfare. Yarmouth, MA: IFAW. 295 pp.

Parsons, E.C.M. & Rawles, C. (2003). The resumption of whaling by Iceland and the potential negative impact in the Icelandic whale-watching market. *Current Issues in Tourism* 6, 444–448.

Pike, D.G., Gunnlaugsson, T., Víkingsson, G.A. & Mikkelsen, B. (2010). Estimates of the abundance of minke whales (*Balaenoptera acutorostrata*) from the T-NASS Icelandic and Faroese ship surveys conducted in 2007. Cambridge: IWC, 12 pp.

Rafnsson, M. & Edvardsson, R. (2009). Foreign Whaling in Iceland Archaeological Excavations at Strákatangi in Hveravík, Kaldrananeshreppi 2009, Field report. Report for Náttúrustofa Vestfjarða and Strandagaldur, 27 pp.

Rasmussen, M.H. (2004). A study of communication and echolocation sounds recorded from free-ranging white-beaked dolphins (*Lagenorhynchus albirostris*) in Icelandic waters. PhD Thesis, University of Southern Denmark, Odense University. 109 pp.

Rasmussen, M.H. & Miller, L.A. (2002). Whistles and clicks from white-beaked dolphins, *Lagenorhynchus albirostris* recorded in Faxaflói Bay. *Aquatic Mammals* 28, 78–89.

Reeves, R.R., Smeenk, C., Kinze, C.C., Brownell, R.L. Jr & Lien, J. (1999). White-beaked dolphin, *Lagenorhynchus albirostris* Gray, 1846. In S.H. Ridgway & R. Harrison

(Eds), *Handbook of Marine Mammals*. San Diego, CA: Academic Press, pp. 1–30.

Revery, F. (2011). Avoidance or indifference – Investigating effects of boat traffic on the behaviour of Humpback whales (*Megapteranovaeangliae*) in Skjálfandi Bay. Student report, University of Iceland.

Salo, K. (2004). Distributions of Cetaceans in Icelandic waters. Master thesis. University of Southern Denmark, Odense, Denmark.

Schaffer, A., Madon, B., Garrigue, C. & Constantine, R. (2009). Avoidance of whale watching boats by humpback whales in their main breeding ground in New Caledonia. International Whaling Commission S/61/WW6.

Scheidat, M., Castro, C. & Gonzalez, J. (2004). Behavioural responses of humpback whales (*Megaptera novaeangliae*) to whalewatching boats near Isla de la Plata, Machalilla National Park, Ecuador. *Journal of Cetacean Research Management* 6, 63–68.

Siglaugsson, P. (2005). *The Whale Meat Market. Study on Current and Possible Markets and Cost of Operations in Minke Whaling.* Reykyavik: GJ Financial Consulting. 15 pp.

Stefánsson, G., Sigurjónsson, J. & Víkingsson, G.A. (1997). On dynamic interactions between some fish resources and cetaceans off Iceland based on a simulation model. *Journal of Northwest Atlantic Fisheries Science* 22, 357–370.

Veal, L. (2010). *Whaling Profitable but Bad for Iceland's Image.* Retrieved from http://ipsnews.net/news.asp?idnews=51067 (accessed 4 July 2012).

Víkingsson, G.A. (2008). Retrieved from http://www.hafro.is/hrefna/pdf/nidurstodur08.pdf (in Icelandic) (accessed 4 July 2012).

Víkingsson, G.A. & Ólafsdóttir, D. (2004). Hnýðingur (White-beaked dolphin). In P. Hersteinsson (Ed.), *Íslensk Spendýr (Icelandic Mammals).* Vaka- Reykjavík: Helgafell, pp. 154–157.

Víkingsson, G.A., Ólafsdóttir, D., Gunnlaugsson, P., *et al.* (2006). Research programme on common minke whales (*Balaenoptera acutorostrata*) in Icelandic waters. A progress report. IWC 2006 SC/58/O20.

Walton, D. (2003). *Iceland divided over whaling.* Retrieved from http://news.bbc.co.uk/2/hi/business/3153591.stm (accessed 4 July 2012).

Ægisson, S., Ásgeir, J. & Hliðberg, J.B. (1997). *Icelandic Whales, Past and Present.* Reykjavik: Mal Og Menning.

8

Iceland and the resumption of whaling

An empirical study of the attitudes of international tourists and whale-watch tour operators

Tommy D. Andersson, Susanna E. Gothall and Béatrice D. Wende

This chapter investigates how the resumption of Icelandic whaling has impacted the whale-watching tourism industry in Iceland viewed both from a demand side and a supply side. The impact on international tourists and domestic whale-watching tour operators' attitudes were subjects of a case study from Iceland in 2007. After an introduction to Icelandic whale-watching and the Icelandic whaling situation, the results of this case study are presented. In the concluding section the development of whale-watching tourism in Iceland from 1995 to 2009 is viewed and analysed through these two perspectives.

Whaling versus the whale-watching tourism industry

For a number of reasons, the coexistence of whaling and whale-watching has been described as incompatible (Hoyt & Hvenegaard, 2002). Removal of whales and disturbances or changes to their regular activities are direct effects of whaling. Whaling will thus reduce the number of existing whales for whale-watching at the same time as it can cause avoidance responses to boats in several ways, such as increases in dive intervals (Baker et al., 1988; MacGibbon, 1991; Janik & Thompson, 1996) and changes in movement and direction (Edds & Mac-Farlane, 1987; Salvado et al., 1992; Bejder et al.,

1999; Nowacek et al., 2001) which are caused by the presence and navigation of vessels on the water in proximity to animals (Constantine, 2001). Other potential implications of the coexistence are negative attitudes towards the destination image by whale-watchers as well as other tourists. There are, however, some proponents of the Icelandic resumption of commercial whaling who say that it is possible for commercial whaling and whale-watching to coexist and that money can be gained from both whaling and whale-watching (e.g. Moyle & Evans, 2001 cited in Parsons & Rawles, 2003). In Iceland and Norway, the whale-watching industry exists side by side with the whaling industry.

Higham and Lusseau (2007) suggest that tourists visiting Iceland and Norway might react in three different ways to the destinations' whaling activities. The first type of response is that tourists accept that whaling and whale-watching coexist; the second type is that whaling, whether commercial, scientific or for sustainable harvest, will be seen by tourists as offensive, which will have a direct negative impact on the whale-watching industry; the third type is that indigenous whaling is seen as an acceptable local cultural expression, which might even enhance the visitors' interests in the destination.

In 2002, after a pause of 14 years, the Icelandic government decided to join the International Whaling Commission (IWC) and to start *scientific*

Whale-watching: Sustainable Tourism and Ecological Management, eds J. Higham, L. Bejder and R. Williams.
Published by Cambridge University Press. © Cambridge University Press 2014.

whaling once again, which caused considerable controversy among the public (Kirby, 2003 cited in Parsons & Rawles, 2003). In 2003, altogether 36 whales were killed in Iceland (Parsons & Rawles, 2003). Four years later, in 2006, Iceland also decided to resume *commercial* whaling.

As whales are an integral part of Iceland's destination image, whaling is seen as a threat to Icelandic tourism (Helgason, 2007) and harms the general image of the island, as most tourism attractions in Iceland are nature-based (Sigursteinsdottír, 2003 in Helgason, 2007). A Gallup poll conducted in 2006 when commercial whaling resumed showed that 48% of the surveyed Icelanders believed that commercial whaling would have a negative impact on the country's tourism industry (IFAW, 2006). According to the Icelandic Tourist Industry Association, whaling might harm the growing whale-watching tourism industry, which is said to 'create a very positive image for Iceland' (CBS News, 2003). Vignir Sigursveinson, operating manager of Elding Whale Watching in Iceland, feared that the whales would be afraid of boats in general due to the resumption of whaling (CBS News, 2003). It is likely that whales cannot distinguish between whale-watching boats and whaling boats and consequently avoid all boats (e.g. Bejder *et al.*, 1999; Constantine, 2001; Nowacek *et al.*, 2001).

The main research question addressed in this empirical study is: how is the whale-watching industry in Iceland affected by the fact that the country allows whaling?

The objectives of the study are therefore to determine the impact whaling has on:
• the attitudes of international tourists to Iceland and
• the attitudes and operations of whale-watching tour operators in Iceland.

After a brief introduction to whale-watching in Iceland, followed by an overview of whaling activities in Iceland, the results of a survey of international tourists arriving in Iceland as well as Icelandic whale-watching tour operators will be presented. The conclusions will address the research question

through these two perspectives, discussing the development of whale-watching tourism in Iceland from 1995 to 2009.

Whale-watching in Iceland

Iceland is one of Europe's most popular whale-watching destinations (IFAW, 2003) and a variety of species can be found in Icelandic coastal waters, including humpback whales (*Megaptera novaeangliae*), minke whales (*Balaenoptera acutorostrata*), blue whales (*B. musculus*), killer whales (*Orcinus orca*), white-beaked dolphins (*Lagenorhynchus albirostris*) and harbour porpoises (*Phocoena phocoena*) (North Sailing, 2003 cited in Parsons & Rawles, 2003). These different species can be seen within only a couple of hours of each other (IFAW, 2003). The high season for whale-watching in Iceland is from April to September (Helgason, 2007). The whale-watching tourism industry in Iceland mainly consists of local small and medium-sized enterprises (Helgason, 2007) and started on a small scale in 1991 with 100 persons whale-watching annually. The industry then only consisted of one whale-watching tour operator (Hoyt, 1995 cited in Fisher, 2007).

Ecotourism and especially whale-watching constitute an increasing source of income for the Icelandic economy (IFAW, 2003). Even though the Icelandic whale-watching industry is only two decades old, in 2007 it was worth around $US 24 million, competing with the turnover from the Icelandic whaling industry at its peak time in the 1980s (Helgason, 2007). 'Tourism is the fastest growing sector in the Icelandic economy where whale-watching is the fastest growing activity' (Gudmundsson, 2007). Whale-watching has thus gained great importance for Iceland. For Iceland generally, and its several isolated communities in particular, whale-watching is therefore of significant value (IFAW, 2003). It is said that the number of new jobs, business spin-offs and social benefits of the whale-watching tourism industry outweigh the social and economic influences that the whaling industry has had and will have in the future (Oddson, 2004 cited in Helgason, 2007).

In Iceland, whale-watching mainly takes place in three areas: the Reykjavík area and the nearby Reykjanes Peninsula area, the north with the fishing towns of Húsavík and Olafsvík, and the Snaefellsnes Peninsula. Húsavík, with a population of approximately 2500 inhabitants, is known as the top spot for whale-watching in Iceland (Arctic Experience, 2007). The main industry of Húsavík used to be fishing; however, currently fishing has been replaced by the tourism industry, specifically the whale-watching tourism industry. Since whale-watching tourism developed in Húsavík, the unemployment rate has decreased, and at the same time the aesthetic appearance of the town has improved (Helgason, 2007).

Whaling in Iceland

Between 1883 and 1990, around 35,000 whales were caught off the coast of Iceland (Altherr, 2003). Whaling in the waters of Iceland stopped in 1989 due to the moratorium on whaling, but at the same time the first preparations were made to recommence whaling (Altherr, 2003). These preparations included the following two developments: the country became a member of CITES (a convention which regulates the international trade in endangered species) and it started to import Norwegian whale products. Iceland also tried to enhance its relationship with Japan, discussing fisheries and whaling issues with the country. At that time, the Icelandic Minister of Fisheries stated the intention to export whale products to Japan in the near future. Iceland rejoined the IWC and in March 2003 Iceland prepared a plan for a two-year so-called research whaling programme, including the killing of 500 whales, which was presented to the IWC.

Kristjan Loftsson, the chief executive of the Icelandic whaling firm Hvalur Hf., strongly believes that to preserve fishing stocks, the number of whales must be reduced (CBS News, 2003). Loftsson has little sympathy for the whale-watching tour operators and claims that since the 1940s, whaling has always been a feasible industry whereas whale-watching companies have sometimes had problems staying in business (CBS News, 2003). Loftsson argues that without the financial support from anti-whaling groups, many whale-watching companies would have been subject to bankruptcy (High North Alliance, 2004). In contrast, anti-whaling groups claim that in Iceland there is more to be earned from whale-watching than whaling (High North Alliance, 2004).

The former Prime Minister of Iceland, Geir H. Haarde (Forsaetisraduneyti, 2007), declared in 2003 that he could not accept that Iceland would not be allowed to utilize whales, as they constitute an important marine resource when other resources in the sea are harvested (CBS News, 2003). The next Prime Minister, Johanna Sigurdadottír, who gained power in 2009, continues the political approach of her predecessor (Mattsson, 2010). Whaling is considered a sensitive topic in Iceland and Icelanders are strongly divided in their opinions towards whaling (Helgason, 2007). According to Helgason (2007), 'In the past, extreme action taken by environmental groups or individuals, have resulted in solidifying Icelanders consensus towards whaling as an icon of independence regardless of pros and cons' (Helgason, 2007: 13). In 2009, the issue of whaling was top of the political agenda when Iceland applied for membership of the European Union (EU Business, 2009). Britain, Germany and France protested against the larger hunting quotas that were introduced in spring 2009 and for this reason, Iceland's future EU membership was jeopardized. These whaling quotas include 150 fin whales and 100 minke whales per year (EU Business, 2009).

With reference to Hoyt and Hvenegaard (2002), no surveys have been conducted to explore whether local whale-watching communities have any preferences towards whaling or whale-watching. Hoyt and Hvenegaard (2002) state that local communities, which are involved in whale-watching tourism, in general are satisfied with whale-watchers and the whale-watching industry, in terms of residents' feelings towards the influence the tourists have on the area. This aspect is interesting, as Iceland practises both whale-watching and whaling.

In 2009, The Icelandic Ministry of Fisheries and Agriculture declared that there will be research on the macroeconomic significance of whaling, which will be conducted by the Economic Institute of the University of Iceland (The Icelandic Ministry of Fisheries and Agriculture, 2009). It also said that they will introduce specific ocean areas for whale-watching where whaling will be forbidden to avoid the overlapping of both activities. To summarize, this report concludes that whaling is economically beneficial for Iceland; however, the conditions for calculations of multiple impact from whaling were considered too vague, and therefore it did not take this aspect into account (*Iceland Review Online*, 2010). However, the Minister of Fisheries and Agriculture at that time, Jón Bjarnasson, 'would not say whether this estimate and the economists' statement that whaling is nationally beneficial are sound enough reasons for it to continue' (*Iceland Review Online*, 2010).

Commercial whaling and scientific whaling

Commercial whaling in Iceland has a long history and had its peak in the early twentieth century, when whale oil was seen as an essential product (Siglaugsson, 2005). Since 1982, more than 100 scientific whaling permits have been issued for a number of countries such as Canada, USA, former USSR, South Africa as well as Japan (Steuer, 2005). Subsequent to 1982, when the IWC decided to suspend commercial whaling, the extent of permits and the number of whales taken began to increase, because some countries used the special permit as an option to circumvent the suspension in commercial whaling (Steuer, 2005).

Between 1986 when the IWC Whaling Moratorium went into effect and 2006, a total of 463 whales were killed in whaling operations for scientific purposes in Iceland (IWC, 2006). Among these, the

North Atlantic fin whales were the major part with 292, followed by 101 North Atlantic minke and 70 North Atlantic sei whales. It is notable that there was no whaling between 1990 and 2003, due to a pause. Since 2003, officially only North Atlantic minke whales have been killed in Iceland, in contrast to 1986–1990, when North Atlantic fin and sei whales were targeted (IWC, 2006). In 2006, Iceland announced that the country would commence commercial whaling consisting of catching nine fin whales and 30 minke whales per year (Black, 2006). According to a representative of the government, 43,600 minke whales and around 25,800 fin whales live in Icelandic waters and therefore the commercial whaling activity is seen as sustainable by the Icelandic government (Black, 2006).

Iceland and the whale meat market

The whale meat market in Iceland, if seen in an international and a domestic context, is very small (Siglaugsson, 2005). When the whale meat from scientific whaling entered the Icelandic market in 2003, Icelanders were no longer used to eating whale meat as this product had not been available during the past 20 years, and the government launched a promotional campaign to increase the demand for whale products on the domestic market. This media campaign included the publication of cooking recipes for whale meat and Icelandic politicians had meals containing whale meat in the presence of the media. Articles in the Icelandic press even argued that the consumption of whale meat is especially healthy. As a result, the consumption of whale meat in Iceland increased in the same year from around 150 to 200 tonnes per year (Altherr, 2003). However, the repeat sale rate was low. It also appeared difficult for Iceland to find an adequate export market owing to the trade restrictions as well as the fact that the market for whale meat is decreasing dramatically. Potential markets for whale meat which were identified by Iceland were China, former USSR, Norway and Japan. As Japan, China and Russia have no CITES import permits, these countries cannot be classified as potential trading partners (Siglaugsson, 2005). In 2009, however, Iceland gained permission to export whale meat to Japan after the 18-year pause (EU Business, 2009).

Iceland and its IWC membership – a difficult relationship

Iceland withdrew from the IWC in 1992 and created an alternative body called NAMMCO, the North Atlantic Marine Mammal Commission, which comprises only states and territories in the North Atlantic area with a strong interest in whaling, including Norway, Greenland and the Faroe Islands (Altherr, 2003). NAMMCO was established in 1992; however, its foundation was unsuccessful and never gained international recognition as a North Atlantic whalers club, which was Iceland's intention. The result was that two invited countries, Canada and former USSR, refused to become members. Norway showed no strong commitment towards it and Denmark did not join the organization at all, and consequently Iceland recognized that NAMMCO had become an international failure (Altherr, 2003).

In 2001, Iceland tried to regain membership of the IWC; however, it stated the condition not to accept the moratorium on commercial whaling (Altherr, 2003). The UK, the USA, New Zealand, Australia as well as Germany and Mexico declared that they were unwilling to accept the lifting of the moratorium (IWC, 2001). The IWC chairman at that time, Bo Fernholm from Sweden, thought that if Iceland's reservation would be accepted, other states who are dissatisfied with the IWC decision would follow Iceland as an example. They would leave and later return without being tied to the decision that was the reason for them to leave IWC in the first place (IWC, 2001). The former IWC chairman Michael Canney stated at the time that 'A convention is there. You join or you don't join. You can't pick the bits you like and leave out the bits you don't like' (IWC, 2001: 6).

IWC membership was denied to Iceland in July 2001 as well as in May 2002 (Altherr, 2003). On 14 October 2002, Iceland was finally allowed to rejoin the IWC; however, the procedure of voting was questionable and the country was even allowed to vote for itself. Afterwards, several countries expressed their controversial feelings that Iceland had rejoined the IWC (Altherr, 2003). The main reason why Iceland wanted to rejoin the IWC was that Japan only allows imports of whale products from IWC member countries (IWC, 2001). As the domestic market for whale meat in Iceland is small, the country hoped to export its whale products to Japan. This is also believed to be why Norway remains a member of the IWC (IWC, 2001).

Data collection

The objectives of the two empirical surveys conducted are to describe the knowledge and attitudes towards whale hunting of:
1. international tourists visiting Iceland; and
2. whale-watching tour operators in Iceland.

These two survey populations were therefore targeted by two samples of respondents. First, international tourists visiting Iceland answered a self-administered survey, which resulted in 160 valid questionnaires collected between 29 June and 9 July 2007 at the following locations: Keflavík International Airport, The Blue Lagoon, Reykjavík Bus Terminal, Reykjavík City Hostel and Reykjavík Harbour. The locations were selected because they are frequented by a large number of international tourists. Attention should be drawn to the fact that the survey of the tourists was conducted in Iceland and therefore excludes those tourists who had chosen not to go to Iceland for various reasons. As the survey was conducted in Iceland, the interviewed tourists were already in the country and therefore had decided to travel to Iceland in spite of the fact that the country conducts whale-hunting. Thus, tourists who decided to boycott the destination because of the whaling situation were not included in the population of this study. Second, four tour operators participated in the survey: Seatours, located in Snaefellsnes Peninsula; Gentle Giants, located in Húsavík; Elding Whale-Watching, located in

Table 8.1 Tourists' attitudes towards whaling (%).

Question	Yes	No	Not sure
Is whale-watching one of the main purposes for your visit to Iceland?	14	85	1
Did you know that Iceland practises whale-hunting before your arrival at this destination?	79	21	0
Would you go on a whale-watching trip in Iceland when you know that the country practises whale hunting?	71	18	11
Taking into consideration the whale-hunting situation in Iceland, would you still consider visiting the destination in the future?	83	7	10
Would you consider whale-watching an animal-friendly activity?	49	21	30
Have you attended a whale-watching trip in Iceland?	41	52	7

Reykjavík; and Sjóferdir Snorra EHF, located in Dalvík, near Húsavík.

Questions asked in the survey as well as answers received are described in Tables 8.1–8.4. The answers were mainly analysed in terms of frequency distributions.

Results from the study of tourists' and tour operators' attitudes

Two separate surveys were conducted in order to examine the whale-watching industry from both the demand side and the supply side. In the first section the attitudes of international tourists will be discussed and in the second section the opinions of representatives from the whale-watching tour operators will be presented.

Tourists' attitudes towards Icelandic whaling

The majority (86%) of the respondents were European citizens and 69% of the respondents travelled to Iceland for the main purpose of leisure. Although 14% of the international tourists to Iceland during summer 2007 travelled with the main purpose of whale-watching, 41% had experience of whale-watching in Iceland. International tourists were well-informed about the whaling situation in Iceland and 79% knew before their arrival that Iceland practises whaling (Table 8.1).

The question 'Would you go on a whale-watching trip in Iceland when you know that the country practises whale hunting?' got a positive response from 71% of the respondents. Female respondents were less willing to consider going on a whale-watching trip in Iceland when they knew about the whaling situation of the country than were male respondents. Answers to the question 'Taking into consideration the whale hunting situation in Iceland, would you still consider visiting the destination in the future?' showed that 83% of respondents would still consider visiting Iceland in the future regardless of the existing whaling situation at this destination. It seems that most of the respondents who answered 'yes' to this question are positive about Iceland as a destination, despite the fact that the country practises whaling. An American tourist expressed his opinion in the following statement; 'I like Iceland, regardless of the whale hunting' (Table 8.2).

Many respondents who would consider visiting Iceland in the future irrelevant of the whaling situation shared the opinion that they do not want to blame the entire country for the whaling situation. Subsequently, the three types of responses suggested by Higham and Lusseau (2007) can be extended and applied specifically to international tourists in Iceland by saying that the first type of response of tourists visiting a whaling nation is that they also do not associate the activity of whaling with either whale-watching or the entire country. This opinion is shown in the following

Table 8.2 Tourists' feelings about Iceland practising whaling.

Please describe your feelings about Iceland practising whaling	Negative feelings	Positive feelings	Neutral feelings	Non-response rate
Number of tourists	74	31	42	13
Percentage of total number of respondents (%)	47	19	26	8

Table 8.3 What is your attitude towards whaling, referring to the different reasons (%)?

Type of whaling	Definitely refuse	Refuse	Neutral	Accept	Definitely accept	No answer
Commercial	46	24	16	9	1	4
Scientific	19	21	26	26	4	4
Aboriginal	21	9	27	32	8	3

statement made by an English tourist: 'I do not associate the whale hunting situation with Iceland as a whole' and another made by a Swedish tourist: 'I do not think there is a relationship between whale-watching and whale hunting'. International tourists accept that whale-watching and whaling coexist.

The second type of response is that whaling, regardless of whether it is commercial, scientific or for sustainable harvest, will be seen by the tourists as offensive and shocking. This will have a direct impact on the whale-watching industry (Higham & Lusseau, 2007). In the case of Iceland, this would mean that tourists are so offended by the whaling that they associate it with the destination image of Iceland and therefore will not return to the destination in the near future. This opinion can be seen in the answers by an English tourist who would not visit Iceland in the future if the country increases its whaling activities: 'Not if whaling increases'. Another German tourist stated that: 'Iceland is not economically dependent on hunting whales; therefore I believe that the destination image of Iceland will be more positive if they protect the whales instead.'

The third type of response is that tourists see the whaling as an integral part of a destination's traditions and culture, no matter if they accept it or not. This opinion can be seen in the following quotations:

Every country has its drawbacks. What about bull-fighting in Spain and fox hunting in England? Travelling to these countries does not mean that I accept the cruelties. (Latvian tourist)

I would not condemn whale hunting, if it is managed in line with international treaties. Whale hunting is part of the culture even if I may not agree with it. (English tourist)

Both tourists see whaling as an integral part of Iceland's culture although they do not agree with it. Another English tourist states that 'I did not come here to preach or tell the locals what to do.' Tourists who see whaling as an integral part of Iceland's culture have the attitude that they do not have the right to tell somebody else what they should do. The attitude towards whaling can be divided according to the following three forms of whaling, as shown in Table 8.3.

If percentages in Table 8.3 for 'Accept' and 'Definitely accept' are added, tourists surveyed accept whaling to large extent (40%) when it is conducted by indigenous peoples. Furthermore, when whaling is conducted for scientific reasons, it is also highly accepted (30%); however, commercial whaling is only accepted by 10% of the respondents (Table 8.3). One Latvian tourist summarizes this finding in the following statement. 'Whaling can be justified only for scientific or aboriginal purposes. Commercial whaling should be prohibited.' Commercial whaling, however, is seen by the tourists as questionable because whale populations are decreasing. A German tourist supports this result by saying: 'In earlier time, whale hunting was used as a source to obtain food and therefore it was acceptable. Nowadays however the whale hunting in Iceland is commercialized and therefore it is not acceptable at all.' The study findings also show that commercial whaling is not accepted for the reason that whale populations are decreasing and Iceland is seen as a civilized country that generates revenue from other commercial activities instead of engaging in whaling. These statements by tourists, which are coherent with the results in Table 8.3, clearly illustrate that tourists are much more prepared to accept whale-hunting when it appears to be on a sustainable and reduced scale.

Results of the survey of the whale-watching tour operators' attitudes

The four tour operators participating in the survey are all small-scale businesses, ranging from 2 to 45 employees and offering whale-watching tours and other types of tours. The approximate number of customers per year varies between 1500 and 60,000. All four tour operators have negative feelings towards whaling for scientific, commercial and indigenous people's purposes; however, for various reasons. The managing director of the company Gentle Giants stated his personal opinion about whaling for indigenous people's purposes as follows: 'Culture is changing from year to year' (Gudmundsson, 2007) and therefore hunting for aboriginal purposes cannot be justified. The whale-watching tour operators think whaling should be prohibited in Iceland and they do not believe in the coexistence of whale-watching and whaling seen in a long-term perspective. Gudmundsson stated his opinion that: 'The question is killing or watching?' (Gudmundsson, 2007).

The whale-watching tour operators are affected, some to a small, others to a large degree by whaling in Iceland. The main concern is that the areas where both the whaling and the whale-watching activities are conducted overlap and on occasions

whaling has actually been carried out in front of whale-watching tourists (Sjóferdir Snorra EHF, 2007). The daily operations of the tour operators have also been affected by the fact that some tour operators have to travel a longer distance to find whales, which results in higher demands of time and operational costs. Despite the fact that only a small number of minke whales are hunted, the Icelandic whaling has an influence on the quality of the tourists' whale-watching experience in Iceland, as some whale species, especially the hunted ones, avoid whale-watching boats. A decrease in spotted whales, especially the hunted species, and altered behaviour of the whales has been observed by the whale-watching tour operators. Gudmundsson (2007) expressed this development as follows: 'There are fewer Minke whales observed nowadays and they have become more shy and careful'. As the whale-watching tour operators assumed, the most curious whales have been killed by whaling boats (Table 8.4). A lower quality of whale-watching experience can be anticipated due to further alteration in whale behaviour if whaling continues.

Whaling also seems to have an impact on the consumer purchase behaviour of the tourists. Some tour operators claim they have observed a decrease in bookings as well as cancellations of whale-watching trips by travel agencies because of whaling. This was especially noticeable immediately following Iceland's resumption of whaling and will be elaborated further in the Discussion section below and illustrated in Figure 8.1.

According to a representative of Sjóferdir Snorra EHF (2007), whaling in Iceland has affected their business and the number of reservations in a negative manner. The representative also noticed that several customers are concerned about the whaling situation in the country. Vigner Sigursveinson (2007), of Elding Whale-Watching, stated that people have contacted the company due to their concern about the whaling situation (Sigursveinson, 2007). In contrast to Elding Whale-Watching, the Sjóferdir Snorra EHF has not noticed that their customers are concerned about the whaling situation. Ragnheidúr Valdemarsdottír, the representative of

Table 8.4 The impacts of whaling for the whale-watching tour operators.

Company name	Whaling in front of the whale-watching boat	Longer distance to find whales	Altered behaviour of whales	Change in number of bookings
Sjóferdir Snorra EHF	Yes	No	Yes, especially the minke whales	Slight change downwards
Seatours	No	No	No	No change
Gentle Giants	No	No	Yes, especially the minke whales	No change
Elding Whale-Watching	No	Yes, definitely the hunted species and especially in the last two years	Whales approach boats less often. Minke whales act more shyly. Curious whales mostly seemed to have been killed	Some change at the beginning of the resumption of whaling

Seatours (2007), said that the company has received letters from concerned customers. With reference to the managing director Stefan Gudmundsson (2007) of Gentle Giants, all the company's customers expressed negative feelings towards whaling.

Tourists want to be informed about the whaling situation, but only one out of four whale-watching tour operators interviewed informs customers onboard the whale-watching boats on a regular basis. Their reasons behind this are that they do not want to ruin the tourists' whale-watching experience by mentioning whaling onboard the whale-watching boats and they do not want to upset their customers by telling them about the Icelandic whaling situation. A statement by a French tourist underlines the importance of being informed by the whale-watching tour operator: 'I would agree to go on a whale-watching trip if the organizer aims at making the tourists and media etcetera aware of the problem of whale hunting.'

Development of spotted whales

One of the whale-watching tour operators (Seatours) gave additional information regarding spotted whales, specifically the different spotted whale

Table 8.5 Spotted whale species by Seatours from 2004 to 2007 (Valdemarsdottír, 2007).

Year	Minke whales spotted in percentage of the tours	Orcas spotted in percentage of the tours	Sperm whales spotted in percentage of the tours
2004	69	43	27
2005	42	35	35
2006	66	28	35
2007	72	20	6

species during recent years (Valdemarsdottír, 2007; Table 8.5). It should be taken into consideration that other species were also seen during this period, but their numbers were not significant. Noticeable is that Seatours had an unusually high number of spotted minke whales in 2007. To be more precise, minke whales were seen in 72% of the whale-watching tours. In 2005, the numbers of spotted minke whales were considerably lower in comparison to the other years.

According to Sjóferdir Snorra EHF (2007), the whales behave more shyly towards the boats since the resumption of whaling. In particular, the behaviour of the Minke whales has altered since whaling of this species recommenced in Iceland. The number of whales seen has changed; nowadays there are more humpback whales compared to before but fewer minke whales (Sjóferdir Snorra EHF, 2007). Gudmundsson of Gentle Giants has noticed the same development, and that minke whales have become more shy and cautious (Gudmundsson, 2007). In contrast, Ragnheidúr Valdemarsdottír of Seatours (2007) observed an increasing number of minke whales, while the numbers of orcas were decreasing. This might be due to the fact that Seatours is located in a different area to the other operators. According to Vigner Sigursveinson (2007) of Elding Whale-Watching, the numbers of spotted whales have decreased over recent years, especially in 2007. However, the reasons for this development have not been proven. Sigursveinson believes it might be the lack of food in the bay or a side effect of the resumption of whaling (Sigursveinson, 2007).

Agreements between whale-watching tour operators and whaling companies

All whale-watching companies interviewed stated that whale-watching and whaling locations sometimes overlap and according to the representative of the whale-watching company Sjóferdir Snorra EHF (2007), a dolphin was once shot in front of their whale-watching tourists. However, there is a silent agreement between the whale-watching companies and the whaling companies not to hunt close to the whale-watching sites even though these sites overlap (Gudmundsson, 2007). With reference to Sigursveinson (2007), the whaling companies have asked the whale-watching companies to draw a map of the whale-watching locations in order to know where the whaling boats should not hunt. Later on, however, they disregarded the silent agreement because the scientific whaling research methods demanded sampling in all areas (Sigursveinson, 2007).

Although whaling impacts on the whale-watching tourism industry, the whale-watching tour operators do not blame the whaling companies or the

local communities for the resumption of whaling. One tour operator tried to explain the locals' positive attitudes by saying that the reason for Icelanders' attitudes is that the country recently gained its independence and therefore people want to assert their independent decisions, without listening to other European countries. The four whale-watching tour operators cooperate with the organization Icewhale, which is a non-governmental organization in Iceland. It represents the opinion of the whale-watching tourism industry in Iceland. The tour operators are in contact with the Icelandic government via Icewhale (Sigursveinson, 2007). None of the whale-watching tour operators are in touch with the whaling companies.

Discussion

The worldwide growing awareness of the environment has placed nature at the top of the agenda and environmental values are gaining increasing importance. With this in mind, it would not be surprising if international tourists react more sensitively towards whaling in Iceland in the near future. Whaling can be perceived to compromise the image of Iceland, and may result in potential travellers preferring another whale-watching destination. The statement below by an American tourist shows the foremost opinion of international tourists visiting Iceland according to the study findings. Although most international visitors that were surveyed in 2007 generally have negative feelings towards whaling, they would still visit the country in the future. The power of influence that the Icelandic whaling has on tourists' willingness to return to the destination in the future can be regarded as relatively small.

While I in general do not support whale hunting, that will not affect my decision to visit Iceland again in the future (American tourist).

The majority of international tourists travelling to Iceland in 2007 did not accept Icelandic whaling; however, they did not show their disagreement by changing their behaviour or actions and therefore their protest can be seen as silent.

The results of the survey describe attitudes of a population of international tourists who visited Iceland in 2007 and were well aware (71% of the respondents) of the fact that Iceland had resumed whaling. This population is interesting to discuss in relation to the three types of responses to whaling that Higham and Lusseau (2007) suggest:

1. Whaling and whale-watching can coexist without one adversely affecting the other;
2. Whaling, whether commercial, scientific or for sustainable harvest, will be seen by tourists as offensive and impossible to reconcile with; and
3. Indigenous whaling is seen as an accepted local cultural expression, which might even enhance the visitors' interests in the destination.

According to these three categories, the second category would not be expected to visit Iceland and is probably poorly represented in the surveyed population of international tourists in Iceland in 2007, but the first as well as the third category may be well represented. An important issue for the whale-watching tourism industry must, then, be how large proportions of the total potential for whale-watching tourism each of these three categories represent.

One month after the resumption of whaling in Iceland in 2003, a drop of 25% in bookings for whale-watching trips was reported by a British tour operator (Williams, 2006). The national airline Icelandair has also experienced cancellations due to the resumption, which implies that whaling is an important issue for the total Icelandic tourism industry and that the resumption of whaling is problematic (Conlin, 2006). Another survey of Icelandic tourists, conducted in 1998, showed that 54% of the respondents anticipated negative impacts on their decision about repeat visits if Iceland recommenced whaling (Fisher, 2007).

There is a tendency that most whale-watching tourists respond negatively to the hunting (Higham & Lusseau, 2007). A study revealed that 91% of whale-watching tourists do not want to visit a country that engages in commercial whaling (Parsons & Rawles, 2003). This demonstrates that the purpose of the whaling truly affects the tourists' attitudes. It implies that aboriginal and scientific whaling is accepted more than the commercial form. Another study has shown that residents from Australia were the ones that were opposed most to whaling (60%), followed by the United States (57%), Germany (54%), and England (43%), whereas the whaling countries Japan and Norway had the lowest numbers; with Japan (25%) and Norway (22%) (Freeman & Kellert, 1992, cited in Hoyt & Hvenegaard, 2002). The opposition to whaling is strong in important source countries for whale-watching tourism to Iceland, such as Germany and England, that generate a considerable number of foreign tourists (Hoyt & Hvenegaard, 2002).

Figure 8.1 shows the numbers of passengers on whale-watching tours in Iceland between 1995 and 2009, ranging from almost nil in 1995 to more than 120,000 in 2009. The dotted line shows, in percentage terms, whale-watching passengers as percentage of total number of international tourists in Iceland ranging from nil to 25% during the same time period. The diagram shows how the number of passengers has developed in relation to the time period when there was a pause in whaling in Iceland (1990–2002) and when scientific whaling (2002–) and commercial (2006–) whaling took place.

Two, seemingly contradictory, pieces of information are conveyed by Figure 8.1. In terms of total number of visitors, there has been a strong and steady growth throughout the period despite the resumption of scientific whaling in 2002 and commercial whaling in 2006. Referring to the typology of Higham and Lusseau (2007), it may be assumed that Iceland has been able to continue attracting 'Type 1' and 'Type 3' whale-watching tourists and it looks as if whaling has not had any severe negative impact on the whale-watching industry.

However, in relative terms, there are clear indications of impacts in 2002 (when scientific whaling resumed) and in 2006 (when commercial whaling resumed). After a continuous growth in relative importance for the tourism industry during 1995–2002, this growth came to a halt in 2002 and levelled

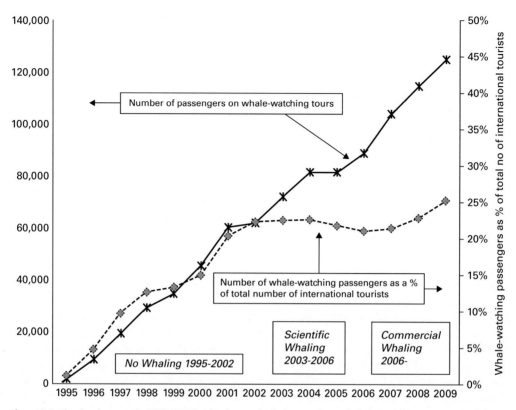

Figure 8.1 The development in 1995–2009 in absolute and relative numbers of whale-watching tourism in Iceland.

out. There were also signs of a relative decrease in importance in 2006. Referring again to the typology of Higham and Lusseau (2007), these results may indicate that 'Type 2' whale-watching tourists is an important market segment that Iceland has lost and that this loss has severely affected the whale-watching tourism industry in Iceland. There may, of course, be other explanations to these changes than the resumption of whaling, such as a lack of capacity in the whale-watching tourism sector, a global saturation of the demand for this type of tourism activity or more attractive services from competitors in other countries. These alternative explanations have not been investigated in this study.

Iceland resumed commercial whaling in 2006 and the full impact that whaling might have on the tourism industry cannot yet be fully estimated, but

it is assumed that tourists will react with objection, resulting in decreasing numbers of whale-watching participants which will impact the whale-watching tour operators' businesses. Whaling might result in tourists only watching some species in the future and this change will impact the whale-watching tour operators on a day-to-day basis. Altered behaviour of the whales has already been observed by the whale-watching tour operators. In the future, whale-watching tour operators may have to concentrate on watching other whale species, even other wildlife species, if Iceland continues whaling.

According to Conlin (2006), boycotting Iceland due to the resumption of whaling might hurt only the tourism industry with its whale-watching companies. Icelandic companies are concerned about the impact of whaling on the Icelandic tourism industry and therefore they are lobbying against it (Conlin, 2006). Parsons and Rawles (2003) found that 79% of whale-watchers in Iceland stated that they would boycott a destination that conducts whaling operations. In addition, 12.4% of visitors said that they would visit a country that conducts whaling; however, they would not participate in a whale-watching trip in this country. This result also confirms the tourists' attitudes of this study. One German tourist believes that 'it is a contradiction that Iceland conducts whale hunting and offers whale-watching tours at the same time.'

All the interviewed whale-watching tour operators are against all three different types of whaling and they do not believe that whale-watching and any type of whaling can coexist in the future. In contrast, the tourists accepted indigenous whaling more than commercial whaling. If commercial whaling continues, it will not only have a negative impact on the destination image of Iceland, but it might also have serious consequences for the whale-watching tourism industry, as commercial whaling is seen as the least-accepted form of whaling among tourists.

Female participants in the survey react more sensitively towards the different forms of whaling and the whaling situation in Iceland in comparison to male respondents. The conclusion can be drawn that female tourists' image of Iceland is likely to be more damaged. In the case study, tourists from Germany and Denmark were most opposed to the whaling situation in Iceland, which implies that these nationalities are influenced the most by the whaling situation. The study findings show that the tourists want to be informed about the whaling situation while attending a whale-watching trip. It can further be concluded that tourists who originate from whaling countries are the ones that are least influenced by the Icelandic whaling situation.

Although whale-watching and whaling are conducted at the same location, most tourists are not confronted with the whaling to the same extent as the whale-watching tour operators are. Most tourists see both activities only from a superficial

perspective and some of them are not even aware of whaling being conducted in Iceland. Tour operators, on the other hand, see both activities as interrelated as they are confronted with the implications of whaling more frequently and they also have deeper background knowledge about the topic.

The importance of tourism for the Icelandic economy is considerable and it is the fastest-growing sector in the Icelandic economy (Gudmundsson, 2007), but whale-watching may no longer be the fastest-growing tourism activity as, according to Gudmundsson (2007), it used to be before Iceland resumed whaling.

The results of this study suggest that since 2002 when Iceland resumed whaling, whale-watching is no longer the fastest-growing activity in the tourism industry. With these facts in mind, it appears problematic that the country attracts tourists with whale-watching as one of its major attractions at the same time as Iceland is hunting whales for scientific and commercial purposes.

Conclusion

The research question addressed in this study was how the whale-watching tourism industry in Iceland has been affected by the fact that the country allows whaling. Whale-watching tour operators all have negative attitudes towards the decision to resume whaling in Icelandic waters. They have noticed not only a drop in the number of customers, but also severe avoidance responses and changes in behaviour by the whales. Most of them prefer not to discuss the Icelandic policy with tourists onboard as it may have a negative impact on the their whale-watching experience.

Tourists seemed to be more ready to accept the situation, and a large majority would still consider visiting Iceland (83%) and also go on a whale-watching trip (71%) even when they are aware of the fact that Iceland allows whaling. Here it must be underlined that the tourist population studied consists of the tourists that came to Iceland in 2007 and thus the tourists who were strongly against visiting

Iceland are therefore less likely to be included in the population sampled.

Tourism statistics also show a strong growth in tourism in Iceland after 2002, when Iceland resumed whaling. The number of tourists participating in whale-watching activities has also grown considerably (as illustrated in Figure 8.1), but in relative terms, there is a noticeable change after 2002, when whale-watching no longer represents a growing tourism activity relative to other activities. Thus the relative growth came to a halt in 2002 when Iceland resumed whaling, and during the period 2001–2009 whale-watching levels out as an activity that 25% of the total number of international tourists visiting Iceland partakes in.

Acknowledgements

This chapter is based on a thesis from the master programme in Tourism and Hospitality Management at the School of Business, Economics and Law at the University of Gothenburg: 'Match Point: Watching versus Catching' by Beatrice Wende and Susanna Gothall available at: http://gupea.ub.gu.se/bitstream/2077/9617/1/2007_26.pdf.

REFERENCES

Altherr, S. (2003). *Iceland's whaling comeback preparations for the resumption of whaling.* Pro Wildlife. Retrieved from: http://www.wdcs.org/dan/publishing.nsf/c525f7df6cbf01ef802569d600573108/b2460680bc28d8f480256d4a0040d97b/$FILE/Iceland-report-english.pdf (accessed 14 September 2007).

Arctic Experience (2007). *Whale-watching in Iceland. Iceland whale-watching.* Retrieved from: http://www.arctic-experience.co.uk/whalewatching/tour.pdf (accessed 10 September 2007).

Baker, C.S., Perry, A. & Vequist, G. (1988). Humpback whales of Glacier Bay, Alaska. *Whalewatcher* Fall, 13–17.

Bejder, L., Dawson, S.M. & Harraway, J.A. (1999). Responses by Hector's dolphins to boats and swimmers in Porpoise Bay, New Zealand. *Marine Mammal Science* 15(3), 738–750.

Black, R. (2006). *Iceland begins commercial whaling.* BBC, 17 October 2006. Retrieved from: news.bbc.co.uk/1/hi/sci/tech/6059564.stm (accessed 28 November 2007).

CBS News (2003). *Whale of a debate in Iceland.* Retrieved from: http://www.cbsnews.com/stories/2003/06/11/world/main558254.shtml (accessed 11 October 2007).

Conlin, J. (2006). The resumption of whaling hurts Icelandic tourism. *New York Times*, 12 November 2006. Retrieved from: http://travel.nytimes.com/2006/11/12/travel/12transwhale.html (accessed 4 September 2007).

Constantine, R. (2001). Increased avoidance of swimmers by wild bottlenose dolphins (*Tursiops truncatus*) due to long-term exposure to swim-with-dolphin tourism. *Marine Mammal Science* 17(4), 689–702.

Edds, P.L. & MacFarlane, J.A.F. (1987). Occurrence and general behaviour of balaenopterid cetaceans summering in the Saint Lawrence Estuary, Canada. *Canadian Journal of Zoology* 65, 1363–1376.

EU Business (2009). *EU membership will kill whale hunting in Iceland: CEO* Retrieved from: http://www.eubusiness.com/news-eu/1245396721.64/ (accessed 8 August 2009).

Fisher, S. (2007). *Whale-watching in Iceland.* Whale and Dolphin Conservation Society. Retrieved from: http://www.wdcs.org/dan/publishing.nsf/c525f7df6cbf01ef802569d600573108/96d7234e5c1bc1d1802568f5004c62de/$FILE/iceland.pdf (accessed 14 September 2007).

Forsaetisraduneyti (2007). Prime Minister's Office. Retrieved from: http://eng.forsaetisraduneyti.is/minister/cv (accessed 12 October 2007).

Gudmundsson, S. (2007). Managing director at Gentle Giants, whale-watching tour company. Telephone interview, 7 November 2007.

Helgason, J.G. (2007). Comparison of two companies in tourism in Northern Iceland – Case studies of innovation systems. The Icelandic Tourism Centre, Háskólinn á Akureyri.

Higham, J.E.S. & Lusseau, D. (2007). Urgent need for empirical research into whaling and whale-watching. *Conservation Biology* 21(2), 554–558.

High North Alliance (2004). *Whale-watching – Recipe for economic disaster.* Retrieved from: http://www.highnorth.no/news/nedit.asp?which=336 (accessed 12 October 2007).

Hoyt, E. & Hvenegaard, G.T. (2002). A review of whale-watching and whaling with applications for the Caribbean. *Coastal Management* 30(4), 381–399.

Icelandic Ministry of Fisheries and Agriculture (2009). *Press Release from The Icelandic Ministry of Fisheries and Agriculture on Whaling,* 18 February. Retrieved from:

http://eng.sjavarutvegsraduneyti.is/news-and-articles/nr/9580 (accessed 8 August 2010).

Iceland Review Online (30 March, 2010). *New report says whaling is beneficial for Iceland.* Retrieved from: http://icelandreview.com/icelandreview/daily_news/?cat_id=16539&ew_0_a_id=360170 (accessed 21 July 2010).

IFAW (2003). A new threat at the IWC. *Our Shared World.* A quarterly publication. Retrieved from: http://www.ifaw.org/ifaw/dfiles/file-75.pdf (accessed 4 September 2007).

IFAW (2006). *Icelanders believe whaling damages tourism.* Press releases. Retrieved from: http://www.ifaw.org/ifaw.org/ifaw/general/default.aspx?oid=198522 (accessed 5 January 2008).

IWC (2001). Are whalers welcome at the IWC? *The International Harpoon.* No 1, 23 July 2001, London. Retrieved from: http://www.highnorth.no/iwc2001/harpoon/Harp1new.pdf (accessed 14 October 2007).

IWC (2006). *Total whales killed in whaling operations since the IWC Whaling Moratorium went into effect.* Retrieved from: http://assets.panda.org/downloads/totalwhaleskilled2006.pdf (accessed 17 September 2007).

Janik, V.M. & Thompson, P.M. (1996). Changes in surfacing patterns of bottlenose dolphins in response to boat traffic. *Marine Mammal Science* 12, 597–602.

MacGibbon, J. (1991). Responses of sperm whales (*Physeter macrocephalus*) to commercial whale watching boats off the coast of Kaikoura. Unpublished Report to the Department of Conservation. Christchurch, New Zealand: University of Canterbury.

Mattsson, B.M. (25 July, 2010). Det är Johannas tid på Island. *Göteborgs-Posten* 152(197), 18.

Nowacek, S.M., Wells, R.S. & Solow, A.R. (2001). Short-term effects of boat traffic on bottlenose dolphins, *Tursiops truncatus*, in Sarasota Bay, Florida. *Marine Mammal Science* 17(4), 673–688.

Parsons, E.C.M. & Rawles, C. (2003). The resumption of whaling by Iceland and the potential negative impact in the Icelandic whale-watching market. *Current Issues in Tourism* 6(5), 444–448.

Salvado, C.A.M., Kleiber, P. & Dizon, A.E. (1992). Optimal course by dolphins for detection avoidance. *Fishery Bulletin* 90, 417–420.

Siglaugsson, P. (2005). The whale meat market – study on current and possible markets and cost of operations in minke whaling. *GJ Financial Consulting June 2005.* Retrieved from: www.natturuverndarsamtok.is/myndir/Whale_meat_report.pdf (accessed 28 November 2007).

Sigursveinson, V. (2007). Operating director at Elding Whale-watching, whale-watching tour company. Telephone interview, 7 November 2007.

Sjóferdir Snorra EHF (2007). Sjóferdir Snorra EHF, whale-watching tour operator. Self-administrated questionnaire, 5 November 2007.

Steuer, K. (2005). *Scientific whaling in the 21st century. A report by WWF June 2005. WWF for a living planet. Science, profit and politics.* World Wide Fund for Nature. Gland. Switzerland. Retrieved from: http://assets.panda.org/downloads/wwfsciwhalingreportfinal.pdf (accessed 14 September 2007).

Valdemarsdottír, R. (2007). Representative of Seatours, whale-watching tour company. Telephone interview, 7 November 2007.

Williams, N. (2006). Iceland shunned over whale hunting. *Current Biology* 16(23), 975–976.

9

Green messengers or nature's spectacle

Understanding visitor experiences of wild cetacean tours

Heather Zeppel and Sue Muloin

Introduction

In western countries, whales and dolphins are iconic wildlife species and have been a key focus of marine conservation efforts since the 1970s. Social values based on conservation influence the type of benefits now sought from marine wildlife interactions, such as the trend towards non-consumptive viewing of wild cetaceans rather than killing whales or dolphins (Frohoff & Packard, 1995; Muloin, 1998; Bulbeck, 1999; Kellert, 1999; Hoyt, 2003; Parsons et al., 2003; Higham & Lusseau, 2004; Corkeron, 2006; Neves, 2010; Brakes & Simmons, 2011), or seeing wild instead of captive dolphins (Hughes, 2001; Bulbeck, 2005). These new environmental and amenity values of cetaceans as 'charismatic mega-fauna' have underpinned the rapid world-wide growth in whale- and dolphin-watching as a marine tourism activity (Hoyt, 2001; Orams, 2005; Cisneros-Montemayor et al., 2010). In 2008, over 13 million people went on whale-watching tours in 119 countries, generating income of US $2 billion in coastal economies (IFAW, 2009). The economic and conservation benefits of cetacean tours are supported by organizations such as the Pacific Whale Foundation, Whale and Dolphin Conservation, and Whales Alive.

Whale-watching is defined as 'the watching of any cetacean in the wild, an activity which is almost invariably conducted from a platform (e.g. ship, cliff or aeroplane)' (Warburton, 1999: 12) and as 'any commercial enterprise which provides for the public to see cetaceans in their natural habitat' (Warburton et al., 2001: 5). Swimming with humpback whales (*Megaptera novaeangliae*) in Tonga (Orams, 2001; Kessler & Harcourt, 2010), dwarf minke whales (*Balaenoptera acutorostrata* subspecies) in Australia (Birtles et al., 2002; Valentine et al., 2004), and wild dolphins (Amante-Helweg, 1996; Orams, 1997a; Hughes, 2001; Luck, 2003; Samuels et al., 2003; Blewitt, 2008; Zeppel, 2009; Draheim et al., 2010) are popular activities in selected locations where this is legally permitted. As a result, there is a growing body of research about visitor experiences of wild cetaceans.

This chapter reviews and evaluates benefits for participants on a range of guided whale and dolphin tours involving watching or swimming encounters, or feeding wild dolphins. The focus is on non-consumptive, free-ranging cetacean tourism where visitors can view, photograph, feed (dolphins), swim with, or assist in research on whales and dolphins in the wild. Most research on cetacean tourism addresses environmental impacts on whales and dolphins, industry compliance with codes of conduct and managing visitor interactions with wild cetaceans. This chapter reviews empirical studies that focus on tourist experiences of whale and dolphin tours in Australia, New Zealand, Scotland, western Canada/USA, Dominican Republic, Oman,

Whale-watching: Sustainable Tourism and Ecological Management, eds J. Higham, L. Bejder and R. Williams.
Published by Cambridge University Press. © Cambridge University Press 2014.

Taiwan and Hong Kong. Much of this research on cetacean tourism is site- or species-specific and limited to one type of encounter (i.e. land-, boat- or water-based). Therefore, there is a need for more systematic, in-depth evaluation of whale and dolphin tourism experiences and educational programmes to identify strategies that increase tourist benefits and knowledge, promoting environmental attitude shifts and lifestyle changes (Samuels *et al.*, 2003). This includes both on-site and longer-term conservation behaviours by visitors that benefit whales and dolphins and help to protect marine ecosystems. Hence, this chapter critically reviews the (1) psychological, (2) emotional, (3) educational, and (4) conservation benefits of visitor participation in cetacean tours. It begins with a discussion of the framework used to review this research about education programmes and visitor experiences on cetacean tours.

Framework for managing cetacean tourism experiences

This chapter follows the framework provided by Orams (1995a, 1995b, 1999) that measures positive changes in both tourists and the marine environment for effective management of cetacean tourism (see Table 9.1). Indicators of tourist benefits from marine animal encounters include enjoyment and learning contributing to pro-environmental attitude and behaviour changes, along with conservation benefits for marine environments and marine wildlife. Indicators of conservation benefits include tourists reducing wildlife disturbance, protecting habitats and aiding the viability of marine ecosystems. The framework by Orams (1999) was based on a previous model of experiential education in whale-watching tours in Hawai'i (Forestell, 1992, 1993). This model focuses on the cognitive states or learning of visitors using interpretation in marine settings to reduce impacts and promote pro-environmental behaviours on whale tours.

Luck (2003, 2008) evaluated the key role of interpretation on swim-with-wild-dolphin tours in New

Table 9.1 Indicators for managing cetacean tourism experiences (Orams, 1995a, 1999).

Tourist

1) Satisfaction/enjoyment (closeness/type of interaction)
2) Education/learning (knowledge)
3) Attitude/belief change
4) Behaviour/lifestyle change

Marine Environment
1) Minimize disturbance
2) Improve habitat protection
3) Contribute to the long-term health and viability of the ecosystem

Adapted from Zeppel and Muloin (2008a).

Zealand, based on models by Forestell (1993) and Orams (1997b). Orams (1999) extended the three-step experiential education sequence of Forestell (1993) into a four-stage sequence of indicators of desirable tourist outcomes from marine education programmes (i.e. satisfaction/enjoyment, education/learning, attitude/belief change, and behaviour/lifestyle change). A study of visitors feeding wild dolphins also adopted a model based on changing attitudes, beliefs, behaviours, and actions through wildlife interaction and interpretation (Mayes *et al.*, 2004). The following sections apply Orams' (1999) key indicators to review the psychological, emotional, education and conservation benefits of visitor participation in wild cetacean tours.

Visitor satisfaction with wild cetacean encounters

A variety of studies have addressed visitor satisfaction with whale-watch experiences in the wild. Seeing whales, their proximity, and whale behaviours are factors determining visitor satisfaction with whale-watching tours, followed by other marine wildlife, the coastal scenery or boat trip, and learning about whales (Duffus & Dearden, 1993; Muloin, 1998; Foxlee, 2001; Andersen & Miller, 2006). The provision of whale information and interpretation is also important. Experiential aspects, such as hearing the Orca (*Orcinus orca*) 'blow', also heightened visitor satisfaction (Andersen & Miller, 2006). Seeing and feeding dolphins, natural scenery, socializing, and a learning experience all contributed to visitor enjoyment of shore-based interactions with wild dolphins in Australia (Mayes *et al.*, 2004; Smith *et al.*, 2006). Overall, seeing whales, being close to nature, the natural habitat and animal behaviour were important for all wild cetacean encounters.

Visitor satisfaction from swimming with dwarf minke whales (*Balaenoptera acutorostrata* subspecies) included closeness, total interaction time, and the number of whales observed during encounters. The best elements were the minke whales (18%) and minke whale interactions (22%), followed by diving (15%) and seeing other marine life (10%) (Valentine *et al.*, 2004). In contrast, Orams (2000) found proximity to the whales (4%) was not a major part of visitor satisfaction compared to whale behaviours and boat features. Muloin (1998) reports visitor satisfaction (78% novices) mainly related to seeing whales in the wild and minimizing impacts through boat approach regulations of staying 100 m away from whales. Furthermore, Muloin (1998) noted a gender difference where women reported a higher level of satisfaction and were more emotional in their responses to whales than were men. Other studies found social aspects, such as spending time with family or friends, added to visitor satisfaction on boat tours viewing whales (Andersen & Miller, 2006) and at wild dolphin feeding sites (Mayes *et al.*, 2004).

A survey at Bunbury in Western Australia found tourists watching dolphins from a boat saw more dolphins and were more satisfied than those swimming with dolphins. Swimmers were unhappy with visibility underwater (37%, 3 m at best), cold water and weather (30%), and lack of proximity to dolphins (17%), with half of those where dolphins came within five metres stating this was still too distant (O'Neill *et al.*, 2004). Tour operation presentations of dolphins as 'sexy' or maternal also

improve tourist satisfaction with dolphin encounters (Besio *et al.*, 2008). The type of in-water dolphin encounter method such as snorkel lines, free swimming, underwater scooters, and boom nets may also affect visitor satisfaction (Zeppel, 2009).

Psychological and emotional benefits of wild cetacean encounters

A body of existing research addresses a range of psychological benefits that visitors may derive from viewing whales. Watching humpback whales (*Megaptera novaeangliae*) at Hervey Bay in Australia provided personal benefits for visitors such as thrills and excitement (whales), tranquility and peacefulness (nature), inspiration, learning and relaxation. Escaping from normal life and enjoying a special experience with family and other people interested in whales were other key benefits (Muloin, 1998). On killer whale or orca (*Orcinus orca*) tours, the main visitor benefits were seeing resident orcas (75%); the boat ride and coastal scenery (66%); learning about marine wildlife (38–42%); and spending time with family and friends (14%) (Andersen & Miller, 2006). The learning benefits of information presented about whales were secondary to the 'mood' benefits of tourists directly experiencing cetaceans and nature (Milstein, 2007, 2008).

Recent studies emphasize the role of emotions and experience intensity in animal encounters (Curtin, 2005; Milstein, 2007, 2008). These studies report intense personal and emotional experiences of visitors with whales and dolphins through close encounters and mediated interactions. The anthropomorphism of whales and dolphins by tour guides and visitors includes the attribution of human emotions such as 'love', along with intelligence and maternal care, during wild cetacean encounters (Bulbeck, 2005; Peace, 2005; Servais, 2005; Besio *et al.*, 2008). There is a recent trend in marine wildlife tourism from viewing marine mammals onboard boats or from land to more active in-water encounters (Orams, 2002; Curtin, 2005). This desire for close personal encounters has prompted strong emotional reactions to wild cetaceans. Tourists on swim-with-dwarf-minke-whale (*Balaenoptera acutorostrata* subspecies) tours elicited emotional responses (10%) such as 'awed', 'humbled/enlightened', 'dream come true', and 'feel closer to whales/nature' (Birtles *et al.*, 2002). Whales approaching whale-watch boats also heightened the emotional affinity and a sense of privilege felt by visitors.

Tourists viewing southern right whales (*Eubalaena australis*) from the Nullarbor cliffs in South Australia felt closer to nature (82%) and spiritually uplifted (32%); while tourists interacting with dolphins at Monkey Mia (Western Australia) reported feelings of affection (50%), being protective (40%), and that the dolphins liked them (25%) (Bulbeck, 1999, 2005). Other general feelings elicited about dolphins at Monkey Mia were fondness/affection (84%), fun/pleasure (58%), peace/tranquility (47%), comradeship/oneness (32%), and a feeling of communication (24%) with the dolphins (Bulbeck, 2005). Hence, for many visitors on cetacean tours, 'the basis of the participants' enjoyment of the experience is aesthetic and emotional, rather than intellectual' (Neil *et al.*, 1996: 186).

Educational benefits of wild cetacean encounters

The educational benefits of whale and dolphin tours include visitor learning, knowledge and information presented about cetaceans and marine environments. These marine education programmes involve talks by tour guides, interpreters and rangers onboard boats or at shorelines; and visitor centres; displays; signs; and brochures. This information covers the biology, ecology and behaviours of cetaceans, best-practice guidelines for cetacean interaction, and threats to cetaceans. Visitors may learn about cetacean biology and marine conservation issues both onshore (Orams, 1995c; Muloin, 1998; Orams & Hill, 1998; Christensen *et al.*, 2009) and onboard whale- and dolphin-watching boats (Birtles *et al.*, 2002; Corkeron, 2004; O'Neill *et al.*,

2004; Andersen & Miller, 2006; Forestell & Kaufman, 2007; Milstein, 2007, 2008; Peake, 2011; Peake *et al.*, 2009; Porter & Kaufman, 2009). This section reviews the education/learning (knowledge), attitude/belief changes, and behaviour/lifestyle changes in visitors on wild cetacean tours.

Education and learning (knowledge)

Educational experiences were important for visitors on dwarf minke whale tours, on swim-with-dolphin tours, and other whale-watching tours. On the Great Barrier Reef, some 14% of visitors highlighted learning about dwarf minke whales (*Balaenoptera acutorostrata* subspecies), marine life on the reef, and the educational experience and research conducted about these whales (Birtles *et al.*, 2002). On whale-watching tours in Scotland, visitors learnt about minke whales (71%), that cetaceans were mammals (36%), and other features such as their anatomy, rarity, intelligence and social nature (8–12%). They also learnt about threats to whales from marine pollution (31%), the impacts of whaling (20%), whales being caught in nets (17%), and from overfishing (15%) (Warburton, 1999; Warburton *et al.*, 2001; Parsons *et al.*, 2003). However, Luck (2003) found that, while most visitors on swim-with-dolphin tours in New Zealand increased their knowledge of dolphins and wildlife (range of 66–69%), only 29% strongly agreed the dolphin tour was an educational experience. One dolphin operator did not have a guide onboard and most visitors wanted more interpretation about dolphins, the marine environment and threats to dolphins (Luck, 2003). In Taiwan, visitors on eco-labelled cetacean tours gained more knowledge of whales, were more likely to learn about marine issues, promote ecotours and support sustainable use of marine resources (Chen, 2011). A survey of 114 dolphin-watchers in Oman on tours with minimal interpretation found no significant difference in the level of knowledge about dolphins held by novice and experienced visitors, or awareness of dolphin watching guidelines (Ponnampalam, 2011).

Visitors on whale-watch tours also wanted more information about the marine environment (Foxlee, 2001). Whale-watching visitors in Queensland (Australia) preferred information about: marine species and their behaviour (98%); conservation issues (91%); and practical information about protecting marine life (94%); while only 26% wanted basic facts (Ballantyne *et al.*, 2009). A survey of 229 shore-based whale-watchers in Oregon (USA) found whale viewers had stronger biocentric values and greater awareness of their impacts on the marine environment and whales after speaking to an on-site guide (Christensen *et al.*, 2007). Whale-watch passengers in Hawai'i ($n = 488$) supported more stringent regulations for boats, sewage disposal at harbours, and on board education (Shapiro, 2006). Whale organizations (e.g. Pacific Whale Foundation, Whale and Dolphin Conservation and Whales Alive) strongly promote the educational and conservation benefits of whale-watching by raising public awareness of whales and marine conservation issues (WDC, 2013; Whales Alive, 2009; PWF, 2013).

In Queensland (Australia), visitor knowledge about dolphins increased by 81% at Tangalooma Resort on Moreton Island and by 47% at the small seaside town of Tin Can Bay after wild dolphin feeding talks (Mayes *et al.*, 2004). The educational benefits for visitors of the dolphin interpretation and feeding interaction programme at Tangalooma Resort have been well documented by Orams (1994, 1995b, 1995c, 1995d, 1996, 1997a, 1997b, 1999; Orams & Hill, 1998). The site includes a Dolphin Education Centre and a ranger giving nightly talks about dolphin biology and behaviour to both dolphin feeders and observers. Learning about the dolphins at Tangalooma motivated Australian tourists to protect marine environments, while Japanese tourists wanted to touch and physically interact with dolphins. Language barriers also impeded the Japanese visitors from understanding the dolphin-feeding programme or from adopting more environmentally responsible behaviours (Takei, 1998). Beasley (1997) found visitors on dolphin tours in Akaroa (New Zealand) and Hong Kong had short-term increases in their knowledge of marine areas.

The quality of interpretation provided about whales and dolphins also influenced conservation outcomes and other environmentally responsible behaviours as reported by visitors. For example, whale-watchers in southern New South Wales (Australia) did not tell others about whale conservation, as this was not covered in the interpretation. An educational brochure given to boat-based whale-watchers in 2003, however, increased visitor knowledge about threats to whales and awareness of whale conservation. Interpretation promoted stronger environmental appreciation in boat-based whale-watchers (Stamation *et al.*, 2007). Only about half of surveyed visitors on two dolphin-watching boats in Port Stephens, New South Wales (Australia) were highly satisfied with interpretation about conserving dolphins (52%) and their ocean environment (42%), indicating a need for improvement in this area. There was greater support for conservation of marine wildlife onboard the boat with a full interpretation of dolphins linked to conservation messages and behaviours (Mayes & Richins, 2009).

A study of interpretation delivered by tour guides onboard eight whale-watching boats in Hervey Bay (Australia) found less than 50% delivered whale conservation messages, while 25% encouraged conservation actions such as joining an environmental or wildlife group; and 12.5% mentioned the impacts of Japanese whaling (Peake, 2011). In addition, a 2005 survey of whale-watch visitors ($n = 1519$) on eight boats in Hervey Bay found that while 22% received general facts about whales, only 30% received a conservation message and 42% received no conservation messages at all from their experience (Peake *et al.*, 2009; Peake, 2011). Older visitors over 41 years of age received more conservation messages (75%) than did younger visitors (45%), while Australian visitors were more receptive (35%) than were international visitors (23%). Other key factors that positively affected conservation messages received by visitors were tour guides suggesting conservation actions, level of visitor satisfaction, and interpretation about whaling and marine conservation (Peake *et al.*, 2009). The Hervey Bay whale-watch operators mainly focused on visitor entertainment and did not see educating tourists about marine conservation issues as part of their role. However, a whale-watching education programme by the Queensland Parks Service expected tour operators to provide conservation messages about whales and the marine environment as part of their permit conditions (Peake, 2011).

In contrast, the Pacific Whale Foundation provides interpretation onboard seven whale-watching boats in Hawai'i based on modelling and reinforcement of desired conservation behaviours (e.g. volunteering) and environmental best practice (i.e. biodiesel, recycling) to minimize human impacts on marine areas (Forestell & Kaufman, 2007). A study of 29 other marine tourism operators in Hawai'i, however, including cetacean tours, found no interpretation delivered about marine degradation, threats to marine life, or suggestions for conservation actions. Interpretation on swim-with-dolphin tours in Hawai'i also did not address preservation of dolphins and their habitat (Wiener *et al.*, 2009). Therefore, visitor support for protection of cetaceans and marine areas depends on conservation education during the experience. However, there is great variability in commitment to and delivery of education programmes during wild cetacean encounters. There is much scope for improvement in the delivery of cetacean conservation messages and minimizing environmental impacts and this is critical to the future of whale-watching tourism.

Attitude and belief change

Several studies suggest that marine wildlife tours with a strong educational focus can change the environmental attitudes and beliefs of visitors. A number of empirical studies have sought to understand changes in the environmental attitudes of visitors on whale and dolphin tours (Table 9.2). On swim-with-dwarf-minke-whale (*Balaenoptera acutorostrata* subspecies) tours, 27% of tourists changed their attitudes to conservation, displaying a greater awareness of whales, marine life, whaling and other human impacts (Birtles *et al.*, 2002). In

Scotland, 46% of visitors thought that whale-watching had a positive impact mainly by increasing visitor awareness of whales (40%) and visitor concern for boat interference (33%) or noise (10%) affecting whale behaviours (Warburton *et al.*, 2001). In the USA, land-based whale-watchers (39%) were more concerned than boat tourists (21%) about the impacts of noise, boats and kayaks on killer whales (*Orcinus orca*) as it affected their own experience (Finkler & Higham, 2004; Higham & Hendry, 2008). A 2005 survey of 229 shore-based whale-watchers in Oregon (USA) at seven 'Whale Watching Spoken Here' sites with guides found that experienced whale viewers had stronger biocentric values and greater awareness of their impacts. Visitor awareness of the behavioural consequences of their daily actions on the marine environment and whales was higher after speaking to an on-site guide (Christensen *et al.*, 2007, 2009). A survey of whale-watchers in British Columbia, Canada, reported a stronger conservation orientation in terms of attitudes towards whale management among advanced (or experienced) whale-watchers compared to intermediate or novice groups (Malcolm & Duffus, 2008). Tourists' desire for close encounters, then, was matched by awareness of human impacts on whales.

Visitors on wild dolphin feeding tours at Tin Can Bay and Tangalooma Resort, Australia, felt more strongly about conservation (81%), the state of marine areas (66%) and helping out with conservation programmes (52%) after their wild dolphin experience. They also disagreed with dolphins in captivity (59%) and Indigenous people hunting dolphins (68% of Australians), while 9% also disagreed with the practice of feeding wild dolphins (Mayes *et al.*, 2004). Visitors on dolphin-watching tours in the Dominican Republic supported education provision (82%), conservation involvement by operators (65%) and minimizing impacts, although 57% wanted to swim with dolphins (Draheim *et al.*, 2010). Education, then, is a key element of managing tourist–wildlife interactions, with the outcome of changing environmental attitudes and the potential to reduce visitor impacts on wildlife and marine

Table 9.2 Changes in environmental or conservation attitudes.

Environmental/conservation attitudes	% Response
Dwarf minke whale tours QLD Australia (Birtles *et al.*, 2002)[a]	
Greater awareness/concern/appreciation of (marine) life/nature	10.3
Increased/reinforced conservation awareness	6.0
Greater awareness/concern/appreciation of whales	5.8
Greater awareness of whaling issues	1.9
Greater awareness (whale guidelines, ecotourism, impacts on whales)	1.8
Increased awareness of human impacts on marine life/nature	1.3
Increased awareness of coastal development & resource exploitation	0.4
Total – YES – Conservation Attitudes	27.7
Killer whale watching USA (Finkler & Higham, 2004)[b]	(land-based /boat-based)
Effects of noise on whales	74/ 54
Power boats placed in the path of whales	73/ 56
Disturbance of whales by (other) power boats	69/ 52
Impacts of kayaks approaching whales	27/ 18
Dolphin feeding QLD Australia (Mayes *et al.*, 2004)[c]	
Felt more strongly about conservation of the environment generally	81
Felt they could make more of a difference to the state of the (marine) environment	66–67
Felt more confident in assisting with conservation programmes	52
Disagreed with indigenous people hunting dolphins (international visitors/Australians)	30/ 68
Disagreed with keeping dolphins in aquariums	59
Disagreed with feeding wild dolphins	9

[a] = Represents total coded themes or elements ($n = 466$), not numbers of respondents. Questionnaires $n = 527$ from 52 trips on 5 dive boats.

[b] = Land-based whale-watch surveys ($n = 327$), boat-based whale-watch surveys ($n = 306$), total questionnaires ($n = 633$).

[c] = Survey of visitors at wild dolphin feedings, $n = 105$ questionnaires (54 at Tangalooma, 51 at Tin Can Bay). Adapted from Zeppel and Muloin (2008a).

areas (Orams, 1995b; Zeppel, 2008; Zeppel & Muloin 2008a, 2008b, 2008c). However, a key issue is sustaining these pro-environmental attitudes in visitors over time in the weeks, months and years after the experience of interacting with wild cetaceans. Sustaining these conservation beliefs over time depends on repeat encounters with wildlife, otherwise a pro-conservation attitude towards marine wildlife may diminish. Cetacean education, then, needs to create longer-term sustainable differences in attitude and beliefs leading to behavioural or lifestyle changes in visitors.

Behaviour and lifestyle changes

Some studies suggest that marine wildlife tours with a strong educational focus can create longer-term behavioural or lifestyle changes in visitors. These behavioural changes include minimizing human impacts on wildlife, donating money to wildlife conservation, and direct actions supporting environmental issues (Moscardo *et al.*, 2004). Table 9.3 presents some reported changes in the personal behaviour of visitors after a wild dolphin feeding encounter. Visitors on wild dolphin feeding tours

Table 9.3 Changes in visitor behaviour after a wild dolphin tour, Australia.

Behaviour or lifestyle changes	% Response
Dolphin feeding QLD (Mayes et al., 2004)[a]	
Remove beach litter that could harm dolphins	75
Assist in the protection of whales and dolphins where possible	64
Decrease their contribution to water pollution	60
Tell others about the need to care more for our oceans and wildlife	56
Become more involved in marine conservation issues	23
Touching dolphins is OK (Tin Can Bay/Tangalooma)	25/ 3
Dolphin education and feeding Tangalooma QLD (Orams, 1996)[b]	(DEP visitors/control)
Get more information on dolphins	41/ 13
Picked up rubbish from beaches	65/ 44
Become more involved in environmental issues	32/ 6
Made a donation to an environmental organization	23/ 11

[a] = Definite responses only, $n = 105$ questionnaires (54 Tangalooma, 51 Tin Can Bay) for visitors feeding wild dolphins.
[b] = DEP = Dolphin Education Program visitors, $n = 104$, Control group (pre-DEP) $n = 110$ (phone interviews).
Adapted from Zeppel and Muloin (2008a).

at Tin Can Bay and Tangalooma (56–75%) stated they would remove beach litter, assist in protecting dolphins, decrease water pollution and tell others about caring for oceans and marine life (Mayes *et al.*, 2004). However, there is often a gap between stated intentions and implementation of conservation behaviours by visitors (Stamation *et al.*, 2007). In regard to onsite visitor behaviours that impact on marine wildlife, 25% of Tin Can Bay's unmonitored visitors thought it was acceptable to touch wild dolphins compared to 3% of visitors at Tangalooma that

learnt about human impacts on dolphins at regulated feeding talks. This reflects the level of cetacean education provided and the monitoring of visitors feeding wild dolphins (Mayes *et al.*, 2004).

At Tangalooma Resort, follow-up phone interviews conducted with visitors two to three months after the dolphin experience also found longer-term changes in environmental behaviours. Visitors that participated in the dolphin education programme actively sought dolphin information, picked up beach rubbish, were more involved in environmental issues, and donated money to environmental organizations (Orams, 1996, 1997b). The Tangalooma education programme also reduced inappropriate onsite behaviours, such as visitors touching dolphins (Orams & Hill, 1998). A US study reported that American college students with higher education or knowledge had a more scientific attitude about dolphins, were more environmentally friendly, and stated they were less likely to engage in harmful activities such as touching, feeding or boating near wild dolphins (Barney *et al.*, 2005). Many studies measure intention to act rather than actual conservation behaviours and rely on self-reporting by visitors, with no longer-term studies of changes.

Conservation benefits of wild cetacean encounters

The conservation benefits gained from wildlife tourism include (1) wildlife management and research; (2) financial support for conservation of species; (3) socioeconomic benefits; and (4) education of visitors potentially leading to more conservation-focused behaviour and support (Higginbottom & Tribe, 2004). The conservation outcomes for marine wildlife and marine environments aim to: (1) minimize disturbance; (2) improve habitat protection; and (3) contribute to the long-term health and viability of ecosystems (Orams,1995a, 1999). Several studies have addressed conservation appreciation and pro-environmental actions by visitors as a consequence of wild cetacean tours

Table 9.4 Conservation actions reported by cetacean tourists.

Conservation appreciation or actions	% Response
Humpback whale-watching, QLD, Australia (Wilson & Tisdell, 2003)[a]	
Take more action to protect whales in Australia	80
Complete worldwide ban on whaling	78
Report stranding of whales and injured/mistreated whales	73
Whale-watching, Scotland, UK (Rawles & Parsons, 2005)[b]	
Regularly recycled items	83
Bought comestic/hygiene items not tested on animals	73
Used energy-saving light bulbs	60
Members of environmental/animal welfare organizations	47
Regularly purchased organic or environmentally friendly products	46
Used other energy-saving devices	42
Voluntary work for environmental/animal welfare organizations	27
Humpback whale-watching, NSW, Australia (Stamation *et al.*, 2007)[c] (Land/Boat)	
Donate to and/or be actively involved in an environmental group	17/23
Recycle	43/48
Choose household products better for the environment	41/48
Avoid putting oil, fat, paint or turpentine down sink or toilet	47/50
Avoid putting litter or detergents into gutters or storm water drains	46/49
Use alternatives to plastic bags when grocery shopping	40/43
Whale-watching/wildlife visitors, QLD, Australia (Ballantyne *et al.*, 2009)	
Recycling, conserving water and energy	80
Purchase environmentally friendly products, pick up other people's litter	37
Participate in clean-up, donate money, voluntary environmental work	6

[a] = 702 questionnaires from whale-watchers in Hervey Bay (Qld).
[b] = 236 questionnaires on whale-watch boats in western Scotland.
[c] = 1037 questionnaires from 6 whale-watching boats and 1569 surveys from land-based whale-watchers in southern NSW.
  Adapted from Zeppel and Muloin (2008a).

(Table 9.4). These studies generally report that visitors on whale-watching tours and at wild dolphin feeding programmes often exhibit a higher level of support for marine conservation in comparison to the general public (Wilson & Tisdell, 2003; Mayes *et al.*, 2004).

Visitors on whale-watching tours in Hervey Bay ($n = 702$) indicated that they would do more to protect whales (80%); supported a global ban on commercial whaling (78%); and would report (73%) stranded whales (Wilson & Tisdell, 2003). The conservation actions of visitors on whale-watching tours in Scotland ($n = 236$) included regular recycling (83%); purchasing products not tested on animals (73%); using energy saving bulbs (60%) and other energy-saving devices (42%); and buying organic/environmentally friendly products (46%). They were also members of (46%), or did voluntary work (27%) for, environmental or animal welfare organizations (Rawles & Parsons, 2005). At Mull in western Scotland, whale-watch visitors ($n = 183$) were more likely to be involved in wildlife-related conservation activities (91%), including volunteer work (18%), and to be members of wildlife and environmental organizations (58%) than other general visitors (Warburton, 1999). In Australia, 30% of visitors on swim-with-dwarf-minke-whale (*Balaenoptera acutorostrata* subspecies) tours were members of wildlife or conservation organizations, while 27% had greater awareness of marine impacts (Birtles *et al.*, 2002).

Ninety visitors surveyed on other Australian whale-watching cruises reported that they conserved water and energy, bought eco-friendly products and practised recycling. They exhibited similar conservation actions up to four months after the whale tour (Ballantyne *et al.*, 2006). Around 70–77% of whale-watch visitors supported conservation education, especially practical ways to help protect wildlife (Ballantyne *et al.*, 2009). Surveys of land-based ($n = 1569$) and boat-based whale-watchers ($n = 1037$) in southern New South Wales (Australia) also recorded 40–50% response rates for environmentally friendly activities such as recycling, proper disposal of liquid wastes, not using plastic bags, and choosing less harmful household products (Stamation *et al.*, 2007). There was a lower response for active involvement or donating to conservation groups (17–23%), although visitors with a higher environmental rating were more likely to support environmental groups. Visitors on boat-based whale-watch trips ($n = 1037$) were also more likely to pick up litter (59%), tell people general facts about whales (70%) and inform others about whale conservation (48%). Boat-based whale-watchers also told others more frequently about whales than land-based watchers (Stamation *et al.*, 2007). However, a follow-up survey six to eight months later of boat and land-based whale-watchers found no overall changes in performing environmentally friendly behaviours as originally stated: 18% stayed the same, 37% increased and 45% decreased (Stamation *et al.*, 2007).

Another study found whale-watch visitors and other wildlife tourists in Queensland (Australia) also supported conservation activities such as recycling, conserving water and energy (80%); picking up litter; and buying green products (37%); with least commitment to donating money, clean up days, or environmental volunteer work (6%). All of these wildlife tourists wanted practical information about aiding marine conservation rather than simply general information on conservation issues (Ballantyne *et al.*, 2009). A 2005 survey of 229 shore-based whale-watchers in Oregon (USA) found whale viewers that talked with on-site guides had greater awareness of human threats to whales and how recycling or reducing pollution can help to protect whales. While agreeing it was important to protect marine areas and whales, few visitors volunteered or donated to environmental causes, yet they did talk to others about marine issues (Christensen *et al.*, 2007, 2009). Consequently, whale organizations support whale tourism as it profiles threats to whales and fosters support for conservation (Hoyt, 2001; IFAW, 2009).

On the other hand, an observational study of 29 marine tourism operators in Hawai'i (USA), including whale and dolphin tours, found that some companies implementing 'green' practices were still waste dumping from holding tanks ($n = 16$, 55%), removing or harassing marine life ($n = 5$, 17%),

or dumping food scraps overboard ($n = 3$, 10%). For example, spinner dolphins (*Stenella longirostris*) were disturbed while resting in shallow bays, numerous boats encircled dolphins, or bow-riding dolphins were steered towards tourists in the water. Tour boats also crowded or pursued humpback whales (*Megaptera novaeangliae*). Only 14% ($n = 4$) of companies recycled materials while just 10% ($n = 3$) collected donations for marine conservation (Wiener *et al.*, 2009). Benefits are limited through these 'depreciative practices' by tour operators and lack of conservation education by guides. However, the Pacific Whale Foundation actively promoted conservation actions and best practices onboard its seven whale-watching boats based in Hawai'i (Forestell & Kaufman, 2007; PWF, 2012).

Dolphin-feeding visitors at Tangalooma (Australia) also recorded a strong commitment towards informing others about conservation (61%); using energy saving devices (44%); donating time/money to wildlife conservation or environmental organizations (22–28%); and joining a dolphin or mammal stranding group (17%). The Tangalooma visitors had a far higher response to all conservation actions, particularly donating time or money or joining organizations, than Tin Can Bay visitors. Almost two-thirds of dolphin visitors (63%) at Tangalooma and Tin Can Bay, however, were not currently involved in environmental organizations or activities (Mayes *et al.*, 2004). Visitors ($n = 195$) on dolphin-watching boats in Port Stephens (NSW) also expressed high to very high support for pro-environmental actions such as removing harmful litter from oceans (60%) and beaches (51%); decreasing personal water pollution (54%); assisting in protection of dolphins (49%); and telling others about caring for marine areas and wildlife (49%). There was moderate support for actions requiring time, money, or effort such as donating money to conservation organizations (37%); getting involved in conservation issues (33%); donating time to wildlife conservation (31%); helping stranded mammals (29%); and joining a conservation group (28%) (Mayes & Richins, 2009).

Conservation messages delivered on guided tours, brochures, and displays were evaluated at

12 marine wildlife attractions in New Zealand (Higham & Carr, 2002, 2003). The two dolphin tours and one whale tour highlighted environmental issues such as hunting and/or protection of marine mammals; the Southern Ocean Whale Sanctuary; recreational set netting killing dolphins; and Maori environmental values. Introduced species, predator eradication and wildlife recovery programmes were also covered. Some 54% of visitors on Dolphin Watch Marlborough, where tourists could participate in collecting and recording data about dolphin sightings and behaviours, stated their wildlife experience had affected their environmental values and actions. Half of all visitors on five tours read conservation publications about dolphins. Interviews with visitors found 30% believed tour operators needed to promote conservation (Higham & Carr, 2002, 2003).

In sum, personal encounters with wild cetaceans linked with on-site education programmes were more likely to generate conservation appreciation or actions. Research evidence indicates that not only are there short-term educational benefits for most visitors, but the pro-environmental attitudes/behaviours do continue several months later (Ballantyne et al., 2006, 2009; Moscardo et al., 2004; Rawles & Parsons, 2005; Warburton, 1999; Wilson & Tisdell, 2003). Nevertheless, changes that require more effort or active intervention such as volunteering and donating money are less likely outcomes in the longer term for many visitors (Christensen et al., 2009; Hughes, 2013; Mayes et al., 2004; Mayes & Richins, 2009; Rawles & Parsons, 2005; Stamation et al., 2007).

Managing visitor experiences on wild cetacean tours

This chapter has reviewed a range of psychological, educational, and conservation benefits for visitors on whale and dolphin tours. The on-site benefits of increased understanding or emotional responses to wild cetacean encounters can lead to off-site benefits such as greater environmental awareness,

supporting nature conservation work and protecting endangered species. Empirical studies of cetacean tourism experiences were assessed against the framework provided by Orams (1995a, 1999), measuring positive changes in both tourists and the marine environment for effective management of cetacean tourism operations and sites. Tourist benefits from cetacean encounters include enjoyment and learning contributing to pro-environmental attitude and behaviour changes, and the longer-term intention to engage in conservation actions that benefit cetaceans and marine environments. Therefore, cetacean tours with a strong educational focus and conservation programme can create attitude, behaviour or lifestyle changes in visitors (Birtles et al., 2002; Ballantyne et al., 2006, 2009). It can also enhance wildlife guardianship (Geneste, 2005), when tour guides link whale and dolphin conservation with marine habitat protection (Milstein, 2008).

Therefore, this review of visitor benefits from guided encounters with wild cetaceans supports the framework of indicators developed by Orams (1999) for managing marine tourism experiences through interpretation leading to improved outcomes for visitors and the marine environment. Key factors for tourist enjoyment and satisfaction are seeing wild cetaceans; their proximity and behaviours; the natural setting; other marine wildlife; and learning about marine life. The psychological benefits of visitors interacting with wild cetaceans include relaxation or stimulation, mood benefits or positive feelings, and learning benefits (Muloin, 1998). The learning benefits obtained from information about cetaceans reinforced the emotional benefits of directly experiencing whales and dolphins in their natural habitat (Milstein, 2007, 2008). Furthermore, there is some evidence that personal benefits and visitor satisfaction with wild cetacean encounters differ according to gender, age or previous experience (Muloin, 1998; Christensen et al., 2007, 2009; Malcom & Duffus, 2008; Peake et al., 2009) as well as the type of wildlife encounter or platform such as boat-based, land-based, or in-water encounters with whales or dolphins (Finkler, 2001; Finkler &

Higham, 2004; Birtles *et al.*, 2002; Finkler & Higham, 2004; Stamation *et al.*, 2007; Higham & Hendry, 2008).

In addition, quality educational experiences are important for visitors to increase their short-term knowledge of whales and dolphins. Cetacean tours with a strong educational focus changed or affected the pro-environmental attitudes, beliefs and behaviour of many visitors (Birtles *et al.*, 2002; Ballantyne *et al.*, 2006, 2009; Orams, 1996; Peake *et al.*, 2009; Rawles & Parsons, 2005; Stamation *et al.*, 2007; Warburton, 1999; Wilson & Tisdell, 2003). Visitors swimming with dwarf minke whales (*Balaenoptera acutorostrata* subspecies) reported that this experience influenced their conservation attitudes (27%), along with educational (14%) and other personal (10%) benefits (Birtles *et al.*, 2002). On whale and dolphin tours with quality educa-tion programmes, tourists changed their attitudes to conservation, displaying a greater knowledge of cetaceans and awareness of threats to marine life. Visitors interacting with dolphins at Tanga-looma also adopted short-term pro-environmental behaviours (up to 4 months later) such as clean-ing up beaches, recycling, and donating money to wildlife groups (Orams, 1996). Close proxim-ity to whales or dolphins during in-water encoun-ters or shore-based feeding interactions with dol-phins magnified these environmental and personal benefits. The level or intensity of the encounter with wild cetaceans needed to change tourist atti-tudes was linked to direct, close contact with ani-mals more than passive viewing from a boat or on land.

The thrilling and exciting aspects of seeing wild cetaceans, however, may overwhelm other conser-vation messages with 'people mistaking the mes-senger for the message, or missing the nature for the whale' (Milstein, 2007: 112). The quality of interpretation about cetaceans, thus, does influence conservation outcomes and other environmentally responsible behaviours as reported by visitors (Bal-lantyne *et al.*, 2006, 2009; Christensen *et al.*, 2007, 2009; Mayes *et al.*, 2004; Mayes & Richins, 2009; Muloin, 1998; Orams, 1996; Peake, 2011; Peake *et al.*,

2009; Rawles & Parsons, 2005; Stamation *et al.*, 2007; Warburton, 1999; Wilson & Tisdell, 2003). A key find-ing of this chapter is that the quality of these edu-cation programmes delivered during wild cetacean encounters is critical to visitor satisfaction and longer-term conservation benefits. In many cases, however, the quality is lacking, or cetacean conser-vation and environmental actions are not included or reinforced enough to visitors.

These psychological, educational, and conserva-tion benefits for visitors on whale and dolphin tours depend on sound management of cetacean encounters and interpretation programmes that integrate knowledge with the emotional aspects of observing cetaceans in the wild. Visitor learning for fun and enjoyment during leisure activities is an important part of tourism experiences and in fostering environmentally sustainable attitudes and behaviour (Ballantyne & Packer, 2005). The bene-fits for participants on wild cetacean tours are real-ized when the affective (emotional) benefits and excitement of seeing whales and dolphins are inte-grated with the cognitive (educational) benefits of learning. Key aspects are the experiential, partic-ipatory, and situational contexts of wild cetacean tours along with tour guides presenting conserva-tion issues and actions to visitors (Peake *et al.*, 2009). Visitor interactions with wild cetaceans that involve making personal connections in a learning context provide a range of psychological/mood and educational benefits. In-water encounters also pro-vide physical and emotional benefits from swim-ming, snorkelling or diving with whales or dol-phins where this is legally permitted. Cetacean tours that increase both environmental awareness and positive feelings, then, are more likely to gener-ate environmental actions resulting in conserva-tion benefits for whales, dolphins and the marine environment.

Sustainable management of wild cetacean tourism, then, requires integrated techniques for conservation education, research and vessel man-agement/interaction regulations (Higham & Bejder, 2008; Stamation, 2008; Higham *et al.*, 2008; Parsons & Rose, 2009; Scarpaci & Parsons, 2012). Research on the critical role of conservation education in wild cetacean encounters is vital as 'sustainable tourism depends on sustainable education' (Andersen & Miller, 2006: 111).

This chapter focused on tourist experiences of free-ranging whales and dolphins mainly in western countries. There is a need for further cetacean tourism research that investigates social and environmental values and other cultural perspectives of whales and dolphins, such as by Asian tourists. The personal benefits of wild cetacean encounters for Asian visitors may differ from benefits sought by western tourists (Takei, 1998; Choi, 2010; Chen, 2011). The environmental attitudes of whale-watch visitors in countries that operate both whale-watching tours and commercial or subsistence whaling such as in Iceland, Norway, Japan, South Korea, Tonga and the Caribbean, need further investigation (Hoyt, 1993; Ris, 1993; Freeman & Kellert, 1994; Orams, 2001; Hoyt & Hvenegaard, 2002; Parsons & Rawles, 2003; Herrera & Hoagland, 2006; Higham & Lusseau, 2007, 2008; Parsons & Draheim, 2009; Choi, 2010). There is also limited research and education about the potential impacts of global climate change on the occurrence and distribution of whales (Lambert *et al.*, 2010). The psychological and personal benefits for staff and operators of wild cetacean tours, driving their involvement in this activity, along with their conservation attitudes, behaviour and consequent influence on education programmes also require further investigation (Geneste, 2005). In addition, longer-term studies need to measure ongoing actual conservation actions of visitors one to five years after wild cetacean interactions, beyond self-reported intentions to act environmentally. While the wildlife experience itself in a scenic natural area may heighten visitor concern and appreciation for cetaceans, behavioural changes may not always follow. A more in-depth evaluation of the content, delivery and impact of interpretation programmes on visitor benefits from wild cetacean tours will identify attitude shifts and ongoing conservation behaviours that ultimately benefit whales and dolphins.

Conclusion

Close personal encounters with cetaceans in the wild provide a range of psychological, emotional, educational and conservation benefits for visitors. These mediated encounters can offer the potential to motivate visitors to respect marine life; foster environmentally responsible attitudes and behaviours; and benefit marine conservation. Marine interpretation programmes that highlight cetacean biology and human impacts also influence visitor attitudes, beliefs and conservation outcomes. Linking affective and cognitive responses to experiencing wild cetaceans increases environmental awareness; changes visitor attitudes; modifies intentions to act pro-environmentally; and fosters conservation appreciation and actions by marine tourists. Personal benefits for visitors depend on the intensity and frequency of tourist encounters with cetaceans as well as the type of learning experience provided by operators. This chapter highlighted the critical importance of interpretation/education programmes delivered to visitors during encounters with wild cetaceans to modify both on-site behaviours and to influence longer-term conservation outcomes for marine wildlife. The challenge ahead lies in achieving demonstrated long-term changes in the way people think and act in relation to the marine environment generally and cetaceans in particular.

Visitors also differ in their desired mix of psychological, educational and conservation benefits from encounters with wild cetaceans, with important contrasts identified in this chapter based on age, gender, platform (i.e. land- or boat-based), type of encounter (i.e. watching, feeding or swimming) and level of previous experience (i.e. novice versus repeat visitor). Experienced whale-viewers had stronger biocentric values and greater overall awareness of their impacts on whales, while novice visitors supported whale-watch regulations to minimize impacts from boats. Therefore, visitor benefits and potential dissatisfaction with wild cetacean encounters need to be carefully managed by the operators of marine wildlife tourism experiences.

The thrilling aspects of cetaceans as 'nature's spectacle' need to be matched by their growing role as 'green messengers' for marine conservation. Wild cetacean tours need to engage visitors and deliver effective conservation messages about marine life and ecosystems, while also cautiously managing the visitor desire for close interactions with whales and dolphins.

REFERENCES

Amante-Helweg, V. (1996). Ecotourists' beliefs and knowledge about dolphins and the development of cetacean ecotourism. *Aquatic Mammals* 22: 131–140.

Andersen, M.S. & Miller, M.L. (2006). Onboard marine environmental education: Whale watching in the San Juan Islands, Washington. *Tourism in Marine Environments* 2(2): 111–118.

Ballantyne, R. & Packer, J. (2005). Promoting environmentally sustainable attitudes and behavior through free-choice learning experiences: What's the state of the game? *Environmental Education Research* 11(3): 281–295.

Ballantyne, R., Packer, J. & Hughes, K. (2006). Designing 'Jonah' experiences: Developing visitor conservation learning through whale watching. In: *Conference Program: The 2nd Australian Wildlife Tourism Conference, 13–15 August 2006*. Perth, WA: Promaco Conventions, p. 28.

Ballantyne, R., Packer, J. & Hughes, K. (2009). Tourists' support for conservation messages and sustainable management practices in wildlife tourism experiences. *Tourism Management* 30(5): 658–664.

Barney, E.C., Mintzes, J.J. & Yen, C.F. (2005). Assessing knowledge, attitudes, and behavior toward charismatic megafauna: The case of dolphins. *The Journal of Environmental Education* 36(2): 41–55.

Beasley, I. (1997). Marine mammal tourism: Educational implications and legislation. Diploma of Wildlife Management Thesis, University of Otago, New Zealand.

Besio, K., Johnston, L. & Longhurst, R. (2008). Sexy beasts and devoted mums: Narrating nature through dolphin tourism. *Environment and Planning A* 40: 1219–1234.

Birtles, A., Valentine, P., Curnock, M., Arnold, P. & Dunstan, A. (2002). *Incorporating Visitor Experiences into Ecologically Sustainable Dwarf Minke Whale Tourism in the Northern Great Barrier Reef*. Townsville: CRC Reef Research Centre.

Blewitt, M. (2008). Dolphin–human interactions in Australian waters. *Australian Zoologist* 34: 197–210.

Brakes, P. & Simmons, M.P. (Eds) (2011). *Whales and Dolphins: Cognition, culture, conservation and human perceptions*. London: Routledge.

Bulbeck, C. (1999). The 'nature dispositions' of visitors to animal encounter sites in Australia and New Zealand. *Journal of Sociology* 35: 129–148.

Bulbeck, C. (2005). *Facing the Wild: Ecotourism, conservation and animal encounters*. London: Earthscan.

Chen, C.L. (2011). From catching to watching: Moving towards quality assurance of whale/dolphin watching tourism in Taiwan. *Marine Policy* 35: 10–17.

Choi, M.A. (2010). Whale-watching or whaling-watching? Contradictory perceptions of whale-watching tourism in Jangsaengpo, South Korea. Masters Thesis. King's College, London.

Christensen, A., Rowe, S. & Needham, M.D. (2007). Value orientations, awareness of consequences, and participation in a whale watching education program in Oregon. *Human Dimensions of Wildlife* 12(4): 289–293.

Christensen, A., Needham, M.D. & Rowe, S. (2009). Whale watchers' past experience, value orientations, and awareness of consequences of actions on the marine environment. *Tourism in Marine Environments* 5(4): 271–285.

Cisneros-Montemayor, A.M., Sumaila, U.R., Kaschner, K. & Pauly, D. (2010). The global potential for whale watching. *Marine Policy* 34(6): 1273–1278.

Corkeron, P.J. (2004). Whale watching, iconography, and marine conservation. *Conservation Biology* 18(3): 847–849.

Corkeron, P.J. (2006). How shall we watch whales? In D.M. Lavigne (Ed.), *Gaining Ground: In pursuit of ecological sustainability*. Guelph, ON: IFAW, pp. 161–170.

Curtin, S. (2005). Nature, wild animals and tourism: An experiential view. *Journal of Sustainable Tourism* 4(1): 1–15.

Draheim, M., Bonnelley, I., Bloom, T., Rose, N. & Parsons, E.C.M. (2010). Tourist attitudes towards marine mammal tourism: An example from the Dominican Republic. *Tourism in Marine Environments* 6(4): 175–183.

Duffus, D.A. & Dearden, P. (1993). Recreational use, valuation and management of killer whales (*Orcinus orca*) on Canada's Pacific coast. *Environmental Conservation* 20(2), 103–117.

Finkler, W. (2001). The experiential impact of whale watching: Implications for management in the case of the San Juan Islands, USA. Unpublished Master of Science Thesis, Department of Marine Science, University of Otago, New Zealand.

Finkler, W. & Higham, J. (2004). The human dimensions of whale watching: An analysis based on viewing platforms. *Human Dimensions of Wildlife* 9(2): 103–117.

Forestell, P.H. (1992). The anatomy of a whale watch: Marine tourism and environmental education. *Journal of the National Marine Educators Association* 10: 10–15.

Forestell, P.H. (1993). If leviathan has a face, does Gaia have a soul? Incorporating environmental education in marine eco-tourism programs. *Ocean & Coastal Management* 20: 267–282.

Forestell, P.H. & Kaufman, G.D. (2007). Speaking from experience: Whale and dolphin watching boats as venues for marine conservation education. In M. Luck, A. Graupl, J. Auyong, M.L. Miller & M.B. Orams (Eds), *Proceedings of the 5th Coastal & Marine Tourism Congress: Balancing Marine Tourism, Development and Sustainability*. Auckland: Auckland University of Technology, pp. 63–67.

Foxlee, J. (2001). Whale watching at Hervey Bay. *Parks and Leisure Australia* 4(3): 17–18.

Freeman, M.R. & Kellert, S.R. (1994). International attitudes to whales, whaling, and the use of whale products: A six country survey. In M.R. Freeman & U.P. Kreuter (Eds), *Elephants and Whales: Resources for whom?* Basel, Switzerland: Gordon & Breach Science Publishers, pp. 293–315.

Frohoff, T.G. & Packard, J.M. (1995). Human interactions with free-ranging and captive bottlenose dolphins. *Anthrozoos* 8(1), 44–57.

Geneste, L. (2005). Hong Kong Dolphinwatch: The evolution of an ecopreneurial business venture. In M. Schaper (Ed.), *Making Ecopreneurs: Developing sustainable entrepreneurship*. Aldersot: Ashgate, pp. 203–213.

Herrera, G.E. & Hoagland, P. (2006). Commercial whaling, tourism and boycotts: An economic perspective. *Marine Policy* 30(3), 261–269.

Higginbottom, K. & Tribe, A. (2004). Contributions of wildlife tourism to conservation. In K. Higginbottom (Ed.), *Wildlife Tourism: Impacts, management and planning*. Altona, Vic: Common Ground/Sustainable Tourism CRC, pp. 99–123.

Higham, J.E.S. & Bejder, L. (2008). Managing wildlife-based tourism: Edging towards sustainability. *Current Issues in Tourism* 11(1): 75–83.

Higham, J.E. & Carr, A.M. (2002). Ecotourism visitor experiences in Aotearoa/New Zealand: Challenging the environmental values of visitors in pursuit of pro-environmental behaviour. *Journal of Sustainable Tourism* 19(4): 277–294.

Higham, J.E. & Carr, A.M. (2003). Sustainable wildlife tourism in New Zealand: An analysis of visitor experiences. *Human Dimensions of Wildlife* 8: 25–36.

Higham, J.E.S. & Hendry, W.F. (2008). Marine wildlife viewing: Insights into the significance of the viewing platform. In J. Higham & M. Luck (Eds), *Marine Wildlife and Tourism Management: Insights from the natural and social sciences*. Wallingford: CABI, pp. 347–360.

Higham, J. & Lusseau, D. (2004). Ecological impacts and management of tourist engagements with marine mammals. In R. Buckley (Ed.), *Environmental Impacts of Ecotourism*. Wallingford: CABI, pp. 171–186.

Higham, J.E.S. & Lusseau, D. (2007). Urgent need for empirical research into whaling and whale watching. *Conservation Biology* 21(2): 554–558.

Higham, J.E.S. & Lusseau, D. (2008). Slaughtering the goose that lays the golden egg: Are whaling and whale-watching mutually exclusive? *Current Issues in Tourism* 11(1): 63–74.

Higham, J.E.S., Bejder, L. & Lusseau, D. (2008). An integrated and adaptive model to address the long-term sustainability of tourist interactions with cetaceans. *Environmental Conservation* 35(4): 294–302.

Hoyt, E. (1993). *Kujira Watching – Whales and Dolphins: Alive and being watched Japanese style*. Bath: Whale & Dolphin Conservation Foundation.

Hoyt, E. (2001). *Whale Watching 2001: Worldwide tourism numbers, expenditures and expanding economic benefits*. Yarmouth Port, MA: IFAW.

Hoyt, E. (2003). Towards a new ethic for watching dolphins and whales. In T. Frohoff & B. Peterson (Eds), *Between Species: Celebrating the dolphin–human bond*. San Francisco, CA: Sierra Club Books, pp. 168–177.

Hoyt, E. & Hvenegaard, G.T. (2002). A review of whale-watching and whaling with applications for the Caribbean. *Coastal Management* 30(4): 381–399.

Hughes, K. (2013). Measuring the impact of viewing wildlife: Do positive intentions equate to long-term changes in conservation behavior? *Journal of Sustainable Tourism* 21(1): 42–59.

Hughes, P. (2001). Animals, values and tourism – Structural shifts in UK dolphin tourism provision. *Tourism Management* 22(4): 321–329.

IFAW (2009). *Whale Watching Worldwide: Tourism numbers, expenditures and expanding economic benefits*. Yarmouth Port, MA: International Fund for Animal Welfare.

Kellert, S.R. (1999). *American Perceptions of Marine Mammals and their Management*. Washington, DC: Humane Society of the United States.

Kessler, M. & Harcourt, R. (2010). Aligning tourist, industry and government expectations: A case study from the swim with whales industry in Tonga. *Marine Policy* 34(6): 1350–1356.

Lambert, E., Hunter, C., Pierce, G.J. & MacLeod, C. (2010). Sustainable whale-watching tourism and climate change: Towards a framework of resilience. *Journal of Sustainable Tourism* 18(3): 409–427.

Luck, M. (2003). Education on marine mammal tours as agent for conservation – But do tourists want to be educated? *Ocean & Coastal Management* 46(9): 943–956.

Luck, M. (2008). Managing marine wildlife experiences: The role of visitor interpretation programmes. In J.E.S. Higham & M. Luck (Eds), *Marine Wildlife and Tourism Management: Insights from the natural and social sciences*. Wallingford: CABI, pp. 334–346.

Malcolm, C. & Duffus, D. (2008). Specialization of whale watchers in British Columbia. In J. Higham & M. Luck (Eds), *Marine Wildlife and Tourism Management: Insights from the natural and social sciences*. Wallingford: CABI, pp. 109–129.

Mayes, G. & Richins, H. (2009). Dolphin watch tourism: Two different examples of sustainable practices and proenvironmental outcomes. *Tourism in Marine Environments* 5(2/3): 201–214.

Mayes, G., Dyer, P. & Richins, H. (2004). Dolphin–human interaction: Pro-environmental attitudes, beliefs and intended behaviours and actions of participants in interpretation programs: A pilot study. *Annals of Leisure Research* 7(1): 34–53.

Milstein, T. (2007). Watching endangered orcas: The role of communication in balancing marine tourism and sustainability. In M. Luck, A. Graupl, J. Auyong, M.L. Miller & M.B. Orams (Eds), *Proceedings of the 5th International Coastal & Marine Tourism Congress: Balancing Marine Tourism, Development and Sustainability*. Auckland: AUT University, pp. 104–113.

Milstein, T. (2008). When whales 'speak for themselves': Communication as a mediating force in wildlife tourism. *Environmental Communication: A Journal of Nature and Culture* 2(2): 173–192.

Moscardo, G., Woods, B. & Saltzer, R. (2004). The role of interpretation in wildlife tourism. In K. Higginbottom (Ed.), *Wildlife Tourism: Impacts, management and planning*. Altona, Vic: Common Ground/Sustainable Tourism CRC, pp. 231–251.

Muloin, S. (1998). Wildlife tourism: The psychological benefits of whale watching. *Pacific Tourism Review* 2: 199–212.

Neil, D.T., Orams, M.B. & Baglioni, A.J. (1996). Effect of previous whale watching experience on participants knowledge of, and response to, whales and whale watching. In K. Colgan, S. Prasser & A. Jeffrey (Eds), *Encounters with Whales 1995 Proceedings*. Hervey Bay, Queensland, 26–30 July 2005. Canberra: Australian Nature Conservation Agency, pp. 182–188.

Neves, K. (2010). Cashing in on cetourism: A critical ecological engagement with dominant E-NGO discourses on whaling, cetacean conservation, and whale watching. *Antipode* 42(3): 719–741.

O'Neill, F., Barnard, S. & Lee, D. (2004). *Best Practice and Interpretation in Tourist/Wildlife Encounters: A wild dolphin swim tour example*. Wildlife Tourism Research Report Series No. 25. Gold Coast: Sustainable Tourism CRC.

Orams, M.B. (1994). Creating effective interpretation for managing interaction between tourists and wildlife. *Australian Journal of Environmental Education* 10(2): 21–34.

Orams, M.B. (1995a). Towards a more desirable form of ecotourism. *Tourism Management* 16: 3–9.

Orams, M.B. (1995b). A conceptual model of tourist-wildlife interaction: The case for education as a management strategy. *Australian Geographer* 27(1): 39–51.

Orams, M.B. (1995c). Development and management of a feeding program for wild bottlenose dolphins at Tangalooma, Australia. *Aquatic Mammals* 21(2): 137–147.

Orams, M.B. (1995d). Using interpretation to manage nature-based tourism. *Journal of Sustainable Tourism* 4(2): 81–94.

Orams, M.B. (1996). Cetacean education: Can we turn tourists into 'greenies'. In K. Colgan, S. Prasser & A. Jeffrey (Eds), *Encounters with Whales 1995 Proceedings*. Hervey Bay, Queensland, 26–30 July. Canberra: Australian Nature Conservation Agency, pp. 167–178.

Orams, M.B. (1997a). Historical accounts of human–dolphin interaction and recent developments in wild dolphin based tourism in Australasia. *Tourism Management* 18: 317–326.

Orams, M.B. (1997b). The effectiveness of environmental education: Can we turn tourists into 'greenies'? *Progress in Tourism and Hospitality Research* 3: 295–306.

Orams, M. (1999). *Marine Tourism: Development, impacts and management.* London: Routledge.

Orams, M.B. (2000). Tourists getting close to whales, is it what whale watching is all about? *Tourism Management* 21: 561–569.

Orams, M. (2001). From whale hunting to whale watching in Tonga: A sustainable future? *Journal of Sustainable Tourism* 9(2), 128–146.

Orams, M.B. (2002). Feeding wildlife as a tourism attraction: A review of issues and impacts. *Tourism Management* 23: 281–293.

Orams, M. (2005). Dolphins, whales and ecotourism in New Zealand: What are the impacts and how should the industry be managed? In C.M. Hall & S. Boyd (Eds), *Nature-based Tourism in Peripheral Areas: Development or disaster?* Clevedon: Channel View, pp. 231–245.

Orams, M.B. & Hill, G.J.E. (1998). Controlling the ecotourist in a wild dolphin feeding program: Is education the answer? *Journal of Environmental Education* 29: 33–38.

PWF (Pacific Whale Foundation) (2013). *Whale or dolphin watch.* Retrieved from http://www.pacificwhale.org/cruises/whale-and-dolphin-watches (accessed 11 April 2013).

Parsons, E.C.M. & Draheim, M. (2009). A reason not to support whaling – A tourism impact case study from the Dominican Republic. *Current Issues in Tourism* 12(4): 397–403.

Parsons, E.C.M. & Rawles, C.J.G. (2003). The resumption of whaling by Iceland and the potential negative impact in the Icelandic whale-watch market. *Current Issues in Tourism* 6(5): 444–448.

Parsons, E.C.M. & Rose, N.A. (2009). Whale watching and the International Whaling Commission: A report of the 2008 whale-watching subcommittee meeting. *Tourism in Marine Environments* 6(1): 51–57.

Parsons, E.C.M., Warburton, C.A., Woods-Ballard, A., *et al.* (2003). Whale-watching tourists in West Scotland. *Journal of Ecotourism* 2(2): 93–113.

Peace, A. (2005). Loving leviathan: The discourse of whale-watching in Australian ecotourism. In J. Knight (Ed.), *Animals in Person: Cultural perspectives on human–animal intimacy.* Oxford: Berg, pp. 191–210.

Peake, S. (2011). An industry in decline? The evolution of whale-watching tourism in Hervey Bay, Australia. *Tourism in Marine Environments* 7(3/4): 121–132.

Peake, S., Innes, P. & Dyer, P. (2009). Ecotourism and conservation: Factors influencing effective conservation messages. *Journal of Sustainable Tourism* 17(1): 107–127.

Ponnampalam, L.S. (2011). Dolphin watching in Muscat, Sultanate of Oman: Tourist perceptions and actual current practice. *Tourism in Marine Environments* 7(2): 81–93.

Porter, B. & Kaufman, M. (2009). Realizing the potential of marine naturalist interpreters as conduits of a mission. In A. Albers & P.B. Myles (Eds), *Proceedings of CMT2009, the 6th International Congress on Coastal and Marine Tourism.* Port Elizabeth: Kyle Business Projects, pp. 288–290.

Rawles, C.J.G. & Parsons, E.C.M. (2005). Environmental motivation of whale-watching tourists in Scotland. *Tourism in Marine Environments* 1(2): 129–132.

Ris, M. (1993). Conflicting cultural values: Whale tourism in Norway. *Arctic* 46(2): 156–163.

Samuels, A., Bejder, L., Constantine, R. & Heinrich, S. (2003). Swimming with wild cetaceans, with a special focus on the southern hemisphere. In N. Gales, M. Hindell & R. Kirkwood (Eds), *Marine Mammals: Fisheries, tourism and management issues.* Collingwood, Vic: CSIRO, pp. 277–303.

Scarpaci, C. & Parsons, E.M.C. (2012). Recent advances in whale-watching research: 2010–11. *Tourism in Marine Environments* 8(3): 161–171.

Servais, V. (2005). Enchanting dolphins: An analysis of human–dolphin encounters. In J. Knight (Ed.), *Animals in Person: Cultural perspectives on human–animal intimacy.* Oxford: Berg, pp. 211–230.

Shapiro, K.R. (2006). Whale watch passengers' preferences for tour attributes and marine management in Maui, Hawai'i. Master of Resource Management Thesis, School of Resource and Environmental Management, Simon Fraser University.

Smith, A., Newsome, D., Lee, D. & Stoeckl, N. (2006). *The Role of Wildlife Icons as Major Tourist Attractions – Case Studies: Monkey Mia Dolphins and Hervey Bay Whale Watching.* Gold Coast: CRC for Sustainable Tourism.

Stamation, K. (2008). Understanding human–whale interactions: A multidisciplinary approach. *Australian Zoologist* 34: 211–224.

Stamation, K., Croft, D., Shaughnessy, P.D., Waples, K.A. & Briggs, S.V. (2007). Educational and conservation value of whale watching. *Tourism in Marine Environments* 4(1): 41–55.

Takei, A. (1998). A cross-cultural assessment of the effectiveness of the interpretive program at the dolphin provisioning program, Tangalooma, Australia. Diploma in Marine Science Thesis, University of Queensland.

Valentine, P.S., Birtles, A., Curnock, M., Arnold, P. & Dunstan, A. (2004). Getting closer to whales – Passenger expectations and experiences, and the management of swim with dwarf minke whale interactions in the Great Barrier Reef. *Tourism Management* 25: 647–655.

Warburton, C.A. (1999). *Marine Wildlife Tourism and Whale-watching on the Island of Mull, West Scotland*. Mull: The Hebridean Whale and Dolphin Trust.

Warburton, C.A., Parsons, E.C.M., Woods-Ballard, A., Hughes, A. & Johnston, P. (2001). *Whale-watching in West Scotland*. Mull: The Hebridean Whale and Dolphin Trust.

WDC (Whale and Dolphin Conservation) (2013). *Whale watching*. Retrieved from http://www.wdcs.org/connect/whale_watch/index.php (accessed 11 April 2013).

Whales Alive (2009). *Whale watching. Whales Alive*. Retrieved from http://www.whalesalive.org.au/whalcwatching.html (accessed 11 April 2013).

Wiener, C.S., Needham, M.D. & Wilkinson, P.F. (2009). Hawaii's real life marine park: Interpretation and impacts of commercial marine tourism in the Hawaiian Islands. *Current Issues in Tourism* 12(5/6): 489–504.

Wilson, C. & Tisdell, C. (2003). Conservation and economic benefits of wildlife-based marine tourism: Sea turtles and whales as case studies. *Human Dimensions of Wildlife* 8(1): 49–58.

Zeppel, H. (2008). Education and conservation benefits of marine wildlife tours: Developing free-choice learning experiences. *The Journal of Environmental Education* 39(3): 3–17.

Zeppel, H. (2009). Managing swim with wild dolphin tourism in Australia: Guidelines, operator practices and research on impacts. In J. Carlsen, M. Hughes, K. Homes & R. Jones (Eds), *Proceedings of the 18th Annual CAUTHE Conference*. Perth, WA: Curtin University of Technology.

Zeppel, H. & Muloin, S. (2008a). Marine wildlife tours: Benefits for participants. In J.E.S. Higham & M. Luck (Eds), *Marine Wildlife and Tourism Management: Insights from the natural and social sciences*. Wallingford: CABI, pp. 19–48.

Zeppel, H. & Muloin, S. (2008b). Education and conservation benefits of interpretation on marine wildlife tours. *Tourism in Marine Environments* 5(2/3): 215–228.

Zeppel, H. & Muloin, S. (2008c). Conservation benefits of interpretation on marine wildlife tours. *Human Dimensions of Wildlife* 13: 280–294.

10

Whale-watching

An effective education programme is no fluke

Genevieve Johnson and Cynde McInnis

Introduction

The ocean is the largest remaining wild place on Earth, and contains populations of cetaceans (whales, dolphins and porpoises) in oceanic, coastal and river systems. Despite comprising 71% of the Earth's surface, less than 1% of the world's oceans and adjacent seas has protected status, compared to approximately 12% of the world's land area (Hoyt, 2005). Today, the oceans and their cetacean inhabitants face multiple human threats – pollution, both chemical (Wise Sr *et al.*, 2009) and noise (Hatch *et al.*, 2012), prey depletion (Bearzi *et al.*, 2005), entanglements in fishing gear (Robbins & Mattila, 2004), hunting (Hovelsrud *et al.*, 2008) and ship strikes (Panigada *et al.*, 2006) are among the most pervasive. Now more than ever, people need to be educated about the oceans and encouraged to act on their behalf.

Humans have always had a strong connection to whales. An evolution in human attitudes and behaviour towards cetaceans has taken us from hunting them to hailing them as icons of marine conservation. Attitudes towards whales, their suffering and their death have changed dramatically over time, with a steep turnaround in the 1970s–1980s (Bearzi *et al.*, 2010). Cetaceans are now renowned for the education opportunities they present as ambassadors, communicating the need for us to protect them and their aquatic habitats. There is no question of the allure of cetaceans and the passion they inspire in most of us. There is also little doubt that watching whales and dolphins in their natural environment offers an ideal platform to educate visitors (Orams, 1997) about the many endangered species of cetaceans, the overall health of their habitats and what can be done by individuals to help. Because of this, we believe that whale-watching is one of the best ways to combat a society increasingly alienated and distanced from nature, and an excellent chance to affect conservation-based behaviour change. The issue is not whether we have the right species or the right experience from which to capture the hearts and minds of visitors, it is are we currently planning and offering effective education programmes onboard whale-watch trips that change not only attitudes toward cetaceans and the oceans, but also behaviour?

In this chapter we emphasize the critical importance of an effective whale-watch education programme in order to promote behaviour change. We draw on existing theoretical models that address behaviour change through whale-watching as well as our own extensive personal experience. We introduce a third model, the McInnis model, that is influenced by and builds on the previous two, while presenting a comprehensive methodology for teaching on whale-watch trips. Based on years of experience at Cape Anne Whale Watch (CAWW) – an education-based whale-watch experience out of Gloucester,

Whale-watching: Sustainable Tourism and Ecological Management, eds J. Higham, L. Bejder and R. Williams.
Published by Cambridge University Press. © Cambridge University Press 2014.

Massachusetts (MA), and its associated research programme – we illustrate why this approach to education is one that will enlighten visitors and encourage behaviour change.

Current status of whale-watch education

Since the late 1980s, whale-watching has grown rapidly with 12% annual growth through most of the 1990s – a rate 3–4 times the growth rate of overall tourism (Hoyt, 2001, 2009; O'Connor *et al.*, 2009). Hoyt (2012) competed a more recent overview and concluded that some 13 million people a year travel to or within 119 countries and overseas territories to go whale-watching, spending more than US$ 2.1 billion (Hoyt, 2009; O'Connor *et al.*, 2009). In all reports, Hoyt noted that education could be improved in almost every location and that education is a key element of high-quality, sustainable whale-watching or 'sustainable marine ecotourism' (Hoyt, 2012).

Education programmes onboard whale-watch expeditions have the potential to be at the forefront of marine conservation efforts worldwide, but how often does our relatively recent conservation mindset translate to a change in behaviour? Does participating in an education-based whale-watch mean that visitors will consume less and recycle more, refrain from using chemicals in the home or only buy seafood that wasn't caught in fishing gear that entangles marine mammals? Will they actively contribute to the preservation of the cetaceans they watched, and most importantly, will their commitment be enduring in the long-term?

Today, education of various kinds is a component of some businesses involved in the global ecotourism phenomenon of whale-watching. However, a broad education programme needs to be a regular part of whale-watching in every community (IFAW, WWF and WDCS, 1997). A 1998 world survey of whale-watch operations found that only 35% of all operators had enlisted naturalists to guide their trips (Hoyt, 1998). It is now generally accepted that ecotourism should include three major components: that it is based on a nature-based attraction in a natural setting, that it is managed in a sustainable manner and that it includes an educational component for the tourists (Valentine, 1990; Weaver, 2001). It seems, however, that there has been little consideration of the educational value of these experiences for the participants. This is somewhat surprising considering that the incorporation of education into eco-tours is held to be an essential component of ecotourism (Weaver, 2001). However, exactly what constitutes 'educational value' is rarely, if ever, defined. Russell and Hodson (2002) highlight the oft-proclaimed educational value of whale-watching, although notably little research exists to support such claims. Orams (1996) argues that it should be the attempt to move eco-tourists from a state where they simply minimize their impact on the host natural environment (attraction) to a state where they actively contribute to the health and viability of that environment. He argues that this transition is (or should be) the purpose of educational services offered as part of the ecotourism experience, and that participants' knowledge, attitudes and intentions to change their behaviour are suitable indicators of this transition.

It would be fair to say that many whale-watch companies, with few notable exceptions, assume that exposure to nature and the transmission of information in an entertaining manner will produce responsible, environmentally aware visitors willing to change their behaviour and contribute to the overall well-being of the marine environment. Russell and Hodson (2002) describe a case study of a Canadian whale-watching experience that showed it did not live up to its potential as a form of critical science education, although the authors believed the goal was worth pursuing. Lück (2003) distributed questionnaires to whale-watch visitors in New Zealand, in which respondents clearly indicated their desire to receive more information about the environment. Changing human behaviour is notoriously difficult, and if the whale-watch industry is to make a positive contribution to the conservation of cetaceans and the marine environment, companies must adopt carefully designed

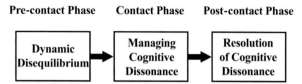

Figure 10.1 Forestell and Kaufman (1990). Education and interpretation model.

educational programmes, which incorporate strategies and methodologies that facilitate behaviour change. Ideally, the inclusion of a recognized education programme should be a condition of permitting to be monitored and assessed when permits are due for renewal. Currently, the vast majority of so-called 'ecotourism' operators are simply offering information and entertainment, not environmental education or interpretation. Ecotourism operators range dramatically in their approach to education. Based on extensive personal observation by the authors during whale-watch trips in Australia; North, South and Central America; Europe and Island nations in the Pacific, Atlantic and Indian Oceans, it has been noted that some companies, while offering natural history information about the species being observed, rarely put the species in the context of their habitat. Threats to whales in general, and specifically to the animals being watched, are inadequately addressed, if at all, and visitors are seldom given tangible opportunities to help. Many owners do not consider the role of the 'naturalist' to be a paid position. In some instances organizations such as the American Cetacean Society (ACS) train volunteers to provide interpretive education, while in others, it is the captain who shares bits of information when and if the opportunity to deviate from their main focus arises. It is both logical and obvious that if an education programme does not deliberately set out to change attitudes and behaviour it is extremely unlikely to do so (Orams, 1997). So how can we isolate and capitalize on the situations that are conducive to environmental learning?

Despite the rapid global growth of whale-watching, and its recognition as an ecotourism activity, limited empirical research has been conducted into how education programmes are planned and executed, and the effectiveness of those education efforts in changing the attitudes and long-term behaviour of whale-watch visitors. The handful of studies that have been conducted (Orams, 1997; Orams & Taylor, 2005; Neil *et al.*, 2003; Stamation *et al.*, 2007) demonstrate a clear link between education programmes and behaviour change, and that whale-watching experiences can improve visitors' knowledge about whales. However, Stamation *et al.* (2007) showed that this increase in knowledge is short-lived, while Bierman (2002b) notes that although seeing whales in their natural environment can be a powerful emotional experience, without a framework within which to support and expand on the experience, its potential long-term benefits decay very quickly.

Existing theoretical models

There are two existing theoretical models that specifically address whale-watch education. The Forestell and Kaufman (1990) model influenced the Orams (1997) model. This chapter presents the McInnis model, which has been influenced by both of these models. The McInnis model combines the breakdown of different stages of the visitor experience (Forestell & Kaufmann, 1990) with the affective and cognitive domain that Orams contributes.

Based on cognitive psychology theory, Forestell and Kaufman (1990) contribute a model for the design and testing of education and interpretation programmes based on field experiences with nature-based tourists in Hawai'i (Figure 10.1). Forestell and Kaufman's work suggests that effective whale-watch experiences should be divided into three stages of information. In the pre-contact stage, 'dynamic disequilibrium' should be created or promoted. This is described as the behavioural response to seeing whales (clapping, cheering, smiling) followed by the desire to ask questions – a 'state of wonder' (Forestell, 1993). Visitors are excited about the pending encounter with whales, feeling a perceived need for information and,

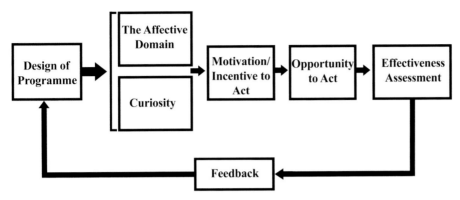

Figure 10.2 Orams (1997). Features of an effective interpretation programme.

consequently, a motivation to learn. The intention of this stage is to create questions in the mind of the visitor prior to any interaction with wildlife. During the contact phase that follows, the 'cognitive dissonance' created during the pre-contact phase should be addressed. 'Cognitive dissonance' refers to an instance when what an individual perceives externally is different from what is known internally. Answers to questions about the animals and their behaviour must be provided in an informed and interesting manner and should be relevant to what the visitor is experiencing. Finally, the post-contact phase or 'resolution of cognitive dissonance' is a time of personal validation. It involves follow-up activities encouraging visitors to incorporate new information into changed behaviour.

Forestell and Kaufman's model was the basis for further development by Orams (1997), who proposed a model based on five phases of the learning process to effect behaviour change (Figure 10.2). The Orams (1996) model incorporates theories that balance both cognitive dissonance and affective domain. Creating questions in people's minds and arousing curiosity is central to the Orams (1996) model. The affective domain engenders participants' emotional involvement through topics such as birth, death or caring. Orams (1997) suggests that by invoking the 'affective domain', programme messages are more effectively 'internalized' and are more likely to be acted upon. The next phase creates incentive to act by outlining specific problems

relevant to the animals being watched, followed by a personalized message offering ways to help with opportunities for immediate action. The final phase involves assessment and feedback.

The development of a revised model: the McInnis model

In recognition of the need for education programmes that encourage long-term behaviour change among whale-watch visitors, we present a model that combines the previous two models to specifically accommodate and address behaviour change as well as implementation and follow-up strategies. Stellwagen Bank off the coast of New England in the United States has long been a global leader in whale-watching. In the mid 1970s New England operators established their own brand of commercial whale-watching with strong scientific and educational components, and paid naturalists on every trip who were often working researchers. Educational programmes to introduce school children to wild cetaceans – begun in southern California by such groups as the American Cetacean Society – were expanded in New England (Hoyt, 2009).

This model was developed as part of McInnis's Master's Program (Bierman, 2002b) and has been used and tested over 2500 trips by McInnis and others, at CAWW in Gloucester, MA since 1999.

Surveys and observations have been used to evaluate the programme. Components of the model have been revised over the years as new observations are made or as new ideas come forth, and development continues. Two PhD research projects have been conducted on participants of this whale-watch programme (Domizi, 2008 and Deissler, 2007).

Theories that inform the McInnis model

The following is an introduction to three theories that inform the development and implementation of the McInnis model.

Transformative learning

Mezirow and Associates (1990) put forth a theory of how adults change – a theory termed transformative learning or perspective transformation (Cranton, 1994). We define transformative learning as an experience resulting in the challenging, evaluating, and reworking of one's values, beliefs and underlying assumptions leading to new and more open perspectives on life. There is a series of five stages that a person goes through in order for this process to be complete (Cranton, 1994; Brookfield, 1987).

1. *Triggering event*: transformative learning begins with either a triggering event or a slow combination of occurrences that are harder to pinpoint. These experiences lead to a disorienting dilemma. People find discrepancies in their perception of how the world should be and how the world really is. (This is another word for cognitive dissonance.)
2. *Self-examination*: the result of this disorienting dilemma is that one identifies and examines his or her assumptions, values and beliefs. This phase involves much self-scrutiny and introspection. It can be difficult, painful, or anxiety-provoking to examine oneself in this way.
3. *Exploration*: by talking with people, reading or investigating, one starts to explore other ways of approaching the world.

4. *New perspective*: once different avenues are explored, a new perspective begins to be formed incorporating what has been learned.
5. *Action*: these new perspectives and assumptions are then incorporated into a person's life in his or her actions and beliefs.

Transformative learning is a long process, and will not be completed on a single whale-watch trip. However, it should be encouraged to begin. While transformative learning has been studied in formal educational settings, its application to environmental education or interpretive settings has only recently begun. Coghlan and Gooch suggest that transformative learning should be applied to volunteer tourism in order to get the desired results of the participants – to be changed by their experience (Coghlan & Gooch, 2011).

Multiple intelligences

Gardner (1983) introduced his theory of multiple intelligences in the book, *Frames of Mind*, where he describes humans as having seven different intelligences. Gardner defines intelligence as 'the ability to solve problems or to create products that are valued in one or more cultural settings' (Gardner, 1999). The criteria of each intelligence is that it has the potential of isolation by brain damage, has an evolutionary history, contains an identifiable core operation or set of operations, and is susceptible to encoding in a symbol system (Gardner, 1999). According to Gardner's theory, every one of us has some level of each of these intelligences. Each intelligence is independent of the others, yet they are sometimes used together to solve problems. The nine recognized intelligences are: (a) qualitative, (b) quantitative, (c) visual–spatial, (d) bodily-kinesthetic, (e) musical, (f) naturalist, (g) interpersonal, (h) intrapersonal, and (i) existential.

Multi-sensory learning

Different parts of our brains are responsible for responding to our various senses (Christie, 2000). When thinking about how people learn, it makes

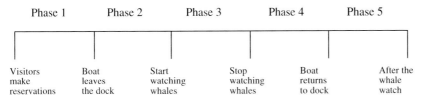

Phase 1	Phase 2	Phase 3	Phase 4	Phase 5
Visitors make reservations	Boat leaves the dock	Start watching whales	Stop watching whales	Boat returns to dock	After the whale watch

Figure 10.3 Phases of the whale-watch experience.

sense to try to physically involve as many parts of the brain as possible. If more parts of the brain were working with the same information, that information would have a greater chance of being understood, remembered, and recalled. Educators would literally be stimulating more neural pathways in the brain (Christie, 2000). This is the concept behind multi-sensory learning. The basic definition is 'the process of using as many of the senses as possible when doing exercises and lessons' (Reading with Phonics, 2002).

Relation to the previous models

Expanding on the Forestell–Kaufman model, we use a five-phase experience based on Koth, Field, and Clark's (1982) description of a recreational experience on a wildlife-viewing trip to Alaska. These phases and the psychological perspective of the visitor at each point are described in detail in the 'Action' section (section 7) (Figure 10.3). These phases create the context in which to apply the model (Figure 10.4). Based on the assumption that natural learning takes place in natural settings (Lave & Wenger, 1991), and teaching for behaviour change has to be planned for (Schell & Schell, 2007), this model illustrates how change can occur on whale-watching trips (Figure 10.4).

It is crucial that all connections be established for behaviour change to occur. There must be an emotional connection to the whales that motivates the visitor to care about the animals and want to act to protect them. Next, an intellectual connection is made between the whales and the environment they live in. Finally, when there is an emotional

response to seeing the animals, as well as a carefully designed education programme in place that facilitates a cognitive connection, a person's attitude toward the animals will be strengthened from these feelings and knowledge. The affective domain, cognitive domain and attitude all contribute to the likelihood of behaviour change occurring. As visitors go on more whale-watches, or have more encounters with the wild world, each experience can reinforce or strengthen their attitude towards the environment and long-term behaviour change is more likely to occur.

From personal experience, it is important to bear in mind that some aspects of the model are difficult to control, such as a person's emotional response. Whales are wild animals and the numbers and species that may be seen on any given trip are not predictable and can impact a person's response to the trip. Bierman found a difference in how people rated the trip when controlling for species seen. When only humpback whales were sighted, passengers rated the trip 4.61 on a scale of 5. When only fin whales were seen, the rating dropped to 4.22 (Bierman, 2003). Likewise, the behaviour of the whales influenced ratings. With more than 10 active humpbacks (feeding or breaching), passengers rated the trip 4.74; with 10 or fewer active humpbacks, ratings were 4.38 (Bierman, 2003). What people observed on the trip influenced how they rated the trips. Although observations do not necessarily correspond to emotional response, it is something to bear in mind.

Another important aspect that educators are likely unaware of is the attitude of the individual visitor coming aboard. It has been said that naturalists are 'preaching to the converted' on whale-watch

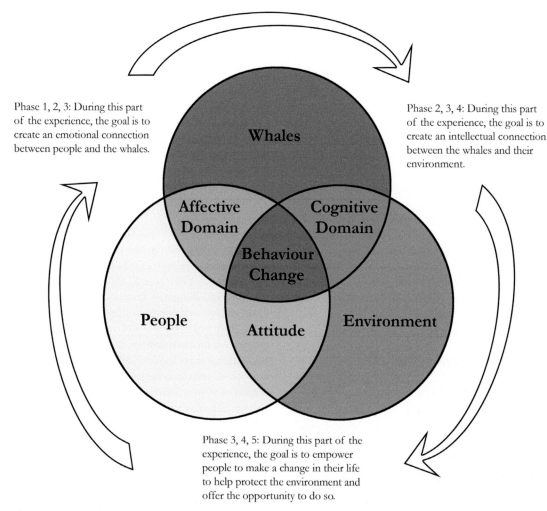

Phase 1, 2, 3: During this part of the experience, the goal is to create an emotional connection between people and the whales.

Phase 2, 3, 4: During this part of the experience, the goal is to create an intellectual connection between the whales and their environment.

Phase 3, 4, 5: During this part of the experience, the goal is to empower people to make a change in their life to help protect the environment and offer the opportunity to do so.

Figure 10.4 The McInnis Model.

trips, but as Forestell points out, and we would agree from personal observation, whale-watch visitors represent a cross-section of the travelling public and they are not necessarily more environmentally aware (Forestell, 1993). Seasonal surveys conducted aboard CAWW vessels have shown there are trends in age, education level and reasons for whale-watching. In autumn and spring, couples aged 30–50 travelling without children and coming to learn about whales are the norm. This is in contrast to summer visitors, who tend to comprise

families on vacation (unpub. data). Even though there may be trends in visitors, it is difficult to assess individual attitudes toward the wild world.

The McInnis model education programme becomes especially important when aiming for the goal of behaviour change because we need to account for the variety of whale experiences as well as the unknown attitudes of our passengers. The approach to education described in this chapter, together with a well-trained and insightful naturalist who is flexible based on whale behaviour and visitor attitude, is more likely to lead to behaviour change in passengers.

What does this model look like? If a visitor has an enjoyable experience watching whales, but is not given insight into behaviours observed or threats to animals in the habitat observed, the experience will not likely impact on future behaviour. If a visitor sees only one whale surface for a few breaths, this is not likely to elicit a strong emotional response, in which case, the cognitive connection will be exceedingly important for behaviour change to be a result of the trip. If a visitor sees multiple whales or one or more active whales, learns about the threats they face and their conservation status, they are likely to have their environmental attitude reinforced or developed, with behaviour change a more likely result.

The theories applied to whale-watching (theories)

Being in the presence of whales can have a powerful impact on the affective and cognitive domain. Strong reactions come from observing gentle interactions, from watching their sheer power as they jump out of the water, or from seeing first-hand how humans are negatively impacting their habitats. Many visitors enthusiastically pay to experience the deeper psychological responses to whales that whale-watching can provide (Muloin, 1998). In the case of humpback whales off New England, it is not uncommon to see an animal with entanglement scars on its body (Robbins & Mattila, 2004). This has the potential to elicit an emotional reaction in visitors, leading to their seeking more information. When interviewing whale-watch participants at Cape Ann Whale Watch, Domizi (2008) found that at a six-month follow-up, visitors often used the same descriptors as they did immediately following the trip to describe the emotional impact of the trip. A variety of encounters with whales can become a triggering event for transformative learning. The triggering events can be either negative or positive (Brookfield, 1987). It could be an animal breaching over and over (positive) or seeing a whale entangled in fishing gear (negative). Each visitor, based on their personal situation and life experiences, will have a different response, and as educators, we might never know when these triggering events occur.

Guidance and critical reflection from educators throughout the trip can take these powerful experiences and help people begin to realize why they are greatly moved by whales. The reason why might be part of their disorienting dilemma. Guiding visitors through the uncovering of their basic assumptions, values and beliefs can begin to happen at various times during the trip. Therefore, by encouraging people to examine themselves, they will be on a path to new discovery and hopefully new perspectives and actions. This is the reasoning behind the circular pattern to the model. Future whale-watch experiences will provide more opportunities for exploration and creation of new perspectives. The more a visitor whale-watches, the more likely he or she will implement long-term behavioural changes.

Implementing the McInnis model (methods)

Aboard whale-watching trips, educators never know who the visitors are going to be until they board the boat. Even at that point, we know very little about them. Bierman (2001) found visitors to CAWW represented 45 of the 50 United States, and 35 countries, as well as all ages, occupations and educational backgrounds. If programme planners keep in mind that visitors represent all multiple intelligence types, can learn through all their senses, educational material on the whale-watch boats can be delivered in ways that stimulate the entire brain and each individual. For example, interpretation while watching whales is conducted in an auditory fashion. Bones, baleen and whale models can be made available for people to touch, see and smell.

How can an educator implement the programme into their whale-watch trip?

Figure 10.5 Intern Lisa Gibbler presents the baleen to passengers. (See colour plate section.)

Interns and volunteers

Internships are increasingly popular with college students and are useful for a variety of reasons. Internships help students gain experience and figure out whether that field is something they want to pursue further. It provides students with an outlet for their knowledge and allows them to sharpen their skills. Both interns and companies benefit from the experience. The companies have an increased staff they do not have to pay. Interns will also be on the whale-watch boat for an extended amount of time (often three months). Not all interns come to the experience knowing that researching whales is what they want to do with their lives. Some just love whales and want to enjoy their summer. This group has the largest potential for change because of their extended experience. (See also 'Education over time'.)

Teaching tools

Teaching tools are objects, papers, technologies – devices that support education. Teaching tools can be thought of as entry points, which are ways to open up conversation, engage and entice people (Gardner, 1999). By creating teaching tools that incorporate the different intelligences and senses, every audience member has the opportunity to engage in learning about whales. For example, orca or sperm whale teeth together with baleen affords passengers the opportunity to touch, see and even smell bio facts (Figure 10.5). A 3D model of the coastline of Massachusetts as well as the ocean floor, and specifically Stellwagen Bank, allows visitors to better understand upwelling and why animals return to the area year after year (Figure 10.6). Playing back audio recordings of the sounds of the whales allows someone with musical intelligence to relate to the whales on a very emotional level.

Interactions or stations

In many cases, the travel time to the sites of animal encounters can be up to an hour or two. This provides an excellent opportunity to interact with the visitors. One way is to set up stations on the boat where people can interact with various teaching tools on their own. Another alternative is to take the teaching tools to the visitors. With teaching tools in hand, educators can gather small groups of people to explain concepts, answer questions and help facilitate visitors' emotional and intellectual connection to the whales. The interaction gives people the opportunity to ask questions about specific topics they are interested in. It also gives educators the opportunity to explore potential disorienting

Figure 10.6 Cynde McInnis holds up the 3D model of Stellwagen Bank. (See colour plate section.)

dilemmas with passengers and help them on their path of transformative learning. By having small groups, or one-on-one conversations, everyone is allowed the opportunity to maximize their learning.

Interpretative commentary

Interpretive commentary occurs during phase 3 and is delivered by a skilled whale-watch educator. There is probably no more important person for a successful whale-watching tour than the naturalist guide. This person is a critical component in promoting behaviour change. Currently, a surprisingly large number of tours rely on the boat captain or boat operator as the sole guide (Hoyt, 1998). This is rarely ideal or suitable in the long term.

Environmental interpretation – an educational effort directed at revealing the relationships and meanings behind aspects of the natural environment (Tilden, 1957) – has a long-standing tradition in the park management field (Knudson *et al.*, 1995). '[Through exhibits, guided tours, and talks] interpretation enhances our understanding, appreciation, and therefore, protection of historic sites and natural wonders' (Beck & Cable, 1998). In 1957, Tilden wrote *Interpreting Our Heritage*, where he outlines six principles, which have been the foundation for the field. In their book, *Interpretation for the 21st Century: 15 guiding principles for interpreting nature and culture*, Beck and Cable build on Tilden's original 6 and expand to 15 principles of interpretation. These should be part of every naturalist's repertoire.

Domizi's (2008) study of visitors to CAWW found most participants indicated 'educators or naturalists were important sources of information as well as pleasurable sources of conversation and social interaction' (p. 198). Bierman (2001) asked CAWW visitors to rate components of the trip on a scale from one to four (one being less important and four being more important to the overall trip). The table below shows the results.

Interpretive commentary	Interaction with educators	Teaching tools	Conservation message	Introductory talk	Harbour tour
3.66	3.48	3.44	3.4	3.37	3.15

In ideal circumstances, Hoyt (2006) suggests a list of tasks that a well-trained, knowledgeable and personable guide (or guiding staff on larger cruise ships) can accomplish on a trip.

- Manage customer care and answer questions before, during and after the trip.
- Give safety briefings before trips and take the lead in introducing passengers to the boat and making them feel comfortable and safe (including dealing with seasickness).
- Introduce passengers to the natural, cultural, geological and oceanographic features of an area.
- Help passengers with photo tips.
- Tell good sea and whale stories and be entertaining.
- Show passengers how to identify individual animals and point out the names, identification details and life histories of individual whales when known.

- Make passengers realize the nature of a wildlife watching trip – nothing is certain and every trip is different; the more time and more trips one takes, the more likely one is to experience extraordinary things (Hoyt, 2006).

The above tasks meet what we consider the minimum requirements of a whale-watch guide. The following tasks are what separate an average guide, and therefore most likely an average experience for the visitor, from a highly skilled naturalist. These tasks must be mastered and are necessary to encourage behaviour change in visitors.

- Become the bridge between the largely urban world of most passengers and the natural world of whales, dolphins and the sea.
- Impart essential take-home conservation messages.
- Help forge the essential link between passengers and the sea and ensure that their first whale-watching trip is a success, no matter how many or even whether whales are seen at all (Hoyt, 2006).

In relation to the McInnis model, we would add:

- Present interpretation in a variety of manners: visual, kinesthetic, auditory, digital.

Follow-up – online and social networks

Follow-up, reinforcing events and opportunities to act are critical for this process of behaviour change (Orams, 1996; Rennie & McClafferty, 1995). When visitors leave the setting of the whale-watch boat, they go back to their lives, and their experience will leave the forefront of their mind. Phase 5 might be one of the most important phases to focus on for long-term change to occur.

Today, whale-watch companies worldwide are increasingly using online media and social networks to keep visitors participating far beyond their 'at sea' whale-watching experience. Through now traditional Internet outlets such as websites, blogs and electronic newsletters, companies publish information and articles about upcoming trips, news, whale-watching regulations and ways people can help. Social networks such as Facebook, Twitter,

Flickr and Pinterest allow visitors to follow a company, preview, or even follow-up by keeping in touch with animals they sighted on their trip. Posting photos, stories about trips, the animals, threats whales face and what can be done to help, can generate bidirectional communications to keep people connected to their experience. This media has the added bonus of being able to be 'shared' with friends from a computer, tablet and mobile phone. In MA, many of the whale-watch companies update their Facebook page on a daily and often per-trip basis, including images of individual animals seen and descriptions of their behaviour, GPS position, status if injured or entangled.

Mobile 'apps' – applications running on mobile phones and tablets – allow participants to become 'citizen scientists' on trips. Citizen science is the involvement of citizens from the non-scientific community in academic research (Trumbull *et al.*, 2000; Lee *et al.*, 2006). Within the past two decades, this use of volunteers in research has proliferated and evolved into 'citizen science' (Catlin-Groves, 2012). The New South Wales Government and National Parks and Wildlife Service (Australia) released a free iPhone app entitled 'Wild about Whales'. Connected to an online research database, people can use the app to collect data by recording their own sightings, view the latest sightings on a map, learn about the species in the area, as well as connect with other whale-watchers. Embedded sensors allow researchers to validate GPS and image data and are now affordable and regularly used by citizens (Catlin-Groves, 2012). This type of social marketing is the modern way of truly sharing experiences with wildlife and in turn, promoting the whale-watch company and even sustainable whale-watching practices. Information that comes from whale-watch participants to their friends increases transparency and pressure on operators to fulfil best practice and offer the best experience (pers. obs.). In a whale-watching management context, marine environmental education complements regulation as a mechanism for changing human conduct and fosters the protection of whales (Andersen & Miller, 2006).

Although the participants in Domizi's (2008) study had no reinforcing events to share at a six-month follow-up and were unable to report gains in knowledge since leaving the boat, they had indicated intent to change immediately following the trip. Bierman (2001) found that at the end of the whale-watch, a little over half of the respondents to the survey had the intention of changing a behaviour – not littering, being aware of household chemicals, recycling, or learning more. Thirty-five percent said they would not change any behaviours, and 14% considered themselves already environmentally conscious. There has been no follow-up with these participants, but it indicates that at the end of a trip, visitors are willing to make a change, and by providing reinforcing events and opportunities to act, change is more likely to occur. Mezirow (Cranton, 1994) notes that transformative learning is a long process and cannot happen in a vacuum. People need to be supported through the process. If change is going to occur after a whale-watch trip, reminders of the participant's affective and cognitive responses to the whales will help provide this support.

Methods and theories in action

How does this look in action? CAWW in Gloucester, MA has used this education programme on over 2500 whale-watch trips since 1999. All naturalists and interns have received the same onboard training over the years and therefore consistency exists.

Phase 1: Anticipation of the event

This occurs from when the visitors sign up for the trip until they board the boat. Excitement and possibly apprehension characterize this phase (Koth, *et al.*, 1982). It is a time where people may seek information, so having it available is important. Upon making a reservation, visitors should be encouraged to go to the website to download information or a species checklist or connect with social networking sites. Once visitors board the boat, signs and menu holders provide oportunities for people to see the species and behaviours in the area. This provides those who enjoy quietly reading with an opportunity to learn about the animals while waiting for the trip to begin. This phase ends with the dock talk (introductory talk), which includes information about safety, where the boat is going, what might be seen, and what themes will be presented throughout the trip.

Phase 2: Travelling to the site

During this phase, it is thought that substantial information can be presented because interest in learning is high (Koth *et al.*, 1982). Information is available in a variety of ways. Signage and menu holders on each table are entry-points for intrapersonal people, linguistic and visual learners. This is the time when interactions with educators take place. With two to four interns or volunteers per trip, each educator is assigned a teaching tool for the day. Some teaching tool suggestions are: baleen and teeth, a 3D model of the ocean floor, close-up photos of whale body parts, and the songs and sounds of cetacean species. A 3D model shows visitors how the bottom topography relates to why whales are on Stellwagen Bank – a visual–spatial intelligence tool. Seeing, smelling and feeling the baleen and teeth helps people understand the two kinds of cetaceans. Baleen and the 3D model were rated the highest of the teaching tools that have been evaluated, by passengers (Bierman, 2001). Mobile tablets such as iPads can also be used as teaching tools for showing multimedia. This phase of the trip lays a foundation for information to be presented more formally upon reaching the whales. A goal is to facilitate the visitor's emotional attachment to and intellectual interest in the whales. As Orams (1996) points out, 'issues which involve humans' affective domain are likely to be those issues that are central to all life. Topics such as reproduction, birth, death, competition and conflict, sickness and social relationships are emotional areas for most humans. Therefore, interest and emotional response to other living things struggling with these same issues is likely.'

Information like this could bring about a disorienting dilemma in a person, so educators could ask questions that might begin the self-examination phase of transformative learning.

Phase 3: 'On-site experience'

Information and interpretation play a more formal role and are presented either over the PA system or more personally depending on the size of the boat. Educators need to remember Beck and Cable's principles of interpretation during this time (1998). For example, one principle is to relate what you are saying to the lives of the visitors. When talking about a whale's blow, relate it to how a visitor can see their breath on a cold day. A humpback's flipper has the same bone structure as a human hand and arm. When choosing which threats to talk about, choose ones that people can help alleviate – entanglements and pollution are two examples.

During the on-site experience, the order in which information is given relates to the McInnis Model on a finer scale. During the first third of the time spent watching whales, point out physical aspects of the whales those visitors can see. The blow, the flippers, the flukes, why they lift them up and dive sequences are topics to mention first to get people excited about their own ability to recognize what they are seeing – exciting the affective domain. The second third of this time is spent talking about meatier topics such as migration, calving, feeding styles, surface activity (if they are doing these things). This is to help people gain knowledge and establish that intellectual connection between whales and the environment in which they live. Move into population levels and threats whales face while concluding their time with whales. This addresses the attitude component of the model to help people connect to the environment and adopt an attitude that helps protect these animals. This time also provides an opportunity to begin the self-examination and exploration phase. Offering tips on what people can do to help the whales gives them insight into what changes they can make in their life that will make a difference.

Phase 4: Travelling from the site

On the ride back to the dock after leaving the whales, the goal is twofold: reinforce, on a more personal basis, information that was given while watching the whales as well as helping individuals make meaning of their experience. During this time it is important to clarify visitor questions and encourage further study of whales, the ocean environment and what changes they might make in their own lives to help protect whales. This may include a specific action, or encouragement of the visitor to follow the whale-watch or a specific whale via social networks. (This will be discussed further later.) On the return trip, teaching tools could include discovery scopes with plankton collected during the trip, games for children and an entanglement teaching. It is important to remember that each person will have a different response to seeing the animals and talking with them on the ride in can be valuable.

When talking with visitors during this phase, ask questions that help people reflect on their experience. Reflection is what might uncover those assumptions that can lead to behaviour change. This is also a time to take advantage of 'teachable moments'. Those encounters that are unplanned and unexpected can have an enormous impact on people. When a whale approaches the boat and rolls over, its eye looking at visitors, this is a teachable moment. Ask people how they feel. What do they think? How might the experience impact that person? Educators can also provide opportunities to act – a petition to sign, membership forms, product suggestions or lists of things to do at home (Orams, 1996).

Phase 5: The 'recollection' phase and follow-up

During this phase, remind visitors of what they saw, reinforce the themes from the trip as well as encourage visitors to make a small change in their life to help protect these animals. Provide opportunities for action and reinforcing events. This is providing some of the support people need to complete the process of transformative learning. Some ideas may include the following.

- Give visitors a pledge sheet. Have them write down the action they will take in the next week/month to help whales and dolphins and put it up on a board on the vessel.
- Give visitors a seafood guide to put in their wallet, along with instructions on how to use it.
- Tell them about the guide and associated app, if applicable, and make sure there are signs around to reinforce the message.
- Remind people to go to the website, sign up for Facebook, Twitter or electronic newsletters.
- Gather email addresses as people make their reservation or on the boat as a way to actively engage them after the experience instead of hoping they come to you.
- Have them sign a more formal campaign or petition or take part in a photo. Providing options for people and challenging them to do something daily/weekly/monthly is important.

This formal wrap up should be done in such a way that everyone could listen – over the PA in the last few minutes, or on the stern before docking. The focus should be on specific behaviours people can adopt in their lives as a result of being informed and inspired on their trip as well as offering ongoing opportunities to connect with their experience and the whales via more whale-watch trips and social networks.

Organizing the trip into five phases helps when putting together an entire programme as it offers a chance to focus energies and establish goals for each section of the experience that will work for individual whale-watches.

Education over time

Remembering that long-term behaviour change does not happen quickly and without support (Cranton, 1994), each exposure people has to whales or the wild world can heighten their awareness and desire to help protect these animals. The more exposure, the more likely a person is to change. Domizi

(2008) found after a single whale-watch that 'short-term learning (immediately following the experience) focused largely on gains in content knowledge, the value of the lived experience, and recognition of the emotional impact of the experience'. Some did mention they had an increased awareness in relation to the oceans. While others had plans to change their behaviours to be more environmentally aware, six months later they reported no sustained behavioural change, likely because there were no reinforcing events, although they did admit to paying more attention to environmental issues, particularly those that related to the ocean.

In addition to the daily whale-watch trips, the University of Georgia has a doctoral-level class called, 'The Whale Class'. This involves graduate students going to CAWW for a week-long class on social learning theory, while participating in six or seven whale-watch expeditions. The McInnis model suggests that more exposure to whales means greater potential for change. In fact, some students have based their PhDs on the whale-watch experience. One student went home and began a recycling and water management programme at the high school where he teaches. One student signed up for an undergraduate degree in ecology as a result of spending her week with the whales. She wrote, 'I have no reason. I have no explanation. I don't need another degree. I don't necessarily want another degree. And I may not get one. I really am much more interested in figuring out how I can be a better human being and how I can pass stuff on to my kids...how I can try to make the world a better place' (Deissler, 2007).

Finally, the interns and volunteers that participate in over 50 whale-watches a summer are provided with repeated exposure to whales, the greatest potential for long-term change. CAWW intern Jill Nielson wrote, 'That summer on the boat was a life changing summer. I thought that I wanted to do research either on whales or other wildlife, but the opportunity to teach gave me a different door to open.' Jill is now teaching seventh-grade science. Ildiko Polyak wrote, 'That summer [that I interned],

my whole life changed. For the first time in my life, my eyes were opened to the growing issues from the impact of humans on nature. The words of the educator made me realize that I could make a difference, but the whales made me realize that I HAD to make a difference. I went on to receive my bachelor's degree in molecular biology and to work in an environmental toxicology lab examining levels of toxins affecting various genes in cetaceans.' Ildiko went on to receive her Master's in infectious disease, focusing on aquatic and marine bacterial pathogenesis. Because of their experience, they will go on to influence others, encouraging change occurs on a broader scale.

Conclusion

At first glance, whale-watching may appear to be an ideal blend of conservation and education. However, upon further investigation, the existing educational benefits are not as straightforward as proponents contend (Russell & Hodson, 2002).

Whale-watching has substantial unrealized potential in terms of the educational opportunities it offers; however, as highlighted by Hoyt (2009), education must be improved and is key to a high-quality whale-watch experience that promotes behaviour change. In this chapter, we evaluated the importance of an effective whale-watch education programme and the critical influence of a trained onboard educator, while highlighting the need for the implementation of effective education programmes throughout the global whale-watch industry. An effective education programme and effective educators are a necessity if long-term behaviour change is to occur as a result of watching whales, and therefore should constitute mandatory components of all whale-watch ventures. An international accreditation programme or standard that requires the implementation of an effective education programme in conjunction with the training of educators and guides could help achieve this outcome.

For any whale-watch education programme to be effective, change needs to be planned for and innovative ways to offer follow-up to visitors must be utilized so they can reconnect with their experiences. The McInnis model is envisioned as a plan to disseminate best practice in whale-watch education by providing a step-by-step, five-phase lesson plan together with application ideas and follow-up activities that harness new technologies that can be tailored specifically to the needs of the whale-watch educator and operator.

It is known that whale-watch companies promote the educational value of whale-watching; yet, we also know that limited research exists to support such claims. A comprehensive review by Zeppel and Muloin (2008) provides insight into gaps in existing research on attitude and behaviour change as a result of whale- and dolphin-watching experiences. Behaviour change studies and their application to environmental education while watching whales have begun only relatively recently and are attracting worldwide interest. Gaps in long-term research specifically related to this topic offer enormous opportunity to further our knowledge on the critically important issue of whale watch education and behaviour change.

REFERENCES

Andersen, M.S. & Miller, M.L. (2006). Onboard marine environmental education: Whale watching in the San Juan Islands, Washington. *Tourism in Marine Environments* 2(2), 111–118.

Bearzi, G., Politi, E., Agazzi, S. & Azzellino, A. (2005). Prey depletion caused by overfishing and the decline of marine megafauna in eastern Ionian Sea coastal waters (central Mediterranean). *Biological Conservation* 127, 373–382.

Bearzi, G., Pierantonio, N., Bonizzoni, S., Notarbartolo di Sciara, G. & Demma, M. (2010). Perception of a cetacean mass stranding in Italy: The emergence of compassion. *Aquatic Conservation: Marine and Freshwater Ecosystems* 20(6), 644–654.

Beck, L. & Cable, T. (1998). *Interpretation for the 21st Century: 15 guiding principles for interpreting nature and culture.* Champaign, IL: Sagamore Publishing.

Bierman, C. (2001). Can Cetacean Education Evolve from Educating Humans? One Approach. 2001 Poster Presentation: 14th Biennial Conference of the Society for Marine Mammalogy. Vancouver, British Columbia, Canada: Society for Marine Mammalogy.

Bierman, C. (2002a). Naturalists as facilitators of environmental responsibility. Culture of Whales Poster Presentation: 8th International Conference of the American Cetacean Society. Seattle: American Cetacean Society.

Bierman, C. (2002b). Whale watch education, a curriculum to foster environmental responsibility. Masters Thesis, Lesley University, Cambridge, MA.

Bierman, C. (2003). Whale Watch Education: What do passengers like most? 2003 Poster Presentation: 15th Biennial Conference of the Society for Marine Mammalogy. Greensboro, NC: Society for Marine Mammalogy.

Brookfield, S.D. (1987). *Developing Critical Thinkers: Challenging adults to explore alternative ways of thinking and acting.* San Francisco, CA: Jossey-Bass Publishers.

Catlin-Groves, C.L. (2012). The citizen science landscape: From volunteers to citizen sensors and beyond. *International Journal of Zoology* Article ID 349630.

Christie, S.B. (2000). The brain: Utilizing multi-sensory approaches for individual learning styles. *Education* 2(1), 327–331.

Coghlan, A. & Gooch, M. (2011). Applying a transformative learning framework to volunteer tourism. *Journal of Sustainable Tourism* 19(6), 713–728.

Cranton, P. (1994). *Understanding and Promoting Transformative Learning: A guide for educators of adults.* San Francisco, CA: Jossey-Bass Publishers.

Deissler, C.H. (2007). Educators' belief change in a situated learning environment. A dissertation submitted to the Graduate Faculty of The University of Georgia. Athens, Georgia. Unpublished data.

Domizi, D.P. (2008). Learning in context: Exploring short- and long-term experiences situated in an informal learning environment. A dissertation submitted to the Graduate Faculty of The University of Georgia. Athens, Georgia. Unpublished data.

Forestell, P.H. (1993). If leviathan has a face, does Gaia have a soul? Incorporating environmental education in marine eco-tourism programs. *Ocean & Coastal Management* 20, 267–282.

Forestell, P.H. & Kaufman, G.D. (1990). The history of whale watching in Hawai'i and its role in enhancing visitor appreciation for endangered species. In M.L. Miller & J. Auyong (Eds), *Proceedings of the 1990 Congress on Coastal and Marine Tourism*, 25–31 May 1990, Honululu, Hawai'i, USA. Newport, OR: National Coastal Resources Research & Development Institute, Vol. 2, pp. 399–407.

Gardner, H. (1983). *Frames of Mind: The theory of multiple intelligences*. New York, NY: Basic Books.

Gardner, H. (1999). *Intelligence Reframed*. New York, NY: Basic Books.

Hatch, L.T., Clark, C.W., Van Parijs, S.M., Frankel, A.S. & Ponirakis, D.W. (2012). Quantifying loss of acoustic communication space for right whales in and around a U.S. national marine sanctuary. *Conservation Biology* 26(6), 983–994.

Hovelsrud, G.K., McKenna, M. & Huntington, H.P. (2008). Marine mammal harvests and other interactions with humans. *Ecological Applications* 18(2, Suppl.), S135–S147.

Hoyt, E. (1998). Watch a whale; learn from a whale: enhancing the educational value of whale watching. In *Proceedings of the 1998 International Forum on Dolphins and Whales*. Muroran, Japan, pp. 5–19.

Hoyt, E. (2001). *Whale Watching Worldwide: Tourism numbers, expenditures, and expanding socioeconomic benefits. A Special Report From the International Fund for Animal Welfare*. Yarmouth Port, MA: IFAW.

Hoyt, E. (2005). *Marine Protected Areas for Whales, Dolphins and Porpoises: A world handbook for cetacean habitat conservation*. London: Earthscan.

Hoyt, E. (2006). *Whale Watching and Marine Ecotourism in Russia*. Chippenham, Wiltshire: Far East Russia Orca Project and WDCS, the Whale and Dolphin Conservation Society, 84.

Hoyt, E. (2009). Whale watching. In W.F. Perrin, B. Würsig & J.G.M. Thewissen (Eds), *Encyclopedia of Marine Mammals*. San Diego, CA: Academic Press, pp. 1219–1223.

Hoyt, E. (2012). *Whale Watching Blueprint – I. Setting up a marine ecotourism operation*. North Berwick, Scotland: Nature Editions.

International Fund for Animal Welfare (IFAW), World Wildlife Fund (WWF) and WDCS, the Whale and Dolphin Conservation Society. (1997). *Report of the International Workshop on the Educational Values of whale Watching*. Provincetown, MA: IFAW, WWF, and WDCS, 40.

Knudson, D.M., Cable, T.T. & Beck, L. (1995). *Interpretation of Cultural and Natural Resources*. State College, PA: Venture Publishing.

Koth, B., Field, D.R. & Clark, R.N. (1982). Cruiseship travelers to Alaska: Implications for onboard interpretation. *The Interpreter* 13(1), 39–46.

Lave, J. & Wenger, E. (1991). *Situated Learning: Legitimate peripheral participation*. Cambridge: Cambridge University Press.

Lee, T., Quinn, M.S. & Duke, D. (2006). Citizen science, highways, and wildlife: Using a web-based GIS to engage citizens in collecting wildlife information. *Ecology and Society* 11(1), 11.

Lück, M. (2003). Education on marine mammal tours as agent for conservation – But do tourists want to be educated? *Ocean & Coastal Management* 46, 943–956.

Mezirow, J. & Associates. (1990). *Fostering Critical Reflection in Adulthood: A guide to transformative and emancipatory learning*. San Francisco, CA: Jossey-Bass Publishers.

Muloin, S. (1998). Wildlife tourism: The psychological benefits of whale watching. *Pacific Tourism Review* 2, 199–213.

Neil, D., Orams, M. & Baglioni, A. (2003). Effect of previous whale watching experience on participants knowledge of, and response to, whales and whale watching. Retrieved from http://www.tangalooma.com/dolphinweb/research/papers/paper3.pdf (accessed 3 August 2010).

O'Connor, S., Campbell, R., Cortez, H. & Knowles, T. (2009). *Whale Watching Worldwide: Tourism numbers, expenditures and expanding economic benefit*. A special report prepared by Economists at Large. Yarmouth, MA: International Fund for Animal Welfare.

Orams, M.B. (1996). An interpretation model for managing marine wildlife–tourist interaction. *Journal of Sustainable Tourism* 4, 81–95.

Orams, M. (1997). The effectiveness of environmental education: Can we turn tourists into 'Greenies'? *Progress in Tourism and Hospitality Research* 3, 295–306.

Orams, M.B. & Taylor, A. (2005). Making ecotourism work: An assessment of the value of an environmental education programme on a marine mammal tour in New Zealand. In C. Ryan, S.J. Page & M. Aicken (Eds), *Taking Tourism to the Limits: Issues, concepts and managerial perspectives – Advances in tourism research*. Oxford: Elsevier Science, pp. 83–98.

Panigada, S., Pesante, G., Zanardelli, M., Capoulade, F., Gannier, A. & Weinrich, M.T. (2006). Mediterranean fin whales at risk from fatal ship strikes. *Marine Pollution Bulletin* 52, 1287–1298.

Reading with Phonics. (2002). Retrieved from http://www.earthmagneticfiled.com/multisens.html (accessed 11 February 2002).

Rennie, L.J. & McClafferty, T. (1995). Using visits to interactive science and technology centers, museums, aquaria, and zoos to promote learning in science. *Journal of Science Teacher Education*, 6(4), 175–185.

Robbins, J. & Mattila, D.K. (2004). Estimating humpback whale (*Megaptera novaeangliae*) entanglement rates on the basis of scar evidence. Report to the National Marine Fisheries Service. Order number 43ENNF030121. 22.

Russell, C.L. & Hodson, D. (2002). Whalewatching as critical science education? *Canadian Journal of Science, Mathematics and Technology Education* 2(4), 485–504.

Schell, J.W. & Schell, B.A. (2007). Communities of practice: A curriculuar model that promotes professional reasoning. In B.A. Schell & J.W. Schell (Eds), *Clinical and Professional Reasoning in Occupational Therapy*. Baltimore, MD: Lippincott, Williams & Wilkins.

Stamation, K.A., Croft, D.B., Shaughnessy, P.D., Waples, K.A. & Briggs, S.V. (2007). Educational and conservation value of whale watching. *Tourism in Marine Environments* 4(1), 41–55.

Tilden, F. (1957). *Interpreting Our Heritage*. Chapel Hill, NC: University of North Carolina Press.

Trumbull, D.J., Bonney, R., Bascom, D. & Cabral, A. (2000). Thinking scientifically during participation in a citizen-science project. *Science Education* 84, 265–275.

Valentine, P.S. (1990). *Nature-Based Tourism: A Review of Prospects and Problems*. Townsville, Australia: Author.

Weaver, D. (2001). Ecotourism. Wild about Whales. Retrieved from http://www.wildaboutwhales.com.au/ (accessed 10 October 2012).

Wise Sr., J.P., Payne, R., Wise, S.S., *et al.* (2009). A global assessment of chromium pollution using sperm whales (*Physeter macrocephalus*) as an indicator species. *Chemosphere* 75, 1461–1467.

Zeppel, H. & Muloin, S. (2008). Education and conservation benefits of interpretation on marine wildlife tours. *Tourism in Marine Environments* 5(2–3), 215–228.

11

What's in it for the whales?

Exploring the potential contribution of environmental interpretation to conservation

Mark Orams, Paul Forestell and Jonathon Spring

Introduction

The question posed in the title to this chapter is deliberately provocative and it is a question which is seldom addressed in the multitude of studies and publications on whale-watching (cetacean-based tourism) world-wide. This is despite the fact it is now widely understood that whale-watching activities (especially those that occur from vessels and in the sea) are not benign (Higham & Bejder, 2008; Parsons, 2012). A wide range of research into the effects of vessel-based watching of whales and dolphins and swimming with wild cetaceans in differing locations and on different species has shown short-term behavioural impacts and, in some cases, long-term detrimental impacts (Wade *et al.*, 2012).[1]

Thus, it is clear that the risk of negative consequences for cetaceans from tourism activity exists (Bejder *et al.*, 2006a), although it is questionable whether the degree of impact is equivalent across species (Wade *et al.*, 2012). On the human side of the relationship a growing body of research has focused on the economic, social and, to a lesser degree, cultural impacts of whale-watching (Scarpaci *et al.*, 2009). It is reported that this tourism activity has grown rapidly (Hoyt, 2001; Parsons, 2012) and that it generates significant economic revenue for operators and destinations which in turn supports communities, cultures and, in some cases, nations (O'Connor *et al.*, 2009). Indeed, this volume has presented summaries of the above in some detail. That is, there are demonstrable negative impacts or, at the least, risks of negative consequences for cetaceans targeted for tourism; and conversely, that there are a range of benefits for the human side of the relationship (economic, social, cultural and personal). Given humans' relationships with natural resources over history this is not

---

[1] Examples of studies showing impacts of tourism on cetaceans include: for the Amazonian boto *Inia geoffrensis* (de sá Alves *et al.*, 2009); for bottlenose dolphins *Tursiops* species (Bejder *et al.*, 2006b; Constantine, 2001; Constantine *et al.*, 2004; Janik & Thompson, 1996; Lusseau, 2003; Lusseau *et al.*, 2006; Miller *et al.*, 2008; Nowacek *et al.*, 2001); for Chilean dolphins *Cephalorhynchus eutropia* (Ribeiro *et al.*, 2005); for common dolphins *Delphinus* species (Neumann & Orams, 2006; Stockin *et al.*, 2008); for dusky dolphins *Lagenorhynchus obscurus* (Barr & Slooten, 1999; Dans *et al.*, 2008; Lundquist & Markowitz, 2009; Würsig *et al.*, 1997; Yin, 1999); for dwarf minke whales *Balaenoptera accutorostrata* (Birtles *et al.*, 2002); for Hector's dolphins *Cephalorhynchus hectori* (Bejder *et al.*, 1999; Green, 2004; Martinez & Orams, 2009; Nichols *et al.*, 2001); for humpback whales *Megaptera novaeangliae* (Baker & Hermann, 1989; Scheidat *et al.*, 2004);

for killer whales *Orcinus orca* (Lusseau *et al.*, 2009; Williams *et al.*, 2006, 2009); for Pacific humpback dolphins, *Sousa chinensis* (Van Parijs & Corkeron, 2001); for sperm whales *Physeter macrocephalus* (Gordon *et al.*, 1992; Richter *et al.*, 2006); for spinner dolphins *Stenella longirostris* (Courbis, 2004; Green & Calvez, 1999); for spotted dolphins *Stenella frontalis* (Ransom, 1998); and for tucuxis *Sotalia fluviatilis* (Carrera *et al.*, 2008).

Whale-watching: Sustainable Tourism and Ecological Management, eds J. Higham, L. Bejder and R. Williams. Published by Cambridge University Press. © Cambridge University Press 2014.

surprising. The Judeo-Christian view of humans having dominion over nature and that nature's purpose is to provide benefit for humankind has become the dominant world-view through the twentieth century (Geisinger, 1999), and has provided a framework for debating the morality of killing whales (Gambell, 1990). Indeed, it is rare that development associated with human activity in the post-industrialized world has benefited nature and natural resources. So, the observation that whale-watching as an activity tends to be dominated by the utilization of cetaceans for human benefit, as opposed to the whale's benefit (Forestell, 2002), is, perhaps, to be expected. This is not, however, how these activities are marketed and presented to the wider public (Corkeron, 2006). Almost universally in the popular press (print and electronic media) cetacean-based tourism activities are held up as examples of enlightened sustainable 'eco-tourism' which is both good for nature and good for tourism (and economies and tourists). For example, New Zealand's largest cetacean-based tour operator 'Whale Watch Kaikoura' has been lauded as an enlightened example of sustainable ecotourism (for example, this business was named Supreme Award winner at the 2009 Responsible Tourism Awards in London), despite evidence that their activities do impact the natural behavioural patterns of the whales they target (see Richter *et al.*, 2006).

In contrast to popular press representations of the apparent enlightened use of whales for tourism, the academic literature has tended to concentrate on the negative impacts of such tourism activities for the whales (Scarpaci *et al.*, 2009; Parsons, 2012). The growing evidence of negative impacts, which has been shown at differing locations and across a range of species (Wade *et al.*, 2012), provides a strong basis for a conclusion that the cetacean-based tourism industry may be unsustainable rather than inherently sustainable. Such a conclusion is reasonable because the findings of research on the effects of cetacean-based tourism are so widespread and the weight of evidence so strong. As a consequence, an important question to address is that

posed in the title to this chapter: what's in it for the whales? How can we justify subjecting them to interactions with tour boats, sea-kayakers, snorkelling tourists, underwater photographers and circling aircraft?

A potential answer may be found in the argument that by valuing cetaceans alive as a tourism attraction there is additional incentive to protect these animals and to ensure their long-term health and viability by also protecting their habitat. For example, the value of whales as a tourism attraction has been used at the International Whaling Commission (IWC) and in countries considering or involved in whaling as an argument for conservation and against whaling practices (Alie, 2008). Similarly, at an individual country level, a similar contention has been made using an economic valuation of whales for tourism as an incentive for their continued protection (eg. Orams, 2001). However, the most commonly touted argument for the 'use' of cetaceans for tourism is in their claimed ability to change people's perceptions, attitudes and behaviour (Anderson & Miller, 2006), either through simple direct exposure to animals in the wild (Payne, 1995), or through the explicit educational activities of tour guides (Forestell, 1990). The idea that whales and dolphins have the ability to influence people's lives and the way they choose to live them is a long-held view (Corkeron, 2004; Forestell & Kaufman, 1990; Orams, 1993; Russell, 1994). It fits within the broader context of nature-based tourism or 'ecotourism' and the opportunity to have tourists, as a consequence of their interaction with nature, become more conservation-minded and actively contribute to the betterment of nature, habitats and ecosystems.

Thus, the overall aim of this chapter is to explore the role and potential benefits of education and interpretation in cetacean-based tourism. More specifically, the chapter will consider whether education programmes could mitigate the short-term impacts of approaches by humans, and enhance the long-term changes in human attitude and behaviour needed to protect marine mammals and their environment.

Ecotourism and wildlife

It has been argued that ecotourism is an attempt to construct a model of tourism that is appropriate for the environment and which contributes to the fiscal and employment needs of the society within which it operates (Fennell, 2003; Wearing & Neil, 1999; Weaver, 2001). An educational element has been noted as a key feature in definitions of ecotourism (Fennell, 2003). Typically, this has been in the form of environmental education information and services provided as part of the ecotour experience (Weiler & Davis, 1993).

Wildlife viewing is often a central feature of ecotourism products. A wildlife tourism product can be considered to be within the parameters of ecotourism if it adheres to the principles of sustainability associated with ecotourism (Fennell, 2003; Weaver, 2001). Wildlife-based ecotourism is therefore viewed as a sustainable use of wildlife for tourism purposes which has, as a central component of the experience, an educational outcome for the tourists (Higginbottom & Tribe, 2004).

Wildlife-based ecotourism occurs in a variety of contexts. These activities can occur in natural habitats or captive environments, such as zoos, or in settings that incorporate both natural and captive environments. However, there has been a dramatic increase in ventures that enable tourists to have a direct experience with wildlife in natural settings (Higginbottom, 2004; Newsome *et al.*, 2005; Tisdell & Wilson, 2004). Wildlife–visitor interactions in natural settings can range from watching wildlife, spotlighting nocturnal wildlife, swimming with marine wildlife, touching wildlife, to feeding wildlife (Higginbottom, 2004; Orams, 2002; Rodger *et al.*, 2007).

Environmental education/interpretation and wildlife-based tourism

Environmental education is generally considered in the context of formal educational activities based on an approved curriculum with pre-determined learning objectives and formal assessment of student performance related to those objectives. It is usually delivered as a lesson or lecture in school, college or university classroom setting or via field-trip based learning experiences. Consequently, while the term environmental education is often used in association with wildlife-based tourism, it is seldom the correct label for what occurs in those settings. Environmental interpretation is a related specialty, but one where the education is less formal and occurs in natural settings where visitors are made aware of the intrinsic values of the resources of the site and its overall relationship to human experience (Ham, 1992). A direct experience may refer to a visit to a nature-based site and/or a wildlife encounter (Forestell, 1992). Non-personal interpretation covers media such as signs, panels, brochures, dioramas and audio-visual material. Personal interpretation describes a direct experience at a site where a designated person (such as a guide) talks to visitors and facilitates learning (Brochu & Merriman, 2002).

Environmental education theory and practices have informed the learning process in environmental interpretation (Butler & Hvenegaard, 2002; Knapp, 2007). Both environmental education and interpretation involve explaining the influence of human behaviour and actions on natural ecosystems. A goal of environmental education is to develop an ethic of responsible environmental behaviour of students and this also is a goal for many environmental interpretation programmes (Benton, 2009; Butler & Hvenegaard, 2002; Knapp, 1994).

The term 'interpretation' was first used in relation to guiding by Enos Mills in the early twentieth century (Beck & Cable, 2002). Beck and Cable (2002) note close parallels between Mills' opinions on guiding and the principles of interpretation developed by Tilden (1977) in his seminal 1957 work. Tilden (1977) viewed interpretation as an art whose aim was to stimulate interest in site resources. An ideal interpretive experience adds to a visitor's appreciation of the resources encountered (McArthur & Hall, 1996). From a management perspective, an early objective of interpretation was to reduce the practice of taking mementos such as stalagmites (Beck & Cable, 2002; Merriman & Brochu, 2002). Since those early days, interpretation has developed into a speciality in its own right and has diversified in terms of its thinking, application and contribution as a management tool (Beck & Cable, 2002).

Environmental interpretation has been advocated as one of a number of complementary strategies that can contribute to a sustainable form of wildlife-based tourism (Ham & Weiler, 2002; Moscardo *et al.*, 2004; Zeppel, 2008). One of the ways interpretation can do this is through the provision of information to the visitor that addresses the implications of the visitor's behaviour on the animals observed. Personal face-to-face interpretation facilitated by a tour guide provides visitors with the opportunity to interact with staff for personally relevant information about what they are experiencing (Brochu & Merriman, 2002; Ryan & Dewar, 1995).

A central tenet in the discourse about interpretation is its relationship with information. Interpretation relies upon information and without information interpretation becomes simply a form of entertainment (Beck & Cable, 2002). However, Tilden (1977: 9) was concerned about the relationship between interpretation and information. This was articulated in the second of his six principles about interpretation: 'Information, as such, is not interpretation. Interpretation is revelation based on information. But they are entirely different things. However all interpretation includes information'.

Tilden (1977) emphasized the distance between interpretation and information due to his concern about the approach of United States National Parks Service staff in their provision of interpretation:

Others were unduly impressed by the word 'education'. The word, coming from well-known educators suggested direct and detailed instruction. Thus, in so many cases that we have observed, the provocation to the visitor to search out meanings for himself, and join the expedition like a fellow discoverer, was sometimes submerged in a high tide of facts, perfectly accurate, perfectly ineffectual. (Tilden, 1977: 36)

As the role of interpretation is seen as provoking the recipient to think or be moved, there is a view that passing of information to the visitor is not enough to act as the sole catalyst for creating any meaningful level of connection with nature.

The relationship between behavioural change and environmental interpretation

Tilden (1977: 32) introduced the idea that 'the chief aim of interpretation is not instruction but provocation'. He differentiated interpretation from guided talks that relayed information through the interpreter's attempt to make a topic relevant to visitors' life experience. Tilden (1977) emphasized the need to create a connection between the visitor and on-site resources through deepening their knowledge of the site. A range of natural resource management agencies have attempted to adopt the approach advocated by Tilden as a tool for both managing visitor behaviour and enhancing visitor experiences (Ham, 2007). For example, the National Parks Service in the United States of America uses interpretation programmes to explain its work, and to foster appreciation and protection of on-site resources (Beck & Cable, 2002). Others have advocated the application of meaningful and high-quality environmental interpretation programmes as an integral part of wildlife-based tourism experiences including whale-watching (e.g. Forestell, 1992; Orams, 1996). They have, however, cautioned that a sound understanding and application of the potential connection between environmental interpretation, wildlife-based tourism and behaviour change in tourists is essential in conducting such activities.

Relevant theory and research into interpretation and behaviour change

Ham and Krumpe (1996) consider behavioural change through interpretation in the context of the theories of Reasoned Action and Planned Behaviour. These theories postulate that behavioural

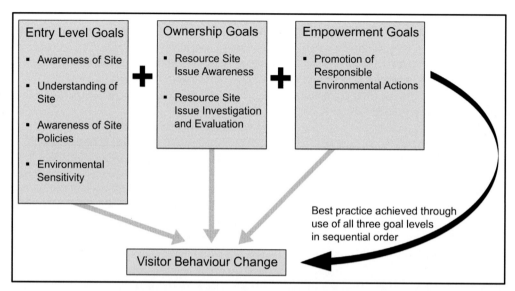

Figure 11.1 Environmental Interpretation Behaviour Change Model (derived from Knapp, 2007: 56).

intention, from where actual behaviour origi-nates, is affected by three constructs: attitude towards behaviour, subjective norm and perceived behavioural control (Ajzen, 1992; Fishbein & Man-fredo, 1992). These three factors are controlled by three respective sets of salient beliefs; behavioural beliefs, normative beliefs, and control beliefs. Con-tent and delivery of content through different inter-pretation media can attempt to change behaviour through thematic interpretation that focuses 'com-munication content on the primary beliefs salient to the targeted behaviour' (Ham & Krumpe, 1996: 18).

Knapp (2007) is cautionary about expectations relating to the role of interpretation in facilitating behavioural change particularly due to the short duration of most interpretation experiences. He suggests a model (Figure 11.1) to develop interpre-tation based on three sets of goals that are sequen-tially hierarchical. These are entry-level goals, such as awareness of the site and environmental sensi-tivity, then ownership goals that allow visitors to consider issues relating to resources at the site, and finally empowerment goals that promote responsi-ble environmental actions. The inclusion of these

goals into any personal interpretation experience would be determined by its duration and available knowledge about the audience.

One of the challenges of personal interpretation is that each individual participating in the trip has a set of beliefs, attitudes and interests that can be fundamentally different to the other people par-ticipating in the trip (Beck & Cable, 2002; Ham, 1992). Thus, attempts to deliver a standardized and scripted interpretive talk or presentation are seldom effective in achieving the desired educational out-comes. This is further compromised by the signifi-cant low conversion rate that occurs post-trip, when stated intentions of participants to take positive action which benefits conservation seldom convert into actual behavioural change (Orams, 1997). This degradation over time in potential positive envi-ronmental actions is analogous to the waning of interest in media subjects, termed the 'issue atten-tion cycle', where a problem 'suddenly leaps into prominence, remains there for a short time, and then – though still largely unresolved – gradually fades from the centre of public attention' (Downs, 1972: 38).

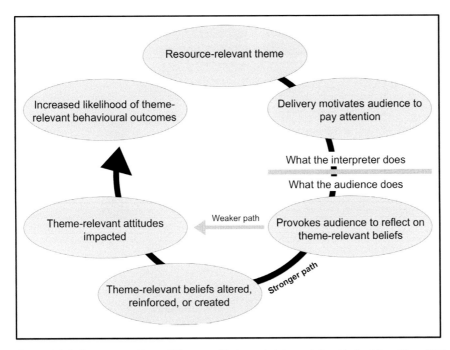

Figure 11.2 Pathways to making a difference with thematic interpretation (derived from Ham, 2007: 47).

Ham (2007) explored how interpreters could make a difference to audiences' perceptions and behaviour. In this regard, he tried to respond to the following key questions, drawing upon a range of literature on cognitive and behavioural psychology. Does increasing knowledge impact on attitudes? Is there any 'mental effort' needed of visitors for attitudinal changes to occur? Do certain attitudes predict certain behaviours? How may interpreters affect visitors' behaviours? The findings from this study highlighted that an increase of 'factual knowledge' (2007: 42) and an ability to evaluate 'good and bad' does not automatically lead to changes in attitude. There is a 'mental effort' needed for attitudinal changes to occur. The 'new knowledge' received must, in some way, provoke a reflection upon visitors' own beliefs. However, in order for the visitor to react, this reflection must be agreed with intrinsically. Strong efforts for long-lasting behavioural effects were generally found to be ineffective. Short, specific messages directed at

desired on-site behaviour were found to be more effective because visitors may implement such actions relatively quickly (e.g. remaining on a designated pathway). Predictions that certain behaviours follow certain attitudes are difficult to make.

Ham (2007) contends that the focus of interpretation should be on attitudes towards a specific behaviour as opposed to general environmental conservation attitudes. Hence, he argues there is the need to target the attitudes 'about the behavior we want them to engage in' (p. 45). Ham (2007) proposes that research should focus on comparing pertinent beliefs of those people enacting a certain attitude with those who do not take action. As a consequence, interpreters may be able to understand differences and may then target audiences particularly with those beliefs held by 'enactors'.

Ham (2007) developed a framework (Figure 11.2) that differentiates between the role of the interpreter (guide) and the audience (visitor or tourist).

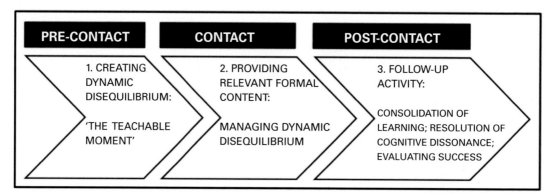

Figure 11.3 Three steps to an effective interpretation programme (after Forestell & Kaufman, 1990; adapted by Orams, 1996).

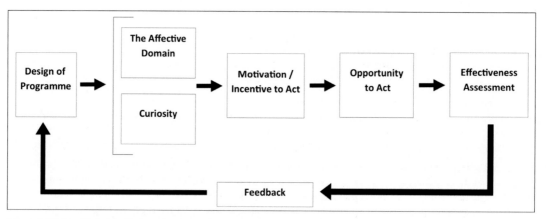

Figure 11.4 The Orams Model for Effective Interpretation (Orams, 1996: 86).

The model proposes that desired behavioural outcomes are best achieved when interpreters provide theme-related, enjoyable, relevant and organized information to audiences. He particularly emphasizes that the 'theme' and its accessibility to the audience are key points for reinforcing or changing attitudes and behaviours and states that 'interpreters who make compelling presentation of strongly relevant themes [beliefs] stand the greatest chance of having enduring impacts on their audiences' (Ham, 2007: 48).

In contrast to Ham's theme-based approach, Lundberg (1997) advocates for thesis-based interpretation whereby an idea, hypothesis or statement that challenges visitors' attitudes or beliefs is used.

This is more consistent with Forestell and Kaufman (1990), who argue that the creation of dynamic disequilibrium within the visitor can provide a 'teachable moment' and that the management of cognitive dissonance can consolidate learning during a guided interpretive experience. Included in the Forestell and Kaufman model (Figure 11.3) is the goal to get tourists to participate in activities that could facilitate the changing of existing attitudes and behaviours (Forestell, 1992).

Orams (1996) extended the Forestell and Kaufman (1990) model and included the creation of an opportunity to act, as well as to use outcomes assessment to improve the programme design (see Figure 11.4). His testing of this model in a cetacean-based

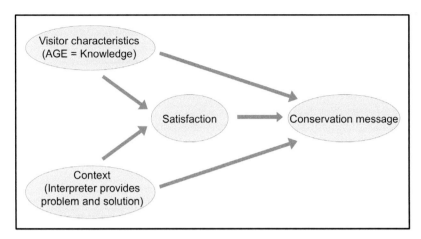

Figure 11.5 The mechanism behind effective communication of conservation messages (derived from Peake *et al.*, 2009: 122).

tourism scenario (Orams, 1997) provided evidence that a proportion of tourists stated they had engaged in some kind of pro-conservation behaviour several months after their experience.

Peake *et al.* (2009) examined the influence of interpreters (tour guides) on the understanding of the conservation message delivered as part of a whale-watching experience in Hervey Bay, Australia. They found that the understanding of the conservation message and related action taken is influenced by three main factors. Broadly, these are first, the age of the respondent (which is correlated with their experience and knowledge); second, the information provided by the tour guide with regard to the specific conservation issue or problem; and third, the provision of specific solutions to the problem outlined, which is a stimulus for visitor empowerment and a sense of responsibility. Peake *et al.* (2009) highlight that it is particularly important that the communication from the tour guide highlights the important role that the visitor (tourist) can play in achieving positive conservation outcomes (see Figure 11.5).

Similarly, Powell and Ham (2008) looked at the influence of an ecotourism operator's interpretation programme on the pro-environmental behaviour among visitors to the Galapagos Islands. Their findings show that well-designed and delivered programmes can facilitate support for conservation activities, may foster pro-environmental attitudes and may generally enhance 'philanthropic behaviour' (p. 474). However, they also point out that knowledge gain is not necessarily always followed by pro-conservation behaviour. Changes may particularly occur if environmental ideas or concepts which visitors already understand ('salient beliefs') are targeted consistently. Moreover, the opportunity to take action (e.g. donation of money) at the time of the tour has greater impact than the opportunity to act after the experience is completed (i.e. solicitation to donate in the future).

Zeppel (2008) summarized the benefits of education and conservation within marine-based settings for both the visitors and the host environment. She showed that empathy gained during wildlife encounters can foster direct changes towards pro-conservation and environmentally sound behaviour (e.g. adherence to guidelines). In addition, Zeppel and Muloin (2008) found that the actual encounter with wildlife when combined with the provision of an interpretive programme can have a cognitive and affective impact on visitors, at least for a short time. On the other hand, Marion and Reid (2007) examined the impact of educational programmes on visitors' environmental attitudes and the implementation of behavioural change. They conclude that many studies find that educational programmes indeed had an effect upon visitors' knowledge but, to a lesser degree, on their subsequent behaviours. They argue that conservation messages should focus on a particular goal and aim to reason within an ecological rather than within a social framework for action-taking. They also state that the timing (on-site), the credibility (and consistency) of the transmitted message, and the means of communication itself, have a major influence. The characteristics of the audience have an influence on the deciphering of the message. Audiences may sometimes be less likely to adhere to desired behaviours because these may contradict what has actually been already experienced in nature. Hence, messages should target specific beliefs instead of specific socio-demographics. Finally, they state that messages should be practical and simple as these seem more effective.

The interpretation models reviewed above provide useful discussion points. Knapp's (2007) model recognizes that there are different levels in visitors' environmental literacy. Incorporating Knapp's (2007) model into an interpretation programme would involve integrating the three sets of goals into either one or a series of interpretive encounters for the same visitor, with key issues being the scale of repeat visitation and the number of interpreters to cater to different visitor needs. Ham's (2007) model allows interpretation programme planners to conceptualize how the interpretive encounter may impact on different visitors and highlights the need for interpretation to be thematic, organized, relevant and enjoyable. Peake *et al.* (2009) and Forestell and Kaufman (1990) emphasize the importance of the guide in provoking the visitor to think more deeply about what they are experiencing and the latter structures the interpretation within the different stages of a wildlife encounter. This is a critical point, because the majority of interpretation theory and models proposed are based on interpretive opportunities that are relatively predictable and allow for the careful design and structuring of the interpretive programme before the experience. Wildlife tourism situations such as whale-watching

are unpredictable in their nature (Forestell, 1992). For example, the size of platform, the sea conditions, visibility, the sighting of whales and whale behaviour (including duration of encounters, proximity and number of whales and surface activity) are highly variable. As a consequence, operators and whale-watch tour guides have to be flexible in their delivery of interpretive material.

Whale-watching, interpretation and conservation outcomes

The application of relevant theory on interpretation, management and behavioural change in the context of cetacean-based tourism experiences has been relatively limited. However, a number of authors have proposed approaches based on the successful utilization of interpretation in other contexts (e.g. Ballantyne & Packer, 2005; Ballantyne *et al.*, 2008; Forestell, 1990; Lück, 2003, 2008; Orams, 1996, 1997; Peake *et al.*, 2009) and others have proposed and tested environmental interpretation programmes based on their own experiences within the industry (e.g. Forestell & Kaufman, 2007).

Marion and Reid's (2007) points in terms of whale-watching, the relevance of the message in relation to the whale encounter, behaviour change, the trip itself and the social setting of the experience are useful issues to explore. The nature of the whale-watching encounter defines the relevance of the message. One of the great challenges facing successful interpretation programmes during marine mammal tourism encounters is the wide range of size and types of platforms and operations (Hoyt, 2005). The diversity of whale-watching platforms and experiences around the world is dramatic (O'Connor *et al.*, 2009), and includes land-based, marine and aerial tours. Marine tours range from four-passenger *jukung* (long, narrow, wooden outrigger-type dug-outs) in Bali, to 400-passenger dedicated whale-watch vessels off the east coast of the USA. With such a wide range of platforms, it becomes very challenging to propose a 'best-practice' model for an interpretation programme. Flexibility has to exist within the structure of the interpretation so that it can accommodate smaller, more one-on-one settings without a formal guide, as well as larger platforms with multiple crew and guides, and an onboard PA system. In general, the overall behaviour of the whale-watch operator underpins the credibility of any conservation-based message. How the operator decides to behave in the presence of the whale and what the operator chooses to provide, emit, dispose, recycle or re-use during and after the trip will likely affect how the operator's patrons respond to calls to change their own behaviour. The whale-watching trip provides an important opportunity to model environmentally responsible behaviour (Forestell & Kaufman, 2007).

In keeping with Knapp's (2007) model, even on very small platforms, operators have the opportunity to promote achievement of 'entry-level goals' by modelling a number of important behaviours that can influence passenger mindsets. These include: managing passenger expectations so that, even if sightings are few or non-existent, the experience is still positive; operating in a respectful and non-intrusive way when the target species is present; avoiding pollution through poorly maintained or operated engines; preventing material from being thrown or blown into the ocean; demonstrating awareness of, and compliance with, management regulations. While it is more challenging on small platforms to promote 'ownership' or 'empowerment' goals, these also can be promoted in creative ways. In almost all whale-watch venues, even when the platforms are very small, there are owner/operator associations that work in conjunction with the broader tourism supply chain (e.g. accommodation, agents and resource management officials) to promote and protect the growth and sustainability of the industry. These organizations are well placed to promote conservation and research messages for tourists before and after the whale-watch as part of the overall tourism experience at a destination.

In venues where larger platforms operate, allowing for more formal interpretation programmes, the interaction between interpreters and tourists occurs within a complex set of social relationships that can include family, friends, strangers, and other employees. It is important therefore, that the quality of the staff and the way the whale-watch business operates is considered. One of the most important contributions that environmental interpretation can make is in the effective training of the whale-watch operators (skippers, guides and other staff) themselves, in the collaborative pursuit of all three levels of goals described by Knapp (2007).

What is understood about the conservation outcomes derived from whale-watching experiences remains limited (Corkeron, 2006; Stamation *et al.*, 2007; Parsons, 2012). While there have been a number of useful advances in developing, testing and refining models directed at prompting behaviour change in tourists via interpretation associated with whale-watch experiences, much remains to be done. Particular priorities for future research are:

1. developing and testing the effectiveness of training programmes for marine mammal tour operators (skippers, guides and other staff);
2. exploring and understanding the specific situations/opportunities where learning for conservation outcomes can be best achieved or created (the 'teachable moment');
3. developing and examining post-visit follow-up activities in order to better understand how conversion rates to more environmentally responsible behaviour can be improved. More specifically, how conservation-oriented behaviour change can be 'cemented' to stop 'back-sliding';
4. exploring how social media and information technology-based mobile devices might be used to enhance environmental interpretation and improve conversion rates to more environmentally responsible behaviour;
5. finally, in a wider sense an important question is: how can high-quality environmental interpretation programmes associated with whale-watching contribute to a wider social move towards more sustainable lifestyles?

It is important that management regimes directed at minimizing the potential negative impacts of cetacean-based tourism also include efforts at maximizing the conservation benefits of these activities. In some places, such an approach is already mandated. For example, New Zealand requires that permitted marine mammal tour operators offer services of 'sufficient educational value' (New Zealand Marine Mammals Protection Regulations, 1992: Section 6h). While it is not clearly defined what this value is, the intent of the regulations is to require marine mammal tourism to contribute to the conservation of the species being targeted (New Zealand Marine Mammals Protection Regulations, 1992). Orams and Taylor (2005) found that an application of an interpretation programme onboard a marine mammal tour operator under this regime did have some longer-term conservation benefits. Most importantly they concluded that:

> The operator surveyed does not conduct their educational programme in order to simply meet a requirement of their marine mammal tourism permit, but that they do so because they appear to strongly believe in taking advantage of the opportunity to maximise the benefit of the tourism experience for their patrons. Moreover, the passion and dedication of the staff onboard the vessel revealed a deep commitment to marine conservation and a genuine compassion for marine mammals and their future. (p. 97)

This point is important. Mandating high-value interpretation programmes in order to achieve conservation outcomes is unlikely to be a successful strategy (Stamation *et al.*, 2007). High-quality interpretation (as with any educational activity) will always be dependent on the passion, abilities and effectiveness of those involved in its planning and delivery. Fortunately, there are many marine mammal tour operators worldwide who remain committed to contributing to the longer-term health and viability of the animals they focus their tours on. What remains to be done is to take environmental interpretation in the marine mammal tourism scenario to a higher level of professionalism whereby resources, curriculum, teacher training, and outcomes assessment approach the levels common in more traditional educational fields. There are a number of very experienced operators who deliver

high-quality interpretation services who can be utilized as 'best-practice' models to help lift the standard of others. In addition, management agencies and relevant non-governmental organizations have an important role to play in facilitating and building the educational capability of marine mammal tour operators and their staff.

Conclusions

The limited number of studies into the effectiveness of environmental interpretation has shown that a carefully designed and implemented programme applied in a cetacean-based tourism setting can achieve some positive environmental outcomes. However, even these programmes show a low conversion rate from intentions to change behaviour to actual behaviour change and longer-term studies reveal significant drop off in those changes in behaviour over the longer term. While such results are sobering, they are not unusual. A wide variety of programmes and significant investment has been made in social marketing campaigns designed to reduce damaging behaviours such as smoking, family violence, driving under the influence of alcohol and the pirating of music and movies (Kotler & Lee, 2008). The wide range of work undertaken in these endeavours reveals the same well-understood phenomenon: that is, human behaviour is notoriously difficult to change.

Despite the growing amount of research that focuses on increasing the effectiveness of environmental interpretation in stimulating more environmentally responsible behaviour in tourists, it is clear that this outcome is ambitious and difficult to achieve (Kollmuss & Agyeman, 2002). Claims and assumptions that cetacean-based tourism experiences on their own (i.e. in the absence of high-quality interpretation programmes) engender more environmentally conscious consumers and contribute to a more sustainable world are, therefore, naïve and reveal an ignorance of the challenges associated with changing human behavioural patterns.

As is often the case when empirical research is undertaken on activities or phenomena that are assumed to have specific actions and related consequences, the research reveals that the assumptions are flawed. This is true for the assumption that cetacean tourists become more environmentally responsible as a consequence of their experiences. Dolphins, whales and porpoises are indeed charismatic and, as evidenced by the massive growth in cetacean-based tourism globally (O'Connor *et al.*, 2009), their attraction for people has spurred the growth of a massive industry. The economic and social benefits are clear. A variety of work has quantified or estimated these consequences for communities and nations (e.g. see Hoyt, 2001). In addition, a wide variety of research has also shown that there are negative consequences for many cetacean species subject to tourism attention (Parsons, 2012). In some cases, these findings have resulted in management agencies changing the number, type, location and timeframes available to tour operators in an attempt to mitigate impacts, but it remains unclear whether such regulation is any more successful than efforts utilizing interpretation programmes in mitigating those impacts (Forestell, 2008).

With regard to the tourists themselves and the consequences of their experiences with cetaceans, there is little empirical research to draw upon. Furthermore, the contention that ecotourism is an active contributor to environmental conservation is also an assumption that has seldom been tested in an empirical sense. What research has shown, not surprisingly, is that cetacean-tourists are generally satisfied with their experiences and that they express a great deal of affection for and interest in cetaceans and conservation. What remains to be shown is what specific, measured and long-term changes in these tourists' behaviour and lifestyle occurs as a consequence of their experiences. We consider this an important priority for further research, or more specifically, how to move from immediate gratification and rapid decay of experience/intentions, to longer-term influences on visitor behaviour post-experience.

It is interesting to consider that for most business operations, even those in the tourism industry, there is a need for that business to invest in the basis for the attraction of customers to that business. For example, Disney World has made massive investment in its theme parks that attract customers (tourists) to their business. In addition to this initial capital investment, the company must maintain their assets, depreciate their value over time and reinvest in upgrading their attractions. However, for cetacean-based tour operators there is no capital investment required in order to create the attraction that is the basis of their business. While such operators will point to their investment in vessels, safety equipment, visitor centres, marketing material and, in some cases, costs of permitting, the primary attraction of their business (whales, dolphins and porpoises) has not required any capital investment on their part. The availability of cetaceans for tourism is not, however, a free public good. Their protection has been hard-fought. In many coastal areas they were brought to the brink of extinction as a consequence of whaling. Non-governmental organizations and conservation-focused groups and individuals spent considerable time, effort and money working to protect cetaceans. This work continues as newer and growing threats of fisheries by-catch, habitat degradation, human-created ocean noise, shipping and vessel strikes and disturbance, and climate change all threaten cetaceans and other marine life.

There is a pronounced need for the cetacean-based tourism industry to consider its obligations in terms of actively contributing to the attraction upon which their business is based. In addition to assuming greater responsibility for ensuring participants achieve the 'entry-level goals' (Knapp, 2007) described above, they should invest in the conservation and welfare of cetaceans. One of the most important ways they can do this is through creating, implementing and measuring the effectiveness of state-of-the-art environmental interpretation efforts as a core aspect of their business operations. Furthermore, their activities should support research directed at measuring the consequence of their 'use' of cetaceans for their clients. Such feedback is a central part of any sound educational process and the findings can be used to further develop, refine and improve the educational services provided as an integral and central part of the tourism experience.

To deliver entertaining information, have satisfied customers and a financially viable tourism business is not enough. When the question 'what's in it for the whales' is put to a cetacean-based tour operator, specific answers which demonstrate the actual benefits for the animals which are the basis of their business are in order. Platitudes, ecotourism awards and assumed conservation benefits are not enough. State-of-the-art, carefully designed, well-delivered and thoroughly researched and tested interpretation programmes can provide critical answers to this fundamentally important question. Enlightened operators will see that, in the growing competitive market for tourism focused on charismatic wildlife, such an approach will be an important competitive advantage that is good for business and has long-term benefits for the whales, dolphins and porpoises as well.

We recognize that it has become passé among many in the marine mammal science community to be critical of the marine mammal tourism industry (Corkeron, 2004, 2006). This is, perhaps, understandable given the wide range of published research that shows that cetacean-based tourism induces changes in natural behaviour patterns of targeted animals (Parsons, 2012). In contrast, there exists little evidence of effective interpretation programmes to either mitigate or justify such effects (Stamation et al., 2007). It is, however, important to recognize that cetaceans respond and adapt to a wide range of pressures and changes in their ecosystems, and not all species are equally impacted by human-generated disturbance (Wade et al., 2012). Changes in behaviour as a response to new stimuli should not be assumed to always equate to negative outcomes. We are also mindful of a number of locations where decades of relatively intensive whale-watching pressure have been coincident with a continued recovery of the targeted populations of cetaceans. For example, in Maui,

Hawai'i, there are now more humpback whales resident during their breeding season than there were 50 years ago (Barlow et al., 2011). A similar pattern of recovery has occurred for the same species on the east coast of Australia (Noad et al., 2011). Off the west coast of North America, grey whale numbers have recovered spectacularly (Punt & Wade, 2010). Thus, intensive and long-term whale-watching activity does not necessarily equate to negative conservation outcomes for whales. The critical challenge now is to maximize the conservation benefits for cetaceans and the environment through the opportunity to effectively educate the millions of tourists who interact with whales and dolphins every year.

REFERENCES

Ajzen, I. (1992). Persuasive communication theory in social psychology: A historical perspective. In M. Manfredo (Ed.), *Influencing Human Behavior*. Champaign, IL: Sagamore, pp. 1–27.

Alie, K. (2008). Whales: More valuable alive than dead? A question for decision makers in eastern Caribbean whale-watching destinations. *Business, Finance & Economics in Emerging Economies* 3(1), 177–190.

Anderson, M.S. & Miller, M.L. (2006). Onboard marine environmental education: Whale watching in the San Juan Islands, Washington. *Tourism in Marine Environments* 2(2), 111–118.

Baker, C.S. & Herman, L.M. (1989). *Behavioral Responses of Summering Humpback Whales to Vessel Traffic: Experimental and opportunistic observations*. Washington, DC: US National Park Service.

Ballantyne, R. & Packer, J. (2005). Promoting environmentally sustainable attitudes and behavior through free-choice learning experiences: What is the state of the game? *Environmental Education Research* 11(3), 281–295.

Ballantyne, R., Packer, J. & Hughes, K. (2008). Tourist support for conservation messages and sustainable practices in wildlife tourism experiences. *Tourism Management* 1–8. DOI: 10.1016/j.tourman.2008.1011.1003.

Barlow, J., Calambokidis, J., Falcone, E., et al. (2011). Humpback whale abundance in the North Pacific estimated by photographic capture–recapture with bias correction from simulation studies. *Marine Mammal Science* 27, 793–818.

Barr, K. & Slooten, E. (1999). *Effects of Tourism on Dusky Dolphins at Kaikoura.* Conservation Advisory Science Notes 229. Wellington: Department of Conservation.

Beck, L. & Cable, T. (2002). *Interpretation for the 21st Century: Fifteen guiding principles for interpretation nature and culture* (2nd edn). Champaign, IL: Sagamore Publishing.

Bejder, L., Dawson, S.M. & Harraway, J.A. (1999). Responses by Hector's dolphins to boats and swimmers in Porpoise Bay, New Zealand. *Marine Mammal Science* 15(3), 738–750.

Bejder, L., Samuels, A., Whitehead, H. & Gales, N. (2006a). Interpreting short-term behavioural responses to disturbance within a longitudinal perspective. *Animal Behaviour* 72, 1149–1158.

Bejder, L., Samuels, A., Whitehead, H., *et al.* (2006b). Decline in the relative abundance of bottlenose dolphins exposed to long-term disturbance. *Conservation Biology* 20(6), 1791–1798.

Benton, G.M. (2009). From principle to practice: Four conceptions of interpretation. *Journal of Interpretation Research* 14(1), 7–31.

Birtles, A., Arnold, P.W. & Dunstan, A. (2002). Commercial swim programs with dwarf minke whales on the northern Great Barrier Reef, Australia: Some characteristics of the encounters with management implications. *Australian Mammalogy* 24(1), 23 38.

Brochu, L. & Merriman, T. (2002). *Personal Interpretation: Connecting your audience to heritage resources.* Fort Collins, CO: National Association for Interpretation.

Butler, J.R. & Hvenegaard, G.T. (2002). Interpretation and environmental education. In P. Dearden & R. Rollins (Eds), *Parks and Protected Areas in Canada: Planning and management* (2nd edn). Ontario, Canada: Oxford University Press, pp. 179–203.

Carrera, M.L., Favaro, E.G.P. & Souto, A. (2008). The response of marine tucuxis (*Sotalia fluviatilis*) towards tourist boats involves avoidance behaviour and a reduction in foraging. *Animal Welfare* 17(2), 117–123.

Constantine, R. (2001). Increased avoidance of swimmers by wild bottlenose dolphins (*Tursiops truncatus*) due to long-term exposure to swim-with-dolphin tourism. *Marine Mammal Science* 17(4), 689–702.

Constantine, R., Brunton, D.H. & Dennis, T. (2004). Dolphin-watching tour boats change bottlenose dolphin (*Tursiops truncatus*) behaviour. *Biological Conservation* 117(3), 299–307.

Corkeron, P.J. (2004). Whale watching, iconography, and marine conservation. *Conservation Biology* 18(3), 847–849.

Corkeron, P.J. (2006). How shall we watch whales? In D.M. Lavigne (Ed.), *Gaining Ground: In pursuit of ecological sustainability.* Guelph: International Fund for Animal Welfare, pp. 161–170.

Courbis, S.S. (2004). Behavior of Hawaiian spinner dolphins (*Stenella longirostris*) in response to vessels/swimmers. Masters Thesis, San Francisco State University.

Dans, S.L., Crespo, E.A., Pedraza, S.N., Degrati, M. & Garaffo, G.V. (2008). Dusky dolphin and tourist interaction: Effect on diurnal feeding behaviour. *Marine Ecology Progress Series* 369, 287–296.

de sá Alves, L.C.P., Andriolo, A. & Orams, M.B. (2009). Feeding Amazonian boto (*Inia geoffrensis*) as a tourism attraction. A path toward tragedy? *Proceedings of the World Congress on Coastal and Marine Tourism.* Port Elizabeth, South Africa. CD-ROM.

Downs, A. (1972). Up and down with ecology – The 'Issue-Attention Cycle'. *The Public Interest*, 28–38.

Fennell, D.A. (2003). *Ecotourism* (2nd edn). London: Routledge.

Fishbein, M. & Manfredo, M. (1992). A theory of behavior change. In M. Manfredo (Ed.), *Influencing Human Behavior.* Champaign, IL: Sagamore, pp. 29–50.

Forestell, P.H. (1990). Marine education and ocean tourism: Replacing parasitism with symbiosis. In M.L. Miller & J. Auyong (Eds), *Proceedings of the 1990 Congress on Coastal and Marine Tourism (25–31 May 1990, Honolulu, Hawaii, USA).* Newport, OR: National Coastal Resources Research & Development Institute, Vol. 1, pp. 35–39.

Forestell, P.H. (1992). The anatomy of a whalewatch: Marine tourism and environmental education. *Current* 11(1), 9–15.

Forestell, P.H. (2002). Popular culture and literature. In W. Perrin, B. Würsig & H.G.M. Thewisson (Eds), *Encyclopedia of Marine Mammals.* San Diego, CA: Academic Press, pp. 957–974.

Forestell, P.H. (2008). Protecting the ocean by regulating whale watching: The sound of one hand clapping. In J.E.S. Higham & M. Lück (Eds), *Maine Wildlife and Tourism Management: Insights from the natural and social sciences.* Wallingford: CABI Publishing, pp. 272–293.

Forestell, P.H. & Kaufman, G.D. (1990). The history of whale watching in Hawaii and its role in enhancing visitor appreciation for endangered species. In M.L. Miller & J. Auyong (Eds), *Proceedings of the 1990 Congress on Coastal and Marine Tourism (25–31 May 1990, Honolulu, Hawaii, USA).* Newport, OR: National Coastal Resources Research & Development Institute, Vol. 2, pp. 399–407.

Forestell, P.H. & Kaufman, G.D. (2007). Speaking from experience: Whale and dolphin watching boats as venues for marine conservation education. In M. Lück (Ed.), *Proceedings of the 5th International Coastal and Marine Tourism Congress (September 11–15, Auckland, New Zealand)*. Auckland, New Zealand: AUT University, pp. 63–67.

Gambell, R. (1990). Whaling – A Christian position. *Science & Christian Belief* 2(1), 15–24.

Geisinger, A. (1999). Sustainable development and the domination of nature: Spreading the seed of the western ideology of nature. *Boston College Environmental Affairs Law Review* 27(1), 43–73.

Gordon, J., Leaper, R., Hartley, F.G. & Chappell, O. (1992). *Effects of Whale-watching Vessels on the Surface and Underwater Acoustic Behaviour of Sperm Whales off Kaikoura, New Zealand*. New Zealand Department of Conservation Science and Research Series 52. Wellington: Department of Conservation.

Green, E. (2004). *Review of the Effects of Whale Watching on Cetaceans and an Assessment of Dolphin Watching Tourism in Canterbury*. Christchurch, New Zealand: Department of Conservation Canterbury Conservancy.

Green, M. & Calvez, L. (1999). Research on Hawaiian spinner dolphins in Kealakekua Bay, Hawaii. In K. Dudzinski, T. Frohoff & T. Spradlin (Eds), *Abstracts: Wild Dolphin Swim Program Workshop*. 13th Biennial Conference on the Biology of Marine Mammals. Maui, Hawaii: The Society of Marine Mammalogy.

Ham, S.H. (1992). *Environmental Interpretation: A practical guide for people with big ideas and small budgets*. Golden, CO: Fulcrum.

Ham, S.H. (2007). Can interpretation really make a difference? Answers to four questions from cognitive and behavioural psychology. In *Interpreting World Heritage 2007: Proceedings of National Association for Interpreters (NAI) International Conference, Vancouver, Canada*. Fort Collins, CO: NAI, pp. 42–52.

Ham, S.H. & Krumpe, E.E. (1996). Identifying audiences and messages for non-formal environmental education – A theoretical framework. *Journal of Interpretation Research* 1(1), 11–23.

Ham, S.H. & Weiler, B. (2002). Interpretation as the centrepiece of sustainable wildlife tourism. In R. Harris, T. Griffin & B. Weiler (Eds), *Sustainable Tourism: A global perspective*. Oxford: Butterworth Heinemann.

Higginbottom, K. (2004). Managing impacts of wildlife tourism on wildlife. In. K. Higginbottom (Ed.), *Wildlife Tourism: Impacts, management and planning*. Altona, VIC: Common Ground Publishing, pp. 211–229.

Higginbottom, K. & Tribe, A. (2004). Contributions of wildlife tourism to conservation. In. K. Higginbottom (Ed.), *Wildlife Tourism: Impacts, management and planning*. Altona, VIC: Common Ground Publishing, pp. 99–123.

Higham, J.E.S. & Bejder, L. (2008). Managing wildlife-based tourism: Edging slowly towards sustainability? *Current Issues in Tourism* 11(1), 75–83.

Hoyt, E. (2001). *Whale watching 2001 – Worldwide tourism numbers, expenditures and expanding socioeconomic benefits*. Yarmouth Port, MA: International Fund for Animal Welfare (IFAW) and the United Nations Environmental Program (UNEP).

Hoyt, E. (2005). *Marine Protected Areas for Whales, Dolphins and Porpoises*. London: Earthscan.

Janik, V.M. & Thompson, P.M. (1996). Changes in the surfacing patterns of bottlenose dolphins in response to boat traffic. *Marine Mammal Science* 12(4), 597–602.

Kotler, P. & Lee, N.R. (2008). *Social Marketing: Influencing behaviours for good*. Los Angeles, CA: Sage.

Knapp, D.H. (1994). Validating a framework for goals for program development in environmental interpretation. Unpublished Doctoral dissertation, Southern Illinois University, Carbondale.

Knapp, D.H. (2007). *Applied Interpretation: Putting research into practice*. Fort Collins, CO: National Association for Interpretation.

Kollmuss, A. & Agyeman, J. (2002). Mind the gap: Why do people act environmentally and what are the barriers to pro-environmental behavior? *Environmental Education Research* 8(3), 239–260.

Lück, M. (2003). Education on marine mammal tours as agent for conservation – But do tourists want to be educated? *Ocean & Coastal Management* 46(9–10), 943–956.

Lück, M. (2008). Managing marine wildlife experiences: The role of visitor interpretation programmes. In J. Higham & M. Lück (Eds), *Marine Wildlife and Tourism Management: Insights from the natural and social sciences*. Wallingford: CABI, pp. 334–346.

Lundberg, A.E. (1997). Towards a thesis-based interpretation. *Legacy* 8(2), 14–17, 30–31.

Lundquist, D.J. & Markowitz, T.M. (2009). Effects of tourism on behaviour and movement patterns of dusky dolphin groups monitored from shore stations. In T.M. Markowitz, S. DuFresne & B. Würsig (Eds), *Tourism Effects on Dusky Dolphins at Kaikoura, New Zealand*. Wellington, New Zealand: Department of Conservation.

Lusseau, D. (2003). Effects of tour boats on the behavior of bottlenose dolphins: Using Markov chains to model anthropogenic impacts. *Conservation Biology* 17(6), 1785–1793.

Lusseau, D., Slooten, E. & Currey, R.J. (2006). Unsustainable dolphin watching activities in Fiordland, New Zealand. *Tourism in Marine Environments* 3, 173–178.

Lusseau, D., Bain, D.E., Williams, R. & Smith, J.C. (2009). Vessel traffic disrupts the foraging behavior of southern resident killer whales *Orcinus orca*. *Endangered Species Research* 6, 211–221.

Marion, J. & Reid, S. (2007). Minimising visitor impacts to protected areas: The efficacy of low impact education programmes. *Journal of Sustainable Tourism* 15(1), 5–27.

Martinez, E. & Orams, M.B. (2009). *Report on the Use of Stones as Auditory Stimulants During Swim-with-dolphins Encounters with the South Island Hector's dolphins (Cepahlorhynchus hectori hectori) in Akaroa Harbour, New Zealand.* Canterbury, New Zealand: Department of Conservation.

McArthur, S. & Hall, C.M. (1996). Interpretation: Principles and practice. In C.M. Hall & S. McArthur (Eds), *Heritage Management in Australia and New Zealand* (2nd ed.). Melbourne: Oxford University Press, pp. 88–106.

Merriman, T. & Brochu, L. (2002). *The History of Heritage Interpretation in the United States.* Fort Collins, CO: National Association for Interpretation.

Miller, L.J., Solangi, M. & Kuczaj II, S.A. (2008). Immediate response of Atlantic bottlenose dolphins to high speed personal watercraft in the Mississippi Sound. *Journal of the Marine Biological Association of the United Kingdom* 88(6), 1139–1143.

Moscardo, G., Woods, B. & Saltzer, R. (2004). The role of interpretation in wildlife tourism. In K. Higginbottom (Ed.), *Wildlife Tourism: Impacts, management and planning.* Altona, VIC: Common Ground Publishing, pp. 231–251.

Neumann, D.R. & Orams, M.B. (2006). The impacts of tourism on common dolphins in the Bay of Plenty, New Zealand. *Aquatic Mammals* 32, 1–9.

Newsome, D., Dowling, R. & Moore, S. (2005). *Wildlife Tourism.* Clevedon: Channel View Publications.

New Zealand Marine Mammals Protection Regulations. (1992). *New Zealand Marine Mammals Protection Regulations.* Wellington, New Zealand: New Zealand Government Printer.

Nichols, C., Stone, G., Hutt, A., Brown, J. & Yoshinaga, A. (2001). *Observations of Interactions between Hector's dolphins (Cephalorhynchus hectori), Boats and People at Akaroa Harbour, New Zealand.* Science for Conservation 178. Wellington, New Zealand: Department of Conservation.

Noad, M., Dunlop, R., Paton, D. & Cato, D. (2011). Absolute and relative abundance estimates of Australian east coast humpback whales (*Megaptera novaeangliae*). *Journal of Cetacean Research Management* (Special Issue 3), 243–252.

Nowacek, S.M., Wells, R.S. & Solow, A.R. (2001). Short-term effects of boat traffic on bottlenose dolphins, *Tursiops truncatus*, in Sarasota Bay, Florida. *Marine Mammal Science* 17(4), 673–688.

O'Connor, S., Campbell, R., Cortez, H. & Knowles, T. (2009). *Whale Watching Worldwide: Tourism numbers, expenditures and expanding economic benefits.* A special report from the International Fund for Animal Welfare, Yarmouth, MA. USA. Prepared by Economists at Large.

Orams, M.B. (1993). The role of education in managing marine wildlife – Tourist interaction. *Proceedings of the 7th National Marine Education Society of Australasia Conference.* Brisbane, Queensland: Marine Education Society of Australasia, pp. 6–11.

Orams, M.B. (1996). An interpretation model for managing marine wildlife–tourist interaction. *Journal of Sustainable Tourism* 4(4), 81–95.

Orams, M.B. (1997). The effectiveness of environmental education: Can we turn tourists into greenies? *Progress in Tourism and Hospitality Research* 3(4), 295–306.

Orams, M.B. (2001). From whale hunting to whale watching in Tonga: A sustainable future? *Journal of Sustainable Tourism* 9(2), 128–146.

Orams, M.B. (2002). Feeding wildlife as a tourism attraction: A review of issues and impacts. *Tourism Management* 23(3), 281–293.

Orams, M.B. & Taylor, A. (2005). Making ecotourism work: An assessment of the value of an environmental education programme on a marine mammal tour in New Zealand. In C. Ryan, S.J. Page & M. Aicken (Eds), *Taking Tourism to the Limits: Issues, concepts and managerial perspectives – Advances in tourism research.* Oxford: Elsevier Science, pp. 83–98.

Parsons, E. (2012). The negative impacts of whale-watching. *Journal of Marine Biology.* doi:10.1155/2012/807294

Payne, R. (1995). *Among Whales.* New York, NY: Bantam Doubleday Dell Publishing.

Peake, S., Innes, P. & Dyer, P. (2009). Ecotourism and conservation: Factors influencing effective conservation messages. *Journal of Sustainable Tourism* 17(1), 107–127.

Powell, R. & Ham, S. (2008). Can ecotourism interpretation really lead to pro-conservation knowledge, attitudes and behavior? Evidence from the Galapagos Islands. *Journal of Sustainable Tourism* 16(4), 467–489.

Punt, A. & Wade, P. (2010). *Population status of the eastern North Pacific stock of gray whales in 2009.* U.S. Department of Commerce, NOAA Tech. Memo. NMFS-AFSC-207, 43 pp.

Ransom, A.B. (1998). Vessel and human impact monitoring of the dolphins of Little Bahama Bank. Masters thesis. San Francisco, CA: San Francisco State University.

Ribeiro, S., Viddi, F.A. & Freitas, T.R.O. (2005). Behavioural responses by Chilean dolphins (*Cephalorhynchus eutropia*) to boats in Yaldad bay, Southern Chile. *Aquatic Mammals* 31(2), 234–242.

Richter, C., Dawson, S. & Slooten, E. (2006). Impacts of commercial whale watching on male sperm whales at Kaikoura, New Zealand. *Marine Mammal Science* 22(1), 46–63.

Rodger, K., Moore, S.A. & Newsome, D. (2007). Wildlife tours in Australia: Characteristics, the place of science and sustainable futures. *Journal of Sustainable Tourism* 15(2), 160–179.

Russell, C.L. (1994). Ecotourism as experiential education? *The Journal of Experiential Education* 17(1), 16–22.

Ryan, C. & Dewar, K. (1995). Evaluating the communication process between interpreter and visitor. *Tourism Management* 16(4), 295–303.

Scarpaci, C., Lück, M. & Parsons, E.C.M. (2009). Recent advances in whalewatching research: 2008–2009. *Tourism in Marine Environments* 6(1), 39–50.

Scheidat, M., Castro, C., Gonzalez, J. & Williams, R. (2004). Behavioural responses of humpback whales (*Megaptera novaeangliae*) to whalewatching boats near Isla de la Plata, Machalilla National Park, Ecuador. *Journal of Cetacean Research and Management* 6(1), 63–68.

Stamation, K., Croft, D., Shaughnessy, P., Waples, K. & Briggs, S. (2007). Educational and conservation value of whale watching. *Tourism in Marine Environments*, 4, 41–55.

Stockin, K.A., Lusseau, D., Binedell, V., Wiseman, N. & Orams, M.B. (2008). Tourism affects the behavioural budget of the common dolphin *Delphinus* sp. in the Hauraki Gulf, New Zealand. *Marine Ecology Progress Series* 355, 287–295.

Tilden, F. (1977). *Interpreting Our Heritage* (3rd edn). Chapel Hill, NC: University of North Carolina.

Tisdell, C. & Wilson, C. (2004). Economics of wildlife tourism. In K. Higginbottom (Ed.), *Wildlife Tourism: Impacts, management and planning.* Altona, VIC: Common Ground Publishing, pp. 145–163.

Van Parijs, S.M. & Corkeron, P.J. (2001). Boat traffic affects the acoustic behaviour of Pacific humpback dolphins, *Sousa chinensis. Journal of the Marine Biological Association of the UK* 81(3), 533–538.

Wade, P., Reeves, R. & Mesnick, S. (2012). Social and behavioural factors in cetacean responses to overexploitation: Are odontocetes less 'resilient' than mysticetes? *Journal of Marine Biology.* doi:10.1155/2012/567276

Wearing, S. & Neil, J. (1999). *Ecotourism: Impacts, potentials and possibilities.* Oxford: Butterworth Heinemann

Weaver, D.B. (2001). *Ecotourism.* Milton, Australia: Wiley.

Weiler, B. & Davis, D. (1993). An exploratory investigation into the roles of the nature-based tour leader. *Tourism Management* 14(2), 91–98.

Williams, R., Lusseau, D. & Hammond, P.S. (2006). Estimating relative energetic costs of human disturbance to killer whales (*Orcinus orca*). *Biological Conservation* 133, 301–311.

Williams, R., Bain, D.E., Smith, J.C. & Lusseau, D. (2009). Effects of vessels on behaviour patterns of individual southern resident killer whales *Orcinus orca. Endangered Species Research* 6, 199–209.

Würsig, B., Cipriano, F., Slooten, L., Constantine, R., Barr, K. & Yin, S. (1997). Dusky dolphins (*Lagenorhynchus obscurus*) off New Zealand: Status of present knowledge. *Reports of the International Whaling Commission* 47, 715–722.

Yin, S.E. (1999). Movement patterns, behaviors and whistle sounds of dolphin groups off Kaikoura, New Zealand. Masters thesis, Texas A & M University.

Zeppel, H. (2008). Education and conservation benefits of marine wildlife tours: Developing free-choice learning experiences. *The Journal of Environmental Education* 39(3), 3–17.

Zeppel, H. & Muloin, S. (2008). Conservation and education benefits of interpretation of marine wildlife tours. *Tourism in Marine Environments* 5(2–3), 215–227.

12

Integrating traditional ecological knowledge and community engagement in marine mammal protected areas

Naomi McIntosh, Kepā Maly and John N. Kittinger

Introduction

Increasing impact on the world's oceans from development, pollution, overharvesting of resources and other human-related causes stems from our failure to recognize that our oceans are not an indestructible and infinite resource. Today, signs of stress and the diminishing health of our oceans are commonplace and globally we have begun to acknowledge that our oceans are in crisis (Pauly *et al.*, 1998; Mora *et al.*, 2009). Marine life and vital coastal habitats are under increasing pressure from overuse, and the cumulative effects of human activities in the ocean reduces its ability to function as a healthy ecosystem (Halpern *et al.*, 2008). The costs of pollution, overfishing, and destruction of habitats threaten local food security, livelihoods, and the health and welfare of human communities reliant on the ocean (Kent, 1997; Bell *et al.*, 2009; Halpern *et al.*, 2012). Current approaches to management, planning and policy have thus far relied too heavily on fragmented and piecemeal governance, rather than systematic, ecosystem-based approaches that are needed to preserve ocean ecosystem health and the resilience of human communities (Crowder *et al.*, 2006; Spalding *et al.*, 2008; Pressey & Bottrill, 2009).

Impacts and depletion of marine mammals is a subset of the global crisis facing our ocean. Persistent and major threats to marine mammals include historical exploitation, habitat loss and by-catch, pollution, ocean noise, and overharvesting (Erbe, 2002; Schipper *et al.*, 2008; Simmonds & Isaac, 2007; McClenachan & Cooper, 2008). Reducing these threats requires a systematic approach to management of marine mammals, with multi-dimensional responses that are both species- and location-specific (Taylor *et al.*, 2000; Marsh *et al.*, 2003).

Protected areas, sustainable ecotourism and customary approaches to stewardship can all be effective strategies to both conserve marine mammals and benefit the economic, cultural, social and recreational values of coastal communities. Across the world, marine protected areas (MPAs) provide a foundation for sustainable recreation and tourism-based activities (Agardy, 1993; Eagles *et al.*, 2002). MPAs attract visitors in much the same way as terrestrial parks (Abate, 2009). Tourists expecting the marine life to be more available and abundant in MPAs than in non-protected areas are drawn to these places, often providing significant economic benefits to local communities (e.g. Oracion *et al.*, 2005). Additionally, tourism in MPAs provides a direct connection between visitors and their host communities, with opportunities for cross-cultural dialogue and learning.

However, managing the relationship between the visitor, host and place requires a thoughtful approach to sustain the cultural, aesthetic and

Whale-watching: Sustainable Tourism and Ecological Management, eds J. Higham, L. Bejder and R. Williams.
Published by Cambridge University Press. © Cambridge University Press 2014.

intrinsic values of a given environment and its resources for both the visitors and the host community. A growing number of local communities where marine mammal protected areas occur are working to strengthen local influence and accountability for the health and long-term sustainability of their natural resources through revitalization of local traditions and resource knowledge. Areas designated to protect whales, dolphins and other marine mammals can be a source of inspiration for local communities to work on collaborative initiatives that minimize the negative impacts of tourism on the environment and revive, revitalize and provide learning opportunities for cultural heritage.

The purpose of this chapter is to provide an overview of how traditional ecological knowledge, customary practices and community engagement can be integrated effectively into sustainable ecotourism in marine mammal protected areas. We draw on two promising case studies to show how ecotourism operations, visitor experiences and host communities can benefit from community engagement and the incorporation of traditional practices, principles and values. Drawing on these examples and relevant literature, we explore opportunities to promote a broader understanding of the importance of these habitats and species, and the role of sustainable ecotourism, marine mammal protected areas, and cultural heritage in engendering a greater sense of responsibility for marine mammal populations. We hope this information will be useful to managers, business owners, researchers and other constituencies seeking to enhance the public awareness of the challenges and opportunities for marine mammal conservation and to those seeking to increase ecotourism's social, cultural, political and economic value within their communities.

Marine mammal protected areas

The designation of marine protected areas has been used as a tool to benefit marine mammals since 1869 with the declaration of the Pribilof Islands as a 'special reservation' to bring regulations to the commercial hunt for northern fur seals (Reeves, 2000). However, it was some time before the first MPA aimed at specifically protecting cetaceans was established in Mexico in 1972 by Presidential decree. Laguna Ojo de Libre National Gray Whale Refuge (Scammon's Lagoon) was designated to protect a prime grey whale (*Eschrichtius robustus*) mating and calving lagoon (Hoyt, 2005).

An MPA is defined by the World Conservation Union (IUCN) as 'any area of intertidal or subtidal terrain, together with its overlying water and associated flora, fauna, historical and cultural features, which has been reserved by law or other effective means to protect part or all of the enclosed environment' (IUCN, 1994; Hoyt, 2005: 18). The concept of an MPA embraces a number of different types of designations and involves a variety of area-based approaches to marine conservation designed to fulfil and address different objectives. Sanctuaries, parks, reserves, preserves, refuges and marine managed areas are all considered to fall within the broad meaning of the term 'marine protected area' (Reeves, 2000). MPAs can be designated as 'no take' areas where extractive activities such as fishing and plant removal are prohibited. However, many MPAs are designed as 'multiple use' areas where specific recreation and commercial uses compatible with the resource protection goals of the area are allowed.

In the USA, MPAs span a range of habitats including the open ocean, coastal area, inter-tidal zones, estuaries and the Great Lakes. They also vary widely in purpose, legal authorities, agencies, management approaches, level of protection and restrictions on human uses. MPAs in the USA have been established at all levels of government and are managed mainly by various federal, state/territories and tribal agencies through legislative acts.

The National Oceanic and Atmospheric Administration's Office of National Marine Sanctuaries (ONMS) serves as the trustee for a system of 14 MPAs that encompass almost 500,000 km^2 (150,000 square miles) of ocean and Great Lakes waters. The system includes 13 marine sanctuary sites and one marine national monument.[1] Marine sanctuaries in the US protected a variety of habitats including deep-ocean gardens, coral reefs, whale migration corridors, deep-sea canyons, seamounts, kelp forests, sea grass beds, historically significant shipwrecks, and other important cultural and archaeological sites. They range in size from 0.66 km^2 (one-quarter square mile) in Fagatele Bay (American Samoa), to more than 362,075 km^2 (139,000 square miles) at the Papahānaumokuākea Marine National Monument, located in the northwest Hawaiian Archipelago (Table 12.1).

  Core programmes managed by the ONMS focus on conservation, scientific research and education and outreach. These programmes include benefits such as support for research and monitoring the condition of important sanctuary resources and the promotion of public understanding and appreciation of the value of sanctuaries and their natural and cultural resources. In effect, US National Marine Sanctuaries are managed areas to promote long-term protection, biological diversity, uses that are sustainable and opportunities that enhance the economic, natural environment and cultural benefits to the community. Most sites within the Sanctuary system have resident or migratory marine mammal populations and contribute to marine mammal conservation by preserving habitat, reducing threats to coastal and marine environments, enhancing appreciation and awareness, and by promoting careful use within these natural protected areas (Reeves, 2000). In addition, some sites such as Gerry E. Studds Stellwagen Bank, Hawaiian Islands Humpback Whale, Channel Islands, Monterey Bay, Gulf of the Farallones and Olympic Coast National Marine

Sanctuaries specifically manage activities through regulations and guidelines that more directly target protection for marine mammals and their habitats within their designated areas.

  The legislative authority of the National Marine Sanctuaries Act can be a useful management tool in non-regulatory ways as well and has the potential to extend, enhance and/or complement other regulatory authorities that protect the marine environment by offering benefits that other programmes do not (Abate, 2009). The place-based management focus of US national marine sanctuaries provides significant opportunities to apply an integrated approach to management that considers the entire ecosystem and diverse human dimensions with ocean environments. Such approaches are often termed 'ecosystem-based management', or EBM, which is defined as 'an integrated approach to management that considers the entire ecosystem, including humans. The goal of EBM is to maintain an ecosystem in a healthy, productive and resilient condition so that it can provide the services humans want and need. EBM differs from conventional approaches that usually focus on a single species, sector, activity, or concern; it considers the cumulative impacts of different sectors' (McLeod *et al.*, 2005). Ecosystem-based management approaches are based on the understanding that protection of a species requires broad collaborative efforts that often require multijurisdictional authorities to work together in a collaborative, adaptive and inclusive decision-making approach.

Traditional ecological knowledge and customary management practices

Many marine mammal protected areas occur in areas where coastal cultures have a long history and association with marine mammals and the environments they occupy. For example, marine mammal protected areas in the Hawaiian Islands, New Zealand, the Arctic, and other parts of the world have been established in areas where local cultures and customs have long-standing relationships

---

[1] The Papahānaumokuākea Marine National Monument, located in the northwestern Hawaiian Islands, is administered jointly by three co-trustees – the Department of Commerce through NOAA's Office of National Marine Sanctuaries and the National Marine Fisheries Service Pacific Islands Regional Office, the Department of Interior through the US Fish and Wildlife Service's Pacific Region National Wildlife Refuge System and Pacific Islands Fish and Wildlife Office and the State of Hawai'i through the Department of Land and Natural Resources Division of Aquatic Resources and Division of Forestry and Wildlife (Kittinger *et al.*, 2010).

Table 12.1 Designation of US Marine Sanctuaries and Area.

US National Marine Sanctuary site	Protected resources	Year designated	Area (km^2)	Area (sq. miles)	Area (sq. nms)	Area (acres)
USS Monitor (Virginia/North Carolina)	Wreck of ironclad Civil War ship	1975	2.33	0.90	0.7	576
Channel Islands (California)	Kelp forests, rocky shores, fisheries, marine mammals, endangered species	1980	3818	1474	1113.1	943,360
Gulf of the Farallones (California)	Coastal beaches, fisheries, endangered whales, estuaries, seabirds	1981	3250	1280	966.3	803,200
Gray's Reef (Georgia)	Patchy limestone reefs, endanged or threatened whales, turtles, fisheries	1981	60	22	16.7	14,720
National Marine Sanctuary of American Samoa – formerly Fagatele Bay National Marine Sanctuary (American Samoa)	Coral tropical reef ecosystem in eroded volcanic crater	1986	35,024	13,523	10,211.5	8,654,720
Cordell Bank (California)	Productive upwelling area around pinnacles, ridges and the bank	1989	1362	529	399.5	336,640
Florida Keys (Florida)	Third largest coral reef system in the world, shallow near-shore habitats, fisheries, endangered species, shipwrecks	1990	9845	3801	2870.2	2,432,640
Flower Garden Banks (Texas)	Three underwater banks of healthy offshore coral reefs, endangered turtles	1992	145	56	42.4	35,840
Monterey Bay (California)	Deep marine canyons, kelp forests, rich fishing grounds, elephant seals, sea otters	1992	13,776	5319	4016.5	3,404,160
Gerry E. Studds Stellwagen Bank (Massachusetts)	Endangered whales and habitat around the sand and gravel bank, shipwrecks, fisheries	1992	2191	846	638.9	541,440
Hawaiian Islands Humpback Whale (Hawai'i)	Endangered whale breeding and calving grounds	1992	3548	1366	1031.3	876,800
Olympic Coast (Washington)	Isolated shoreline, kelp forests, offshore seabird colonies, shipwrecks	1994	8259	3189	2407.9	2,040,960
Thunder Bay (Michigan)	Two centuries of shipwrecks, maritime heritage sites	2000	1160	448	338.4	286,720
National Marine Sanctuaries			82,440	31,854	24,053	20,371,776
Papahanaumokuakea Marine National Monument (Hawai'i)	Coral reefs, shipwrecks and maritime heritage sites, deep water around uninhabited chain of small islands and atolls, seabirds	2006	362,062	139,793	105,560	89,467,520
National Marine Sanctuary System		1972	444,502	171,647	129,613	109,839,296

Source: Office of National Marine Sanctuaries.

with marine mammal species. These customs derive from historical human–environment interactions, and have evolved in a place-based manner into situated knowledge systems often called traditional or local ecological knowledge (Kittinger *et al.*, 2013). Traditional or local ecological knowledge (TEK or LEK) is defined as a cumulative system of knowledge, practice, and belief, which evolves through adaptation and is perpetuated through intergenerational cultural transmission (Berkes, 1999). In many parts of the world, TEK systems form the basis for customary management practices that are designed to regulate the use, access and transfer of resources (Cinner & Aswani, 2007). Although TEK/LEK systems and customary management practices often have historical origins, it is critical to emphasize that customary management systems are dynamic and continually evolving through adaptive processes, including the introduction, loss and syncretization of knowledge (Berkes *et al.*, 1998).

There are many different paths to achieving an EBM approach to managing marine mammal protected areas, and in areas with indigenous cultures and communities, TEK/LEK and customary management practices can play a major role in achieving EBM. The influence of conservation education, for example, within marine mammal protected areas can be substantial in increasing stakeholder awareness and helping promote responsible behaviour towards the environment and the local cultures (Leisher *et al.*, 2012; Gibson & Puniwai, 2006). Efforts to improve public appreciation and enhance awareness through education and outreach initiatives have successfully fostered a sense of respect, interest and cooperation among various members of the public in marine protected areas generally (Leisher *et al.*, 2012), including marine mammal protected areas such as the Hawaiian Islands Humpback Whale National Marine Sanctuary (Hoyt, 2011). Similarly, many whale-watching destinations around the world have received credit for positive benefits to host communities economically, scientifically and in terms of education and conservation (Hoyt, 2001).

An essential factor for achieving such success is ensuring quality standards are integrated as part of the whale-watch experience. In many nations and regions around the world, whale-watching is managed through both regulations and guidelines and enhanced by incorporating education and outreach components. In New Zealand, for example, national policy directs commercial whale-watching operations to ensure no adverse effects on marine mammal species and encourages positive benefits for conservation, management and protection. In addition, specific guidelines recommend whale-watch operations provide sufficient educational value to participants or to the public (IFAW, 2000). Among a list of educational values identified during a 1997 IFAW Workshop on the Educational Values of Whale Watching, whales were noted to be emblems for promoting awareness of endangered species and habitat protection and whale-watching recognized as an opportunity for people across all ages and cultures to become familiar and involved with environmental issues and conservation efforts on a personal, local, regional, national and international level (IFAW, WWF & WDCS, 1997). However, research has also indicated that increasing responsible behaviour as a result of marine mammal tourism experiences is particularly difficult and although the awareness and knowledge of tourists about marine mammals was significantly modified, there was no significant improvement in the conservation values of participants detected in these experiences (Orams, 2002).

To maximize the education benefits of whale-watching, Hoyt (2001) asserts it would be worthwhile to explore how whale-watching can be managed to diminish pressure on the coastal and ocean habitats within communities by creating complementary efforts to promote education and awareness to conserve cetaceans in a way that recognizes the economic, cultural, social and recreational needs of the place and its people, thereby reducing the costs and increasing the benefits. There is a wide variety of information that can be shared and presented during a whale-watching excursion including ecosystem information focusing on the reasons for the presence of whales in a particular location, information about other species present and also on geographical features of the habitat. In addition, historical information can also be provided to educate people about the history and significance of the coastal cultures in the area (IFAW, WWF & WDCS, 1997), and information about the local cultural connections and traditions associated with the species. For example, in Hawai'i and Oceania, whales and other marine megafauna such as turtles have special significance in Native Hawaiian culture (Tsuha, 2008; Rudrud, 2010). Similarly, Arctic communities share a close relationship with marine mammals and these species can constitute a large portion of the subsistence diet for these communities (Lovecraft & Meek, 2010). Marine mammal species can carry special cultural significance and associated TEK systems and also represent key ecological services provided by coastal ecosystems to communities (Dale & Armitage, 2011; Lovecraft & Meek, 2010).

There exists potential to improve whale-watching tourism by seeking opportunities within marine mammal protected areas to create holistic learning opportunities that include not only biological information about the species and its environment, but also cultural knowledge, indigenous values and customary practices associated with marine mammals. Such approaches have significant opportunity to advance public understanding of the species and to advance understanding about the importance of cultural knowledge of the people connected to the place. Culture-based education approaches can be an effective mechanism for increasing cultural understanding and respect for the health and long-term sustainability of natural resources within marine mammal protected areas. Higgins-Desbiolles (2009), for example, examined the role of ecotourism by some indigenous communities to teach indigenous values in the hope of fostering transformations in ecological conscientiousness. She contends the value of this knowledge to ecologists and others who seek to obtain this knowledge in order to inform contemporary planning and environmental management. Stewart-Harawira (2005) suggests that because some aspects of traditional knowledge remain fixed, they provide a framework by which new experiences and situations are

understood and become remade giving them meaning in our time. She argues that the study of culture and ethnicity is vitally important in developing pedagogies for better ways of being in the world and indigenous cultural knowledge is profoundly relevant to this endeavour. Indigenous ways of retaining traditional values while engaging with demands of contemporary living could serve as models of more sustainable living for non-indigenous societies (Kikiloi, 2010).

There are various pathways toward integrating culture-based education approaches to foster more environmentally responsible behaviour (Thaman, 1993, 2002). Engaging in activities to promote shared responsibility and inspire positive actions that supports benefits for the entire community and directs attention to both the natural and cultural environment of the host community offers many inspirational possibilities. Below, we describe two examples of culture-based programmes in existing whale-watching tourism operations. These case studies are meant to illustrate the potential of such approaches in broadening the scope of whale-watching operations to include local communities and cultures.

Whale Watch Kaikoura, New Zealand

There have been several published case studies documenting the success and challenges of tourism development associated with Whale Watch Kaikoura in New Zealand (Orams, 2002; Curtin, 2003; Hoyt, 2007; Department of Sustainability, Environment, Water, Population and Communities, 2009). This operation was formed in 1987 and flourished, subsequently transforming Kaikoura from a small fishing and farming town to an international tourist destination (see Chapter 22; Horn *et al.*, 1998). Here, we focus on highlighting Kaikoura as an example that illustrates how TEK and customary practices have been successfully integrated into an ecotourism programme.

Kaikoura, a small coastal community located on the northeastern coast of New Zealand's South Island, is known internationally for its dramatic scenery and unique oceanic features. Lying close to the shore is the Kaikoura underwater canyon, which stretches 60 km long and 1200 m deep and is home to a diversity of marine mammals, including sperm whales that are rarely found close to the shore. Kaikoura canyon currently has no protective status; however, there are plans in Kaikoura to create a marine reserve, a marine mammal sanctuary and to achieve world heritage status. Kaikoura itself is a place of special significance to the indigenous Māori of New Zealand, who have inhabited the area for around 1000 years (Orams, 2002). Traditionally in the Māori worldview, all life is connected. Like other indigenous cultures, Māori believe there is a deep kinship between humans and the natural world. People are not viewed as being superior to the natural order; they are understood to be part of it (Te Ahukaramu, 2009). *Kaitiakitanga* is the Māori way of expressing this connection. *Kaitiakitanga* means guardianship and protection and is a term that is widely used throughout the environmental field in New Zealand. In this context, *Kaitiakitanga* today expresses traditional ideas and values offering guidance in response to current issues and challenges. It emphasizes that people's interactions with other species should be responsible, respectful and conducted properly.

Today, Whale Watch Kaikoura is the only operator permitted by the New Zealand government to provide boat-based whale-watch tours in Kaikoura. According to Kauahi Ngapora, the company's Chief Operating Officer, 'Whale Watch Kaikoura is a company committed to providing a world-class whale watching experience while carefully managing the use of a rare natural resource within a unique environment' (Reeves, 2009). As a Māori-owned company, Whale Watch Kaikoura is dedicated to providing employment for Kaikoura Māori and also assumes the role of *Kaitiaki* (guardian) of the natural environment. In this way, Whale Watch Kaikoura embraces the philosophy that people, the land, the sea and all living things are connected as one. To fulfil their *Kaitiaki* responsibilities, Whale Watch Kaikoura supports a range of community projects to ensure the benefits of tourism are further

spread throughout the community. Whale Watch invests in the future of the community by providing work experience to students from both the high school and the Kaikoura Centre for Continuing Education, they donate money to local schools to help students in need and offer reduced rates to school and community groups to participate in whale-watching (Horn *et al.*, 1998; Hoyt, 2005). Thus, Whale Watch Kaikoura financially supports education initiatives and helps members of the community to become stronger ambassadors for the natural resources within the area. In addition, Whale Watch Kaikoura demonstrates their commitment to the care and protection of the environment by working to minimize waste, promote recycling, use eco-friendly products, reduce energy consumption, minimize impacts to wildlife and provide financial support for marine mammal research conducted in Kaikoura. A key factor of this case study is that Whale Watch Kaikoura recognizes that the health of the community is reflected in its environment and vice versa. The organization works to incorporate management practices to help increase the benefits of whale-watching to the host community while also working to reduce the potential costs of whale-watching on the environment. Furthermore, by adopting indigenous values, such as *Kaitiaki* (guardian) as a guiding principle, Whale Watch Kaikoura demonstrates an understanding of what is needed to harmonize the preservation of valuable tourism resources with opening them up to the public.

Trilogy Excursions, Maui, Hawaiian Islands

Trilogy Excursions is another example of how whale-watching operators can create sustainable livelihoods while still protecting marine mammal populations and giving back to communities to benefit the economic, cultural, social and recreational values of the area. Trilogy Excursions is a family-owned and operated sailing company established in 1973. The company is based in Lahaina, Maui and provides ocean adventure tours on the Hawaiian islands of Maui and Lāna'i. The company offers a variety of tours including sailing, snorkelling, SCUBA diving, whale-watching and sunset sails, and many of the company's excursions occur in the Hawaiian Islands Humpback Whale National Marine Sanctuary (HIHWNMS). According to Jim Coon, Chief Executive Officer, the company is serious about its responsibility to support the local community on Maui and Lāna'i. Each year, Trilogy hosts a turkey giveaway on the island of Lāna'i to celebrate the national Thanksgiving holiday and to give back to the community. The resident population on Lāna'i is less than 2000, and the giveaway seeks to provide over 1000 families with food for the holiday. Trilogy has also been able to partner with other organizations to offer other community services during the giveaway such as free health screening, children's environmental educational activities and opportunities to evaluate and assess needs within the community, providing a cost-efficient way for these organizations to perform outreach to a large number of people in the community. The company prides itself in being able to attract staff with good moral ethics by incorporating strong family values into the workplace.

Trilogy crewmembers also started the Blue 'Āina Campaign, a monthly reef cleanup that targets local surfing spots and reflects the Native Hawaiian concept of *malama 'āina*, a cultural concept that promotes taking care and giving back to the land and sea. These concepts are also core concepts of the HIHWNMS, a long-standing marine mammal protected area that is increasingly engaged with local communities and indigenous principles of management in its transition toward an ecosystem-based management approach (HIHWNMS Sanctuary Advisory Council, 2011, 2012). Trilogy's Blue 'Āina Campaign has raised over US$350,000 to support local non-profit organizations. Each month, Trilogy donates its boats, employees donate their time, and community volunteers pay a small fee to participate in the reef cleanup. Proceeds benefit a different local non-profit organization each month. Trilogy also supports programmes to teach students sailing skills. The company employs Hawaiian cultural practitioners to provide training for all of their

staff to learn about Hawaiian culture and values. In 2011, Trilogy became one of three organizations to be inaugurated to be gold certified by the Hawai'i Ecotourism Association. Qualifying organizations must, among other things, contribute to the local communities in which they operate and support conservation outcomes of local community-based environmental initiatives.

In a traditional Hawaiian context, nature and culture are one and the same, there is no division between the two. The wealth and limitations of the land and ocean resources gave birth to, and shaped the Hawaiian worldview. The *'āina* (land), *wai* (water), *kai* (ocean), and *lewa* (sky) were the foundation of life and the source of the spiritual relationship between people and their environs. Every aspect of life, whether in the sky, on land or of the waters was believed to have been the physical body-forms assumed by the creative forces of nature, and the greater and lesser gods and goddesses of the Hawaiian people. Respect and care for nature, in turn, meant that nature would care for the people (to access the rich collection of resources on traditional knowledge and historical accounts, see Maly *et al.*, 2004). In this respect, Trilogy strives to incorporate many aspects of Native Hawaiian and local (kama 'āina) culture in their operations and business ethic.

Both Trilogy and Kaikoura Whale Watch provide strong examples that a company's success is not solely based on business profits. It's about setting a high-quality standard and incorporating business practices that nurture responsive actions to benefit the health and welfare of community. It's about perpetuating the understanding that all things are interconnected and the need to support initiatives that promote and foster stewardship of natural environments.

In the words of Kupuna (Elder) Daniel K. Kaopuiki, Sr,[2] '*Maika'i ka hana a ka lima, 'ono no ka 'ai a ka waha!*' (When the hands do good work, the mouth is satisfied with good food!) This expression brings things down to the basic premise that

[2] Daniel K. Kaopuiki, Sr. (1890–1983), island of Lāna'i.

Table 12.2 Keys to success in community engagement and collaborative relationships with stakeholders.

Work with integrity	Many successful operations have overarching visions or principles that direct the organization to work within ethical guidelines, with integrity and a focus on community engagement. Provide an environment that reinforces healthy values and behaviours
Develop trust and work to form meaningful, long-term partnerships	Trust is built up through reciprocity and long-term engagement and relationship-building. Sustainable ecotourism programmes should seek to engage often and early with community leaders, organizations and stakeholder groups, in order to promote the uniqueness of each culture and place
Follow through on promises with substantial action	Organizations that under-promise and over-deliver develop lasting relationships with communities and stakeholders. Programmes that seek opportunities to support community activities and events (e.g. beach cleanups, restoration projects) may have better relationships with stakeholders and community institutions
Integrate local heritage into programmes	Organizations that integrate local cultural heritage and traditional knowledge into their programmes may have better relationships with local communities
Give back to communities and the environment	Healthy resources and healthy communities promotes stronger local businesses. Organizations that support efforts to provide services that promote wellness in communities may benefit from these efforts. Organize and encourage participation in regular community-wide events to celebrate and promote interaction and communication among all members of the community
Help people connect to place	Organizations with a focus on place, and not just the animals that inhabit a local environment, may have better revenue streams, and better relationships with communities through incorporation of place-based cultural and traditional knowledge. Preserving and protecting the environment and resources goes hand-in-hand with history, culture and place

working with integrity and caring for the natural systems about us means that nature will in-turn care for (nurture) us.

Table 12.2 suggests some basic foundational guidelines being offered based on the principle that each one of us has the responsibility to nurture culturally healthy and responsive citizens who contribute to the growth and harmony of the community. These may be used by managers, business leaders and other constituencies seeking to enhance opportunities for marine mammal conservation and to those seeking to increase ecotourism's social, cultural, political and economic value within their communities.

Conclusion

Transitioning toward ecosystem-based management of the oceans requires a deeper commitment to understanding coastal cultures and their complex relationships with coastal environments (Kittinger *et al.*, 2012; Samonte *et al.*, 2010). Communities with long historical and cultural relationships with marine mammals often have associated traditional knowledge systems and associated customary practices. The collective knowledge, experience and customs of such communities have significant opportunity to enrich whale-watching ecotourism operations, both within and outside of marine mammal protected areas. Ecotourism operations can also benefit local communities directly, through business models or approaches that espouse local social norms and cultural customs, and which seek to give back to the environments and communities that support these businesses. Drawing on examples from New Zealand and Hawai'i, we suggest pathways for more meaningful engagement with local communities and cultures, which hold much promise for promoting sustainable, long-term relationships between people and marine mammals.

REFERENCES

Abate, R.S. (2009). Marine protected areas as a mechanism to promote marine mammal conservation: International and comparative law lessons for the United States. *Oregon Law Review* 88, 255–309.

Agardy, M.T. (1993). Accommodating ecotourism in multiple use planning of coastal and marine protected areas. *Ocean & Coastal Management* 20, 219–239.

Bell, J.D., Kronen, M., Vunisea, A., *et al.* (2009). Planning the use of fish for food security in the Pacific. *Marine Policy* 33, 64–76.

Berkes, F. (1999). *Sacred Ecology: Traditional ecological knowledge and resource management.* Philadelphia, PA: Taylor & Francis.

Berkes, F., Folke, C. & Colding, J. (1998). *Linking Social and Ecological Systems: Management practices and social mechanisms for building resilience.* Cambridge: Cambridge University Press.

Cinner, J.E. & Aswani, S. (2007). Integrating customary management into marine conservation. *Biological Conservation* 140, 201–216.

Crowder, L.B., Osherenko, G., Young, O.R., *et al.* (2006). Resolving mismatches in U.S. ocean governance. *Science* 313, 617–618.

Curtin, S. (2003). Whale watch in Kaikoura: Sustainable destination development? *Journal of Ecotourism* 2(3), 173–195.

Dale, A. & Armitage, D. (2011). Marine mammal co-management in Canada's Arctic: Knowledge co-production for learning and adaptive capacity. *Marine Policy* 35, 440–449.

Department of Sustainability, Environment, Water, Population and Communities, Australian Government (2009). Conservation and values – global cetacean summary report, 60 pp. Canberra: Australian Government.

Eagles, P.F.J., McCool, S.F. & Haynes, C.D. (2002). Sustainable tourism in protected areas: Guidelines for planning and management. Best Practice Protected Area Guidelines Series No. 8. Prepared for United Nations Environment Programme, World Tourism Organization and IUCN – The World Conservation Union, Gland, Switzerland.

Erbe, C. (2002). Underwater noise of whale-watching boats and potential effects on killer whales (*Orcinus orca*), based on an acoustic impact model. *Marine Mammal Science* 18, 394–418.

Gibson, B. & Puniwai, N. (2006). Developing an archetype for integrating Native Hawaiian traditional knowledge with earth system science education. *Journal of Geoscience Education* 54, 287–294.

Halpern, B.S., Walbridge, S., Selkoe, K.A., *et al.* (2008). A global map of human impact on marine ecosystems. *Science* 319, 948–952.

Halpern, B.S., Longo, C., Hardy, D., *et al.* (2012). An index to assess the health and benefits of the global ocean. *Nature* DOI: 10.1038/nature11397.

Higgins-Desbiolles, F. (2009). Indigenous ecotourism's role in transforming ecological consciousness. *Journal of Ecotourism* 8(2), 144–160.

HIHWNMS Sanctuary Advisory Council. (2011). *Report of the Ecosystem Protections Working Group For the Sanctuary Advisory Council.* Honolulu: Hawaiian Islands Humpback Whale National Marine Sanctuary (HIHWNMS).

HIHWNMS Sanctuary Advisory Council. (2012). Aloha ʻĀina: A Framework for Biocultural Resource Management in Hawaiʻi's Anthropogenic Ecosystems. Proceedings from a Technical Expert Workshop: Organized by the Native Hawaiian and Research Committees of the Hawaiian Islands Humpback Whale National Marine Sanctuary Advisory Council, 5–6 July 2012, Hawaiian Islands Humpback Whale National Marine Sanctuary, Honolulu.

Horn, C., Simmons, D.J. & Fairweather, J.R. (1998). *Evolution and Change in Kaikoura: Responses to tourism development.* TREC Report no. 6. Lincoln, New Zealand: Lincoln University, Tourism Research and Education Centre.

Hoyt, E. (2001). *Whale Watching 2001: Worldwide tourism numbers, expenditures, and expanding socioeconomic benefits.* Yarmouth Port, MA: International Fund for Animal Welfare, pp. i–vi; 1–158.

Hoyt, E. (2005). *Marine Protected Areas for Whales, Dolphins, and Porpoises: A world handbook for cetacean habitat conservation.* London and Sterling, VA: Earthscan.

Hoyt, E. (2007). *A Blueprint for Dolphin and Whale Watching Development.* Washington, DC: Humane Society International, 32 pp.

Hoyt, E. (2011). *Marine Protected Areas for Whales, Dolphins, and Porpoises: A world handbook for cetacean habitat conservation and planning* (2nd edn). London and New York: Earthscan.

IFAW (2000). *Report of the Workshop on the Legal Aspects of Whale Watching.* Chile: Punta Arenas, 52 pp.

IFAW, WWF, and WDCS (1997). *Report of the International Workshop on the Educational Values of Whale Watching.* Provincetown, MA: IFAW, 44 pp.

IUCN (1994). *Guidelines for Protected Area Management Categories* CNPPA with assistance of WCMC, IUCN, Gland, Switzerland, and Cambridge, 261 pp.

Kent, G. (1997). Fisheries, food security, and the poor. *Food Policy* 22, 393–404.

Kikiloi, K. (2010). Rebirth of an archipelago: Sustaining a Hawaiian cultural identity for people and homeland. *Hūlili: Multidisciplinary Research on Hawaiian Well-Being* 6, 73–114.

Kittinger, J.N., Dowling, A., Purves, A.R., Milne, N.A. & Olsson, P. (2010). Marine protected areas, multiple-agency management, and monumental surprise in the Northwestern Hawaiian Islands. *Journal of Marine Biology* 2011, 1–17. Article ID 241374. http://www.hindawi.com/journals/jmb/242011/241374/.

Kittinger, J.N., Finkbeiner, E.M., Glazier, E.W. & Crowder, L.B. (2012). Human dimensions of coral reef social–ecological systems. *Ecology and Society* 17(4), 17.

Kittinger, J.N., Cinner, J.E., Aswani, S. & White, A.T. (2013). Back to the past? Integrating customary practices and institutions into co-management of small-scale fisheries. In J.N. Kittinger, L. McClenachan, K.B. Gedan, & L. K. Blight (Eds), *Marine Historical Ecology in Conservation: Applying the past to manage for the future.* Berkeley, CA: University of California Press.

Leisher, C., Mangubhai, S., Hess, S., *et al.* (2012). Measuring the benefits and costs of community education and outreach in marine protected areas. *Marine Policy* 36, 1005–1011.

Lovecraft, A.L. & Meek, C.L. (2010). The human dimensions of marine mammal management in a time of rapid change: Comparing policies in Canada, Finland and the United States. *Marine Policy* 35, 427–429.

Maly, K. & Maly, O. Kumu Pono Associates LLC. (2004). Ka Hana Lawai`a a me nā kai ‘Ewalu. Summary of detailed findings from research on the history of fishing practices and marine fisheries of the Hawaiian Islands. The Nature Conservancy, Honolulu, HI. Retrieved from http://www.ulukau.org/elib/collect/maly3/index/assoc/D0.dir/book.pdf

Marsh, H., Arnold, P., Freeman, M., *et al.* (2003). Strategies for conserving marine mammals. In N. Gales, M. Hindell & R. Kirkwood (Eds), *Marine Mammals: Fisheries, tourism and management issues.* Victoria, Australia: CSIRO, pp. 1–19.

McClenachan, L. & Cooper, A.B. (2008). Extinction rate, historical population structure and ecological role of the Caribbean monk seal. *Proceedings of the Royal Society B: Biological Sciences* 275, 1351–1358.

McLeod, K.L., Lubchenco, J., Palumbi, S.R. & Rosenberg, A.A. (2005). Scientific consensus statement on marine ecosystem-based management. Signed by 221 academic scientists and policy experts with relevant expertise and published by the Communication Partnership for Science and the Sea. Retrieved from http://compassonline.org/?q=EBM, 1–21 pp.

Mora, C., Myers, R.A., Coll, M., *et al.* (2009). Management effectiveness of the world's marine fisheries. *PLoS Biol* 7, e1000131.

Oracion, E.G., Miller, M.L. & Christie, P. (2005). Marine protected areas for whom? Fisheries, tourism, and solidarity in a Philippine community. *Ocean & Coastal Management* 48, 393–410.

Orams, M.B. (2002). Marine ecotourism as a potential agent for sustainable development in Kaikoura, New Zealand. *International Journal of Sustainable Development* 5(3), 338–352.

Pauly, D., Christensen, V., Dalsgaard, J., Froese, R. & Torres Jr, F. (1998). Fishing down marine food webs. *Science* 279, 860–863.

Pressey, R.L. & Bottrill, M.C. (2009). Approaches to landscape- and seascape-scale conservation planning: Convergence, contrasts and challenges. *Oryx* 43, 464–475.

Reeves, R.R. (2000). *The Value of Sanctuaries, Parks, and Reserves (Protected Areas) as Tools for Conserving Marine Mammals.* Final report to the Marine Mammal Commission, contract number T74465385. Bethesda, MD: Marine Mammal Commission, 50 pp.

Reeves, R.R. (Ed.) (2009). *Proceedings, First International Conference on Marine Mammal Protected Areas (ICMMPA), Maui, Hawai'i, 30 March–3 April 2009,* 120pp.

Rudrud, R. (2010). Forbidden sea turtles: Traditional laws pertaining to sea turtle consumption in Polynesia (Including the Polynesian Outliers). *Conservation and Society* 8, 84–97.

Samonte, G., Karrer, L.B. & Orbach, M. (2010). *People and Oceans: Managing marine areas for human well-being*. Science and Knowledge Division, Conservation International, Arlington, VA. Retrieved from http://www.conservation.org/Documents/CI_MMAS_Science-to-Action_People_and_Oceans.pdf.

Schipper, J., Chanson, J.S., Chiozza. F., *et al.* (2008). The status of the world's land and marine mammals: diversity, threat, and knowledge. *Science* 322, 225–230.

Simmonds, M.P. & Isaac, S.J. (2007). The impacts of climate change on marine mammals: Early signs of significant problems. *Oryx* 41, 19–26.

Spalding, M.D., Fish, L. & Wood, L.J. (2008). Toward representative protection of the world's coasts and oceans – Progress, gaps, and opportunities. *Conservation Letters* 1, 217–226.

Stewart-Harawira, M. (2005). Cultural studies, indigenous knowledge and pedagogies of hope. *Policy Futures in Education* 3(2), 153–163.

Taylor, B.L., Wade, P.R., De Master, D.P. & Barlow, J. (2000). Incorporating uncertainty into management models for marine mammals. *Conservation Biology* 14, 1243–1252.

Te Ahukaramu, Charles Royal. (2009). Kaitiakitanga – guardianship and conservation, Te Ara – the Encyclopedia of New Zealand, update 25 September 2011. Retrieved from www.teara.govt.nz/en/kaitiakitanga-guardianship-and-conservation.

Thaman, K.H. (1993). Culture and the curriculum in the South Pacific. *Comparative Education* 29, 249–260.

Thaman, K.H. (2002). Shifting sights: The cultural challenge of sustainability. *International Journal of Sustainability in Higher Education* 3, 233–242.

Tsuha, K. (2008). Pae Palaoa, Pae Koholā: An Amalgamation and Articulation of Hawaiian Cultural Insight for Whale Strandings in the Maui County. For Office of the Mayor, County of Maui.

Part III

Ecological dimensions of whale-watching

13

Understanding the ecological effects of whale-watching on cetaceans

Fredrik Christiansen and David Lusseau

Introduction

Whale-watching is a potentially sustainable use of cetaceans and an economically viable alternative to whaling and has become a major contributor to the tourism sector of many countries (Hoyt, 2001; O'Connor *et al.*, 2009). Whale-watching also has the potential to improve people's attitude toward the marine environment, and promote public awareness and support for the conservation issues that targeted species face (Duffus & Dearden, 1993). However, whale-watching can put cetaceans at risk of being harassed and injured by an unknown number of unpredictable impacts which can pose a risk to the viability of the targeted population, as well as the whale-watching industry itself.

Reported effects of human disturbance on cetaceans cover a range of taxa, including many odontocete species and several species of mysticetes (see Chapter 16). Even though many studies have shown that whale-watching can cause both short- (Nowacek *et al.*, 2001; Williams *et al.*, 2002b; Lusseau, 2003a; Christiansen *et al.*, 2010) and long-term negative effects on cetaceans (Bejder *et al.*, 2006; Fortuna, 2006; Lusseau *et al.*, 2006b), few studies have focused on explaining the underlying cause, or ecological and evolutionary mechanisms for these effects (Frid & Dill, 2002). Understanding how human interactions affect wildlife is crucial for the sustainable management of any nature-based tourism activity. This chapter addresses the ecological foundations of whale-watching disturbance on cetaceans. It gives an overview of the documented impacts of whale-watching on cetaceans and compares this to observations of natural predation. It then tries to explain how whale-watching is perceived by the animals to understand the underlying ecological and evolutionary basis for these responses. It goes on to discuss different factors that are likely to influence the response of animals to whale-watching. We then discuss the long-term effects of whale-watching on cetaceans by following the mechanistic link between behavioural effects and vital rates within an energetic framework. Ecological and biological constraints to the ability of cetaceans to cope with disturbance are discussed as well as their implication for long-term vital rates.

The effects of human disturbance on cetaceans

Human disturbance is defined by Frid and Dill (2002) as a deviation in an animal's behaviour from patterns occurring without human influences (see Blanc *et al.*, 2006 for other definitions of disturbance). A large number of studies exist that show that cetaceans are disturbed by whale-watching, even though this is a non-consumptive activity

Whale-watching: Sustainable Tourism and Ecological Management, eds J. Higham, L. Bejder and R. Williams.
Published by Cambridge University Press. © Cambridge University Press 2014.

(Tremblay, 2001). Frid and Dill (2002) explain this by arguing that animals perceive humans as predators and respond to human presence in the same way as in the presence of natural predators. Many cetacean populations do face predation risk from a variety of predators (Visser, 1999; Heithaus, 2001; Ford et al., 2005) and it has been proposed to be the basis for migratory behaviour in mysticetes (Corkeron & Connor, 1999). Therefore, cetaceans have evolved a variety of behavioural strategies to respond to perceived predation risk and it is within this context that they respond to the introduction of novel stimuli (i.e. whale-watching boats) that can be hazardous.

Short-term effects

Short-term behavioural responses towards whale-watching boats are often used to look at the effects of human disturbance on cetaceans, as they are often considered the most direct and hence sensitive measurement of the effects of human disturbance, as well as being relatively easy to measure in the field (Beale & Monaghan, 2004a). Cetaceans respond to whale-watching boats by altering their behaviour in a vertical or horizontal plane, or both. Vertical avoidance means that cetaceans spend less time at the surface, by increasing the depth and duration of dives (Nowacek et al., 2001; Lusseau, 2003b). Horizontal avoidance can be expressed by an increase in swimming speed and frequency of heading changes, to outrun or outmanoeuvre (confuse) the whale-watching boats (Williams et al., 2002b; Lundquist et al., 2012).

Because the vertical and horizontal movement of cetaceans is closely related to the activity they are engaged in (e.g. socializing animals spend more time close to the surface than when travelling), different activity states will differ in their susceptibility to whale-watching boats. Thus, whale-watching boats have been shown to affect the activity state of cetaceans (Lusseau, 2003a; Christiansen et al., 2010; Lundquist et al., 2012). Whale-watching also influences the trade-off processes between perceived predation risk and environmental and social factors, which affects distribution patterns (Allen & Read, 2000; Lusseau, 2005) and group dispersion (Bejder et al., 1999; Nowacek et al., 2001). Finally, both the presence and the noise level of whale-watching boats can affect the vocalization patterns of cetaceans (NRC, 2005; Jensen et al., 2009). Many of the behavioural responses mentioned above have also been observed between cetaceans and natural predators (Visser, 1999; Heithaus, 2001; Ford et al., 2005).

Long-term effects

Because cetaceans are long-lived and relatively hard to study, only three studies to date, all on odontocetes, have been able to show long-term effects of whale-watching on cetacean vital rates (Bejder et al., 2006; Fortuna, 2006; Lusseau et al., 2006b). Two of these identified a decrease in female reproductive success to be the causal factor behind the long-term effects, one resulting in a decrease in population size (Lusseau et al., 2006b) and the other in a shift in relative abundance between two areas (Bejder et al., 2006). For mysticetes, Weinrich and Corbelli (2009) found no long-term effects of whale-watching boats on humpback whales' (Megaptera novaeangliae) calving rate or calf survival. However, this could be because of a lack of proper controls. Therefore, to date, the long-term effects of whale-watching on mysticetes are unknown.

Non-visible effects

Another important aspect of human disturbance on wildlife is that animals can be negatively affected by a disturbance without showing any visible changes in behaviour (Beale & Monaghan, 2004b). Studies of birds have shown that animals can still show physiological responses, stress, to human disturbance without showing behavioural responses (Fowler, 1999; Beale & Monaghan, 2004b) and such responses can reduce breeding success (Carney & Sydeman, 1999). Very little is known about how stress is expressed in cetaceans, and how it can be linked to short-term behavioural responses to disturbance. Wild dolphins showed elevated concentrations of cortisol and aldosterone, two stress hormones, following the encirclement by capture nets (Aubin *et al.*, 1996). While stress can be beneficial in some situations, severe, prolonged or cumulative stress can lead to disease and negative effects on survival and reproductive success (Fair & Becker, 2000). In captive bottlenose dolphins (*Tursiops aduncus*) stress resulting from social instability was shown to contribute to documented mortalities and illnesses (Waples & Gales, 2002).

Explaining the effects of human disturbance on cetaceans

Predation risk

Predation is an important factor shaping the lives of K-selected species (Heithaus *et al.*, 2009). Apart from predators having direct consumptive effects on a population, predators can also have indirect non-consumptive effects in the form of predation risk ; Figure 13.1). Because animals are unaware of actual predation risk, their behavioural decisions will be based on the perceived predation risk. The cost of predation risk can be seen as a trade-off between investments in anti-predator activities (e.g. avoidance, vigilance or group formation) and other fitness enhancing activities (e.g. foraging, mating, parental care), to reduce predation risk (Brown, 1999). As several factors influence perceived predation risk (Figure 13.1), there are also several strategies that animals can use to minimize it, which can vary between species and geographical areas. What all animals have in common, however, is that this trade-off can never be one-sided. If the trade-off is biased towards fitness-enhancing activities, and the investments in anti-predator activities is low, this will lead to an increase in predation rate for the animals which will have direct negative effects on population dynamics. On the other hand, if investment in anti-predator activities is too high, other fitness-enhancing activities will suffer which can lead to negative effects on animal survival and

reproductive success (Creel *et al.*, 2007), and indirect negative effects on population dynamics (Gill & Sutherland, 2000).

Alternatively, high investment in anti-predator activities can lead to a reduction in physiological state (body condition), which might also lead to an increase in predation rate (Sinclair & Arcese, 1995) and direct negative effect on population dynamics. Therefore, the indirect effects of predation, in the form of predation risk, might in fact be greater than the direct effects of predation through consumption (Preisser *et al.*, 2005), especially in cases when the predator also preys on other species or are consuming a relatively small number of animals of the target population (Wirsing, 2007). In the Greater Yellowstone Ecosystem, USA, wolf (*Canis lupus*) presence led to a decline in the recruitment of elk (*Cervus canadensis*) calves in the area. The decline was shown not to be a direct effect of predation by the wolves, but an indirect effect of predation risk which caused a decline in progesterone levels of female elks when they engaged in anti-predator behaviours (Creel *et al.*, 2007).

Predation risk in the absence of predation

Cetaceans often respond to human disturbance in the same way they respond to natural predators (see above). The non-consumptive effects can cause the targeted animals to divert time and energy away from other fitness enhancing activities, as human disturbance can from an evolutionary perspective be likened to predation risk (Frid & Dill, 2002). But how can whale-watching be perceived as predation risk by cetaceans, when it poses such a small risk of injury? Dolphins have been shown to be able to distinguish between different species of natural predators (Irvine *et al.*, 1973), so it should be expected that cetaceans can differentiate between a natural predator and a whale-watching boat. Anti-predator responses are general towards certain stimulus (e.g. rapidly approaching objects) and not predator-specific (Frid & Dill, 2002). Therefore, rather than the object itself, the behaviour of the object is what triggers the anti-predator response

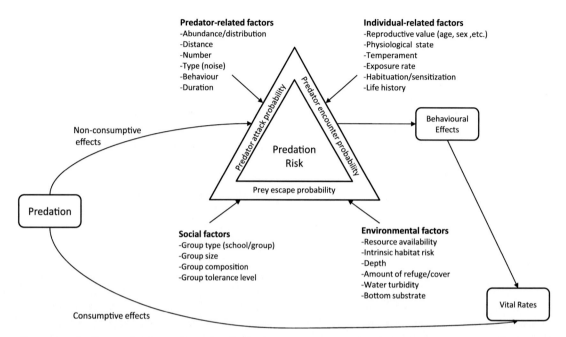

Figure 13.1 The direct and indirect effects of predation on cetaceans. The perceived predation risk of animals is influenced by a number of factors that coexist in complex interactions with each other. Predation risk is made up of three components: the predator encounter probability (the probability that an animal encounters a predator), the predator attack probability (the probability that the animal is attacked when encountered) and the prey escape probability (the probability that the animal can escape an attack from a predator (Hugie & Dill, 1994). For the full mechanistic link between behavioural effects and vital rates, see Figure 13.2.

and animals make no distinction whether the stimulus comes from a natural predator or a non-lethal whale-watching boat, even if the latter source is new to the evolutionary history of the animals, as long as the stimuli that elicit the response are similar (Frid & Dill, 2002). This also illustrates the fact that animals make their decisions based on the perceived predation risk and not the actual predation risk, as the latter cannot be known by the animals.

Further, it is not necessary that predation consumes a high number of prey or that capture attempts occur often (high predator attack probability) or have a high probability of success (low prey escape probability) for a predator to cause a strong response of the targeted animals in the form of predation risk (Preisser *et al.*, 2005). Dolphins in

Shark Bay, Australia, showed strong anti-predator behaviours during time periods when tiger sharks (*Galeocerdo cuvier*) were present in the area, despite dolphins only constituting a minor component of the sharks' diet (Heithaus & Dill, 2002). Further, in cases where hunting by humans has represented a historical real threat to some species over evolutionary time (i.e. commercial whaling), disturbance stimuli and true predation stimuli might be indistinguishable from the perspective of the animals (Frid & Dill, 2002).

Cetaceans are K-selective species, and therefore their life history strategy aims to maximize fitness by increasing survival probability rather than fecundity (Stearns, 1992). To maximize survival, cetaceans have evolved to minimize predation risk at all costs. Therefore, to maximize fitness, cetaceans will over-estimate risks rather than under-estimating them, as the individual costs of anti-predator behaviours are much smaller than the potential cost of predation itself (death) (Bouskila & Blumstein, 1992). This means that whale-watching, even though it is non-consumptive, has the same potential to cause strong responses of the targeted animals as natural predation and that habituation to non-lethal disturbance in general is unlikely. Further, cetaceans might actually learn to associate the distant sound of whale-watching boats with the fast unpredictable behaviour of closer boats, and repeated exposure to whale-watching boats might even lead to a decrease in the threshold level at which a response is triggered (Frid & Dill, 2002), following the concept of sensitization (Bejder *et al.*, 2009).

That cetaceans over-estimate risks further means that the presence of a stimulus (i.e. the presence of whale-watching boats), rather than its intensity (i.e. number of boats), will be the main trigger of a response. This was shown for elks in the Greater Yellowstone Ecosystem, USA, which responded strongly to the mere presence of wolves in the area by altering their habitat use on a very short time scale (<2 hours) and at a relatively large spatial scale (radius >3 km; (Creel *et al.*, 2005). Silent kayaks elicited similar responses as power vessels for killer whales (*Orcinus orca*), showing that boat presence and behaviour rather than boat type was the main source of the disturbance (Williams *et al.*, 2011). Finally, given the life history strategy of cetaceans, it can also be expected that potential long-term effects of whale-watching will be on female reproductive success rather than adult survival, which the few long-term studies on odontocetes confirms (Bejder *et al.*, 2006; Lusseau *et al.*, 2006b; see above).

Factors influencing the effect of human disturbance on cetaceans

The perceived predation risk, and hence the impact of human disturbance, of an animal is influenced by a number of factors (Figure 13.1), of which the relative importance varies between species, populations and geographical areas. Some of these factors relate to the predator (i.e. whale-watching boat), while others relate to the individual being targeted. Predation risk is further influenced by the social structure of the species and the environment in which the targeted animal lives. All these factors influence predation risk, and will hence affect the anti-predator response of an animal, rather than predation rate per se, and the same is therefore expected towards human disturbance (Frid & Dill, 2002).

Predator-related factors

The different components of predation risk are affected by different factors related to the predator, and hence whale-watching boats (Figure 13.1). The predator encounter probability is affected by the abundance and distribution of the predator and is generally proportional to the predator density in an area (Heithaus *et al.*, 2009). Thus, the encounter probability, or exposure rate, of whale-watching boats will influence the perceived predation risk of cetaceans, which will be determined by the number of whale-watching boats in an area as well as the temporal and spatial overlap between whale-watching activities and the distribution of the targeted population. The predator attack probability, as well as the prey escape probability, is affected by predator type (e.g. stealth versus chase, diurnal versus nocturnal, etc.; Heithaus *et al.*, 2009), distance and number (e.g. group size; Beale & Monaghan, 2004b), duration and behaviour (Frid & Dill, 2002). Similarly, the perceived predation risk of cetaceans towards whale-watching boats is influenced by a number of variables. The response is inversely correlated with the distance to the nearest boat (Nowacek *et al.*, 2001; Stamation *et al.*, 2010). The response of cetaceans has also been shown to be affected by the number of boats (Stensland & Berggren, 2007; Williams & Ashe, 2007).

Predation risk can then be understood as a function of the number of predators and the distance to them. The two parameters are inversely correlated, meaning that several predators at a far distance might trigger a similar response as a single predator at a closer distance (Beale & Monaghan, 2004b). The response of an animal is not triggered by the distance to the predator per se, but is initiated when the rate of change of the perceived distance to an approaching object exceeds a certain threshold (Frid & Dill, 2002). The rate of change of the perceived distance will increase if the size of the stimuli is larger and the approach is faster and more direct (the predator becomes visible at a greater rate). Following this theory, both the size of the whale-watching boat, as well as its speed and behaviour are likely to influence the perceived predation risk of the targeted animals. Also the time that the stimulus remains at closer distance to an animal will influence the perceived predation risk (Frid & Dill, 2002). Support can be found in the scientific literature which shows that the response of cetaceans to whale-watching boats is influenced by vessel type and behaviour, the latter being further divided into angle of approach, speed, duration of interaction and movement predictability (consistency) (Bejder et al., 1999; Nowacek et al., 2001; Williams et al., 2002a).

Even though some scientists consider sound to be the primary reason for responses of cetaceans to whale-watching boats (NRC, 2005), other studies indicate that it is far less important and that many other factors are involved (Williams et al., 2011). In pinnipeds, the flushing distance (seals entering the water) of seals in the presence of boats at a haulout was shown to be higher for kayaks and canoes than motor boats (Henry & Hammill, 2001). Apart from being silent, kayaks and canoes generally moved slower, had a lower silhouette, and stayed for a longer time close to the haulout than the motor boats. Thus the behaviour of the kayaks and canoes resembled that of a stealthy predator, which might be the reason why they were perceived by the seals as a higher predation risk compared to the noisier, more easily detectable and therefore more predictable behaviour of motor boats. This suggests that the behaviour, or more precisely the predictability, of the boats is the key component for a successful encounter between cetaceans and whale-watching boats (Williams et al., 2011).

Individual-related factors

It is likely that individual animals differ in their sensitivity, and thus response, to human disturbance as they live different lives, have different personalities, play different roles in their populations and vary in their exposure rate to disturbances (Beale & Monaghan, 2004a; Blanc et al., 2006; Réale et al., 2010; Figure 13.1). The trade-off process between predation risk and other activities is likely to be strongly influenced by the animal's reproductive value (the expected future fitness; Lima, 1998), which in turn is influenced by the age- (Constantine, 2001), sex- (Lusseau, 2003b) and reproductive class (Nowacek et al., 2001). The body condition, or physiological state, of an animal will also influence the response towards human disturbance (Beale & Monaghan, 2004a; Réale et al., 2007). Healthy animals or animals in high-quality (resource-rich) habitats are likely to have sufficient energy reserves or opportunities to acquire energy to afford the costs of avoiding a disturbance by changing their behaviour than animals in poorer conditions or habitats which are more constrained by current requirements and cannot afford to change their behaviour (e.g. stop feeding).

Thus, an animal that is seemingly unaffected by whale-watching boats might actually be in such a poor physiological state that ecological (nutritional) constraints prevent it from engaging in antipredator behaviours (Beale & Monaghan, 2004a). Poor physiological state may be due to natural (e.g. seasonal/migrational) or anthropogenic factors (e.g. pollution, hunting, whale-watching disturbance), or a combination of both. Individuals within a population can also have different temperaments, which are consistent behavioural differences between individuals resulting from differences in reproductive value and not caused by age-, sex- or reproductive class or physiological condition (Réale et al., 2007, 2010).

Different temperaments have different intrinsic predation risks associated with them (Jones & Godin, 2010). For example, some individuals within a population might be bolder and devote more attention to foraging than anti-predator vigilance, while others are more cautious and devote more time to being vigilant. Bolder animals will have a higher foraging efficiency, but also a higher predation risk (Bell & Sih, 2007), while cautious animals will reduce the predation risk at the cost of lost foraging opportunities (Jones & Godin, 2010). While natural predation would favour cautious individuals (Bell & Sih, 2007; Réale *et al.*, 2010), whale-watching will favour bold individuals, as the perceived predation risk does not involve an actual risk of predation (there is no benefit with being more vigilant in the absence of a predator). Whale-watching may therefore work as a selective force working in the opposite direction of natural predation. Thus the response of cetaceans towards whale-watching boats will vary between animals within a population and even for the same individual at different times because the ecological constraints that the animal is subjected to might vary temporarily (e.g. time periods with high prey availability versus low prey availability).

Social factors

The prey escape probability, and hence predation risk, is also influenced by the escape tactics of the targeted animals, which is strongly influenced by factors related to the species social structure (Frid & Dill, 2002; Figure 13.1). Cetaceans can form non-mutualistic groups, called *aggregations* or *schools*, which are formed because of non-social factors (e.g. high food availability) and provide no larger benefit to the members of the group. Alternatively, they can form mutualistic groups, called *groups*, which are based on social factors and which can provide benefits to the members and therefore last for months to decades (Connor & Norris, 1982). In cetaceans, animals from different species can also come together and form what is known as *mixed-species groups* (Stensland *et al.*, 2003). In a predation context, schools can have the benefit of a dilution effect, which decreases an individual's chance of being captured by a predator, and therefore the risk of predation. Forming groups is a way for animals to reduce predation risk once a predator is present by collective vigilance, collective defence and dilution of risk (Stensland *et al.*, 2003; Creel & Winnie, 2005).

Predation risk is believed to be one of the main factors (together with food availability) for group formation in cetaceans (Norris & Dohl, 1980). Further, the group size of cetaceans is generally the result of a trade-off between predation risk and intraspecific competition for food (food availability) (Heithaus & Dill, 2002). Therefore, cetaceans living in fission–fusion societies (e.g. dolphins) adjust their group size depending on the predation risk and their current activity. For group-living cetaceans it is also likely that the tolerance level, and hence response, of an animal to human disturbance will be influenced by the tolerance level of the other individuals in the group (e.g. an animal with a very low tolerance level triggers an anti-predator response of the entire group at a very low predation risk). Therefore, group formation might reduce predation risk on an individual level, but it might increase energetic costs associated with avoidance tactics on a population level (Blanc *et al.*, 2006).

Cetacean calves are physiologically limited in their ability to dive and swim fast (Mann & Smuts, 1998), which can make them more vulnerable to predators. Because a calf is very much dependent on its mother, these physiological limitations will also set limits to the mother's behaviour and also her associations with other conspecifics. These limitations will influence both the calf's and the mother's exposure and vulnerability to predators, and thus the perceived predation risk. In Moreton Bay, Australia, humpback dolphin (*Sousa chinensis*) groups containing pairs of females with calves whistled significantly more (to maintain close contact) than those without, following periods of high boat noise from transiting vessels, suggesting that the perceived predation risk caused by the vessel noise is higher for females with calves than other animals

(Van Parijs & Corkeron, 2001). Because females with calves and groups containing females with calves are more sensitive to interactions with whale-watching boats (Stensland & Berggren, 2007; Stamation *et al.*, 2010), it is not hard to imagine that predation risk in turn can have negative effects on the relationship between mothers and offspring, and therefore the reproduction success of the population (Frid & Dill, 2002).

In the presence of a predator, a mother can either decide to stay with her offspring, which will increase its chances of survival (decrease predation risk) but might decrease the chances of survival of the mother (increase predation risk). Alternatively, the mother can decide to leave her offspring, which will increase the chances of survival of the mother (decrease predation risk) but decrease the chances of survival of the abandoned offspring (increase predation risk) (Edwards, 2002; Frid & Dill, 2002). This trade-off will be influenced by the reproductive value of the mother (Frid & Dill, 2002) and thus the life history of the species. Human disturbance on cetacean females with calves has been shown to trigger the same trade-off process. Chases by tuna purse-seiners in the Eastern Tropical Pacific Ocean frequently lead to the separation between spotted dolphin (*Stenella attenuata*) mothers and calves (Edwards, 2002). The mothers tried to remain with the adult group of dolphins, in order to increase their own survival probability, instead of remaining with their relatively slow-swimming calves, which would have increased the calves' survival probability. Above this, avoidance behaviours caused by whale-watching boats will increase the energetic expenditure for both the mother and the calf. If the energetic increase for the mother is sufficiently large, she might need to redistribute energy from lactation into maintaining her own homeostasis, leading to a decrease in energy acquisition for the calf.

Environmental factors

Resource availability has a strong influence on animal decision-making under predation risk. In general, animals try to match their distribution to that of their prey and avoid areas where predation risk is high (Lima & Dill, 1990). Habitat selection is thus based on a trade-off between predation risk and resource availability in a way that minimizes the ratio between mortality risk and energy intake. An animal is more likely to keep foraging in an area with high predation risk if the prey availability in the area is higher, and the rewards of continuing foraging is enough to meet the costs of the high predation risk, compared to an area with lower prey availability, where the benefits of continued foraging will be low in comparison to the risk of remaining there (Heithaus & Dill, 2006).

Animal escape tactics, influenced by the landscape features (e.g. depth, turbidity, bottom substrate, number of refuges/cover, etc.) can also have a strong influence on predation risk by affecting the prey escape probability (Heithaus *et al.*, 2009). Habitat attributes that modify the probability that the animal will be captured and killed during an encounter with a predator are called the intrinsic habitat risk (Heithaus & Dill, 2006). Therefore, within a habitat, predation risk is influenced by both the probability of encountering a predator and the intrinsic habitat risk. The intrinsic habitat risk of a particular area might be so low (e.g. due to high escape probabilities) that even though the predator encounter probability is relatively high in that area compared to surrounding areas, the animals reduce their overall predation risk by residing there, which paradoxically results in animals selecting areas where the relative predator density is highest (Heithaus *et al.*, 2009).

That predation risk influences habitat selection indicates that disturbance by whale-watching boats can lead to changes in cetacean habitat use, which in turn can lead to increased intraspecific competition or an increase in predation rate by natural predators, which could have negative effects on population dynamics (Gill & Sutherland, 2000). For example, whale-watching activities might cause cetaceans to spend more time in low-quality habitats, resulting in increased intraspecific competition for food, as observed against

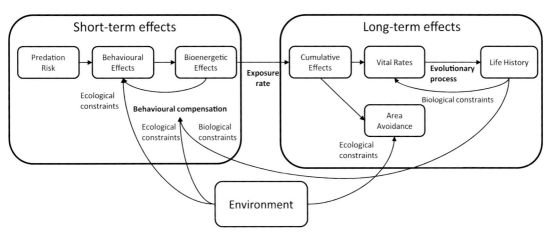

Figure 13.2 A conceptual diagram linking short-term effects of whale-watching to long-term effects. The effects, as well as the ability to compensate for an effect, are constrained by the life history of the species as well as the environment in which it lives.

natural predators (Wirsing *et al.*, 2007). As cetaceans cannot distinguish between perceived predation risk coming from a non-lethal whale-watching boat and that resulting from actual predation (Frid & Dill, 2002), whale-watching activities could also displace animals from relatively safer habitats to areas with higher rates of natural predation, without the animal necessarily becoming aware of the difference.

Estimating the costs of human disturbance on cetaceans

Linking behavioural effects to vital rates

The behavioural response of cetaceans towards whale-watching boats is the result of a trade-off process between the costs of the perceived predation risk (the avoidance behaviour) and the benefits from remaining engaged in a particular activity or occupying a certain area. The trade-off process is greatly influenced by the ecological conditions of the area occupied, which is likely to vary both spatially and temporally, leading to a complex response of the threshold level at which a behavioural

response is triggered. While we have discussed the link between predation risk and behavioural effects, for management it is important to understand how short-term behavioural effects translate into long-term biologically significant effects on individual vital rates (survival and reproductive success), and ultimately population dynamics (NRC, 2005; Réale *et al.*, 2007; Figure 13.2).

Energetic effects

Anti-predator activities all have energetic costs associated with them, either in themselves (e.g. increasing metabolic rate) or by taking time away from other energy-enhancing activities (e.g. foraging), or both (Williams *et al.*, 2006). Predation risk can thus be seen as a trade-off between energy-consumptive activities and energy-acquiring activities. Time allocation patterns (activity budgets) can be used to access the energetic costs of predation risk, and thus human disturbance, on cetaceans by comparing the time an animal allocates to different energy-relevant activities, such as energy acquisition (e.g. foraging), energy consumption (e.g. travelling) and energy conservation (e.g. resting) in the presence and absence of whale-watching boats (Williams *et al.*, 2006; Wirsing *et al.*, 2007). Energetic expenditure can increase if animals increase the swimming speed (Williams & Noren, 2009) and/or the frequency of evasive behaviours in the presence of whale-watching boats, while energy acquisition can decrease if foraging behaviour is reduced or halted (Williams *et al.*, 2006). Some studies show that this latter effect is probably more likely to lead to long-term consequences than the former (Williams *et al.*, 2006).

Cumulative effects

So far we have most particularly discussed the effect of individual interactions between whale-watching boats and cetaceans. Such short-term effects alone are unlikely to lead to long-term energetic effects for animals (Figure 13.2). For example, an animal might be able to compensate for an increase in energetic demands, or a decrease in energy acquisition, by increasing feeding and/or resting time when the disturbance is absent (Blanc *et al.*, 2006). A migrating baleen whale targeted by whale-watching boats for a few hours while on its migratory route is therefore unlikely to be significantly affected by the interaction. However, if a disturbance becomes more frequent, the animal might no longer be able to compensate for the energetic costs, suggesting that the effects of whale-watching are cumulative rather than catastrophic. A resident population of dolphins targeted by whale-watching boats several hours per day throughout the year is therefore more likely to suffer biologically significant effects from the interactions (Lusseau *et al.*, 2006b).

Whale-watching activities can also target individuals within a population disproportionately, with a preference for certain areas, behaviours and/or age-classes. Therefore, the overall exposure rate of an animal to whale-watching activities needs to be estimated in order to assess the long-term effects on bioenergetics and consequently vital rates (Constantine, 2001; Lusseau, 2003a). If whale-watching causes a long-term decline in population size the cumulative exposure rate for the remaining animals will increase, which in turn will lead to a vicious

cycle that can lead to either extinction (Lusseau *et al.*, 2006a) or desertion of the area.

Area avoidance

If the perceived predation risk caused by whale-watching activities is high, the only way for cetaceans to deal with the increased energetic costs of anti-predator activities might be to switch from short-term behavioural avoidance strategies to a long-term area avoidance strategy (Lusseau, 2005; Figure 13.2). Area avoidance means that the animals either temporarily move away from an area during periods when tourism intensity is high, or leave an area entirely to move to another area where they may remain (Bejder *et al.*, 2006). Area avoidance differs from changes in habitat use because the former removes animals from the local breeding population through emigration, whereas the latter results in changes in the population distribution, but the local breeding population remains the same (however, maybe not the effective population size).

As for habitat selection under predation risk, there is a likely cost–benefit consideration behind an area avoidance decision. An area avoidance tactic will be triggered when the costs of remaining in a disturbed area exceed the benefits of staying there. The factors that determine the outcome of such a trade-off include the quality (resource availability, etc.) of the area that is currently being occupied, the quality and distance to other suitable areas, the relative predation risk between the two areas, the relative intraspecific competition between the two areas as well as the investment already made in the current area (Gill *et al.*, 2001). The perceptual range of the species, which is the distance from which an individual can perceive these factors, will set the limits to these decisions (Zollner & Lima, 1997). An animal might also remain in a disturbed area simply because no alternative habitats are available. This might result in negative effects on individual vital rates and even population size, because the high level of predation risk effectively decreases the perceived quality of the habitat (even though the actual quality of the habitat remains the same; Frid & Dill, 2002). Therefore, even if animals remain in an area where whale-watching is taking place, it may be that animals are affected by whale-watching activities (Christiansen *et al.*, 2010) but that ecological constraints prevent them from leaving the area.

Ecological and biological constraints to compensation

Environmental variability will set constraints to the ability of animals to compensate for the energetic costs of human disturbance (Figure 13.2), by constraining their time activity budget (see Chapter 17). For example, if prey is distributed heterogeneously in an area and whale-watching interactions occur in a spot (or during a time period) where prey availability is high, an avoidance response (e.g. change in activity or habitat use) caused by whale-watching will carry a much higher cost than in areas (or during time periods) where prey availability is low. Animals living in areas where prey, or any other vital resource, is distributed heterogeneously will therefore be more constrained in their ability to compensate for whale-watching interactions than animals in areas where prey is distributed homogeneously and where they have more opportunities to compensate for a loss in energy caused by whale-watching.

Apart from the environmental variability setting ecological constraint on the ability of animals to compensate for whale-watching interactions, the life history of a species further sets biological constraints to compensation (Figure 13.2). Life history describes how an animal allocates resources to reproduction throughout its lifetime to maximize lifetime reproductive output given its survival probability (Stearns, 1992). A distinction is made between income breeders, which use current energetic income to finance reproduction, and capital breeders, which use previously stored energy to finance reproduction (Stephens *et al.*, 2009). The two reproductive strategies have different energetic costs associated with them. For example, the

amount of energy that income breeder females can invest in their offspring is directly related to how much prey they can consume during the lactation period and therefore they try to maximize the rate of energetic intake during the lactation period (Costa, 1993). However, an increase in the rate of energy intake also leads to an increase in foraging (and metabolic) effort. An income breeder strategy is therefore energetically relatively costly and consequently only exists in areas that are highly productive where prey is concentrated and predictable (Costa, 1993).

Capital breeders can separate lactation from foraging so females are able to accumulate the energy needed for lactation over a longer time period. This results in a lower daily energy requirement to finance reproduction. This strategy provides more flexibility to acquire energy and to invest energy in reproduction, which enables capital breeders to utilize highly dispersed, patchy and unpredictable prey resources (Costa, 1993). The capital breeding strategy is thus more economical. However, as body reserves are limited, this also limits the total amount of energy that the mother can invest in her offspring. In contrast, income breeders have a less energy-efficient reproductive strategy, but have no limit to the amount of energy they can invest in their offspring, as they can increase their net energy intake by increasing their foraging effort (Costa, 1993).

Most mysticete species undertake long-distance seasonal migrations between high-latitude feeding grounds and low-latitude breeding grounds and thus lactation is separated from foraging both spatially and temporally, meaning that they need to rely on body reserves to nurse their calves in the low productive breeding grounds (Kasuya, 1995). Odontocetes, on the other hand, generally have a relatively restricted habitat and are resident in the sense that they both feed and breed within the same area. The flexibility of capital breeding in acquiring energy and investing it in reproduction enables mysticetes to utilize crustaceans and schooling fish which are abundant seasonally far away from the breeding grounds, while income breeding odontocetes need to rely on locally abundant and predictable resources of marine mammals or fish (Kasuya, 1995).

Mysticetes that have a capital breeding strategy are more likely to be severely affected by whale-watching boats disrupting energy acquisition in their feeding areas (Stephens *et al.*, 2009), because the foraging success in feeding grounds will set the limits to the amount of energy that can be transferred to their offspring in the breeding grounds, and thus the reproductive success. Conversely, because cetaceans are K-selective species, added energetic expenditures on the breeding grounds will also lead to re-routing of energy from lactation to homeostasis (to increase female survival probability), which will ultimately also lead to a decrease in calving success, and reproductive success. Odontocete income breeders, on the other hand, are likely to be more vulnerable to whale-watching boats while breeding (Stephens *et al.*, 2009) as their foraging success at that time is directly linked to their reproductive success. Again, female odontocetes will sacrifice fecundity for survival, by allocating more energy to maintaining their own homeostasis, and thus survival probability, instead of investing it in lactation (reproduction) during interactions with whale-watching boats. Further, capital breeders are more likely to be able to cope with an unpredictable environment shaped by whale-watching boats than income breeders which are not as flexible in respect to energy acquisition and investment in offspring (Stephens *et al.*, 2009).

Management implications

The effect of whale-watching activities on cetaceans can be minimized if the different factors that influence predation risk (Figure 13.1) are taken into consideration when creating guidelines or regulations for whale-watching. Apart from regulating the behaviour of the whale-watching boat itself (e.g. distance and speed), interactions with particular age-, sex- and reproductive classes or group compositions of cetaceans can be avoided (e.g. mother and calves), as well as interactions in areas where animals are likely to be more sensitive to disturbances (e.g. shallow areas or foraging habitats). Knowledge about the ecological and biological constraints to a population should be used to regulate not only the overall exposure rate (daily interaction time), but also where and when interactions should take place, so that the overall time that whale-watching boats can interact with cetaceans (and consequently the economic profit) can be maximized without causing long-term negative effects on cetacean vital rates.

Many of these aims can be achieved by the designation of marine protected areas (Hoyt, 2005; also see Chapter 19). Marine protected areas (MPAs) have been shown to be effective in protecting cetaceans from direct mortalities, such as fisheries by-catch (Gormley *et al.*, 2012); however, with regard to non-lethal effects their design is often inadequate (Williams *et al.*, 2009; Ashe *et al.*, 2010). This chapter highlights many of the factors that need to be taken into account when designing MPAs for cetaceans to minimize the effect of non-consumptive activities. Some studies have already demonstrated that this can be done successfully. Ashe *et al.* (2010) used behavioural data of killer whales to identify which activities were most sensitive to human disturbance, and then evaluated how different habitats were used for these activities, to propose priority habitats to be protected. In a similar way, Williams *et al.* (2009) used behavioural information to evaluate the effectiveness of a voluntary no-entry reserve for killer whales by assessing habitat preference in a behavioural context and the population-level implications of that preference given threats from human activities. Thus, incorporating behavioural data into the selection of MPAs, as well as wildlife management in general, is crucial to minimize the non-lethal effects of whale-watching on cetaceans.

Conclusions

Fundamental to understanding the effect of whale-watching on cetaceans is understanding that animals have evolved anti-predator responses that are general, and therefore make no distinction whether the stimulus that triggers a behavioural response comes from a natural predator or a non-lethal whale-watching boat. Cetaceans will therefore over-estimate risks, making habituation very unlikely to develop. Further, because animals make their behavioural decisions based on the perceived predation risk and not the actual predation risk, the perceived predation risk can often be high in situations where the actual predation risk is low or non-existent. The fact that the indirect effects of predation, in the form of predation risk, can exceed the direct effects emphasizes the need for wildlife managers to take non-lethal effects of human disturbance seriously. Documented long-term negative effects of whale-watching activities on cetacean vital rates further support this statement.

This chapter shows that the perceived predation risk, and thus avoidance responses, are influenced by a number of factors related to the disturbance stimuli, the targeted individual, the social context the animal lives in and the environment, and that these factors coexist in complex interactions with each other. Understanding how short-term behavioural effects of whale-watching activities translate into long-term biologically significant effects on individual vital rates, and ultimately population dynamics is of central importance for managers. Time allocation patterns (activity budgets) relating to different energy-relevant activities can be used to link behavioural changes to bioenergetics. Knowledge of the overall exposure rate of individual whales to whale-watching activities can then be used to assess long-term effects on bioenergetics and consequently individual vital rates. Knowledge of the species life history and the environment in which it lives is necessary to evaluate biological and ecological constraints on compensation, and thus vulnerability.

This information is critical to the effectiveness of operator guidelines and the designation of MPAs. Furthermore, wildlife managers must understand, for example, that in mysticete species long-term effects are not readily detectable, but could affect reproductive success months in the future, at locations thousands of kilometres away. Knowledge of the mechanistic link between behavioural effects of whale-watching and vital rates is fundamental to this understanding, and more research should be directed towards these critical links.

REFERENCES

Allen, M.C. & Read, A.J. (2000). Habitat selection of foraging bottlenose dolphins in relation to boat density near Clearwater, Florida. *Marine Mammal Science* 16, 815–824.

Ashe, E., Noren, D.P. & Williams, R. (2010). Animal behaviour and marine protected areas: Incorporating behavioural data into the selection of marine protected areas for an endangered killer whale population. *Animal Conservation* 13, 196–203.

Aubin, D.J.S., Ridgeway, S.H., Wells, R.S. & Rhinehart, H. (1996). Dolphin thyroid and adrenal hormones: Circulating levels in wild and semidomesticated *Tursiops truncatus*, and influence of sex, age and season. *Marine Mammal Science* 12, 21–13.

Beale, C.M. & Monaghan, P. (2004a). Behavioural responses to human disturbance: A matter of choice? *Animal Behaviour* 68, 1065–1069.

Beale, C.M. & Monaghan, P. (2004b). Human disturbance: People as predation-free predators? *Journal of Applied Ecology* 41, 335–343.

Bejder, L., Dawson, S.M. & Harraway, J. (1999). Responses by Hector's dolphins to boats and swimmers in Porpoise Bay, New Zealand. *Marine Mammal Science* 15, 738–750.

Bejder, L., Samuels, A., Whitehead, H., *et al.* (2006). Decline in relative abundance of bottlenose dolphins exposed to long-term disturbance. *Conservation Biology* 20, 1791–1798.

Bejder, L., Samuels, A., Whitehead, H., *et al.* (2009). Impact assessment research: Use and misuse of habituation, sensitisation and tolerance in describing wildlife responses to anthropogenic stimuli. *Marine Ecology Progress Series* 395, 177–185.

Bell, A.M. & Sih, A. (2007). Exposure to predation generates personality in threespined sticklebacks (*Gasterosteus aculeatus*). *Ecology Letters* 10, 828–834.

Blanc, R., Guillemain, M., Mouronval, J., *et al.* (2006). Effects of non-consumptive leisure disturbance to wildlife. *Revue d'Ecologie (la Terre et la Vie)* 61, 117–133.

Bouskila, A. & Blumstein, D.T. (1992). Rules of thumb for predation hazard assessment: Predictions from a dynamic model. *American Naturalist* 139, 161–176.

Brown, J.S. (1999). Vigilance, patch use and habitat selection: Foraging under predation risk. *Evolutionary Ecology Research* 1, 49–71.

Carney, K.M. & Sydeman, W.J. (1999). A review of human disturbance effects on nesting colonial waterbirds. *Waterbirds* 22, 68–79.

Christiansen, F., Lusseau, D., Stensland, E. & Berggren, P. (2010). Effects of tourist boats on the behaviour of Indo-Pacific bottlenose dolphins off the south coast of Zanzibar. *Endangered Species Research* 11, 91–99.

Connor, R.C. & Norris, K.S. (1982). Are dolphins reciprocal altruists? *The American Naturalist* 119, 358–374.

Constantine, R. (2001). Increased avoidance of swimmers by wild bottlenose dolphins (*Tursiops truncatus*) due to long-term exposure to swim-with-dolphin tourism. *Marine Mammal Science* 17, 689–702.

Corkeron, P.J. & Connor, R.C. (1999). Why do baleen whales migrate? *Marine Mammal Science* 15, 1228–1245.

Costa, D.P. (1993). The relationship between reproductive and foraging energetics and the evolution of the Pinnipedia. *Symposia of the Zoological Society of London* 66, 293–314.

Creel, S. & Winnie, J.A. (2005). Responses of elk herd size to fine-scale spatial and temporal variation in the risk of predation by wolves. *Animal Behaviour* 69, 1181–1189.

Creel, S., Winnie, J., Maxwell, B., et al. (2005). Elk alter habitat selection as an antipredator response to wolves. *Ecology* 86, 3387–3397.

Creel, S., Christianson, D., Liley, S. & Winnie Jr, J.A. (2007). Predation risk affects reproductive physiology and demography of elk. *Science* 315, 960.

Duffus, D.A. & Dearden, P. (1993). Recreational use, valuation, and management of killer whales (*Orcinus orca*) on Canada's Pacific Coast. *Environmental Conservation* 20, 149–156.

Edwards, E.F. (2002). Behavioral contributions to separation and subsequent mortality of dolphin calves chased by tuna purse-seiners in the Eastern tropical Pacific Ocean. National Marine Fisheries Service, NOAA. Administrative Report: LJ–02–28.

Fair, P.A. & Becker, P.R. (2000). Reviews of stress in marine mammals. *Journal of Aquatic Ecosystem Stress and Recovery* 7, 335–354.

Ford, J.K.B., Ellis, G.M., Matkin, D.R., et al. (2005). Killer whale attacks on minke whales: Prey capture and antipredator tactics. *Marine Mammal Science* 21(4), 603–618.

Fortuna, C.M. (2006). Ecology and conservation of bottlenose dolphins (*Tursiops truncatus*) in the North-Eastern Adriatic Sea. PhD Thesis, University of St Andrews, UK.

Fowler, G.S. (1999). Behavioral and hormonal responses of Magellanic penguins (*Spheniscus magellanicus*) to tourism and nest site visitation. *Biological Conservation* 90, 143–149.

Frid, A. & Dill, L.M. (2002). Human caused disturbance stimuli as a form of predation risk. *Conservation Ecology* 6(1): 11.

Gill, J.A. & Sutherland, W.J. (2000). Predicting the consequences of human disturbance from behavioural decisions. In L.M. Gosling & W.J. Sutherland (Eds), *Behaviour and Conservation*. Cambridge: Cambridge University Press, pp. 51–64.

Gill, J.A., Norris, K.S. & Sutherland, W.J. (2001). Why behavioural responses may not reflect the population consequences of human disturbance. *Biological Conservation* 97, 265–268.

Gormley, A.M., Slooten, E., Dawson, S., et al. (2012). First evidence that marine protected areas can work for marine mammals. *Journal of Applied Ecology* 49, 474–480.

Heithaus, M.R. (2001). Predator–prey and competitive interactions between sharks (order *Selachii*) and dolphins (suborder Odontoceti): A review. *Journal of Zoology (London)* 253, 53–68.

Heithaus, M.R. & Dill, L.M. (2002). Food availability and tiger shark predation risk influence bottlenose dolphin habitat use. *Ecology* 83, 480–491.

Heithaus, M.R. & Dill, L.M. (2006). Does tiger shark predation risk influence foraging habitat use by bottlenose dolphins at multiple spatial scales? *Oikos* 114, 257–264.

Heithaus, M.R., Wirsing, A.J., Burkholder, D., et al. (2009). Towards a predictive framework for predation risk effects: The interaction of landscape features and prey escape tactics. *Journal of Animal Ecology* 78, 556–562.

Henry, E. & Hammill, M.O. (2001). Impact of small boats on the haulout activity of harbour seals (*Phoca vitulina*) in Métis Bay, Saint Lawrence Estuary, Québec, Canada. *Aquatic Mammals* 27, 140–148.

Hoyt, E. (2001). *Whale Watching 2000: Worldwide tourism numbers, expenditures, and expanding socioeconomic benefits.* Yarmouth Port, MA: International Fund for Animal Welfare.

Hoyt, E. (2005). *Marine Protected Areas for Whales, Dolphins and Porpoises: A world handbook for cetacean habitat conservation.* London: Earthscan.

Hugie, D.M. & Dill, L.M. (1994). Fish and game – A game theoretic approach to habitat selection by predators and prey. *Journal of Fish Biology* 45(Suppl.A), 151–169.

Irvine, A.B., Wells, R.S. & Gilbert, P.W. (1973). Conditioning an Atlantic bottlenosed dolphin to repel various species of sharks. *Journal of Mammalogy* 54, 503–505.

Jensen, F.H., Bejder, L., Wahlberg, M., *et al.* (2009). Vessel noise effects on delphinid communication. *Marine Ecology Progress Series* 395, 161–175.

Jones, K.A. & Godin, J.G.J. (2010). Are fast explorers slow reactors? Linking personality type and anti-predator behaviour. *Proceedings of the Royal Society B* 277, 625–632.

Kasuya, T. (1995). Overview of cetacean life histories: An essay in their evolution. In A.S. Blix, L. Walløe & Ø. Ulltang (Eds), *Whales, Seals, Fish and Man.* Amsterdam: Elsevier Science, pp. 481–498.

Lima, S.L. (1998). Nonlethal effects in the ecology of predator–prey interactions. *BioScience* 48, 25–34.

Lima, S.L. & Dill, L.M. (1990). Behavioral decisions made under the risk of predation: A review and prospectus. *Canadian Journal of Zoology* 68: 619–640.

Lundquist, D., Sironi, M., Würsig, B., *et al.* (2012). Response of southern right whales to simulated swim-with-whale tourism at Península Valdés, Argentina. *Marine Mammal Science* 29, 24–45.

Lusseau, D. (2003a). Effects of tour boats on the behavior of bottlenose dolphins: Using Markov chains to model anthropogenic impacts. *Conservation Biology* 17, 1785–1793.

Lusseau, D. (2003b). Male and female bottlenose dolphins *Tursiops* spp. have different strategies to avoid interactions with tour boats in Doubtful Sound, New Zealand. *Marine Ecology Progress Series* 257, 267–274.

Lusseau, D. (2005). Residency pattern of bottlenose dolphins *Tursiops* spp. in Milford Sound, New Zealand, is related to boat traffic. *Marine Ecology Progress Series* 295, 265–272.

Lusseau, D., Maersk Lusseau, S., Bejder, L. & Williams, R. (2006a). *An individual-based model to infer the impact of whalewatching on cetacean population dynamics.* The Scientific Committee of the International Whaling Commission. Document: SC/58/WW7.

Lusseau, D., Slooten, E. & Currey, R.J.C. (2006b). Unsustainable dolphin-watching tourism in Fiordland, New Zealand. *Tourism in Marine Environments* 3, 173–178.

Mann, J. & Smuts, B.B. (1998). Natal attraction: Allomaternal care and mother–infant separations in wild bottlenose dolphins. *Animal Behaviour* 55, 1097–1113.

NRC. (2005). *Marine Mammal Populations and Ocean Noise: Determining when noise causes biologically significant effects.* Washington, DC: The National Academies Press.

Norris, K.S. & Dohl, T.P. (1980). The structure and function of cetacean schools. In L.M. Herman (Ed.), *Cetacean Behavior: Mechanisms and functions.* New York, NY: Wiley, pp. 211–261.

Nowacek, S.M., Wells, R.S. & Solow, A.R. (2001). Short-term effects of boat traffic on bottlenose dolphins, *Tursiops truncatus*, in Sarasota Bay, Florida. *Marine Mammal Science* 17, 673–688.

O'Connor, S., Campbell, R., Cortez, H. & Knowles, T. (2009). *Whale Watching Worldwide: Tourism numbers, expenditures and expanding economic benefits.* Yarmouth Port, MA: International Fund for Animal Welfare.

Preisser, E.L., Bolnick, D.I. & Benard, M.F. (2005). Scared to death? The effects of intimidation and consumption in predator–prey interactions. *Ecology* 86, 501–509.

Réale, D., Reader, S.M., Sol, D., *et al.* (2007). Integrating animal temperament within ecology and evolution. *Biological Reviews* 82, 291–318.

Réale, D., Dingemanse, N.J., Kazem, A.J.N. & Wright, J. (2010). Evolutionary and ecological approaches to the study of personality. *Philosophical Transactions of the Royal Society B* 365, 3937–3946.

Sinclair, A.R.E. & Arcese, P. (1995). Population consequences of predation-sensitive foraging: The Serengeti wildebeest. *Ecology* 76, 882–891.

Stamation, K.A., Croft, D.B., Shaughnessy, P.D., *et al.* (2010). Behavioral responses of humpback whales (*Megaptera novaeangliae*) to whale-watching vessels on the southeastern coast of Australia. *Marine Mammal Science* 26, 98–122.

Stearns, S.C. (1992). *The Evolution of Life Histories.* Oxford: Oxford University Press.

Stensland, E. & Berggren, P. (2007). Behavioural changes in female Indo-Pacific bottlenose dolphins in response to boat-based tourism. *Marine Ecology Progress Series* 332, 225–234.

Stensland, E., Angerbjörn, A. & Berggren, P. (2003). Mixed species groups in mammals. *Mammal Review* 33(3), 205–223.

Stephens, P.A., Boyd, I.L., McNamara, J.M. & Houston, A.I. (2009). Capital breeding and income breeding: Their meaning, measurement, and worth. *Ecology* 90, 2057–2067.

Tremblay, P. (2001). Wildlife tourism consumption: Consumptive or non-consumptive? *International Journal of Tourism Research* 3, 81–86.

van Parijs, S.M. & Corkeron, P.J. (2001). Boat traffic affects the acoustic behaviour of Pacific humpback dolphins, *Sousa chinensis. Journal of the Marine Association U.K.* 81, 533–538.

Visser, I.N. (1999). A summary of interactions between orca (*Orcinus orca*) and other cetaceans in New Zealand waters. *New Zealand Natural Sciences* 24, 101–112.

Waples, K.A. & Gales, N. (2002). Evaluating and minimising social stress in the care of captive bottlenose dolphins (*Tursiops aduncus*). *Zoo Biology* 2, 15–26.

Weinrich, M. & Corbelli, C. (2009). Does whale watching in Southern New England impact humpback whale (*Megaptera novaeangliae*) calf production or calf survival? *Biological Conservation* 142, 2931–2940.

Williams, R. & Ashe, E. (2007). Killer whale evasive tactics vary with boat number. *Journal of Zoology* 272, 390–397.

Williams, R. & Noren, D.P. (2009). Swimming speed, respiration rate, and estimated cost of transport in adult killer whales. *Marine Mammal Science* 25(2), 327–350.

Williams, R., Bain, D.E., Ford, J.K.B. & Trites, A.W. (2002a). Behavioral responses of male killer whales to a 'leapfrog-ging' vessel. *Journal of Cetacean Research and Management* 4, 305–310.

Williams, R., Trites, A.W. & Bain, D.E. (2002b). Behavioural responses of killer whales (*Orcinus orca*) to whale-watching boats: Opportunistic observations and experimental approaches. *Journal of Zoology (London)* 256, 255–270.

Williams, R., Lusseau, D. & Hammond, P.S. (2006). Estimating relative energetic costs of human disturbance to killer whales (*Orcinus orca*). *Biological Conservation* 113, 301–311.

Williams, R., Lusseau, D. & Hammond, P.S. (2009). The role of social aggregations and protected areas in killer whale conservation: The mixed blessing of critical habitat. *Biological Conservation* 142, 709–719.

Williams, R., Ashe, E., Sandilands, D. & Lusseau, D. (2011). *Stimulus-dependent response to disturbance affecting the activity of killer whales.* The Scientific Committee of the International Whaling Commission. Document: SC/63/WW5.

Wirsing, A.J., Heithaus, M.R., Frid, A. & Dill, L.M. (2007). Seascapes of fear: Evaluating sublethal predator effects experienced and generated by marine mammals. *Marine Mammal Science* 24, 1–15.

Zollner, P.A. & Lima, S.L. (1997). Landscape level perceptual abilities in white-footed mice: Perceptual range and the detection of forested habitat. *Oikos* 80, 51–60.

14

Whale-watching and behavioural ecology

Rochelle Constantine

Introduction

Animal societies are structured by behavioural events that influence interactions between individuals and their environment. Visual, acoustic and chemical cues are important in signalling a variety of social cues such as readiness to mate, agonistic interactions and position in a hierarchy. Social mammals, such as cetaceans, have a complex social structure that primarily uses visual and acoustic cues to govern interactions between individuals. Whitehead (2007) hypothesized that long-lived animals that are dependent on other populations to survive, and are exposed to greater environmental variation over longer time scales, will develop social learning strategies. This is particularly likely in the marine environment with top predators such as cetaceans. It is challenging to understand cetacean behaviour as the animals spend considerable periods of time below the sea surface, but research with good experimental design, that uses systematic methods and hypothesis testing, can answer questions about cetacean behaviour and in turn the effects of human activities on cetaceans (Mann, 1999; Bejder & Samuels, 2003).

As a result of longitudinal studies, in many cases now spanning in excess of 25 years – for example on Shark Bay, Western Australia and Sarasota Bay, Florida bottlenose dolphins (*Tursiops* spp.), killer whales (*Orcinus orca*) in British Columbia, Canada and humpback whales (*Megaptera novaeangliae*) in the USA – we now have an understanding of the roles of males and females in mating behaviour and the rearing of offspring (Mann *et al.*, 2000). The long period of maternal dependency in odontocetes, in particular, appears to largely serve as time to learn how to function in their complex society (Whitehead & Mann, 2000). This learning can involve horizontal transmission from mother to calf but also vertical transmission through interaction with other individuals in the population. The period of calf dependency varies considerably in cetaceans, with baleen whale calves weaned at around six months of age and odontocetes weaned from eight months to six years of age, although there are of course exceptions to these (Whitehead & Mann, 2000). The period of calf dependency appears to be linked to the social complexity of the species, with more solitary animals exhibiting high levels of group fission–fusion and generally having shorter periods of calf dependency. For migratory whales we know that calves learn their migratory patterns from their mothers and these are timed with environmental changes on the feeding and breeding grounds (Bannister, 2009).

Acoustic communication between individuals is often complex and requires learning that continues throughout the life of some species. For example, bottlenose dolphin signature whistles are extremely complex when compared to other whistles they

Whale-watching: Sustainable Tourism and Ecological Management, eds J. Higham, L. Bejder and R. Williams.
Published by Cambridge University Press. © Cambridge University Press 2014.

produce (Janik *et al.*, 1994; Watwood *et al.*, 2005) and it is now known that they convey individual identity information in these signature whistles that maintain cohesion between individuals (Janik *et al.*, 2006). In killer whales (Deecke *et al.*, 2000; Yurk *et al.*, 2002) and sperm whales (*Physeter macrocephalus*) (Rendell & Whitehead, 2003; Marcoux *et al.*, 2006), vocal clans can be identified where communication is transmitted not only from mother to offspring but also between individuals within a social group. In humpback whales, once calves learn their migratory path from their mother they return to the area they were born, where males develop a complex song that develops over time (Tyack, 2000). Song types are specific to breeding grounds (Winn *et al.*, 1981), but these can be influenced by newcomers to an area (Noad *et al.*, 2000). The lek-type behaviour of males, where song is believed to be used to attract mates and/or socially ordered other males, is vitally important during the short breeding season when many males are competing for only a few females in oestrous (Darling & Bérube, 2006; Clapham, 2009). While these examples cover only a few species, no doubt in time and with further research we will understand more species' social structure in greater depth.

The reasons for smaller-scale movements of cetaceans in coastal waters are often difficult to ascertain unless linked to a seasonal influx of preferred prey as seen in killer whales and bottlenose dolphins (Simulä *et al.*, 1996; Wilson *et al.*, 1997), or short-term environmental changes such as tidal flows (Shane, 1990; Ingram & Rogan, 2002). Seasonal changes in water temperature often drive the change in cetaceans' use of coastal waters and this is likely to be linked to prey distribution. More recently, models have been developed to incorporate multiple variables such as predation risk, competition and prey availability to understand changes in dolphin spatial use (Heithaus & Dill, 2006; Torres *et al.*, 2008). These models better reflect the complexity of habitat use by cetaceans which are no doubt influenced by the social structure of the population. Modelling allows us to incorporate the effect of human activities on cetacean habitat

use, but we are still frequently challenged by a lack of knowledge about the animals' behaviour and spatial distribution prior to human disturbance, which can limit the conclusions from research.

Whale-watch tourism has existed for over 50 years, but since the 1980s has grown at a rapid pace and shows no signs of slowing down worldwide (O'Connor *et al.*, 2009). Initially, no thought was given to the potential effects of boats and swimmers around whales as it was seen as a viable alternative to hunting whales, which was a considerable threat to populations in the 1960s. However, since that time, and with the rapid growth of the industry, questions have been raised by scientists and managers about the effects of tourism on cetaceans and the motivations of whale-watch participants (Orams, 2000; Corkeron, 2004; Higham & Lusseau, 2007). The issues around whether whale-watching constitutes a benign activity have been debated and the over-simplistic view of whale-watching being a better alternative to generate revenue than hunting whales is not relevant given our current knowledge (Corkeron, 2004; Higham & Lusseau, 2007). The rapid growth of the industry, no matter how well-intentioned, provides a challenge for scientists trying to understand cetacean behaviour without anthropogenic influences, especially in coastal waters. Typically, studies involve land- or boat-based methods where data are collected on the behaviour, interactions and location of a single individual alone or part of a group (focal animal follows), or data are collected on all members of a group (focal group follows) (Mann, 1999; Bejder & Samuels, 2003; Whitehead, 2004), but technology is playing a greater role in cetacean research. The increasing use of satellite telemetry has meant larger-scale spatial distribution can be studied over several months, which is useful for migrating whales. For shorter-term tracking devices, such as the suction-cup attached d-tag (Johnson & Tyack, 2003), extremely detailed data can be collected on animals' movements and their acoustic environment. Remote acoustic monitoring can be used to track animals' use of an area (e.g. Rayment *et al.*, 2009) and acoustic recordings help us understand the effects of boat noise on cetacean acoustic behaviour (Van Parijs & Corkeron, 2001; Foote *et al.*, 2004; Jensen *et al.*, 2009). No doubt the advances in technology combined with observational research will allow a better understanding of how whale-watching affects cetacean behaviour.

This chapter will highlight research conducted on understanding the effects of whale-watch tourism on cetacean behavioural ecology. It begins with a review of short- and long-term behavioural studies conducted to date, a brief summary of the energetic costs to cetaceans and impacts on their acoustic behaviour. Where possible, I will point out broad themes in changes in behaviour that are consistent across all species and discuss the long-term consequences of human activities on cetaceans. Finally, I briefly discuss the management responses to research findings to date and the challenges with managing the industry in light of new research findings.

Short-term and long-term behavioural studies

One of the difficulties for cetacean researchers and managers is inferring how short-term behavioural changes as a result of vessel or swimmer presence could have long-term population-level effects (Lusseau & Bejder, 2007). Cetaceans are slow-breeding and long-lived, so the population-level effects of behavioural changes take many years to be revealed, if at all. For whale populations recovering from years of hunting, the effects of tourism may be masked by the increasing number of whales in recovering populations, but this increase could be greater if whale-watching were not occurring. Further complicating behavioural studies is the fact that no pre-tourism data are available in the majority of cases of cetacean populations now exposed to tourism, which means we have no baseline data to compare changes in the population structure, range or behaviour prior to tourism beginning. Another confounding factor is where boat-based studies are the only option, as the research platform itself

has some effect on cetacean behaviour. The long-term interpretations from short-term behavioural data are difficult, but with careful research design, boat handling and robust analyses, the effect of the research platform can be quantified (e.g. Nowacek *et al.*, 2001; Lusseau, 2003; Constantine *et al.*, 2004).

One of the best examples of the value of a long-term data set comes from Shark Bay, Western Australia, where bottlenose dolphins, one of the most frequently watched dolphin species throughout the world, have been studied for almost 30 years. The comprehensive knowledge about individuals' residency patterns, calving rates, relatedness and associations (e.g. Connor *et al.*, 1999; Mann *et al.*, 2000; Krützen *et al.*, 2002) meant that the onset of dolphin-watching tourism could be evaluated against a pre-tourism data set. Bejder *et al.* (2006a) showed that dolphins in the area with the lowest level of dolphin-watch boat visits showed a greater response to boat presence than dolphins with higher levels of dolphin-watch boat exposure. If this study had been a short-term project, it is possible that the results could have been interpreted that frequently exposed dolphins had become habituated or tolerant of boat-presence. However, examination of these findings using the long-term data set found a significant decline, with around one in seven dolphins being displaced from the area of high boat exposure (Bejder *et al.*, 2006b).

It must be noted that this high level of exposure occurred when the population was exposed to two dolphin-watch boats rather than one boat, a very small number compared to the majority of areas where dolphin-watch tourism exists. Bejder *et al.* (2006b) concluded that the decline was due to displacement from the area by more sensitive individuals, as an adjacent control site showed a slight increase in dolphin numbers. Similar displacement behaviour was observed in bottlenose dolphins in Milford Sound, New Zealand where intense periods of boat traffic displaced the dolphins to the entrance to the fiord (Lusseau, 2005). The Fiordland bottlenose dolphins live near the extreme of the species range and some individuals have small home ranges, so the energetic and social consequences of limiting access to a critical habitat is of considerable concern (Lusseau *et al.*, 2006).

Displacement behaviour has been reported in other species, but may be an option afforded only to those individuals with the ability to leave, e.g. individuals in good condition, ability to access resources such as mates and food, or a position in a hierarchy that allows them to defend a new territory (Gill *et al.*, 2001). Individuals that do not leave an area when disturbed may actually be those with the fewest options, so our interpretation of any apparent lack of response indicating habituation may in fact be incorrect (Gill *et al.*, 2001; Beale & Monaghan, 2004). Those individuals that are not displaced may be more vulnerable than those that leave. If the disturbance happens in a critical habitat, such as humpback whale breeding grounds, the animals may have no other choice but to stay, and the long-term consequences of this are currently unknown. One of the more interesting findings from the Shark Bay, Western Australia research was the decline in reproductive success of females exposed to the highest levels of boat exposure (Bejder, 2005). The exact reasons for this decline are unknown, but the increased fission and fusion of groups when a boat was within 50 m of the focal group may have exposed calves to greater levels of predation, or it could be that stress levels are increased in these most vulnerable animals, resulting in decreased reproductive success. This is a well-documented phenomenon in many animals ranging from primates (olive baboons, *Papio anubus*) to reptiles (green anole, *Anolis carolinensis*), where some manage to overcome stress-related infertility but others do not (Wingfield & Sapolsky, 2003). Measuring stress in wild cetaceans is difficult (Hogg *et al.*, 2009), but should definitely be investigated when trying to understand how short-term behavioural responses may have long-term biologically significant consequences.

Another long-term study that highlights the fact that small changes in behaviour should not be ignored is research on the southern British Columbia, Canada–northwest USA populations of killer whales. Early research in British Columbia

showed that killer whales increased their swimming speed when vessels were within 400 m of them (Kruse, 1991). Since that time, researchers have built on our understanding of killer whale responses to boats through controlled exposure experiments which allowed a break-down of which variables affected the animals. We know that when approached either from the side or in their path of travel, killer whales increase the tortuosity of their path – that is, they deviate from their pre-boat approach path of travel more frequently in an attempt to distance themselves from the boat (Williams *et al.*, 2002a, 2002b). This is not an uncommon finding in behavioural studies of cetaceans, e.g. bottlenose dolphins (Janik & Thompson, 1996; Bejder *et al.*, 2006a), humpback whales (Scheidat *et al.*, 2004), sperm whales (Richter *et al.*, 2006) and dusky dolphins (*Lagenorhynchus obscurus*) (Yin, 1999), and it is clear that boat approaches change cetaceans' swimming behaviour. The level of horizontal avoidance by killer whales increased as the boat approached closer than 100 m, but as Williams *et al.* (2002b) noted, their response was not consistent as the number of boats increased. Experimental approaches by multiple vessels found that killer whales' path of travel becomes more linear once there are around three boats within 1000 m of the group and the avoidance tactic of shifting their path of travel as a way of avoiding a few boats is abandoned (Williams & Ashe, 2007). This research is important in our interpretation of killer whale behaviour where intense whale-watching activity occurs.

The examples above are from populations of animals that have high levels of year-round residency in a relatively small area. In contrast, migratory whales that spend only a few months on their breeding or feeding grounds are exposed to very different levels of tour-boat exposure. A recent study, using 25 years of data on humpback whales on their feeding grounds off the New England coast, USA has shown that females' reproductive success was not affected by whale-watching (Weinrich & Corbelli, 2009). Humpback whales have strong matrilineal fidelity to their breeding and feeding grounds and their need to access abundant, reliable sources of prey is a strong factor in whales returning to the same feeding grounds (Clapham, 2009). For mothers whose nutritional demands are considerable when lactating, the benefits of a reliable prey source may outweigh the effects of boat presence. Weinrich and Corbelli (2009) conclude that the effects of boat presence may well be masked by the quality of the feeding grounds for these whales. Animals need to make energetic trade-offs between the level of disturbance and the benefit to staying in the area of disturbance, and for large whales where feeding grounds are not uniformly distributed, the choices may be limited.

In some cases short-term behavioural research forms the baseline study upon which further work is examined. One example of this is the response by bottlenose dolphins to swim-with-dolphin tour participants in the Bay of Islands, New Zealand. The research conducted between 1994 and 1998 found in the first year of study, it was clear that dolphins avoided swimmers placed in their path of travel (Constantine, 2001). This avoidance rate increased by 27% at the end of the study, and avoidance doubled when swimmers entered the water when the dolphins were swimming around the boat. When swimmers were placed in the water off to the side of the dolphins, the avoidance rate was initially low (10%) and this decreased to 1% by the end of the study. Even in this short time-frame it was clear that dolphins responded by avoiding placements that did not give them the choice to approach swimmers, and as time progressed, further exposure of this population to swim attempts led to higher levels of swimmer avoidance, suggesting that they had become sensitized to swimmers. This study, while short-term, still allowed us to understand the effects of swim-with-dolphin tours on bottlenose dolphins.

One way to help interpret short-term responses is to measure the magnitude of the response and exposure levels of individuals to the source of disturbance (Lusseau & Bejder, 2007). In the example above, it was clear early on in the study that the dolphins avoided swimmer placement that did not offer them a choice to approach. There was no doubt that swim attempts are a stressor to these dolphins, with low numbers of individuals interacting with swimmers (average = 19.3% of the group), and the majority of the individuals that did interact were juveniles (average = 67.5%; Constantine, 2001), an age-class that spends a lot of time learning how to function in society pre-adulthood (Connor et al., 2000). Avoidance behaviour increased with time, but the biological implications of this intermittent but persistent stressor are currently unknown. There was a 7.5% annual rate of decline in the population size of bottlenose dolphins using the Bay of Islands, New Zealand between 1997 and 2005 (Tezanos Pinto et al., 2013) and investigations about what might be driving this decline are underway. Changes in calving rates, the number of stranded and animals caught in fisheries and environmental parameters such as sea-surface temperature have all been ruled out as explanatory variables (Nathan, 2010). Given the findings of Bejder et al. (2006a), boat and swimmer interactions must be considered as one of the variables in understanding this change in abundance, but a lack of data on dolphin movements outside the bay, a possible indicator of displacement, means spatio-temporal changes are not able to be determined. Research showed that the dolphins in the Bay of Islands change their behaviour in the presence of boats (Constantine et al., 2004), as seen elsewhere. As this population is not resident in the Bay of Islands, instead having a larger home range (Constantine, 2002), we may not be able to fully understand the reason for the decline.

While research with control and impact sites, before–during–after studies and long time-frames is rarely possible, there are clear and persistent patterns across a number of species of whales and dolphins of short-term responses to the presence of boats. Examples include grouping together more tightly (Bejder et al., 1999, 2006a; Barr & Slooten, 1999), changes in speed and direction of travel (Nowacek et al., 2001; Williams et al., 2002a, 2002b; Scheidat et al., 2004; Stensland & Berggren, 2007; Gulesserian, 2009; Stamation et al., 2010), changes in activity budgets (Acevedo, 1991; Corkeron, 1995; Lusseau, 2003; Constantine et al., 2004; Stockin *et al.*, 2008; Steckenreuter *et al.*, 2011) and altered respiration and surfacing patterns (Stone *et al.*, 1992; Blane & Jaakson, 1995; Janik & Thompson, 1996; Nowacek *et al.*, 2001; Hastie *et al.*, 2003). Many of these strategies are classic animal responses to predation risk (Frid & Dill, 2002). Even though vessels and swimmers are typically non-lethal disturbances, the animals respond to the perceived risk relative to the benefits of remaining in the presence of that risk (Frid & Dill, 2002). The risk–disturbance hypothesis predicts that frequent response to perceived threat can cause a decline in an individual's fitness and, ultimately, population size and habitat use. The consequences of a change in habitat use (Allen & Read, 2000; Bejder *et al.*, 2006a; Lusseau & Higham, 2004; Steckenreuter *et al.*, 2012) could leave animals with less access to resources or increased predation risk. While it is difficult to quantify the myriad behavioural, environmental and anthropogenic disturbances, modelling the effects of multiple sources of change will enable us to better understand the importance of short-term behavioural changes on long-term population-level impacts.

Energetic consequences

Cetacean behavioural studies frequently collect data on events such as leaps, fish chases and pectoral slaps, which are then classified into behavioural states such as travel, socializing, rest and foraging (e.g. Würsig & Würsig, 1980; Shane, 1990; Connor *et al.*, 2000). How animals spend their day is one important way of measuring the effects of change in their environment. The use of energy for behaviours such as socializing must be weighed up against the ability to catch enough prey and gain adequate rest. If animals are not well fed or rested, then they are more vulnerable to predation, less likely to compete for access to mates or adequately care for calves. There are many studies that show that whale-watching changes whale and dolphin behaviour. Of particular concern are those that have documented reduced resting behaviour (e.g.

Lusseau, 2003; Constantine *et al.*, 2004; Steckenreuter *et al.*, 2011), changes in foraging (e.g. Williams *et al.*, 2006; Stockin *et al.*, 2008) and increased travel, which is most likely energetically expensive in particular for females with calves and whales outside their feeding grounds (e.g. Scheidat *et al.*, 2004; Stensland & Berggren, 2007).

The energetic consequences of frequent interruptions to normal behaviour patterns as a result of anthropogenic activities, is of concern. Many whale-watch industries have multiple trips spread throughout a day, but this may not be the best strategy to minimize impact on the animals. In the Bay of Islands, New Zealand there is typically only one group of dolphins present on any given day, and it takes only one tour-boat to stop them from resting (Constantine *et al.*, 2004), so the staggered arrival of tour-boats throughout the day resulted in a decline in resting overall. Whether they compensated for this reduction later in the day is unknown, as research on their nocturnal behaviour was not possible. Williams *et al.* (2006) calculated the energetic cost of a change in killer whale activity budgets due to boat presence to be around 3–4%. Of greater concern was the 18% reduction in energy intake due to lost feeding opportunities in the presence of boats. This is the first study to examine the energetic costs of boat presence and further work should be undertaken, especially for more vulnerable animals like females with calves, noting the decline in reproductive success for Shark Bay, Western Australia females in high disturbance areas (Bejder, 2005), and migratory whales away from their feeding grounds.

Typically, cetaceans exist in an environment with patchy prey distribution. For whales with separate breeding and feeding grounds, the 3–4 months of intense foraging activity (when resident at feeding grounds) are crucial to the whales' surviving the rest of the year off stored energy sources. At breeding grounds where mating and calving occurs, energy output must be balanced with the need to return to their distant feeding grounds. In contrast, non-migratory cetaceans must continuously find prey sources within their home-range. The energetic demands change seasonally with cooler winter waters requiring more energetic output to survive. There is also an additional cost to females when they are gestating or lactating. During the period of calf dependency, mothers are the primary caregiver of their offspring (Whitehead & Mann, 2000), although we know in the case of sperm whales other members of the group may help care for calves (Whitehead, 1996). The increased level of care by females with calves (e.g. Mann & Smuts, 1998; Corkeron & Connor, 1999), combined with the demands of lactation result in considerable demand on females and subsequently this is reflected in their behavioural responses to anthropogenic disturbance.

Female bottlenose dolphins in Zanzibar, Tanzania respond to boats by increasing the amount of travel (Stensland & Berggren, 2007). This type of horizontal avoidance strategy was preferred by females in Doubtful Sound, New Zealand until interactions became intrusive and then females shifted to vertical avoidance strategies by increasing the duration of dives (Lusseau, 2003). Controlled-approach experiments showed that experienced mothers with dependent calves had longer dive times in the presence of boats (Nowacek et al., 2001). This may be a successful strategy to remove the calf from potential harm as they are often less capable in the water (Mann & Smuts, 1998) and have limited dive capability due to limited lung capacity. It is for this reason that younger cetaceans are more likely to be near to the surface, making them more vulnerable to vessel strike (Wells & Scott, 1997; Laist et al., 2001; Panigada et al., 2006; Douglas et al., 2008). In humpback whales on their migration path, pods containing calves were more sensitive to boat presence than non-calf pods (Stamation et al., 2010), which aligns with the protective role of mothers over their offspring.

Acoustic effects on behaviour

Many cetacean species use acoustic communication to maintain group cohesion or spatial distribution, and acquire information about conspecifics and their environment (Tyack, 2000). Underwater noise can impair hearing, either temporarily or permanently, and impede effective communication between individuals (Richardson et al., 1995). Measuring underwater sound and cetacean acoustics is a challenging area as there are many factors that must be measured, such as water depth, substrate type, a species' acoustic range, vessel engine type and speed, to name a few. Recently, researchers have begun to model the potential impacts of noise and their masking effects on cetacean communication (e.g. Erbe, 2002; Jensen et al., 2009).

Research has shown that vessel engine noise can disturb the acoustic behaviour of several species of cetacean and there is evidence that boat noise can mask communication between delphinids. Killer whales in Washington State (USA) waters increased their call duration in the presence of boats by 15% after a 5-fold increase in the number of boats accompanying the whales (Foote et al., 2004). This response once a certain threshold was reached is a common theme in behavioural changes in response to anthropogenic disturbance. Erbe (2002) modelled the effects of vessel noise on killer whales and estimated that fast vessels masked calls at 14 km and could cause a temporary threshold shift in hearing after 30–50 min at 450 m. Prolonged exposure to slow- or fast-moving vessels, as experienced by killer whales in the Washington State, USA–southern British Columbia, Canada region could cause a permanent shift in hearing, which is of considerable concern as these whales rely heavily on vocalizations to maintain their society (Deecke et al., 2000; Yurk et al., 2002). Belugas (*Delphinapterus leucas*) in the St Lawrence River increased the frequency range of their vocalizations once vessels were in close proximity and changed their calling rates as vessels approached within 1 km (Lesage et al., 1999). These effects lasted longer when slow-moving ferries were passing through their habitat, which suggests that as long as the source of disturbance is present, it poses a disturbance to the whales. Bottlenose dolphin communication range was reduced by 26% within 50 m by a small boat travelling at 5 knots in shallow water (Jensen *et al.*, 2009). The same level of noise and speed reduced pilot whale communication by 58% in deeper waters (Jensen *et al.*, 2009). Humpback dolphins (*Sousa chinensis*) increased their rate of whistles once boats had passed by within 1.5 km of the group (Van Parijs & Corkeron, 2001) and bottlenose dolphins increased their whistle rates in the presence of dolphin-watch vessels (Scarpaci *et al.*, 2000).

Given the importance of vocal communication between individual cetaceans to maintain cohesion, provide information about prey sources and attract mates (e.g. Janik, 2000; Tyack, 2000) any disruption to this aspect of their behavioural ecology is concerning. The risk of a temporary or permanent threshold shift in hearing is too great in these animals that rely on sound to survive and maintain healthy societies.

Management responses

Scientific input into management decisions about how to most effectively manage whale-watch operations has been met with a variety of outcomes. In general, self-regulation through measures such as voluntary codes of conduct or guidelines has been found to have low levels of compliance, even when they knew they were being monitored (e.g. Scarpaci *et al.*, 2003; Duprey *et al.*, 2008; Wiley *et al.*, 2008; Schaffar *et al.*, 2010). The use of legally binding regulations allows a degree of control over the growth of an industry (e.g. New Zealand), but only if the agency limits the number of permits and enforces the regulations. Unfortunately, action by management agencies is often lacking as this can be expensive and time-consuming. The management framework with the highest likelihood of an increase in protection for the animals is where operators are permitted to run tours. The most effective example to date is from Western Australia, where the government withdrew one of two dolphin-watch permits in Shark Bay when it was shown that there was a significant impact from these tours (Ministry Media Statement, 2006). In New Zealand, changes

to operators permitting conditions, such as limiting interaction time, restrictions in swimmer placements and restricted periods of time when tours can operate, have occurred in several locations. In some cases changes are implemented that do not effectively protect the animals from exposure to tourism (e.g. Hartel, 2010; Steckenreuter *et al.*, 2012). This highlights the importance of ongoing evaluation of the efficacy of protection measures and the need for adaptive management plans (Constantine & Bejder, 2008).

Conclusion

A significant volume of research now exists that provides increasingly detailed insights into how and to what extent the presence of vessels and swimmers alters the behavioural ecology of cetaceans. All species of cetaceans live in complex societies that rely on a fine balance of finding sufficient prey, competing for access to mates, raising offspring and avoiding predation. However, it remains very much the case that we have a limited and incomplete understanding of the long-term effects of tourism for the majority of species and populations exposed to whale-watching. For large migratory whales, for example, the energetic output required when away from their feeding grounds is immense and we have no idea what the effects of increased vigilance and travel speeds as a result of whale-watch boat presence means to their energetic demands. What is known from long-term studies on killer whales and bottlenose dolphins is that disruptions to their behaviour, energetics, social structure and habitat use can result in biologically significant consequences and this is cause for considerable concern for all cetaceans exposed to tourism.

There are similar and repetitive themes in the short-term responses to boat and swimmer presence by several cetacean species, but more than enough information for managers to act by enforcing particular types of boat handling and swimmer placement. Unfortunately, this often does not happen, and even when some form of agreement exists to regulate boat behaviour it is frequently ignored

(Scarpaci *et al.*, 2003; Duprey *et al.*, 2008; Constantine & Bejder, 2008). Managers need to work closely with researchers and industry to form an effective adaptive management plan and respond rapidly when issues arise (Constantine & Bejder, 2008).

There is discussion about the limitations of studies that quantify the short-term changes in behaviour of cetaceans. Many management or funding agencies do not have the time to wait for long-term studies before acting. This poses a dilemma for behavioural ecologists who know that a pre-tourism data set, followed by carefully designed research testing hypothesis using multiple scenarios, is the best way to advise managers on how to minimize the effect of whale-watching on target populations. While some researchers are taking new approaches to testing hypotheses about human disturbance (e.g. Dans *et al.*, 2012; Thorne *et al.*, 2012), many short-term studies underway are just repeating the same work conducted elsewhere and finding the same results. We need to move beyond this. I suggest that future research needs to take a multi-disciplinary approach to creating models that anticipate the long-term consequences of repetitive exposure to acute stressors to cetacean populations. There are already some collaborative projects underway such as the Population Consequences of Acoustic Disturbance (National Research Council, 2005) that are trying to address the biological consequences of anthropogenic noise and the International Whaling Commission is encouraging research on the effects of whale-watching on the recovery of whale populations. The use of tracking technologies such as AIS have proven effective in monitoring shipping compliance with ship-strike reduction measures (e.g. Vanderlaan & Taggart, 2009; Wiley *et al.*, 2011) and could be used effectively to monitor adherence to restricted areas or operating times. The combined use of technology and innovative analysis solutions are important to ensure the future welfare of populations exposed to tourism.

Whale-watching is not a benign activity, but it can be conducted in a manner that has less impact on the animals' behaviour. Research provides insights into cetacean behavioural changes as a result of human effects and in many cases provides solutions that allow better management of whale-watch industries. The challenge for scientists now is to determine the best ways of evaluating long-term predictors of impacts before they occur so managers can act before it is too late. This requires regulations that are enforced, ongoing evaluation of the efficacy of changes and continued partnerships between all stakeholders. This needs to happen now, otherwise situations as seen in Shark Bay may become a frequent occurrence.

REFERENCES

Acevedo, A. (1991). Interactions between boats and bottlenose dolphins, *Tursiops truncatus*, in the entrance to Ensenada De La Paz, Mexico. *Aquatic Mammals* 17, 120–124.

Allen, M.C. & Read, A.J. (2000). Habitat selection of foraging bottlenose dolphins in relation to boat density near Clearwater, Florida. *Marine Mammal Science* 16, 815–824.

Bannister, J.L. (2009). Baleen whales (Mysticetes). In W.F. Perrin, B. Würsig & J.G.M. Thewissen (Eds), *The Encyclopedia of Marine Mammals* (2nd eds), San Diego, CA: Academic Press, pp. 80–89.

Barr, K. & Slooten, L. (1999). *Effects of Tourism on Dusky Dolphins at Kaikoura*. Conservation Advisory Science Notes 229. Wellington, New Zealand: Department of Conservation, 28 pp.

Beale, C.M. & Monaghan, P. (2004). Behavioural responses to human disturbance: a matter of choice? *Animal Behaviour* 68, 1065–1069.

Bejder, L. (2005). Linking short and long-term effects of nature-based tourism on cetaceans. PhD Thesis, Dalhousie University, Canada. 158 pages.

Bejder, L. & Samuels, A. (2003). Evaluating the effects of nature-based tourism on cetaceans. In N. Gales, M. Hindell & R. Kirkwood (Eds), *Marine Mammals: Fisheries, tourism and management issues*. Melbourne, VIC: CSIRO Publishing, pp. 229–256.

Bejder, L., Dawson, S.M. & Harraway, J.A. (1999). Responses by Hector's dolphins to boats and swimmers in Porpoise Bay, New Zealand. *Marine Mammal Science* 15, 738–750.

Bejder, L., Samuels, A., Whitehead, H. & Gales, N. (2006a). Interpreting short-term behavioural responses to disturbance within a longitudinal perspective. *Animal Behaviour* 72, 1149–1158.

Bejder, L., Samuels, A., Whitehead, H., *et al.* (2006b). Decline in relative abundance of bottlenose dolphins exposed to long-term disturbance. *Conservation Biology* 20, 1791–1798.

Blane, J.M. & Jaakson, R. (1995). The impact of ecotourism on the St. Lawrence beluga whales. *Environmental Conservation* 21, 267–269.

Clapham, P.J. (2009). Humpback whale (*Megaptera novaeangliae*). In W.F. Perrin, B. Würsig & J.G.M. Thewissen (Eds), *The Encyclopedia of Marine Mammals* (2nd edn). New York, NY: Academic Press, pp. 582–585.

Connor, R.C., Heithaus, M.R. & Barre, L.M. (1999). Superalliance of bottlenose dolphins. *Nature* 397, 571–572.

Connor, R.C., Wells, R.S., Mann, J. & Read, A.J. (2000). The bottlenose dolphin: Social relationships in a fission–fusion society. In J. Mann, R.C. Connor, P.L. Tyack & H. Whitehead (Eds), *Cetacean Societies*. Chicago, IL: University of Chicago Press, pp. 91–126.

Constantine, R. (2001). Increased avoidance of swimmers by wild bottlenose dolphins (*Tursiops truncatus*) due to long-term exposure to swim-with-dolphin tourism. *Marine Mammal Science* 17, 689–702.

Constantine, R. (2002). The behavioural ecology of bottlenose dolphins (*Tursiops truncatus*) of northeastern New Zealand: A population exposed to tourism. PhD Thesis, University of Auckland, New Zealand. 195 pages.

Constantine, R. & Bejder, L. (2008). Managing the whale- and dolphin-watching industry: Time for a paradigm shift. In J. Higham & M. Lück (Eds), *Marine Wildlife and Tourism Management*. Wallingford: CABI Publishers, pp. 319–334.

Constantine, R., Brunton, D.H. & Dennis, T. (2004). Dolphin watching tour boats change bottlenose dolphin (*Tursiops truncatus*) behaviour. *Biological Conservation* 117, 299–307.

Corkeron, P.J. (1995). Humpback whales (*Megaptera novaeangliae*) in Hervey Bay, Queensland: Behaviour and responses to whale-watching vessels. *Canadian Journal of Zoology* 73, 1290–1299.

Corkeron, P.J. (2004). Whale watching, iconography, and marine conservation. *Conservation Biology* 18, 847–849.

Corkeron, P.J. & Connor, R.C. (1999). Why do baleen whales migrate? *Marine Mammal Science* 15, 1228–1245.

Dans, S.L., Degrati, M, Pedraza, S.N. & Crespo, E.A. (2012). Effects of tour boats on dolphin activity examined with sensitivity analysis of Markov Chains. *Conservation Biology* 26, 708–716.

Darling, J.D. & Bérube, M. (2006). Interactions of singing humpback whales with other males. *Marine Mammal Science* 17, 570–584.

Deecke, V.B., Ford, J.K.B. & Spong, P. (2000). Dialect change in resident killer whales: Implications for vocal learning and cultural transmission. *Animal Behaviour* 60, 629–638.

Douglas, A.B., Calambokidis, J., Raverty, S., Jeffries, S.J., Lambourne, D.M. & Norman, S.A. (2008). Incidence of ship strikes of large whales in Washington State. *Journal of the Marine Biological Association of the United Kingdom* 88, 1121–1132.

Duprey, N.M.T, Weir, J. & Würsig, B. (2008). Effectiveness of a voluntary code of conduct in reducing vessel traffic around dolphins. *Ocean & Coastal Management* 51, 632–637.

Erbe, C. (2002). Underwater noise of whale-watching boats and potential effects on killer whales (*Orcinus orca*), based on an acoustic impact model. *Marine Mammal Science* 18, 394–418.

Foote, A.D., Osbourne, R.W. & Hoelzel, A.R. (2004). Whale-call response to masking boat noise. *Nature* 428, 910.

Frid, A. & Dill, L. (2002). Human-caused disturbance stimuli as a form of predation risk. *Conservation Ecology* 6(1), 11.

Gill, J.A., Norris, K. & Sutherland, W.J. (2001). Why behavioural responses may not reflect the population consequences of human disturbance. *Biological Conservation* 97, 265–268.

Gulesserian, M. (2009). Anthropogenic interactions of the Group V humpback whales (*Megaptera novaeangliae*) migrating past Sydney, Australia. MPhil Thesis, Macquarie University, Australia. 129 pages.

Hartel, E.F. (2010). Habitat use by bottlenose dolphins (*Tursiops truncatus*) in the Bay of Islands, New Zealand. MSc Thesis, University of Auckland, New Zealand. 101 pages.

Hastie, G.D., Wilson, B., Tufft, L.H. & Thompson, P.M. (2003). Bottlenose dolphins increase breathing synchrony in response to boat traffic. *Marine Mammal Science* 19, 74–84.

Heithaus, M.R. & Dill, L.M. (2006). Does tiger shark predation risk influence foraging habitat use by bottlenose dolphins at multiple spatial scales? *Oikos* 114, 257–264.

Higham, J.E.S. & Lusseau, D. (2007). Urgent need for empirical research into whaling and whale watching. *Conservation Biology* 21, 554–558.

Hogg, C.J., Rogers, T.L., Shorter, A., Barton, K., Miller, P.J.O. & Nowacek, D. (2009). Determination of steroid hormones in whale blow: It is possible. *Marine Mammal Science* 25, 605–618.

Ingram, S.N. & Rogan, E. (2002). Identifying critical areas and habitat preferences of bottlenose dolphins *Tursiops truncatus*. *Marine Ecology Progress Series* 244, 247–255.

Janik, V.M. (2000). Food related bray calls in wild bottlenose dolphins (*Tursiops truncatus*). *Proceedings of the Royal Society London B* 267, 923–927.

Janik, V.M. & Thompson, P.M. (1996). Changes in surfacing patterns of bottlenose dolphins in response to boat traffic. *Marine Mammal Science* 12, 597–602.

Janik, V.M., Dehnhardt, G. & Todt, D. (1994). Signature whistle variations in a bottlenosed dolphin, *Tursiops truncatus*. *Behavioural Ecology and Sociobiology* 35, 243–248.

Janik, V.M., Sayigh, L.M. & Wells, R.S. (2006). Signature whistle shape conveys identity information to bottlenose dolphins. *Proceedings of the National Academy of Sciences* 103, 8293–8297.

Jensen, F.H., Bejder, L., Wahlberg, M., Aguilar Soto, N., Johnson, M. & Madsen, P.T. (2009). Vessel noise effects on delphinid communication. *Marine Ecology Progress Series* 395, 161–175.

Johnson, M.P. & Tyack, P.L. (2003). A digital acoustic recording tag for measuring the response of wild marine mammals to sound. *IEEE Journal of Oceanic Engineering* 28, 3–12.

Kruse, S. (1991). The interactions between killer whales and boats in Johnstone Strait, B.C. In K. Pryor & K.S. Norris (Eds), *Dolphin Societies: Discoveries and puzzles*. Berkeley, CA: University of California Press, pp. 149–159.

Krützen, M., Sherwin, W.B., Connor, R.C., *et al.* (2002). Contrasting relatedness patterns in bottlenose dolphins (*Tursiops* sp.) with different alliance strategies. *Proceedings of the Royal Society London B*, DOI 10.1098/rspb.2002.2229.

Laist, D.W., Knowlton, A.R., Mead, J.G., Collet, A.S. & Podesta, M. (2001). Collisions between ships and whales. *Marine Mammal Science* 17, 35–75.

Lesage, V., Barrette, C., Kingsley, M.C.S. & Sjare, B. (1999). The effect of vessel noise on the vocal behavior of belugas in the St. Lawrence River estuary, Canada. *Marine Mammal Science* 15, 65–84.

Lusseau, D. (2003). Effects of tour boats on the behaviour of bottlenose dolphins: Using Markov chains to model anthropogenic impacts. *Conservation Biology* 17, 1785–1793.

Lusseau, D. (2005). Residency pattern of bottlenose dolphins, *Tursiops* spp., in Milford Sounds, New Zealand, is related to boat traffic. *Marine Ecology Progress Series* 295, 265–272.

Lusseau, D. & Bejder, L. (2007). The long-term consequences of short-term responses to disturbance experiences from whalewatching impact assessment. *International Journal of Comparative Psychology* 20, 228–236.

Lusseau, D. & Higham, J.E.S. (2004). Managing the impacts of dolphin-based tourism through the definition of critical habitats: The case of bottlenose dolphins (*Tursiops* spp.) in Doubtful Sounds, New Zealand. *Tourism Management* 25, 657–667.

Lusseau, D., Slooten, L. & Currey, R.J.C. (2006). Unsustainable dolphin watching tourism in Fiordland, New Zealand. *Tourism in Marine Environments* 3, 173–178.

Mann, J. (1999). Behavioral sampling methods for cetaceans: A review and critique. *Marine Mammal Science* 15, 102–122.

Mann, J. & Smuts, B.B. (1998). Natal attraction: allomaternal care and mother–infant separations in wild bottlenose dolphins. *Animal Behaviour* 55, 1097–1113.

Mann J., Connor, R.C., Tyack, P.L. & Whitehead, H. (2000). *Cetacean Societies*. Chicago, IL: University of Chicago Press.

Marcoux, M., Rendell, L. & Whitehead, H. (2006). Indications of fitness differences among vocal clans of sperm whales. *Canadian Journal of Zoology* 84, 609–614.

Ministry Media Statement. (2006). Retrieved from http://www.mediastatements.wa.gov.au/media/media.nsf/news/958A19167C70F7934825719900206D69. West Australian Government.

Nathan, E. (2010). Investigating the decline of the bottlenose dolphins population of the Bay of Islands. Final Unpublished Report, Faculty of Science Summer Studentship, University of Auckland, New Zealand.

National Research Council. (2005). *Marine Mammal Populations and Ocean Noise: Determining when noise causes biologically significant effects*. Washington, DC: US National Academy of Sciences.

Noad, M.J, Cato, D.H., Bryden, M.M., Jenner, M-N. & Jenner, C.S. (2000). Cultural revolution in whale songs. *Nature* 408, 537.

Nowacek, S.M., Wells, R.S. & Solow, A.R. (2001). Short-term effects of boat traffic on bottlenose dolphins, *Tursiops truncatus*, in Sarasota Bay, Florida. *Marine Mammal Science* 17, 673–688.

O'Connor, S., Campbell, R., Cortez, H. & Knowles, T. (2009). *Whale Watching Worldwide: Tourism numbers, expenditures and expanding economic benefits*. Yarmouth Port, MA: International Fund for Animal Welfare.

Orams, M.B. (2000). Tourists getting close to whales, is it what whale-watching is all about? *Tourism Management* 21, 561–569.

Panigada, S., Pesante, G., Zanardelli, M., Capoulade, F., Gannier, A. & Weinrich, M.T. (2006). Mediterranean fin whales at risk from fatal ship strikes. *Marine Pollution Bulletin* 52, 1287–1298.

Rayment, W., Dawson, S. & Slooten, L. (2009). Trialling an automated passive acoustic detector (T-POD) with Hector's dolphins (*Cephalorhynchus hectori*). *Journal of the Marine Biological Association of the United Kingdom*, doi:10.1017/S0025315409003129.

Rendell, L. & Whitehead, H. (2003). Vocal clans in sperm whales (*Physeter macrocephalus*). *Proceedings of the Royal Society London B* 270, 225–231.

Richardson, W.J., Greene Jr, C.R., Malme, C.I. & Thomson, D.H. (1995). *Marine Mammals and Noise*. San Diego, CA: Academic Press.

Richter, C., Dawson, S. & Slooten, E. (2006). Impacts of commercial whale watching on sperm whales at Kaikoura, New Zealand. *Marine Mammal Science* 22, 46–63.

Scarpaci, C., Bigger, S.W., Corkeron, P.J. & Nugegoda, D. (2000). Bottlenose dolphins (*Tursiops truncatus*) increase whistling in the presence of 'swim-with-dolphin' tour operations. *Journal of Cetacean Research and Management* 2, 183–185.

Scarpaci, C., Dayanthi, N. & Corkeron, P.J. (2003). Compliance with regulations by 'swim-with-dolphins' operations in Port Phillip Bay, Victoria, Australia. *Environmental Management* 31, 342–347.

Schaffar, A., Garrigue, C. & Constantine, R. (2010). Exposure of humpback whales to unregulated whalewatching activities in their main reproductive area in New Caledonia. *Journal of Cetacean Research and Management* 11, 147–152.

Scheidat, M., Castro, C., Gonzalez, J. & Williams, R. (2004). Behavioural responses of humpback whales (*Megaptera novaeangliae*) to whale watch boats near Isla de la Plata, Machalilla National Reserve, Ecuador. *Journal of Cetacean Research and Management* 6, 63–68.

Shane, S.H. (1990). Behavior and ecology of the bottlenose dolphin at Sanibel Island, Florida. In S. Leatherwood & R.R. Reeves (Eds), *The Bottlenose Dolphin*. San Diego, CA: Academic Press, pp. 245–265.

Similä, T., Holst, J.C. & Christensen, I. (1996). Occurrence and diet of killer whales in northern Norway: Seasonal patterns relative to the distribution and abundance of Norwegian spring-spawning herring. *Canadian Journal of Fisheries and Aquatic Sciences* 53, 769–779.

Stamation, K.A., Croft, D.B., Shaunessey, P.D., Waples, K.A. & Briggs, S.V. (2010). Behavioral responses of humpback whales (*Megaptera novaeangliae*) to whale-watching vessels on the southeastern coast of Australia. *Marine Mammal Science* 26, 98–122.

Steckenreuter, A., Harcourt, R. & Möller, L. (2011). Distance does matter: Close approaches by boats impede feeding and resting behaviour of Indo-Pacific bottlenose dolphins. *Wildlife Research* 38, 455–463.

Steckenreuter, A., Harcourt, R. & Möller, L. (2012). Are Speed Restriction Zones an effective management tool for minimizing impacts of boats on dolphins in an Australian marine park? *Marine Policy* 36, 258–264.

Stensland, E. & Berggren, P. (2007). Behavioural changes in female Indo-Pacific bottlenose dolphins in response to boat-based tourism. *Marine Ecology Progress Series* 332, 225–234.

Stockin, K.A., Lusseau, D., Binnedell, V., Wiseman, N. & Orams, M.B. (2008). Tourism affects the behavioural budget of the common dolphin *Delphinus* sp. in the Hauraki Gulf, New Zealand. *Marine Ecology Progress Series* 355, 287–295.

Stone, G.S., Katona, S.K., Mainwaring, A., Allen, J.M. & Corbett, H.D. (1992). Respiration and surfacing rates of fin whales (*Balaenoptera physalus*) observed from a lighthouse tower. *Reports of the International Whaling Commission* 42, 739–745.

Tezanos-Pinto, G., Constantine, R., Brooks, L., *et al.* (2013). Decline in local abundance of bottlenose dolphins (*Tursiops truncatus*) in the Bay of Islands (New Zealand). *Marine Mammal Science*, doi:10.i111/mms.12008.

Thorne, L.H., Johnston, D.W., Urban, D.L., *et al.* (2012). Predictive modeling of spinner dolphin (*Stenella longirostris*) resting habitat in the main Hawaiian Islands. *PLoS ONE* 7, e43167. doi:10.1371/journal.pone.0043167.

Torres, L.G., Read, A.J. & Halpin, P. (2008). Fine-scale habitat modeling of a top marine predator: Do prey data improve predictive capacity? *Ecological Applications* 18, 1702–1717.

Tyack, P. (2000). Functional aspects of cetacean communication. In J. Mann, R.C. Connor, P.L. Tyack & H. Whitehead (Eds), *Cetacean Societies*. Chicago, IL: University of Chicago Press, pp. 270–307.

Vanderlaan, A.S.M. & Taggart, C.T. (2009). Efficacy of a voluntary area to be avoided to reduce risk of lethal vessel strikes to endangered whales. *Conservation Biology* 23, 1467–1474.

Van Parijs, S.M. & Corkeron, P.J. (2001). Boat traffic affects the acoustic behaviour of Pacific humpback dolphins, *Sousa chinensis*. *Journal of the Marine Biology Association, UK* 81, 1–6.

Watwood, S.L., Owen, E.C.G., Tyack, P.L. & Wells, R.S. (2005). Signature whistle use by temporarily restrained and free-swimming bottlenose dolphins, *Tursiops truncatus*. *Animal Behaviour* 69, 1373–1386.

(a)

Figure 3.1a Exemplifying the diversity of human–cetacean interactions within a time and place, common bottlenose dolphins (*Tursiops truncatus*) forage on fish escaping from trawl nets during winch-up off northwestern Australia (photo: S. Allen).

(b)

Figure 3.1b A dolphin caught in a trawl net during the same fishing trip (photo: S. Allen). These dolphins follow trawlers around for days and weeks at a time, benefitting from the concentrated food source, but independent observer reports suggest that around 50 dolphins are caught per annum (Allen & Loneragan, 2010). Underwater video footage taken inside actively fishing trawl nets suggests that by-catch is under-estimated, as some dolphins fall out of the net before being landed (Jaiteh *et al.*, 2013). Dolphins, like all marine mammals, are protected in Australian waters.

(a)

(b)

Figure 3.2a,b Shaped to evoke association with diving whales and dolphins, a single-edged slate knife that was likely to have been used for flensing fish and marine mammals by communities of northern Norway in the Stone Age (photo: A. Icagic/Tromsø University Museum). (b) A mosaic floor section depicting a dolphin from a third-century AD Roman villa in the south of England (photo: D. Allen, Wolf Design).

Figure 3.3b The image of a whale carved into a boab (*Adansonia gregorii*) on the Kimberley coast of northwestern Australia. Humpback whales were, and still are, revered here, while dugongs were, and still are, hunted for consumption by Aboriginal people. Given the boab's longevity, this carving could be decades, hundreds or even over 1000 years old (photo: S. Allen/permission by D. Woolagoodja, senior custodian of the Dambimangari/Worwoorra People).

Figure 3.3c Cetacean researchers conducting shore-based observations of whales from the Big Island of Hawai'i (photo: S. Allen).

Figure 3.3d Researchers conducting boat-based photographic identification of dolphins in southwestern Australia (photo: S. Allen).

(a)

Figure 3.4a A dolphin calf, habituated to close interactions with humans through provisioning and a swim-with-dolphins programme, with fishing line entanglement. This entanglement prevented suckling, resulting in slow starvation (photo: Bunbury Dolphin Discovery Centre).

Figure 3.4b Recreational boaters breaching minimum approach distance regulations to Indo-Pacific bottlenose dolphins, Rottnest Island, Western Australia. After repeated breaches, some individuals started tail slapping and then the resting group moved away into an area where boats could not approach (photo: S. Allen).

(b)

(a)

(b)

Figure 3.5a,b (a) Indo-Pacific bottlenose dolphins leaping in the pressure wave off the bow of a ship leaving the Swan River, Western Australia (photo: S. Allen) (b) Further upstream on the very same day, an emaciated resident dolphin with gross skin lesions and fishing line entanglement perishes adjacent to the city of Perth (photo: S. Allen).

Figure 10.5 Intern Lisa Gibbler presents the baleen to passengers.

Figure 10.6 Cynde McInnis holds up the 3D model of Stellwagen Bank.

(b)

(c)

Figure 21.2b–c As part of the 'Dolphins for Development' Project, infrastructure at the Kampi viewing site was constructed in late 2004. Infrastructure included: (b) no swimming in Kampi Pool signs; and (c) an entrance booth, which was located at the entrance to the Kampi viewing site.

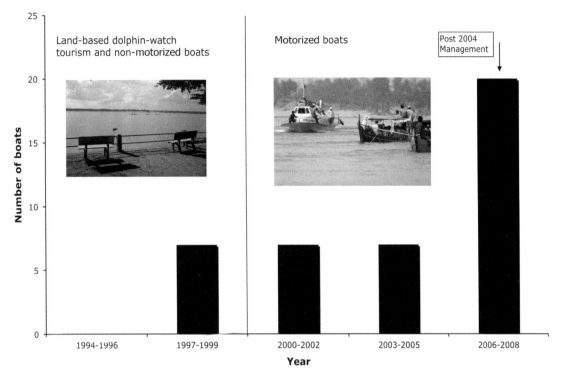

Figure 21.3 Bar graph showing the increase in number of dolphin-watching boats operating at Kampi Pool from 1997 to 2008 (non-motorized boats to the left of the vertical line and motorized boats to the right of the vertical line). Prior to 1997, only land-based observations were undertaken at the Kampi viewing site. Non-motorized boats were used in combination with land-based observations from 1997 to 2000 (left image). Post-2000, all dolphin-watch tourism boats were motorized with shelters for tourists (right image).

(a)

Figure 21.4a Dolphin-watch tourism has expanded quickly at Kampi Pool, and the majority of the local community now receives limited benefit from the dolphin-watch tourism. (a) During holiday periods there are many boats in the pool at one time. As a result of the new Commission for Dolphin Conservation, the number of boats at the dolphin-watch site increased from 7 in 2005 to 20 in 2007.

(b)

Figure 21.4b shows the small size of Kampi Pool, where almost all of the pool is visible in the image (2 km^2).

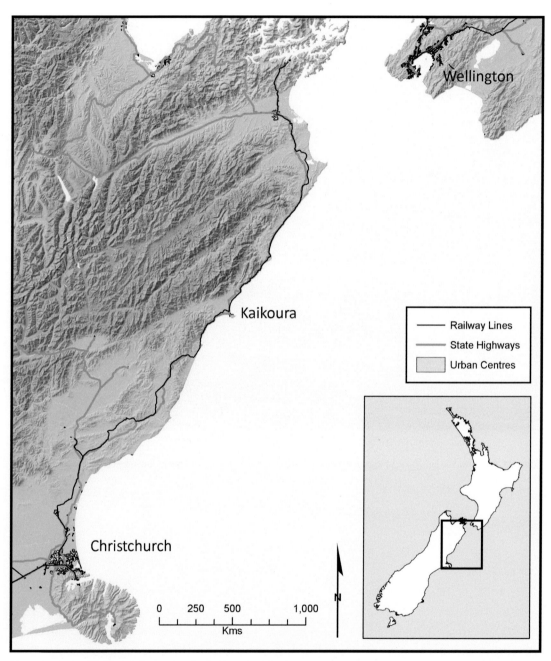

Figure 22.1 Location Map: Kaikoura, South Island, New Zealand.

Weinrich, M. & Corbelli, C. (2009). Does whale watching in Southern New England impact humpback whale (*Megaptera novaeangliae*) calf production or calf survival? *Biological Conservation* 142, 2931–2940.

Wells, R.S. & Scott, M.D. (1997). Seasonal incidence of boat strikes on bottlenose dolphins near Sarasota, Florida. *Marine Mammal Science* 13, 475–481.

Whitehead, H. (1996). Babysitting, dive synchrony, and indications of alloparental care in sperm whales. *Behavioural Ecology and Sociobiology* 38, 237–244.

Whitehead, H. (2004). The group strikes back: Follow protocols for behavioral research on cetaceans. *Marine Mammal Science* 20, 664–670.

Whitehead, H. (2007). Learning, climate and the evolution of cultural capacity. *Journal of Theoretical Biology* 245, 341–350.

Whitehead, H. & Mann, J. (2000). Female reproductive strategies of cetaceans: Life histories and calf care. In J. Mann, R.C. Connor, P.L. Tyack & H. Whitehead (Eds), *Cetacean Societies*. Chicago, IL: University of Chicago Press, pp. 219–246.

Wiley, D.N., Moller, J.C., Pace III, R.M. & Carlson, C. (2008). Effectiveness of voluntary conservation agreements: Case study of endangered whales and commercial whale watching. *Conservation Biology* 22, 450–457.

Wiley, D.N., Thompson, M, Pace III, R.M. & Levenson, J. (2011). Modeling speed restrictions to mitigate lethal collisions between ships and whales in the Stellwagen Bank National Marine Sanctuary, USA. *Biological Conservation* 144, 2377–2381.

Williams, R. & Ashe, E. (2007). Killer whale evasive tactics vary with boat number. *Journal of Zoology* 272, 390–397.

Williams, R., Bain, D.E., Ford, J.K.B. & Trites, A.W. (2002a). Behavioral responses of male killer whales to a 'leapfrogging' vessel. *Journal of Cetacean Research and Management* 4, 305–310.

Williams, R.M., Trites, A.W. & Bain, D.E. (2002b). Behavioural responses of killer whales (*Orcinus orca*) to whale-watching boats: Opportunistic observations and experimental approaches. *Journal of Zoology, London* 256, 255–270.

Williams, R., Lusseau, D. & Hammond, P.S. (2006). Estimating relative energetic costs of human disturbance to killer whales (*Orcinus orca*). *Biological Conservation* 133, 301–311.

Wilson, B., Thompson, P.M. & Hammond, P.S. (1997). Habitat use by bottlenose dolphins: Seasonal distribution and stratified movement patterns in the Moray Firth, Scotland. *Journal of Applied Ecology* 34, 1365–1374.

Wingfield, J.C. & Sapolsky, R.M. (2003). Reproduction and resistance to stress: When and how. *Journal of Neuroendocrinology* 15, 711–724.

Winn, H.E., Thompson, T.J., Cummings, W.C., *et al.* (1981). Song of the humpback whale – Population comparisons. *Behavioural Ecology and Sociobiology* 8, 41–46.

Würsig, B. & Würsig, M. (1980). Behavior and ecology of the dusky dolphin, *Lagenorhynchus obscurus*, in the South Atlantic. *Fishery Bulletin* 77, 871–890.

Yin, S.E. (1999). Movement patterns, behaviors, and whistle sounds of dolphin groups off Kaikoura, New Zealand. MSc Thesis, Texas A&M University, U.S.A. 107 pages.

Yurk, H., Barrett-Lennard, L., Ford, J.K.B. & Matkin, C.O. (2002). Cultural transmission within maternal lineages: vocal clans in resident killer whales in southern Alaska. *Animal Behaviour* 63, 1103–1119.

15

Energetic linkages between short-term and long-term effects of whale-watching disturbance on cetaceans

An example drawn from northeast Pacific resident killer whales

David E. Bain, Rob Williams and Andrew W. Trites

Introduction

Many studies have demonstrated short-term behavioural responses by whales and dolphins in the presence of vessels, but the population-level implications of such changes are poorly understood (Lusseau, 2003, 2004; Bejder *et al.*, 2006a; Lusseau & Bejder, 2007). One means for developing such an understanding is to use a modelling framework such as the Population Consequences of Acoustic Disturbance (PCAD) model. PCAD identifies four levels at which data can be collected, and allows for estimates of modelling parameters at one level to be based on measured data at another level (National Research Council, 2005).

The first level contains short-term behavioural responses, such as those that have been the typical focus of studies on effects of whale-watching. Effects vary within and between species, and include changes in respiration patterns, surface active behaviours, swimming velocity, vocal behaviour, activity state, inter-individual spacing, wake riding, approach and avoidance, and displacement from habitat. Collisions may result in injury or death (Wells & Scott, 1997; Laist *et al.*, 2001). More detailed reviews of vessel effects can be found in Lien (2001) and Ritter (2003).

Fewer studies have addressed the second functional level effects of vessels on whales – quantifying how behavioural responses can affect critical life functions, such as feeding and breeding (Williams *et al.*, 2006; Lusseau *et al.*, 2009; Aguilar Soto *et al.*, 2006). A number of studies have shown vessels to displace whales from habitat, but it is not clear whether it negatively affects feeding behaviour (e.g. Morton & Symonds, 2002; Olesiuk *et al.*, 2002; Richardson *et al.*, 1995a). Even fewer studies have considered the third and fourth functional level effects – namely, the outcome of these disruptions of normal behaviour in the form of altered birth and death rates, and the translation of such changes in the vital rates of individuals to population-level effects (Bejder *et al.*, 2006a,b).

In this chapter, we use data from fish-eating or 'resident' form of killer whales (*Orcinus orca*; Ford *et al.*, 2000) to illustrate how the PCAD framework can be employed to describe the relationship between short-term behavioural responses and population-level consequences. Numerous mechanisms may contribute to population-level consequences. These include stress (Wasser *et al.*, 2010; Ayres *et al.*, 2012), toxin exposure (Ross *et al.*, 2000; Lachmuth *et al.*, 2011), vessel collision (Carretta *et al.*, 2010; Ford *et al.*, 2000), energy expenditure (Williams & Noren, 2009), energy acquisition (Williams *et al.*, 2006; Lusseau *et al.*, 2009), separation of mothers from calves, and disruption of mating (National Research Council, 2003). For the purpose of illustrating PCAD, we chose to limit our quantitative consideration to the energetic

Whale-watching: Sustainable Tourism and Ecological Management, eds J. Higham, L. Bejder and R. Williams.
Published by Cambridge University Press. © Cambridge University Press 2014.

consequences of short-term disturbances on population numbers. Specifically, we were interested in cases in which human-caused disturbance could alter the balance between energy intake and expenditure. Whale-watching disturbance has been shown to increase energy expenditure for a number of cetacean species (from increased travelling time or speed, breaching, etc.) or decrease energy intake (e.g. by reducing time spent foraging; Williams *et al.*, 2006; Lusseau *et al.*, 2009).

We chose to treat energetic consequences as mathematically equivalent to changing intraspecific competition. That is, we combined levels 3 and 4 of the PCAD model into a single level after recognizing that intraspecific competition may result in the individual suffering a change in vital rates being different than the individual exposed directly to the disturbance.

We focused on resident killer whales for several reasons. First, they are extensively watched in the wild (see Figure 15.1 for a map of locations where the bulk of this work has been done). Prior to 1980, recreational whale-watching with northern residents was incidental to recreational boating or an opportunistic portion of commercial cruises. Then, Stubbs Island Charters began offering cruises in British Columbia where the primary focus was whale-watching. Initially, this was a small portion of their business. However, as interest grew, and road access to the north end of Vancouver Island improved, it became a more regular part of their business, eventually becoming a primary focus and growing to include a second vessel. By the mid-1980s, other companies in the region were also regularly offering whale-watching cruises. Subsequently, the geographic scope increased, with additional ports at Kelsey Bay and Campbell River serving passengers seeking to spend part of a day with northern resident killer whales (Jim Borrowman, pers. comm.). Scientific interest in the effects of whale-watching began in the early 1980s with work by Kruse (Kruse, 1991; Williams *et al.*, 2002a,b, 2009a,b; Lusseau *et al.*, 2009).

Commercial whale-watching began a few years earlier in the Southern Community of killer whales.

The commercial whale-watch fleet grew slowly until the late 1980s, grew rapidly through the mid-1990s, and then levelled off (Koski, 2011). Ports regularly allowing passengers to see southern resident killer whales now extend from Seattle, WA to Vancouver, BC, through the San Juan Islands to Vancouver Island, and to the Olympic Peninsula.

Research and commercial whale-watching developed cooperatively, with researchers making it easier for commercial whale-watchers to find whales, and commercial whale-watchers providing logistical support and data to researchers (e.g. Hauser *et al.*, 2007). This cooperative relationship facilitated incorporation of scientific results into the whale-watch industry's own guidelines for self-regulation. Points in these guidelines included not approaching within 100 m (200 m for transients), a no-go zone in the Robson Bight (Michael Bigg) Ecological Reserve, the elimination of leap-frogging behaviour, and reduced operating speeds within 400 and 800 m of killer whales (Bain, 2001). Some of these guidelines were later adopted by one or more governments; in some cases government regulations remain less strict, while in other cases governments imposed stricter regulations (e.g. prohibition of parking in the path of killer whales, and setting a minimum approach distance of 200 m for residents).

In addition to studying the effects of vessels on killer whales in the wild, killer whales held in captivity have been available for detailed studies of their energetics (Kriete, 1995; Kastelein *et al.*, 2000, 2003; Kasting *et al.*, 1989), hearing (Bain & Dahlheim, 1994; Hall & Johnson, 1972; Szymanski *et al.*, 1999), and click production (Schevill & Watkins, 1966; Bain, 1986). Killer whales have also been the subject of a long-term study on their population dynamics (Olesiuk *et al.*, 1990, 2005). In addition, the primary prey of resident killer whales are commercially important, and hence have been extensively studied as well (Hilborn *et al.*, 2012; Ford *et al.*, 2005; Heise *et al.*, 2003; Hanson *et al.*, 2010; Baird & Hanson, 2004). Thus, there are extensive data to parameterize the first, second and fourth levels of the PCAD model, as well as data on mechanisms for how changes at one level may impact another.

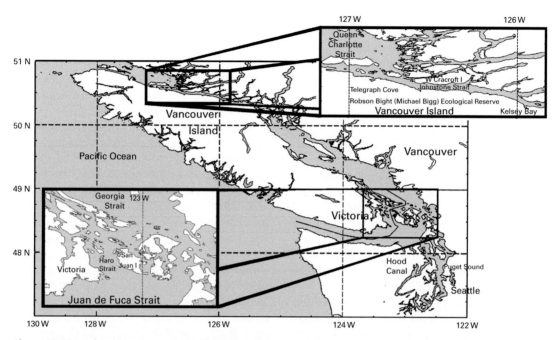

Figure 15.1 Map showing core areas for resident killer whale research. Northern resident killer whale research was centred in Johnstone Strait. Shore-based studies of the effects of whale-watching were conducted from West Cracroft Island, near the Robson Bight Michael Bigg Ecological Reserve. Southern resident killer whale research was centred in Haro Strait. Shore-based studies of the effects of whale-watching were conducted from San Juan Island.

The PCAD approach allows population growth to be projected for an undisturbed population, and to contrast that to the growth of populations experiencing varying degrees of disturbance. Any differences would represent the population-level consequences of disturbance. This modelling framework also allows interaction effects to be examined between disturbance and other factors that affect population growth, such as prey availability (Ward *et al.*, 2009; Ford *et al.*, 2010).

The resident killer whale case study

Under good conditions, resident killer whale populations have increased at a rate of 2.6% per year (Olesiuk *et al.*, 2005). However, the southern resident population of killer whales has failed to achieve this rate of growth since the early 1990s, and in fact declined from a high of 98 individuals in early

1995 to a low of 79 in 2001 (Fisheries and Oceans Canada, 2008; Krahn *et al.*, 2004). The Center for Whale Research reported the 1 July 2012 population was 82 (http://whaleresearch.com/research.html). The causes of the decline have not been confirmed, but may be related to a decline in food availability, and exposure to toxic chemicals, whale-watching and vessel traffic in general (Baird, 2001; Krahn *et al.*, 2004).

Williams *et al.* (2002a,b) and Williams and Ashe (2007) have demonstrated experimentally that whale-watching boats affected the swimming behaviour of 'northern resident' killer whales. Males travelled approximately 13% further when approached by a boat following whale-watching guidelines than when travelling unaccompanied, and females changed direction more from one pair of surfacings to another when accompanied by the experimental boat than when on their own

(Williams *et al.*, 2002a). When followed by a boat that 'leapfrogged', the male whales travelled approximately 17% further than in the absence of boats (Williams *et al.*, 2002b). As a result, it took more energy for whales to travel from one place to another when accompanied by boats than under control conditions, although the metabolic cost of swimming in killer whales is relatively low across a wide range of swimming speeds (Williams & Noren, 2009). Williams *et al.* (2006) estimated that the energetic cost of meeting an activity budget in the presence of boats was only 3–4% higher than that under control conditions. In the presence of boats, however, whales reduced their time spent feeding, which could have resulted in a substantial (18%) decrease in energy intake (Williams *et al.*, 2006). In similar work on 'southern resident' killer whales, Lusseau *et al.* (2009) showed a 25% reduction in the proportion of time spent feeding in the presence of boats.

Bain and Dahlheim (1994) demonstrated that killer whale hearing ability was impaired by masking noise such as vessel noise. A consequence of this is that active space (the range over which biologically important signals are functional) is reduced in the presence of noise (Miller, 2000; Erbe, 2002; Jensen *et al.*, 2009; Clark *et al.*, 2009). In general, the ability of a killer whale to generate echolocation clicks limits the source level, and should be independent of the level of ambient noise. Target strength is a property of the prey, and should also be independent of noise. Thus, a whale in an increasingly noisy environment must be closer to its prey to detect it. Holt (2008) estimated that a boat at cruising speed 400 m from a whale would reduce its echolocation range by 90%. Reducing active space, in turn, should reduce foraging efficiency (Clark *et al.*, 2009). Temporary and permanent threshold shifts are additional mechanisms that may reduce active space (Au *et al.*, 1999; Erbe, 2002).

Increasing the energetic requirements of individuals within the population and reducing effective prey availability are equivalent to reducing the carrying capacity in food-limited populations. Olesiuk *et al.* (2005) produced an equation relating population size and carrying capacity to population growth rate. This model was based on the concept that as populations increase, there is stronger competition for resources such as food, which in turn slows population growth (Gilpin *et al.*, 1976).

Whale-oriented vessel traffic has been monitored for many years (Koski, 2011; Osborne, 1999), as has the population of southern resident killer whales since the mid-1970s (e.g. Bigg *et al.*, 1990; Olesiuk *et al.*, 1990). In addition, the portion of the range of southern resident killer whales that is shared extensively with whale-watching boats has been documented (Osborne, 1999). These data form a basis for testing the strength of relationships between the magnitude of whale-watching and population trends and distribution.

Vessels have been seen accompanying whales passing through a study site off the west side of San Juan Island approximately 90% of the time during daylight hours (Lusseau *et al.*, 2009; Williams *et al.*, 2009a). Combining this estimate with the observation that whales have been regularly seen from April to October suggests that vessels accompanied whales approximately 25% of the time during the year (i.e. 50% of the time during the 6-month whale-watching season).

The following considers the possibility that whale-watching has negatively affected the population dynamics of southern resident killer whales. Specifically, we examined whether whale-watching could affect population dynamics by requiring greater energy expenditure and/or reducing foraging efficiency. To accomplish this, we estimated the increase in energy expenditure due to whale-watching, and estimated reductions in foraging efficiency due to vessel noise. We also developed a model to relate changes in energy balance to changes in population dynamics, and tested whether a significant correlation exists between exposure to whale-watching and population trends. We then assessed whether the model was consistent with (1) the increase in population size in the presence of vessel traffic prior to 1995, (2) the apparent absence of change in range, and (3) pod-specific population trends and patterns of exposure to whale-watching and other vessel traffic.

Methods

Changes in energy expenditure due to whale-watching

We assumed that killer whales that swam 13% or 17% further along a circuitous path would have less time to rest than killer whales that were not being followed by a vessel and swam along a relatively straight-line path. This increase in swimming distance would result in an approximate 13% or 17% increase in energy consumption (Waite, 1988; Kriete, 1995). We multiplied the increased energy expenditure by the proportion of time that whales were exposed to whale-watching to estimate the total increase in energy expenditure (approximately 50% of the time during 50% of the year, or 25% overall over the course of a year).

Changes in energy acquisition due to noise from whale-watching

We used the sonar equation to convert dB of masking or reduced sensitivity due to temporary or permanent threshold shifts to change in detection range. For echolocation at maximum range,

$$DT = SL - 2TL + TS - NR,$$

where DT is the detection threshold, SL is the source level, TL is one-way transmission loss, TS is target strength, and NR is received noise.

At short range in deep water, one-way transmission loss can be approximated by the formula $TL = 20 \log (R)$, where R is the transmission distance (Au, 1993). A correction to this equation for directivity is needed, but insufficient data are available to make this correction quantitatively (Bain & Dahlheim, 1994). Fortunately, the correction for directivity is likely to be the same for any given noise source, independent of absolute noise level. Thus, the equation for passive detection becomes

$$DT = SL - TL - NR.$$

As can be seen from the sonar equation, an increase in noise will result in a decrease in tolerable transmission loss, and hence detection range

will decline. In the absence of wind and current, natural ambient noise can be as low as 20 dB re 1 $\mu Pa^2/Hz$ at 20 kHz (Richardson et $al.$, 1995b). Ambient noise from wind, currents and non-whale-oriented traffic in Haro Strait was typically 50 dB re 1 $\mu Pa^2/Hz$ at 20 kHz (Bain, 2002). This corresponds to an outboard engine operating at high speed at a distance of several kilometres. Thus, noise from whale-watching vessels above this level will increase masking and reduce echolocation range. Received level decreases approximately 6 dB with each doubling of distance. That is, the target will need to be twice as close to offset the masking noise of 12 dB (i.e. 6 dB less loss on the way to the target, and 6 dB less loss on the way back to the whale). For passive listening, two halvings of distance will be required (i.e. the source will need to be four times closer). As a result, for killer whales using passive listening to locate prey (e.g. 'transient' – mammal-hunting killer whales: Barrett-Lennard et $al.$, 1996), the change in distance producing the transmission loss required to offset noise would be approximately twice as large as for whales using echolocation (resident killer whales). Furthermore, whales that use social facilitation in foraging (e.g. resident killer whales (Ford & Ellis, 2005), humpback whales (D'Vincent et $al.$, 1985)) will face the restrictions in range due to masking noise experienced in passive listening.

We assumed that normal foraging efficiency equated to that of a whale with normal hearing in quiet conditions, and expressed foraging efficiency impaired by noise as a percentage of normal foraging efficiency. We also assumed that foraging killer whales ensonified a 'tube' surrounding their travel path, with the radius of the tube reflecting the detection range of prey (active space).

Whether a-priori knowledge of prey distribution affected the ratio of prey detected in noisy conditions to quiet conditions was treated in four ways (see Figure 15.2).

First, the $fixed$-$location$ $model$ assumed that prey were in a fixed location known to the whales (e.g. a particular territory within a reef, as would be the case for many bottomfish species). Whales

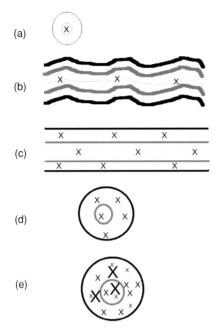

Figure 15.2 Search patterns by killer whales for prey (X) in the water column. (a) known prey location; (b) unknown prey location along a known line; (c) unknown prey location within a plane with search within the plane; (d) unknown location within a plane with search perpendicular to the plane; (e) no information on prey location. X's within the grey boundaries represent prey that would be detected even when hearing is impaired. X's between the grey and black boundaries represent prey that would be missed due to hearing impairment. In (e), large and small X's also represent prey that would be missed.

successfully located prey when they arrived within detection range of this location.

The second model, the *linear search model*, assumed that prey occurred along a line (e.g. a depth contour along the bottom), but at an unknown position. Whales successfully located prey when they arrived within the detection range of the prey.

Our third model, the *within-plane search model*, assumed that prey items lay in a plane (e.g. at a fixed depth or along the bottom), but at an unknown position. Whales were assumed to travel in the plane of the prey. In this model, prey within detection

range of passing whales were detected, but prey sufficiently distant from the paths of whales were undetected. A variation of this is the *across-plane search model*, which assumes that prey are at an unknown position in the plane (e.g. along the mouth of a channel or a current shear). Whales are assumed to travel perpendicular to the prey plane, and successfully locate prey only when they penetrate the plane within detection range of the prey (see Jefferson, 1987).

Our fourth and final model, the *volumetric search model*, assumed that prey could be anywhere in the water column. Whales only located prey if they passed within the detection range of this location. A species such as Chinook salmon may be distributed in a way that requires volumetric searches (Candy & Quinn, 1999).

Population dynamics

Population growth of southern resident killer whales was modelled with the generalized logistic equation (Gilpin *et al.*, 1976) using parameters calculated by Olesiuk *et al.* (2005):

$$\frac{\Delta N}{\Delta t} = 0.026N\left(1 - \left(\frac{N}{100}\right)^{40}\right),$$

where N is the population size, ΔN is the change in population per unit time (Δt), the intrinsic rate of increase of 2.6% per year, and the shape parameter z was 40. Population carrying capacity was estimated to be 100 individuals based on the size of the southern resident population when it peaked around 1960 and again in the mid-1990s (Bain & Balcomb, 1999).

Relationship between fleet size and whale population changes

We performed four statistical analyses to assess whether there was a significant relationship between fleet size and changes in southern resident killer whale population size. The first two analyses used data from 1977 to 2001 (Koski, 2011), while the latter two used only data from five years before the 1996 peak in 1 July population size to five years following the peak (1991–2001).

For each time-frame (i.e. 1977–2001 or 1991–2001), we evaluated two time lags. The first pair of analyses assumed that overall exposure to whale-watching was related to fleet size in that year. That is, we tested whether fleet size predicted population change over the year following exposure. The second pair of analyses assumed that overall exposure to whale-watching was related to consumer demand for whale-watching services. We assumed that operators based their fleet sizing decisions on the number of trips made in the previous year. Thus we tested whether fleet size predicted population change over the year preceding exposure to that particular fleet, as both may have been based on whale-watching activity in the previous season. In summary, we tested for correlation on the assumptions that

1. fleet size determined the amount of exposure to vessels that whales experienced in a given year, and
2. the amount of exposure to vessels that whales experienced in a given year was determined by the same factors as fleet size the following year.

We calculated correlation coefficients for each of the four conditions, as well as the probability that a correlation of that magnitude or higher could occur by chance.

To visualize the data, we plotted the actual values of the variables, as well as moving averages of both variables.

Olesiuk *et al.* (1990) reported 502 encounters with southern residents during the course of their research: 311 sightings of J Pod, 240 of K Pod, and 198 of L Pod (more than one pod was present on many occasions). We used these values as a rough estimate of the relative exposure of the three pods to whale-watching traffic.

Sensitivity

To determine the population-scale effect of whale-watching, we compared the projected population growth in the presence of whale-watching to what

growth would have been in its absence. As some parameters were estimated somewhat arbitrarily, and others were best estimates with broad confidence intervals, we calculated relative population growth rates for a variety of values to determine the sensitivity of the analysis to the values employed.

Results

Changes in energy expenditure due to whale-watching

The extra energy expenditure due to a whale being followed by a single vessel operating in accordance with whale-watch guidelines would total 3.25% over the course of the whale-watching season. A whale being followed by a leapfrogging vessel would expend about 4.25% more energy over the course of a season. This is consistent with the findings of Williams *et al.* (2006) that the energetic cost of meeting an activity budget in the presence of boats was 3–4% higher than during no-boat conditions.

Changes in energy acquisition due to noise from whale-watching

The relative ranges at which whales can detect prey are shown in Figure 15.3 for a variety of detection thresholds related to noise or threshold shifts. Figure 15.3 also shows corresponding reductions in area and volume remaining in the active space.

Population dynamics and sensitivity

Annual changes in the number of killer whales that would be expected at different population sizes according to the generalized logistic model are shown in Figure 15.4 for a range of shape parameters ($z = 1$–40). In general, all of the models predicted a constant per-capita growth rate at low population sizes, and a rapid slowing of growth as the population exceeded 90% of carrying capacity. The models also predicted a loss of 1–4 individuals per year if the population exceeded carrying capacity

Relative Prey Availability Due to Noise-Induced Threshold Changes

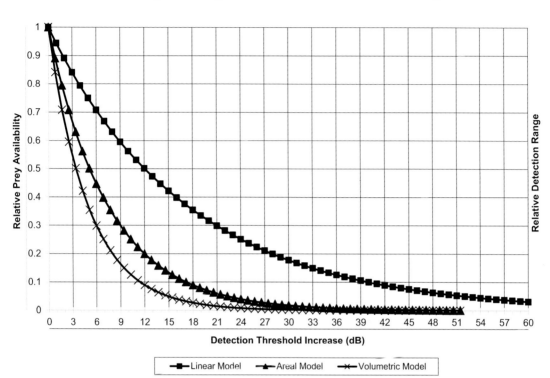

Figure 15.3 Detection range and detection efficiency for killer whales when hearing is impaired. The upper curve shows the relationship between relative detection range and magnitude of hearing impairment. This is also the curve for relative prey detection efficiency for the search within the plane model. The middle curve shows relative prey detection efficiency for the perpendicular-to-prey-plane search model. The lower curve shows the relative prey detection efficiency for the volumetric search model. It is important to note that small elevations in detection thresholds (e.g. 3 dB) can have large effects on the proportion of prey that remain detectable. Points of interest include 6 dB (proposed PTS), 12 dB (proposed TTS), and 30 dB (typical ambient above sea state 0).

by up to 10%. Thus, the most dynamic changes in effects on killer whale numbers appear to occur as they approach or exceed carrying capacity.

The projected energetic effect of whale-watching on killer whale numbers is shown in Figures 15.5 and 15.6. Assuming that population growth follows current patterns (i.e. $z = 40$) and whale-watching increases the energy needs of killer whales by 3% (to compensate for increased swimming and decreased foraging averaged over the course of a year), the model predicted a net loss of one individual per year as the population approached

K (Figure 15.5). In other words, increasing the energy requirements of a population of killer whales due to whale-watching would mean that an environment that once supported 100 killer whales could now support only 97 whales, and equilibrium would be restored over about a three-year period. An increase of 5% would lower carrying capacity to 95 whales, and result in an initial compensation of about two individuals per year. Population-level impacts appeared to be negligible when the population was below 90% of K. Varying z (the shape parameter) or the assumed energetic cost for

Annual Population Growth

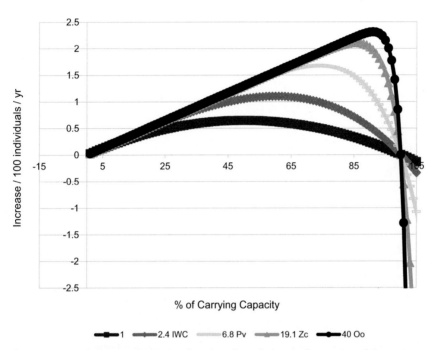

Figure 15.4 Population growth rates as a function of population size for a variety of shape parameters.

whales that were disturbed altered the population response by varying amounts (Figures 15.5 and 15.6). However, all realistic scenarios suggest that responses are negligible until the population is near its carrying capacity.

Relationship between fleet size and whale population changes

Numbers of commercial whale-watching vessels and annual changes in numbers of southern resident killer whales from 1977 to 2001 are shown in Figure 15.7. Regressing vessels on whales (with various time lags) yielded significant correlations. However, correlations over the long term (1977–2001) were weak ($r^2 = 0.18$, $p < 0.05$ for fleet size leading whale change; and $r^2 = 0.24$, $p < 0.01$ for fleet size following whale population change). In contrast, correlations over the more recent 1991–2001

period were stronger ($r^2 = 0.52$, $p < 0.01$, for fleet size leading whale population change; and $r^2 = 0.70$, $p < 0.001$, for fleet size following whale population change). This is consistent with the model result that population-level responses are only to be expected when the population is near carrying capacity.

Discussion of the killer whale case study

Changes in energy expenditure due to whale-watching

The increases in energy expenditure and reduction in prey capture due to whale-watching are probably small – of the order of 10–20%. The cumulative effect will be smaller still (our best estimate is that it is on the order of 3–4%), as not all whales are watched all

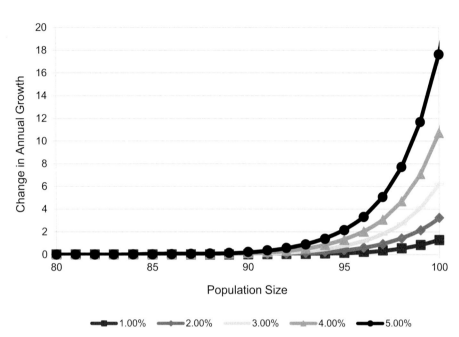

Figure 15.5 Population-scale effects for a variety of total energetic impacts (1–5%) when the shape parameter $z = 40$. Note that the curve at 3% is the best estimate for a whale-watching fleet that follows guidelines, while the curve at 4% is the best estimate for a fleet that continually speeds up to 'leapfrog' whales' predicted paths.

of the time. However, commercial whale-watching is now becoming a year-round industry, and public sighting networks have also allowed recreational whale-watchers to take advantage of viewing opportunities year round.

The commercial fleet has reached a size where all members of the southern resident community can be approached closely at the same time. In the absence of management intervention (e.g. regulation and/or boater education programmes), there is the potential for the frequency of interactions between boats and whales to increase. Although habituation might be expected to reduce or eliminate responses of whales to boats, Williams and Ashe (2007) were able to demonstrate behavioural changes of the magnitude assumed here in northern residents that had over 30 seasons of experience with researchers, commercial photographers,

and other whale-oriented vessels, including > 20 seasons of commercial whale-watching. Similarly, Williams *et al.* (2009a) and Lusseau *et al.* (2009) found changes of this magnitude in southern residents after over 30 seasons of exposure to research vessels and 25 seasons of exposure to commercial whale-watching.

Changes in energy acquisition due to noise from whale-watching

Small increases in detection threshold resulted in a dramatic reduction in the ability of killer whales to detect prey. Au (1993) suggested that echolocation clicks could be detected reliably at about 10 dB above ambient noise, which for a killer whale corresponds to the signal-to-noise ratio at auditory threshold at Sea State 0 (Szymanski *et al.*,

---

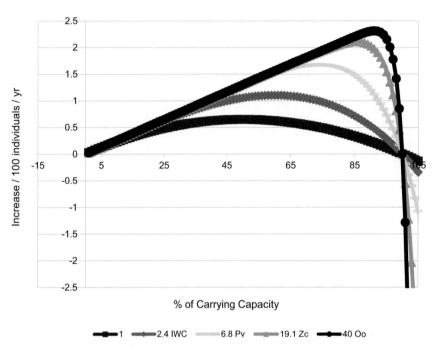

Figure 15.6 Population growth when total energetic impact on carrying capacity is 3% and the shape parameter z ranges from 1 to 40.

1999). Additional noise, whether from natural or man-made sources, would reduce the range of echolocation detection.

A 20 dB increase in noise corresponds to approximately a 3-fold decrease in detection range for echolocation (Figure 15.3). A 60 dB increase in noise corresponds to approximately a 30-fold decrease in detection range. Received noise levels from benign whale-watching (a single outboard-powered vessel that is 100 m to the side, travelling at approximately the same speed as the whale) is on the order of 105–110 dB re 1 µPa, with power spectral densities of approximately 70–80 dB re 1 µPa²/Hz at 20 kHz (Richardson *et al.*, 1995b). Thus, benign whale-watching likely results in masking on the order of 20–30 dB in excess of normal levels of ambient noise.

Temporary threshold shifts due to whale-watching are unlikely to be large enough to exceed the effects of masking. Similarly, permanent threshold shifts are likely to be small relative to temporary threshold shifts and masking. However, temporary threshold shifts may be important in quiet water, where a 12 dB threshold shift would result in a 2-fold decrease in detection range. Similarly, a permanent threshold shift of 6 dB would reduce detection range in quiet water to 70% of the optimum.

Masking will only reduce detection range while whales are in the presence of boats, and is likely to occur for approximately 25% of the year (or 50% of the time during the core whale-watching season). Where temporary threshold shifts occur, the effect might persist for up to 24 hours after

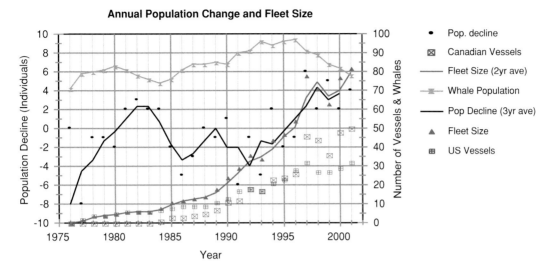

Figure 15.7 Relationship between fleet size and killer whale population changes. 1 July counts of southern resident population size are shown (Krahn *et al.*, 2004). The total number of commercial vessels actively engaged in whale-watching, along with the number based in American and Canadian ports, are shown (after Koski, 2011). Annual changes in whale population size are plotted in the year of the latter count. A three-year moving average of annual whale population change is plotted along with a two-year moving average of total fleet size. Fleet size is used as an index of exposure to whale-watching, although other factors not represented in this graph that affect overall exposure may include: efficiency of whale-watch operators in locating whales; hours per day spent with whales; number and type of engines employed, operating speed, distance, orientation and relative position. Note the tight fit of smoothed whale population change with smoothed fleet size beginning in the early 1990s. Also note that the number of vessels in the commercial whale-watching whale fleet exceeded the number of whales in the population in 2001 (although typically, not all vessels operated simultaneously).

exposure to noise if the duration of the effect in killer whales is the same as that in humans (Erbe, 2002). Because whales may only have 9–12 hours between bouts of whale-watching, this effect might be nearly continuous for half the year. Alternatively, the effect may disappear almost completely within an hour, as is the case with bottlenose dolphins (*Tursiops truncatus*; Nachtigall *et al.*, 2003). Temporary threshold shifts of such a short duration would only slightly increase the effect of noise relative to masking alone. In contrast, permanent threshold shifts would reduce detection range and would be a problem year-round.

The implications for foraging efficiency of reducing active space depend on the foraging tactics that killer whales use to locate prey (Figure 15.1).

Assuming that prey are in a fixed location known to the whales (fixed-location model), killer whales will find prey with the same efficiency regardless of whether their echolocation ability is impaired. Similarly, whales should also find prey with the same efficiency regardless of whether their echolocation ability is impaired if they are on the same path (linear search model) and the whale travels faster than the fish and retains minimal navigation and sensory capability. However, detection efficiency will be impaired if the probability of detecting prey is proportionately related to detection range or the square of the detection range (within- and across-plane search models, respectively). Detection efficiency will be further impaired by noise if prey are randomly located in the water column (volumetric search model), in which case detection efficiency is proportional to the cube of the reduced detection range.

Holt (2008) cited noise levels recorded from a fixed location along San Juan Island. Average mid-day levels were about 12 dB above midnight levels at 15 kHz, suggesting echolocation range would likely be reduced by a factor of at least 3–5 off San Juan Island (average noise received by whales was likely higher than the average received at the fixed site). In the absence of knowledge about prey distribution, this would result in a reduction of available prey by over 95%. Under the planar model (across-plane approach), available prey could be reduced by ~90–95%. Even with approximate knowledge of prey location (planar model, within-plane approach), most prey that would be detected in quiet water could be missed. With outboard-powered whale-watching vessels, even under present guidelines, there would be losses of similar magnitude relative to the already reduced level from other sources of ambient noise (i.e. total reductions in excess of 99%).

It should be pointed out that directional hearing capabilities might reduce the magnitude of the effect of noise on prey detection range relative to that calculated above. Vessels operating to the side or behind whales may have a masking effect that is 10–20 dB lower than the same noise source in front of whales (Bain & Dahlheim, 1994). There is also potential for temporal pattern processing to improve extraction of signal from noise (Szymanski et al., 1998). Even with these corrections, there is still the potential for decreasing foraging efficiency due to noise from increasing whale-watch activity to be more important than changes in foraging efficiency due to changing prey abundance. An 80% reduction in effective prey availability due to noise from whale-watching would be the right order of magnitude to account for the decline in southern residents during the 1990s. This corresponds to an increase in noise received by killer whales of only 9 dB if they have no a-priori knowledge of prey locations.

Population dynamics and sensitivity

The impact of whale-watching on the modelled population dynamics was very sensitive to the shape parameter. When the shape parameter used was 1 (i.e. Maximum Net Productivity Level (MNPL) at 50% of K), the population-scale effects of changes in effective carrying capacity were limited. That is, it would take the population a long time to reach equilibrium with the new carrying capacity. As a result, the change in population growth in any given year would be small.

There were two obvious changes in population-scale effects as the shape parameter increased. The rate of population growth remains near the intrinsic rate of increase even when the population was well above 50% of K. Furthermore, the per capita growth rate remains near the intrinsic rate of increase at higher population sizes as the shape parameter increases (i.e. the production curve became more asymmetrical, with MNPL occurring at higher percentages of K as the shape parameter increases – the reason for the name 'shape parameter'). With a shape parameter of 40 as estimated by Olesiuk et al. (2005), for example, changes in effective carrying capacity had essentially no effect on population growth rate as long as the population remained below 90% of both the old and new values of K. In contrast, the rate of population growth changed very rapidly when the population was near K. That is, small changes in K produce large changes in population growth rate when the population is near K. Populations that are over K rapidly fall below equilibrium.

The second obvious effect of the shape parameter was that the maximum rate of increase in absolute population size increased with increasing shape parameter. For example, the maximum rate of increase for a species with a shape parameter of 1, intrinsic rate of increase of 2.6%, and a K of 100 would be less than one individual per year. When the population was at 105% of carrying capacity, the rate of decrease would be close to 0. In contrast, employing the observed shape parameter for resident killer whales of 40 put the maximum growth rate at over two individuals per year, and the decline of a population at 105% of K was over 16 individuals per year.

The implication of these two observations is that as the shape parameter increases, the potential magnitude of effects of changes in carrying capacity becomes smaller for populations well below carrying capacity, and becomes larger when the population is near K.

Relationship between fleet size and whale population changes

The sensitivity analysis performed above suggests that population growth rate will be independent of K when the population is depleted, as was the case for southern residents at the end of collections for public display (Bigg & Wolman, 1975; Bain & Balcomb, 1999). Once the population recovered, it would become very sensitive to changes in K. As a result, if carrying capacity is affected by whale-watching as suggested above, changes in population growth rate should not have correlated with changes in levels of whale-watching traffic when the population was small. Instead, it should have been tightly correlated with them when the population was large.

Similarly, when the size of the whale-watch fleet was small, the maximum possible cumulative effect would have been small. As fleet size increased, the potential cumulative effect increased, and hence one would expect stronger correlations when the fleet was large than when it was small.

This is the pattern that was observed. The correlation for all years was weak, and whale population size did not track fleet size when both were small. However, the correlation was strong when both were large.

Despite the consistency in the observed trends, we are unable to state whether the relationship between fleet size and population growth rate was causal or whether both parameters were correlated with causal factors that we have not considered.

We have interpreted our data as implying that time spent whale-watching determines fleet size in subsequent years (i.e. consumer demand determines the number of trips and funds available to purchase and maintain vessels in future years), but this hypothesis needs testing.

Population growth in the presence of whale-watching

With relatively healthy fish stocks and depleted numbers of killer whales (due to past collections for public display), the southern resident population was probably well below carrying capacity in the 1970s. According to our model, whale-watching was unlikely to have had any population-scale effect during that period. In the 1980s to the early 1990s, the population may have tracked a fluctuating carrying capacity as fish abundance varied. While there was potential for impact, the small size of the fleet meant that total impact was probably small. As the amount of whale-watching increased through the 1990s, the magnitude of the change in energy balance due to whale-watching may have exceeded the magnitude of the change in energy balance due to changes in fish abundance. If so, this would account for the correlation between fleet size and changes in population size observed over that decade. Our model indicates that if the effect on the population is large, missed prey due to noise is probably a much more important mechanism than excess energy expenditure. The correlation between fleet size and whale population trends merits careful evaluation.

Smoothing fleet size and change in whale numbers by three-year intervals suggests a possible relationship between the two, beginning in the early 1990s (Figure 15.7). While this may be nothing more than coincidence, the smoothing eliminates noise associated with year-to-year variability in calving intervals (a three-year cycle) and possibly whale-watching business decisions.

The energetic consequences of whale-watching that we propose are only relevant when the population is food-limited (whether by reduction in food stocks (Washington Department of Fisheries *et al.*, 1993; Allendorf *et al.*, 1997) or the accessibility of these stocks is limited by disturbance). Other factors associated with whale-watching, such as stress or exposure to burned and unburned fuel, may also inflict population-scale effects in addition to the energetic consequences.

Displacement as a response to noise

Although Morton and Symonds (2002) found that noise excluded northern residents from peripheral habitat near a core area over a period of years, this does not imply that whales will always move in response to disturbance. Our models illustrate that relatively small impacts on effective fish availability can produce population-scale effects. However, population-scale effects of degraded habitat due to whale-watching in core areas (Hauser *et al.*, 2007) could be smaller than the effects of moving to habitat where prey density is lower and optimal foraging tactics may be unknown to the whales (e.g. Juan de Fuca Strait). This contrasts with the case described by Morton and Symonds (2002), in which nearby, familiar areas with similar prey densities (or perhaps higher prey densities, but with more intraspecific competition; Nichol & Shackleton, 1996) were available.

It is important to identify other sources of the decline that may act independently of or synergistically with whale-watching. For example, Ylitalo *et al.* (2001) found lipid concentrations in blubber samples collected from many individuals ranged by a factor of eight (from 7% to 59%). Some of this variation was likely due to methodological issues (e.g. the location on the body of the sample and handling and storage practices). However, some of the variation may have reflected real variation in fat reserves. Because a high proportion of lipids are in the blubber (Borrell *et al.*, 1995), total body concentrations of toxins may vary substantially with energy balance. Thus toxin concentrations could increase as a whale loses weight, even if total body burden remains constant.

Ross *et al.* (2000) found high PCB levels in southern residents, and Reddy *et al.* (2001) found PCB levels tended to be higher in mothers of captive *Tursiops* calves that died as neonates than in mothers of calves that survived the neonatal period. This indicates that reduction in lipid reserves due to impaired energy balance could negatively impact calf survival both through nutritional and toxicological mechanisms, and this may be related to the 50% reduction in calf recruitment rate in L Pod observed over the 1990s (Bain & Balcomb, 1999).

Future work on resident killer whales

Our study has several applications. First, focusing on a subset of the possible consequences of whale-watching provides a minimum estimate of the population-scale effects of whale-watching. Second, the quantitative model we constructed allows the impacts of a variety of scenarios to be evaluated (e.g. different levels of whale-watching – hours per day, days per year, noise exposure or impacts at different population levels, interaction effects of whale-watching and other factors such as prey availability). In addition, the consequences of mathematically equivalent factors can be considered (e.g. reduction in food availability due to fisheries- or climate-induced reductions in salmon stocks would be mathematically equivalent to a reduction in food availability due to noise).

Additional studies should be pursued to test our model predictions, such as that of Lusseau *et al.* (2009), who observed a reduction in foraging effort in the presence of whale-watching vessels. Additional work could address whether there is a detection probability threshold (prey items/unit volume within the masked detection range; Au *et al.*, 2004) for abandoning foraging on optimal foraging grounds that corresponds to this shift in behaviour. Our research could undertake a detailed analysis of whale-watch operators' logs to determine time spent with whales to rigorously assess whether the correlation between fleet size and whale population changes could reflect a causal relationship.

Another study should measure actual noise exposure to refine our estimates of acoustic impact on foraging efficiency (see Griffin & Bain, 2006, and Holt, 2008, as initial efforts toward this). Studies of foraging tactics used by killer whales would allow a better assessment of the importance of reduced prey detection ranges (see Baird & Hanson, 2004). Studies of the distances between whales cooperatively foraging (Ford & Ellis, 2005) successfully and ambient noise whales would also be of interest. Other studies of threshold shifts, directional hearing and masking using captive killer whales would also be valuable (in addition to Bain & Dahlheim, 1994).

Testing whether the correlation between prey availability and population growth (Ward *et al.*, 2009; Ford *et al.*, 2010) is stronger when corrected for whale-watching activity would be valuable. It may also be worthwhile to consider whether it is the timing or total amount of whale-watching that is important. For example, whale-watching could increase the duration of exposure to relatively high toxin concentrations if it slows the rate of weight gain as prey becomes seasonally more abundant. In addition, individuals expend far more energy when growing rapidly or lactating than at other stages of their life cycle (Kriete, 1995), so impairment of energy balance during these periods may have greater survival consequences than at other stages.

There is no doubt that more data are needed to determine whether the *actual* impact of whale-watching exceeds acceptable levels (0.2% of population size: Carretta *et al.*, 2010). The models developed here indicate there is potential for whale-watching to have important effects – mediated by energetic mechanisms – on population growth rates. The work of Ward *et al.* (2009) and Ford *et al.* (2010) suggest resident killer whale populations are near carrying capacity, so energetic effects are likely to be important to population growth rates.

Another area of work would be to develop a total ocean noise budget that includes noise from non-whale-oriented vessel traffic. It will be important to determine the conditions under which noise received from whale-watching vessels is exceeded

by noise from other vessels. This will have important implications for regulating whale-watching (there may be minimal value to requiring whale-watching vessels to keep noise exposure below the actual ambient). It will also have implications for considering whether regulating non-whale-oriented traffic (e.g. commercial vessels operating within shipping lanes in critical habitat, commercial fishing, and other recreational traffic) is likely to promote recovery. If so, management actions specific to other vessels could complement whale-watching-directed actions (Clark *et al.*, 2009).

Finally, additional research should address to what degree killer whales have mechanisms to partially overcome effects of noise. For example, Au *et al.* (2004) found that the high-frequency component of echolocation clicks likely returned the strongest echo from salmon, and Bain and Dahlheim (1994) found masking noise was less effective when it came from the side of or behind the whale than when it came from the same direction as a sound in front of the whale. However, because masking occurs up to at least two octaves above the maximum frequency in masking noise (Bain & Dahlheim, 1994), and vessels produce noise to over 20 kHz (Williams *et al.*, 2002b), masking will be an issue to over 80 kHz, a frequency at which hearing sensitivity is declining and that is above the most powerful part of the click. Further, high frequencies are directional, so even though the 40–60 kHz portion of clicks will be less masked than the 20 kHz portion, the high-intensity portion of the echolocation beam at these high frequencies are less likely to reach the fish than the 20 kHz portion. High frequencies also attenuate faster than lower frequencies. This suggests that killer whales may have the ability to overcome masking by increasing the high-frequency content of clicks, but is likely to be limited in its effectiveness.

Holt *et al.* (2008) found killer whales were able to increase the amplitude of calls in the presence of noise. However, the ability to increase intensity above average is presumably present even in the absence of noise when needed for long-range communication. That is, it remains to be determined whether the maximum ranges at which it is adaptive to communicate and echolocate are less than the maximum range at which successful communication and echolocation are possible in the presence of vessel noise.

Applications to other populations and species

Any attempt to model how cetacean populations respond to disturbance hinges on knowing how cetacean populations behave generally. Our PCAD model shows that the value of the population shape parameter z influences the likelihood that disturbance will lead to population-level consequences. Although the shape parameter is essential for computing population-level effects, we are not aware of it having been published for other cetacean species. The population shape parameter, z, has been reported for some harbour seal (*Phoca vitulina*) populations in the northeast Pacific: $z = 6.77$ for Strait of Georgia (Olesiuk, 2009); $z = 2.43$ and 1.75 for inland and coastal areas of Washington State, respectively (Jeffries *et al.*, 2003). Also, a value of $z = 19.01$ has been reported for pup production of California sea lions (*Zalophus californianus*; Carretta *et al.*, 2010).

There are a limited number of cetacean species (perhaps grey whales, bowhead whales, southern right whales and bottlenose dolphins) for which precise time series of abundance estimates are available that span moderate to high fractions of carrying capacity. Even in such cases, constructing a PCAD model may be more valuable in identifying data gaps than actually performing calculations, as interspecific variability in components of the PCAD model is unknown. However, the killer whale values we present could be used as defaults to make rough estimates while species-specific data are gathered. It should be noted that the default value of z used in baleen whale models in the International Whaling Commission is 2.39, but this does not appear to have been estimated empirically (May, 1980).

The sensitivity analysis shows that for species with a smaller shape parameter than that found for killer whales, the pattern of the onset of population-level changes would be different. Measurable changes would occur at lower population levels relative to carrying capacity in populations with small shape parameters, but the magnitude of changes in these populations near carrying capacity would be smaller than that in killer whales. Because the current carrying capacity must be used, historical estimates are of limited value in habitats that have changed significantly over time (e.g. changes to prey populations due to harvest or habitat degradation).

Only 0.5–2.5% of the daytime observations of 'southern resident' killer whales during the whale-watch season (from 2003 to 2005) occurred in the complete absence of vessels. The time without vessels within 1000 m ranged from 9.9% to 24.3%, 400 m ranged from 19.0% to 43.3%, and the time whales spent without any vessels violating the 100 m guideline then in effect was 71.2–76.7%. The nearly continuous presence of vessels was due to sport fishers present from first light, research vessels from early morning, and commercial whale-watchers from around 9 a.m. until sunset. At night, there is still significant freight traffic, although it is not whale-oriented. This is an issue for conducting control-exposure experiments, of course, but also raises concern about the broader tendency for commercial whale-watching activities to focus, for logistical reasons, on critical habitat areas (Williams *et al.*, 2009b).

Although researchers may be permitted to closely approach killer whales, they rarely do so (~10% of vessels observed within 100 m of whales were research vessels; Bain, 2007). Close approaches occur when researchers approach for photo-identification (Ford *et al.*, 2000), biopsy darting (Krahn *et al.*, 2007), breath sampling (Raverty *et al.*, 2007), or satellite (Andrews *et al.*, 2008) and other tagging (Baird *et al.*, 2005). Prey and faecal sampling may also require close approaches (Ayres *et al.*, 2012), although the use of dogs to locate scat, and the application of DNA techniques to identify prey in scat (Hanson *et al.*, 2010) have reduced the frequency of such approaches. Behavioural and acoustical studies are best conducted without close approaches to minimize the influence of the research vessel on measured behaviour.

As southern resident killer whales approached carrying capacity, population growth slowed due to an increase in non-calf mortality. Subsequently, calf recruitment to six months of age declined as well, resulting in a net decline during the last half of the 1990s (see data in Ford *et al.*, 2005). However, we would not be surprised if other mechanisms for population change are observed in other species.

Sensitivity analyses are useful to assist making management decisions. We have provided a mathematical framework for calculating likely outcomes of various management options. For example, managers could try to strike a balance between costs to the population of whale-watching, and the benefits resulting from changes in human behaviour once passengers return home. Managers could also calculate whether allowing whale-watching poses a risk to the survival of the population. However, it is important to emphasize that the models we outlined above have not been tested, so it is unknown whether they *actually* apply to killer whales, much less other species. However, we feel that our approach provides a quantitative framework to explore how changes in energy balance could lead to population-level consequences, and in the process could provide some critical insights into management issues and identifies additional data required to resolve important management questions.

Finally, we would like to emphasize that we have only addressed energetic mechanisms. As such, our findings provide a minimum estimate of impact. Other mechanisms, such as stress, exposure to toxins, and vessel collisions may be sufficiently important to further impair population growth. Stress may impair the immune system, making individuals more vulnerable to disease (Ayres *et al.*, 2012; Rosalind *et al.*, 2011). Exposure to toxins may also impair the immune system, increasing disease risk further (de Swart *et al.*, 1996). Toxins may also impair reproduction and calf survival (Reddy

et al., 2001). Toxins released from vessels include unburned fuel and exhaust (Lachmuth *et al.*, 2011). Although vessel collisions are rare, and many are survivable, mortalities associated with vessel collisions have been reported (Ford *et al.*, 2000). Synergistic effects of multiple mechanisms also need to be considered. For example, impaired energy balance leads to reduced lipid storage. Disease may lead to temporary cessation of food intake, resulting in further weight loss. Thus, body condition at the onset of disease and the time it takes to overcome pathogens together determine whether the disease is fatal. Similarly, when toxins impair reproduction, but are not fatal to the mother, she continues to compete with other whales in the population for food. Whether this influences population dynamics depends on the nutritional state of the rest of the population.

Concluding remarks

The core objective of efforts to understand and mitigate the effects of whale-watching has been to ensure the survival of populations. The first rigorous studies in the 1980s simply demonstrated that short-term behavioural responses existed. Managers, both in the private sector (whale-watch operators) and public sector responded with simple guidelines. Later studies showed some practices elicited stronger responses than others, and managers again responded with more effective guidelines to preclude activities that elicited strong evasive responses from killer whales. Recent work has shown that these guidelines are insufficient to prevent effects likely to be directly related to fitness (e.g. a reduction in foraging activity and increases in energy expenditure). As managers contemplate how to respond to such results, models such as ours can be constructed to address how short-term effects might relate to population dynamics, and whether data support a correlation between the quantity of exposure and shifts in population growth and decline. As better parameter estimates become available to inform these simple bioenergetics and population dynamics models, they should provide sufficient accuracy and precision to determine whether the effects of vessel traffic are sufficient to reduce the probability that the populations will survive in the long term (i.e. do they exceed potential biological removal (PBR); Wade & Angliss, 1997), or do the cumulative effects of whale-watching and other human activities put the population in jeopardy of imminent extinction?

The resident killer whales of the northeast Pacific represent an exceedingly data-rich case study of behaviour and population biology, but they also represent an interesting study in management. The degree to which management is or is not precautionary determines how much evidence of population-level effects is needed before management actions are imposed. And as some of the science that has been done on this population is applied to other cetaceans, it is important to spell out another lesson learned – namely, that it is essential to specify quantitative management objectives that identify how much of an effect managers are willing to tolerate (limits of acceptable change: Higham *et al.*, 2008; PBR: Wade & Angliss, 1997; and uncertainty: Taylor *et al.*, 2000). Otherwise, the science can and will continue without serving any practical purpose as long as the whales persist.

Acknowledgements

This chapter represents a synthesis of work conducted over the last 20 years, so we owe thanks to many individuals and organizations. Model development was supported primarily by the Orca Relief Citizens Alliance, with additional support from Six Flags Marine World Vallejo. Field studies that contributed to this work were supported by the National Science Foundation, National Marine Fisheries Service, Minerals Management Service, the US Geological Survey, BC Parks, Fisheries and Oceans Canada, and The Jane Marcher Foundation. We also thank whale-watch operators who made our experiments possible, particularly Jim Borrowman and Bill MacKay. Helpful comments on drafts of this manuscript were provided by Mark Anderson, Mike Bennett, Chris Clark, Brad Hanson, Rich Osborne and Bob Otis. We thank Ken Balcomb, Peter Olesiuk, Rich Osborne, Bob Otis and Jodi Smith for providing access to unpublished data. And finally, we would like to thank the many people who provided assistance in the field, not only for their hard work, but also for thoughtful discussions we have had with them over the years.

REFERENCES

Aguilar Soto, N., Johnson, N., Madsen, P.T., Tyack, P.L., Bocconcelli, A. & Borsani, J.F. (2006). Does intense ship noise disrupt foraging in deep-diving Cuvier's beaked whales (*Ziphius cavirostris*)? *Marine Mammal Science* 22(3), 690–699.

Allendorf, F.W., Bayles, D., Bottom, D.L., *et al.* (1997). Prioritizing Pacific salmon stocks for conservation. *Conservation Biology* 11, 140–152.

Andrews, R.D., Pitman, R.L. & Balance, L.T. (2008). Satellite tracking reveals distinct movement patterns for Type B and Type C killer whales in the southern Ross Sea, Antarctica. *Polar Biology* 31, 1461–1468.

Au, W.W.L. (1993). *The Sonar of Dolphins*. New York, NY: Springer.

Au, W.W.L., Nachtigall, P.E. & Pawloski, J.L. (1999). Temporary threshold shift in hearing induced by an octave band of continuous noise in the bottlenose dolphin. *Journal of the Acoustical Society of America* 106, 2251.

Au, W.W.L., Ford, J.K.B., Horne, J.K. & Newman-Allman, K.A. (2004). Echolocation signals of free-ranging killer whales (*Orcinus orca*) and modeling of foraging for chinook salmon (*Oncorhynchus tshawytscha*). *Journal of the Acoustical Society of America* 56, 1280–1290.

Ayres, K.L., Booth, R.K., Hempelmann, J.A., *et al.* (2012). Distinguishing the impacts of inadequate prey and vessel traffic on an endangered killer whale (*Orcinus orca*) population. *PLoS ONE* 7(6): e36842. doi:10.1371/journal.pone.0036842.

Bain, D.E. (1986). Acoustic behavior of *Orcinus*: Periodicity, sequences, correlations with behavior, and an automated technique for call classification. In B. Kirkevold & J. Lockard (Eds), *Behavioral Biology of Killer Whales*. New York, NY: Liss, pp. 335–371.

Bain, D.E. (2001). Noise-based guidelines for killer whale watching. Paper submitted to the Wildlife Viewing Workshop. Vancouver, BC.

Bain, D.E. (2002). Acoustical properties of pingers and the San Juan Island commercial gill net fishery. NMFS Contract Report 40ABNF701651 (unpublished). 14 pp. Available from National Marine Fisheries Service, 7600 Sand Point Way, NE 98115, USA.

Bain, D.E. (2007). The relative importance of different vessel types in the immediate vicinity of southern resident killer whales. NMFS Contract Report Number AB133F-F-04-SE-1272. 24 pp.

Bain, D. & Balcomb, K.C. (1999). Population trends of southern resident killer whales (*Orcinus orca*) from 1960–1999 (unpublished). Status Review Group Workshop, December 1999, Wailea, Maui, HA.

Bain, D.E. & Dahlheim, M.E. (1994). Effects of masking noise on detection thresholds of killer whales. In T.R. Loughlin (Ed.), *Marine Mammals and The Exxon Valdez*. New York, NY: Academic Press, pp. 243–256.

Baird, R.W. (2001). Status of killer whales, *Orcinus orca*, in Canada. *Canadian Field-Naturalist* 115, 676–701.

Baird, R.W. & Hanson, M.B. (2004). Diet studies of 'southern resident' killer whales: prey sampling and behavioral cues of predation. NMFS Contract Report No. AB133F-03-SE-1070. 12 pp.

Baird, R.W., Hanson, M.B. & Dill, L.M. (2005). Factors influencing the diving behaviour of fish-eating killer whales: Sex differences and diel and interannual variation in diving rates. *Canadian Journal of Zoology* 83, 257–267.

Barrett-Lennard, L.G., Ford, J.K.B. & Heise, K.A. (1996). The mixed blessing of echolocation: Differences in sonar use by fish-eating and mammal-eating killer whales. *Animal Behaviour* 51, 553–565.

Bejder, L., Samuels, A., Whitehead, H., *et al.* (2006a). Decline in relative abundance of bottlenose dolphins (*Tursiops* sp.) exposed to long-term disturbance. *Conservation Biology* 20(6), 1791–1798.

Bejder, L., Samuels, A., Whitehead, H. & Gales, N. (2006b). Interpreting short-term behavioural responses to disturbance within a longitudinal perspective. *Animal Behaviour* 72, 1149–1158.

Bigg, M.A. & Wolman, A.A. (1975). Live-capture killer whale (*Orcinus orca*) fishery, British Columbia and Washington, 1962–1973. *Journal of the Fisheries Research Board of Canada* 32, 1213–1221.

Bigg, M.A., Olesiuk, P.F., Ellis, G.M., Ford, J.K.B. & Balcomb, K.C. (1990). Social genealogy of resident killer whales (*Orcinus orca*) in the coastal waters of British Columbia and Washington State. *International Whaling Commission Special Issue* 12, 383–405.

Borrell, A., Bloch, D. & Desportes, G. (1995). Age trends and reproductive transfer of organochlorine compounds in long-finned pilot whales from the Faroe Islands. *Environmental Pollution* 88, 283–292.

Candy, J.R. & Quinn, T.P. (1999). Behavior of adult chinook salmon (*Onchorhynchus tshawytscha*) in British Columbia coastal waters determined from ultrasonic telemetry. *Canadian Journal of Zoology* 77, 1161–1169.

Carretta, J.V., Forney, K.A., Lowry, M.S., *et al.* (2010). U.S. Pacific Marine Mammal Stock Assessments: 2009. U.S. Department of Commerce, NOAA Technical Memorandum NMFS-SWFSC-453. 336 pp.

Clark, C.W., Ellison, W.T., Southall, B.L., *et al.* (2009). Acoustic masking in marine ecosystems as a function of anthropogenic sound sources. IWC/SC/61/E10. Submitted to Scientific Committee, International Whaling Commission. 19 pp.

D'Vincent, C.G., Nilson, R.M. & Hanna, R.E. (1985). Vocalization and coordinated feeding behavior of the humpback whale in southeastern Alaska. *The Scientific Reports of the Whales Research Institute No.* 36, 11–47.

de Swart, R.L., Ross, P.S., Vos, J.G. & Osterhaus, A.D. (1996). Impaired immunity in harbour seals (*Phoca vitulina*) exposed to bioaccumulated environmental contaminants: Review of a long-term feeding study. *Environmental Health Perspectives* 104, 823–828.

Erbe, C. (2002). Underwater noise of whale-watching boats and potential effects on killer whales (*Orcinus orca*), based on an acoustic impact model. *Marine Mammal Science* 18, 394–418.

Fisheries and Oceans Canada. (2008). *Recovery Strategy for Northern and Southern Resident Killer Whales (Orcinus orca) in Canada. Species at Risk Recovery Series.* Ottawa: Fisheries and Oceans Canada. ix + 81 pp.

Foote, A.D., Osborne, R.W. & Hoezel, A.R. (2004). Whale-call response to masking boat noise. *Nature* 428(29 April), 910.

Ford, J.K.B. & Ellis, G.M. (2005). Prey selection and food sharing by fish-eating 'resident' killer whales (*Orcinus orca*) in British Columbia. Can. Sci. Advisory Sec. Res. Doc. 2005/041. 34 pp.

Ford, J.K.B., Ellis, G.M. & Balcomb, K.C. (2000). *Killer Whales*. Vancouver: UBC Press.

Ford, J.K.B., Ellis, G.M. & Olesiuk, P.F. (2005). Linking prey and population dynamics: Did food limitation cause recent declines of 'resident' killer whales (*Orcinus orca*) in British Columbia? Can. Sci. Advisory Sec. Res. Doc. 2005/042. 31 pp.

Ford, J.K.B., Ellis, G.M. & Balcomb, K.C. (2010). Linking killer whale survival and prey abundance: Food limitation in the oceans' apex predator? *Biology Letters* 6, 139–142.

Gilpin, M.E., Case, T.J. & Ayala, F.J. (1976). Theta-selection. *Mathematical Biosciences* 32, 131–139.

Griffin, R.M. & Bain, D.E. (2006). Sound exposure of southern resident killer whales in the southern strait of Georgia. Contract Report to the National Marine Conservation Area Feasibility Study. Vancouver, BC. 23 pp.

Hall, J. & Johnson, C.S. (1972). Auditory thresholds of a killer whale *Orcinus orca* Linnaeus. *Journal of the Acoustical Society of America* 51, 515–517.

Hanson, M.B., Baird, R.W., Ford, J.K.B., *et al.* (2010). Species and stock identification of prey consumed by endangered southern resident killer whales in their summer range. *Endangered Species Research* 11, 69–82.

Hauser, D.D., Logsdon, M.G., Holmes, E.E., VanBlaricom, G.R. & Osborne, R.W. (2007). Summer distribution patterns of southern resident killer whales *Orcinus orca*: Core areas and spatial segregation of social groups. *Marine Ecology Progress Series* 351, 301–310.

Heise, K., Barrett-Lennard, L.G., Saulitis, E., Matkin, C. & Bain, D. (2003). Examining the evidence for killer whale predation on Steller sea lions in British Columbia and Alaska. *Aquatic Mammals* 29(3), 325–334.

Higham, J.E.S., Bejder, L. & Lusseau, D. (2008). An integrated and adaptive management model to address the long term sustainability of tourist interactions with cetaceans. *Environmental Conservation* 35(4), 294–302.

Hilborn, R., Cox, S.P., Gulland, F.M.D., *et al.* (2012). *The Effects of Salmon Fisheries on Southern Resident Killer Whales: Final Report of the Independent Science Panel.* Prepared with the assistance of D.R. Marmorek and A.W. Hall, ESSA Technologies Ltd., Vancouver, B.C. for National Marine Fisheries Service (Seattle. WA) and Fisheries and Oceans Canada (Vancouver. BC). xv + 61 pp. + Appendices.

Holt, M.M. (2008). Sound exposure and Southern Resident killer whales (*Orcinus orca*): A review of current knowledge and data gaps. U.S. Dept. Commer., NOAA Tech. Memo. NMFS-NWFSC-89, 59 p.

Holt, M.M., Noren, D.P., Veirs, V., Emmons, C.K. & Veirs, S. (2008). Speaking up: Killer whales (*Orcinus orca*) increase their call amplitude in response to vessel noise. *Journal of the Acoustical Society of America* 125, EL27–EL32.

Jefferson, T.A. (1987). A study of the behavior of Dall's porpoise (*Phocoenoides dalli*) in the Johnstone Strait, British Columbia. *Canadian Journal of Zoology* 65, 736–744.

Jeffries, S., Huber, H., Calambokidis, J. & Laake, J. (2003). Trends and status of harbor seals in Washington State: 1978–99. *Journal of Wildlife Management* 67, 207–218.

Jensen, F.H., Bejder, L., Wahlberg, M., Johnson, M., Aguilar Soto, N. & Madsen, P.T. (2009). Masking of delphinid communication by small vessels. *Marine Ecology Progress Series* 395, 161–175.

Kastelein, R.A., Walton, S., Odell, D., Nieuwstraten, S.H. & Wiepkema, P.R. (2000). Food consumption of a captive female killer whale (*Orcinus orca*). *Aquatic Mammals* 26, 127–131.

Kastelein, R.A., Kershaw, J., Berghout, E. & Wiepkema, P.R. (2003). Food consumption and suckling in killer whales. *International Zoo Yearbook* 38, 204–218.

Kasting, N.W., Adderley, S.A.L., Safford, T. & Hewlett, K.G. (1989). Thermoregulation in beluga (*Delphinapterus leucas*) and killer (*Orcinus orca*) whales. *Physiological Zoology* 62, 687–701.

Koski, K.L. (2011). 2010 Soundwatch Program annual contract report. Contract Number: AB133F-07-CN-0221 Tasks 2.1.2 & 2.1.3. 75 pp.

Krahn, M.M., Ford, M.J., Perrin, W.F., *et al.* (2004). Status review of Southern Resident killer whales (*Orcinus orca*) under the Endangered Species Act. US Department of Commerce, Vol. NMFS-NWFSC-62.

Krahn, M.M., Hanson, M.B., Baird, R.W., *et al.* (2007). Persistent organic pollutants and stable isotopes in biopsy samples (2004/2006) from Southern Resident killer whales. *Marine Pollution Bulletin* 54, 1903–1911.

Kriete, B. (1995). Bioenergetics in the killer whale, *Orcinus orca*. PhD dissertation (unpublished). University of British Columbia, Vancouver, BC.

Kruse, S. (1991). The interactions between killer whales and boats in Johnstone Strait, B.C. In K. Pryor & K.S. Norris (Eds), *Dolphin Societies: Discoveries and puzzles*. Berkeley, CA: University of California Press, pp. 148–159.

Lachmuth, C.L., Barrett-Lennard, L.G., Steyn, D.Q. & Milsom, W.K. (2011). Estimation of southern resident killer whale exposure to exhaust emissions from whale-watching vessels and potential adverse health effects and toxicity thresholds. *Marine Pollution Bulletin* 62(4), 792–805.

Laist, D.W., Knowlton, A.R., Mead, J.G., Collet, A.S. & Podesta, M. (2001). Collisions between ships and whales. *Marine Mammal Science* 17, 35–75.

Lien, J. (2001). The conservation basis for the regulation of whale watching in Canada by the Department of Fisheries and Oceans: A precautionary approach. Can. Tech. Rep. Fish. Aquat. Sci. 2363: vi + 38 p.

Lusseau, D. (2003). The effects of tour boats on the behavior of bottlenose dolphins: Using Markov chains to model anthropogenic impacts.*Conservation Biology* 17, 1785–1793.

Lusseau, D. (2004). The hidden cost of tourism: Detecting long-term effects of tourism using behavioral information. *Ecology and Society* 9(1), 2 [online] URL: http://www.ecologyandsociety.org/vol9/iss1/art2/.

Lusseau D. & Bejder, L. (2007). The long-term consequences of short-term responses to disturbance: Experiences from whalewatching impact assessment. *International Journal of Comparative Psychology* (Special Issue) 20, 228–236.

Lusseau, D., Bain, D.E., Williams, R. & Smith, J.C. (2009). Vessel traffic disrupts the foraging behavior of southern resident killer whales *Orcinus orca*. *Endangered Species Research* 6, 211–221.

May, R.M. (1980). Mathematical models in whaling and fisheries management. In G.F. Oster (Ed.), *Some Mathematical Questions in Biology*, Vol. 13. Providence, RI: American Mathematical Society.

Miller, P.J.O. (2000). Maintaining contact: Design and use of acoustic signals in killer whales, *Orcinus orca*. PhD dissertation (unpublished). MIT-WHOI.

Morton, A.B. & Symonds, H.K. (2002). Displacement of *Orcinus orca* (L.) by high amplitude sound in British Columbia, Canada. *ICES Journal of Marine Science* 59, 71–80.

Nachtigall, P.E., Pawloski, J.L. & Au, W.W.L. (2003). Temporary threshold shifts and recovery following noise exposure in the Atlantic bottlenosed dolphin (*Tursiops truncatus*). *Journal of the Acoustical Society of America* 113, 3425–3429.

National Research Council. (2003). *Ocean Noise and Marine Mammals*. Washington, DC: National Academies Press. 192 pp.

National Research Council. (2005). *Marine Mammal Populations and Ocean Noise: Determining when noise causes biologically significant effects*. Washington, DC: National Academies Press.

Nichol, L.M. & Shackleton, D.M. (1996). Seasonal movements and foraging behaviour of northern resident killer whales (*Orcinus orca*) in relation to the inshore distribution of salmon (*Oncorhynchus* spp) in British Columbia. *Canadian Journal of Zoology* 74, 983–991.

Olesiuk, P.F. (2009). An assessment of population trends and abundance of harbour seals (*Phoca vitulina*) in British Columbia. Canadian Science Advisory Secretariat Research Document 2009/105, 163 pp. Available from http://www.dfo-mpo.gc.ca/CSAS/Csas/publications/resdocs-docrech/2009/2009_105_e.pdf.

Olesiuk, P.F., Bigg, M.A. & Ellis, G.M. (1990). Life history and population dynamics of resident killer whales (*Orcinus orca*) in the coastal waters of British Columbia and Washington State. *International Whaling Commission Special Issue* 12, 209–243.

Olesiuk, P.F., Nichol, L.M., Sowden, M.J. & Ford, J.K.B. (2002). Effect of the sound generated by an acoustic harassment device on the relative abundance and distribution of harbor porpoises (*Phocoena phocoena*) in retreat passage, British Columbia. *Marine Mammal Science* 18, 843–862.

Olesiuk, P.F., Ellis, G.M. & Ford, J.K.B. (2005). Life history and population dynamics of northern resident killer whales (*Orcinus orca*) in British Columbia. Canadian Science Advisory Secretariat Research Document 2005/045. http://www.dfo-mpo.gc.ca/csas/Csas/DocREC/2005/RES2005_045_e.pdf.

Osborne, R.W. (1999). A historical ecology of Salish Sea 'Resident' killer whales (*Orcinus orca*): With implications for management. PhD dissertation, University of Victoria, Victoria, BC.

Raverty S., Zabek, E., Schroeder, J.P., Wood, R., Bain, D.E. & Cameron, C.E. (2007). Preliminary investigation into the microbial culture and molecular screening of exhaled breaths of southern resident killer whales (*Orcinus* sp) and pathogen screening of the sea-surface microlayer (SML) and sub-surface water samples in Washington State. Poster presented to the Int. Assoc. Aquatic Anim. Med. Conf.

Reddy, M.L., Reif, J.S., Bachand, A. & Ridgway, S.H. (2001). Opportunities for using Navy marine mammals to explore associations between organochlorine contaminants and unfavorable effects on reproduction. *Science of the Total Enivronment* 274, 171–182.

Richardson, W.J., Finley, K.J., Miller, G.W., Davis, R.A. & Koski, W.R. (1995a). Feeding, social and migration behavior of bowhead whales, *Balaena mysticetus*, in Baffin Bay vs. the Beaufort Sea – Regions with different amounts of human activity. *Marine Mammal Science* 11, 1–45.

Richardson, W.J., Greene, Jr, C.R., Malme, C.I. & Thomson, D.H. (1995b). *Marine Mammals and Noise*. San Diego, CA: Academic Press.

Ritter, F. (2003). *Interactions of Cetaceans with Whale Watching Boats – Implications for the Management of Whale Watching Tourism*. Berlin: M.E.E.R. e.V., 91 pp.

Rosalind, M.R., Parks, S.E., Hunt, K.E., *et al*. (2011). Evidence that ship noise increases stress in right whales. *Proceedings of the Royal Society B* 279, 2363–2368.

Ross, P.S., Ellis, G.M., Ikonomou, M.G., Barrett-Lennard, L.G. & Addison, R.F. (2000). High PCB concentrations in free-ranging Pacific killer whales, *Orcinus orca*: Effects of age, sex and dietary preference. *Marine Pollution Bulletin* 40, 504–515.

Schevill, W. & Watkins, W. (1966). Sound structure and directionality in *Orcinus* (killer whale). *Zoologica* 51, 71–76.

Szymanski, M.D., Supin, A.Ya., Bain, D.E. & Henry, K.R. (1998). Killer whale (*Orcinus orca*) auditory evoked potentials to rhythmic clicks. *Marine Mammal Science* 14, 676–691.

Szymanski, M.D., Bain, D.E., Kiehl, K., Henry, K.R., Pennington, S. & Wong, S. (1999). Killer whale (*Orcinus orca*) hearing: Auditory brainstem response and behavioral audiograms. *Journal of the Acoustical Society of America* 106, 1134–1141.

Taylor, B.L., Wade, P.R., DeMaster, D.P. & Barlow, J. (2000). Incorporating uncertainty into management models for marine mammals. *Conservation Biology* 14, 1243–1252.

Wade, P. & Angliss, R. (1997). Guidelines for assessing marine mammal stocks: The GAMMS Workshop. NOAA Technical Memorandum. NMFS-OPR-12. 65 pp. Available from Alaska Fisheries Science Center, 7600 Sand Point Way, NE, Seattle, WA 98115, USA.

Waite, J.M. (1988). Alloparental care in killer whales (*Orcinus orca*). Masters thesis (unpublished). University of California at Santa Cruz, Santa Cruz, CA.

Ward, E.J., Holmes, E.E. & Balcomb, K.C. (2009). Quantifying the effects of prey abundance on killer whale reproduction. *Journal of Applied Ecology* 46, 632–640.

Washington Department of Fisheries, Washington Department of Wildlife and Western Washington Treaty Indian Tribes. (1993). 1992 Washington State Salmon and steelhead stock inventory. Available from Washington Department of Fish and Wildlife, Fish Program, 600 Capitol Way North, Olympia, WA 98501–1091, USA.

Wasser, S.K, Azkarate, J.C., Booth, R.K., *et al*. (2010). Noninvasive measurement of thyroid hormone in feces of a diverse array of avian and mammalian species. *General and Comparative Endocrinology* 168, 1–7.

Wells, R.S. & Scott, M.D. (1997). Seasonal incidence of boat strikes on bottlenose dolphins near Sarasota, FL. *Marine Mammal Science* 13(3), 475–480.

Williams, R. & Ashe, E. (2007). Killer whale evasive tactics vary with boat number. *Journal of Zoology* 272, 390–397.

Williams, R. & Noren, D. (2009). Swimming speed, respiration rate and estimated cost of transport in killer whales. *Marine Mammal Science* 25, 327–350.

Williams, R., Bain, D.E. & Trites, A.W. (2002a). Behavioural responses of killer whales (*Orcinus orca*) to whale-watching boats: Opportunistic observations and experimental approaches. *Journal of Zoology (London)* 256, 255–270.

Williams, R., Bain, D.E., Trites, A.W. & Ford, J.K.B. (2002b). Behavioural responses of male killer whales to a 'leapfrogging' vessel. *Journal of Cetacean Research and Management* 4, 305–310.

Williams, R., Lusseau, D. & Hammond, P. (2006). Estimating relative energetic costs of human disturbance to killer whales (*Orcinus orca*). *Biological Conservation* 133, 301–311.

Williams, R., Bain, D.E., Smith, J.C. & Lusseau, D. (2009a). Effects of vessels on behaviour patterns of individual southern resident killer whales *Orcinus orca*. *Endangered Species Research* 6, 199–209.

Williams, R., Lusseau, D. & Hammond, P.S. (2009b). The role of social aggregations and protected areas in killer whale conservation: The mixed blessing of critical habitat. *Biological Conservation* 142, 709–719.

Ylitalo, G.M., Matkin, C.O., Buzitis, J., *et al*. (2001). Influence of life-history parameters on organochlorine concentrations in free-ranging killer whales (*Orcinus orca*) from Prince William Sound, AK. *Science of the Total Environment* 281, 183–203.

16

Ecological constraints and the propensity for population consequences of whale-watching disturbances

David Lusseau

Introduction

Interactions between boats and cetaceans influence the behaviour of animals. Cetaceans will tend to evade boat interactions and these evasion tactics will lead to altered activity budgets (Williams *et al.*, 2002, 2006; Nowacek *et al.*, 2001, Lusseau, 2003, 2004). In some instances, those influences can have severe impacts for individuals and their populations (Lusseau, 2004, 2005a; Williams *et al.*, 2006; Lusseau *et al.*, 2006; Bejder *et al.*, 2006). However, we currently have no way to determine a priori when boat-interaction effects will lead to biological impacts and when those will have population consequences. This lack of predictive power means that we lack the scientific foundations to manage the current rapid growth of the whale-watching industry and therefore the industry's sustainability is being questioned (Corkeron, 2004; Higham & Lusseau, 2007). In this chapter, I develop a conceptual understanding of how the natural constraints that populations face can be critical factors increasing the propensity for population consequences of whale-watching disturbances. This conceptualization is leading to the development of a number of predictions that I encourage the scientific community to test. Each of these hypotheses is highlighted in their relevant section and I also bring them together in the conclusion.

Cetaceans are exposed to various ecological constraints throughout their lives. Some of those emerge from their life-history strategies (Figure 16.1). For example, baleen whale species migrate between foraging and nursing/breeding grounds and therefore have drastically different activity budgets in different oceanic regions based on their current needs (e.g. maximizing their own energy intake or maximizing the body condition of reared calves through energy transfer). The reason for these migrations, in all contrasting whale migration hypotheses, is linked to matching behavioural needs to suitable habitat (Corkeron & Connor, 1999). Other constraints come from ecosystem factors. For example, perceived predation risk is an important factor shaping habitat use in cetaceans (Heithaus & Dill, 2002; Wirsing *et al.*, 2008). Individuals assess the costs and benefits of an area based on what can be earned from using it in relation to the risks to which they are exposed when resident there (Lima & Zollner, 1996). Also, different populations have different prey species which, importantly, vary in their predictability and heterogeneity in spatial and temporal distribution and abundance. Both of these examples illustrate how existing ecological variables can constrain the time–activity budget of individuals. It is in this ecological context that boat interactions are introduced to cetacean populations and therefore it is valuable to explore how known

Whale-watching: Sustainable Tourism and Ecological Management, eds J. Higham, L. Bejder and R. Williams.
Published by Cambridge University Press. © Cambridge University Press 2014.

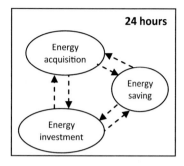 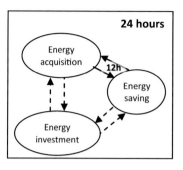 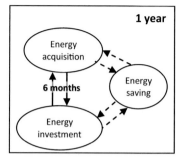

Typical coastal delphinid   **Spinner dolphin-like**   **Typical baleen whale**
                                    pelagic delphinid

Figure 16.1 Examples of the way activity budgets are managed at varying temporal scale by three types of cetacean with varying life history strategies (typical coastal delphinid, spinner dolphin-like pelagic delphinid, and migratory baleen whales). Dotted arrows represent variable transition probabilities while solid arrows represent fixed transition probabilities. So baleen whales have two main states at an annual scale: energy acquisition (feeding) and energy investment (breeding). Within these primary states, whales partition their daily/monthly budget between acquisition and saving or investment (reproduction and growth) and saving. Similarly, spinner dolphins (*Stenella longirostris longirostris*) have two main states at a daily scale: energy acquisition (feeding) and energy saving (resting). Within these primary states, dolphins partition their hourly budget between acquisition and investment or saving and investment.

effects of boat interactions, an added constraint on individuals, can interact with these contextual pressures and under which conditions such ecological interactions will be more likely to lead to biologically significant impacts.

Large-scale: migration and differential sensitivity

Wintering grounds

Migrating whales face different energetic pressures depending on whether they are in their feeding grounds or nursing areas. In nursing grounds, mothers need to maximize energy investment to their calf to ensure their survival. Calf survival is one of the reasons believed to drive migration to locations where predation risk is lower (Corkeron & Connor, 1999). In addition, individuals spend time investing in reproduction in those areas in a number of ways depending on mating strategies (Connor *et al.*, 2000; Whitehead & Mann, 2000).

Mating tactics in baleen whales range from a reliance on sperm competition to something like lekking and greater female choice. This variation in strategies profoundly influences the energetic budget of both sexes during the reproduction stage. By contrast, when in their feeding grounds, individuals maximize energy intake to restore/increase body condition and store energy for migration, mating and nursing. The energy demands of the often vast migrations between feeding and nursing grounds should be understood to represent a genuine and regularly recurring (annual) threat to individual survival. There is a clear dichotomy in the energetic, and resulting activity, budget of baleen whales, with prolonged periods spent intensively investing in energy storage and reproduction in spatially distant locations. Both of these sequential periods of activity are critical to ensuring individual and population survival (Figure 16.1).

For boat interactions to affect the vital rates of individual baleen whales, they would have to significantly disrupt investment in nursing/ breeding grounds and energy intake in feeding grounds.

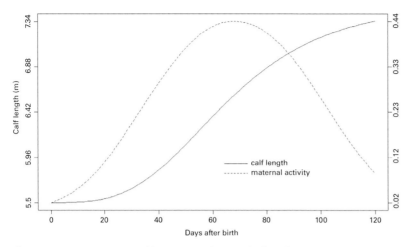

Figure 16.2 A representation of how maternal activity budget changes throughout the nursing season in relation to calf growth in baleen whales: maternal activity predictions, solely for illustrative purposes, emerging from fitting generalized linear models to southern right whale, *Eubalaena australis*, data on activity budget variation in relation to the season at Peninsula Valdés (Thomas & Taber, 1984) overlaid with a plot of the Gompertz growth curve estimated using photogrammetry for southern right whale calf growth at the same location (Whitehead & Payne, 1981). Here I present the proportion of time mothers spent resting as lactation takes place during this activity state (Thomas & Taber, 1984), note the striking variation throughout the season coinciding with the peak in calf growth, the time during which calves will need the most energetic investment from their mothers.

It is possible to envisage that energy investment, such as calf lactation, will be easy to compensate for mothers as lost lactation opportunities due to whale-watching induced displacement can be compensated at a later date. However, lactation does not occur at a constant rate through the development of the calf during the nursing season (Thomas & Taber 1984; e.g. Figure 16.2). Therefore, if tourism interactions are concentrated over the critical period during which calves are trying to maximize growth, these could lead to compromised calf growth, as has been observed in other taxa (McClung *et al.*, 2004), ultimately leading to increased calf mortality or decreased survival probability of those calves as they mature (McMahon *et al.*, 2000). To a lesser extent, mothers will also have to compensate for the added cost of evasive tactics (e.g. increase time spent travelling instead of resting) by balancing the investment they make of their *finite* energy reserves on the nursing grounds by diverting energy that would have gone towards lactation to maintain their homeostasis and ensure survival (i.e. ensure that

they can successfully negotiate the return migration to highly productive feeding grounds). As boat interaction increases, more energy will be diverted away from lactation to maintain the mother's homeostasis and hence can further jeopardise calf survival.

Summering grounds

In contrast, whales will take all opportunities available to them to gain energy when in their foraging grounds. Therefore, any foraging disruption jeopardizes the final body condition that a whale will be able to achieve during a given feeding season. We can therefore envisage that such effects are more likely to lead to impacts on vital rates. Of course, there is already some heterogeneity in the foraging success of whales when they are at feeding grounds and therefore migratory strategies have evolved to ensure that individuals can maximize survival probability given this heterogeneity in what can be achieved. For example, whales may be able to prolong the time they spend in the feeding grounds in order to compensate for lower foraging success, or they may skip migration altogether (Corkeron & Connor, 1999; Craig & Herman, 1997). In other words, there has been strong evolutionary selection pressure to ensure that the survival of individuals is resilient to variation in foraging success at feeding grounds. Energy intake in foraging grounds is a key factor affecting migrating whale survival probability. Those species are K-selected, they are long-lived and slow-reproducing; hence, their behavioural strategies will be geared to maximize survival probability at all costs.

Hypothesis 1a: Counterintuitively, it may be less likely for boat disruption on foraging grounds to lead to effects on vital rates, because the vital rate that could be affected has been subject to evolutionary selection to be resilient to heterogeneity of ecological conditions by the life-history of the species.

Whales still need to accumulate a minimum amount of energy over the season of higher productivity in foraging grounds in order to survive. Boat interactions not only disrupt energy intake, but also increase energy expenditure, as whales swim away from boats or engage in evasive tactics (see Chapters 14 and 16). While the foraging disruption is more energetically demanding (Williams *et al.*, 2006), the added expenditures will increase energetic needs and therefore can send individuals into an energetic deficit spiral as the number of interactions while whales are foraging increases. Therefore, we should expect to observe a stepwise reaction of survival probability to boat interactions exposure: individual whales will be able to cope with boat interactions on foraging grounds up to a certain threshold due to their evolutionary selected behavioural strategies. However, past a certain unknown threshold, individuals will not be able to cope with their metabolic demands and survival rates will drop markedly.

Hypothesis 1b: This non-linear change in vital rates in response to repeated exposure to boat interactions will be a function of the productivity of the foraging ground and the heterogeneity in prey availability.

Habitat preferences and behavioural heterogeneity

Boat interactions can displace cetaceans from preferred habitats (Lusseau, 2005a; Bejder *et al.*, 2006). When boat density is such that engaging in evasive tactics will only result in a new boat interaction, some individuals prefer to leave an area altogether (Lusseau, 2004). The threshold at which this shift from short-term avoidance tactics to long-term area abandonment occurs depends on the costs and benefits of staying in the disturbed area compared to moving elsewhere (Rosenzweig, 1981). Prey availability and predation risk influence habitat use in cetaceans, the latter probably playing a predominant role (Wirsing *et al.*, 2008; Heithaus & Dill, 2006). Individuals spend more time in locations where they are more likely to capture prey and less likely to be detected by predators. Habitat selection can therefore be understood in terms of energetic costs and benefits in the context of survival. Individuals will spend time in areas that maximize energy intake, minimize energy expenditure and maximize possibilities of energy investment, without jeopardizing survival probability. As an environment becomes more risky, individuals will abandon prey-rich habitat in order to minimize mortality risks (Heithaus & Dill, 2002).

Under these premises, boat interaction-induced habitat abandonment can emerge from two different mechanisms. First, evasive tactics can become too costly compared to the energetic gain that can be made in that area and therefore the habitat is deemed degraded by boat interactions. Second, boat interactions can be acting as a perceived risk (i.e. a risk that boats can cause death, for example by colliding with the animals). In that case, individuals can abandon an area even if the area's benefits outweigh the added energetic costs of dealing with boat interactions. The consequence of both mechanisms is the same: individuals are displaced from favoured areas. The number of times these displacements occur, the type of habitats from and to which individuals are displaced, and the likelihood that individuals can compensate for these displacements will influence the consequences of these boat-induced effects. The preference for boats to navigate to areas where cetaceans are more likely to be seen is a feedback mechanism acting on this impact (see Chapter 19). There are strong economic incentives for boats to have non-random habitat use as cetacean sighting is directly linked to economic viability. Moreover, tourist satisfaction levels are linked with increased likelihood of observing aerial displays or specific behavioural events (Lusseau, 2005b; Anwar *et al.*, 2007). Hence, boats will not only interact with cetaceans in their preferred habitat, but also they will favour areas used for non-travelling activity states.

Cetaceans residing year-round in the same home range can develop heterogeneity in space use. For example, there can be heterogeneity in ease of prey capture (Hastie *et al.*, 2004), or predators may not be able to access some refugium (Heithaus & Dill, 2006; Lusseau & Higham, 2004). These activity hotspots can therefore receive a disproportionate level of boat interactions simply because of socioeconomic incentives (Lusseau & Higham, 2004). Resident cetaceans are then pushed to use less-suitable habitat for various activities (e.g. rest or forage in more risky areas or forage in less-productive areas). There are different consequences for disrupting different activity states (Williams *et al.*, 2006).

Hypothesis 2: The spatial ecology context in which disturbances occur will influence the likelihood that a given level of cumulative exposure will have population-level consequences.

Individual exposure and population exposure

Across species, individual cetaceans tend to have overlapping home ranges. Therefore, groups of individuals (e.g. social units in delphinids) will tend to have similar boat exposure. The likelihood that boat interactions will cause population-level consequences depends not only on their ability to elicit impacts on vital rates, but also on the proportion of the population exposed to levels of interactions that can cause vital rate alterations. So, for example, while whale-watching is having serious impacts on bottlenose dolphins (*Tursiops* sp.) in Shark Bay (Australia) (Bejder 2005; Bejder *et al.*, 2006), the proportion of individuals affected in this population is small because activities are concentrated in a small portion of the population's home range. Therefore, there is limited potential for current whale-watching activities to influence the growth rate of that population (Lusseau & Bejder, 2007). By contrast, whale-watching operations in Doubtful Sound (New Zealand) are pervasive throughout the home-range of the resident bottlenose dolphin population. Hence, not only are individuals prevented from engaging in habitat abandonment like their counterparts in neighbouring Milford Sound (Lusseau, 2005a), but also all individuals in the population are exposed to similar levels of cumulative exposure. In that case influences on vital rates of individuals will be similar on the whole population and the population's growth rate will be threatened (Currey *et al.*, 2009a).

Constraints on compensation

I have alluded a number of times now to the issue of compensation. Some populations are exposed to very high levels of behavioural disturbance yet they are seemingly not affected. On the other hand, some populations are exposed to very low level of disturbance and those appear to cause serious alterations of population trajectory. For example, in Sarasota Bay, FL, USA, a bottlenose dolphin (*Tursiops truncatus*) encounters a boat every 6 min, but there are no apparent effects of these interactions on its vital rates (Buckstaff, 2004). In Doubtful Sound, a bottlenose dolphin (*Tursiops truncatus*) encounters a boat every hour and the disturbance this causes can be linked to decreased calving success (Lusseau *et al.*, 2006). This is seemingly completely illogical.

However, these two populations are exposed to very different ecological conditions. In Sarasota Bay, the dolphin population occupies an area where food availability is homogeneous in both time and space (Barros & Wells, 1998). By contrast, the Doubtful Sound population relies on patchy resources that can be serially depleted (Lusseau & Wing, 2006). So in Sarasota Bay, a dolphin can move away from a perturbed area and will not have to travel far to resume its activities. On the other hand, in Doubtful Sound, dolphins will have to undergo prolonged episodes of travelling in order to be able to resume activities after being temporarily displaced from an area (Lusseau & Higham, 2004).

Natural experiments help us confirm this hypothesis. Toxic algal bloom events in Sarasota Bay greatly reduced prey availability to dolphins over the period 2003–2007 (Gannon *et al.*, 2009; McHugh *et al.*, 2011). During these events, female bottlenose dolphins appeared to have insufficient energy available to live in their habitat and rear calves successfully (McHugh *et al.*, 2011; Powell & Wells, 2011). A number of cetacean populations have evolved foraging strategies to make use of 'ephemeral' prey species. Fish-eating killer whales along the western USA and Canada coast are typical examples, as they rely on pulses of salmon returning to their rivers to breed (Ford *et al.*, 2009). In this situation, the timeframe over which the whales can compensate for the added energetic demands of whale-watching is limited to when the prey is available and foraging disruption can therefore lead to complete loss of critical foraging opportunities. We can conclude that habitat costs can only be met if prey availability is sufficient to compensate these costs; that is, fuel the added energetic expenditures they require.

These concepts of habitat heterogeneity can be conceptualized using the notion of environmental noise colour (Vasseur & Yodzis, 2004). In this instance, 'noise' is simply a way to define random variation in the environment and defining its colour is a way to encapsulate its variability. The variance is the same at all scales for a white noise while long period cycles dominate a red noise (Vasseur & Yodzis, 2004). Whether it is on a spatial or temporal scale, environmental noise will affect the ability of cetaceans to compensate for the costs associated with whale-watching operations.

We can predict that a population exposed to redder environmental noise will have fewer opportunities to compensate for disturbances and therefore disturbances of these populations will be more likely to lead to population-level consequences (Figure 16.3). In addition, the more the variance in boat interactions (on a spatial and/or temporal scale) synchronizes with the environmental variance, the more likely it is to lead to population-level consequences. In a correlated system, boats are more likely then to interact with the animals while they are engaged in sensitive activities and the opportunities for them to compensate for these disruptions are sparse. Economic, social and ecological feedback mechanisms force the variance of these two components of the whale-watching–cetacean socio-ecological system to synchronize at the evolutionary scale (see Chapter 19).

Hypothesis 3:   A population exposed to redder environmental noise (Vasseur & Yodzis, 2004) will have fewer opportunities to compensate for disturbances and therefore disturbances of these populations will be more likely to lead to population-level consequences.

Hypothesis 4:   Socio-ecological feedback mechanisms forcing a synchronization of the variance in boat interactions (on a spatial and/or temporal scale) with the environmentally induced variance in the animal's activities will increase the propensity for population-level consequences.

Schooling benefits

Delphinids are group-living species. Individuals draw advantages from being with conspecifics in schools: it increases prey capture per capita and decreases predation rate per capita (Lusseau *et al.*, 2004). Therefore, it is primordial for individuals to stay in schools as long as school membership benefits outweigh its costs (primarily competition). The

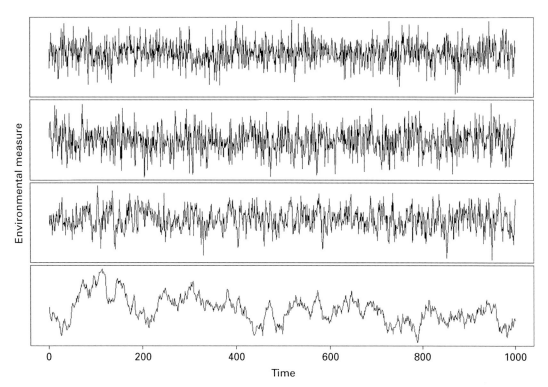

Figure 16.3 Variation in the autocorrelation of environmental noise can be described by the frequency of its variance spectrum and its associated spectral exponent which can be likened to frequencies on the colour spectrum of light. Here I show noise 'colour' varying from white (top, typical of terrestrial systems) to red (bottom, typical of marine systems) in simulated time series of an environmental factor. Opportunities to take advantage of an ecological factor are fewer and further apart in a redder environment and therefore the propensity to compensate for lost opportunities is lower.

marked effect of schooling behaviour on decreased predation risk is a clear advantage to maximize immediate survival chances. Therefore, schooling costs will have to be much higher in populations that are exposed to predation for individuals to leave the school. We can argue that for these K-selected species immediate survival benefits will always outweigh future energetic investment and therefore mothers will maintain school membership if they perceive their immediate survival to be threatened to the detriment of their energetic allocation to their calves (Bell, 1980). They will therefore engage in evasive tactics even if they cannot meet the costs of the resulting altered energetic budget and will reallocate energy for lactation towards maintenance.

I have argued previously that female dolphins will maintain school membership in order to avoid predation risks, real or perceived, even if this action pushes them towards an energetic budget that is too expensive for them to continue investing into lactation (Lusseau, 2004; Lusseau *et al.*, 2006). We still debate whether the tactics cetaceans use to evade boat interactions can be interpreted as those animals perceiving boat interactions as predation risk (Frid & Dill, 2002; Beale & Monaghan, 2004b). Regardless of this, populations that are exposed to real predation risk (e.g. from sharks or killer whales) will still engage in these decision-making processes about school membership during a boat interaction. Staying behind, or not responding to boat interactions as conspecifics do, will increase the risk of predation for individuals after school fission. Evading tactics lead animals both to spend more time in energetically demanding activities and in some situations to forgo foraging (Lusseau, 2004; Williams *et al.*, 2006). Such altered activity budgets can be too expensive for lactating females that will then reabsorb energy they were to invest in lactation to maintain their homeostasis. The perceived survival benefits of maintaining school membership during evading tactics outweigh this cost on reproduction.

Hypothesis 5: The choice of lactating females to maintain school membership benefits may push them to follow suboptimal, expensive, activity budgets leading them to reabsorb energy they were to invest in lactation to maintain their homeostasis.

Conceptualizing the relationship between ecological constraints, boat interactions and population consequences

Boat interactions are an environmental variable that can influence the fitness of targeted individual whales and dolphins. If there is scope for it, individuals can adapt, because of either the resilience or the elasticity of their behavioural ecology. However, individuals do not perform activities at random. The more constraining their habitat already is on the timing or localization of these activities, the less likely they are to be able to cope with interactions without those interactions affecting their vital rates (Figure 16.4). Then, individuals will sacrifice investments that are not essential to their survival, including lactation (hence, decreased calving success) and male reproductive investment.

We can conceptualize when such disturbances are more likely to cause problems by understanding the context in which disturbances occur. For example, a food-limited delphinid population that relies on highly heterogeneously distributed prey items (Figure 16.3, red prey availability variation)

will have more difficulties in compensating the cost of disturbance avoidance and therefore these disturbances will be more likely to cause population-level impacts. In this case, disturbed, and subsequently lost, foraging opportunities cannot be replaced. Also, increasing the energetic costs of a lactating whale on the nursing ground by increasing the amount of time she has to spend travelling to avoid boats will also affect her calving success, as she will not be able to invest as much energy in her calf growth. This ecological tug-of-war happens within an evolutionary context. Cetaceans are K-selected species (Stearns, 1977), which means that the life-history strategy of individuals relies on maximizing adult survival probability. There has therefore been selection pressure on behavioural strategies during the course of their evolutionary history to ensure that the effects of ecological variation on survival probability could be minimized. Targeted boat interactions, whale-watching, are newly introduced ecological constraints that are dealt with within this evolutionary context and therefore we can anticipate that its effects will be more noticeable on other vital rates than adult survival probability.

Environmental variability is a force that influences population dynamics (Petchey *et al.*, 1997). It can be characterized by the frequency of its variance spectrum, which is often described by the associated colour in the light spectrum for such frequency (Vasseur & Yodzis, 2004). This variability, and its associated predictability, is a key factor in the behavioural ecology of disturbances because it drives the ability of individuals to compensate for perturbed opportunities. As long as individuals can compensate for the various added costs associated with disturbances, those are unlikely to alter their vital rates. It is worth noting that while whale-watching will not have any population-level consequences in this case, it can have ecosystem-level consequences as cetaceans have to consume more prey in order to fuel the costs of whale-watching. Therefore, whale-watching development requires careful planning as it can have ecosystem-wide consequences and impact other human activities such as fishing. The more individuals are constrained

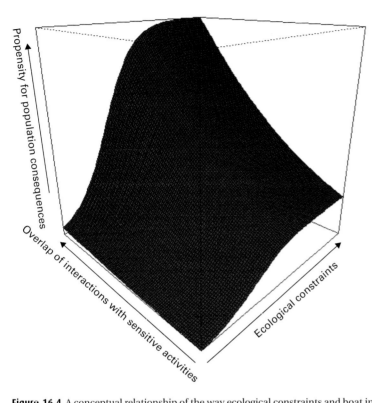

Figure 16.4 A conceptual relationship of the way ecological constraints and boat interactions interact to lead to greater propensity for population consequences of whale-watching disturbances. The likelihood of population consequences depends on (i) whether ecological constraints will allow individuals to compensate for lost activity opportunities caused by boat interactions, (ii) the propensity for boat interactions to cause activity disruption given the activity in which individuals engage, and (iii) the likelihood that boat interactions will occur during disturbance-sensitive activities.

in time and space by other ecological conditions, the less likely they will be to have opportunities to compensate for the effects of whale-watching and therefore impacts on population dynamics will arise more immediately.

Hypothesis 6: Marine environments tend to be redder than terrestrial ones and therefore I hypothesize that behavioural disturbances may be more likely to lead to population-level consequences in marine species than in their terrestrial counterparts.

We have the ability to test this hypothesis by comparing the effects of wildlife tourism on cetaceans and on their phylogenetically close terrestrial relatives (i.e. other ungulates).

In all of the situations described in this chapter, individuals were assumed to be able to afford to engage in a behavioural response to disturbances. As we have seen, in some cases those individuals may not choose to do so, because the cost of responding (i.e. changing activity or leaving an area) may outweigh its benefits. However, there are also situations in which individuals cannot afford to account for this newly introduced ecological factor. Individuals in poor body condition, for example, may have to take the risk of being exposed to a stimulus that elicits predation-risk like psychological and physiological responses in the same way that they may have to occupy poor habitat in other instances (Beale & Monaghan, 2004a). The stress consequences of such exposure are only starting to be explored in other species and can themselves lead to influences on vital rates through, for example, lowered immuno-competence (Ellenberg *et al.*, 2007; Müllner *et al.*, 2004). Further research into the potential for physiological-scale responses to cause population-level consequences remains an urgent priority.

In summary, functional relationships exist between the amount of time cetaceans spend interacting with boats and the likelihood of influences on vital rates given their life-history strategies and ecological conditions. The outstanding question on which we now have to concentrate is the definition of the threshold in cumulative exposure at which vital rate effects are triggered.

Management implications: a first foray into preventive care

Importantly, acknowledging the ecological context of whale-watching disturbances also helps us guide the management of whale-watching activities. An impact assessment should first consider the ability that the targeted population will have to compensate for any disturbances caused by interactions. This will involve defining the ecological landscape of the population such as prey availability and predation risk. Then it is required to understand the spatial and temporal variability in these factors. Finally, we need to determine any intrinsic constraints individuals are facing. These may be caused by their life-history tactics, or the phase of their life cycle (e.g. migrating whales on foraging or nursing grounds).

In addition to guiding impact assessment, these principles can also guide management actions. For example, if a population is spatially constrained, e.g. relying on specific foraging hotspots (Hastie *et al.*, 2004; Hauser *et al.*, 2007), then a spatial management plan of whale-watching disturbances would be sensible (e.g. area closure; Lusseau & Higham, 2004). If a population is temporally

constrained, e.g. time–activity budgets driven by daily cycle of prey availability (Benoit-Bird & Au, 2003), then such a management plan should accommodate temporal variation and include time closures. Most of our management actions are guided by legislations that require targeted populations to remain in favourable conservation status (Mehtälä & Vuorisalo, 2007). The considerations we highlight here provide a means to consider the needs for cumulative exposure quotas in order to minimize the risks to population growth rate.

Conclusion

The population consequences of whale-watching disturbances emerge from the interaction between several functional mechanisms which can be seen as defining a complex adaptive system. Such systems often can exist in several states depending on the conditions to which they are exposed (Folke *et al.*, 2004; Scheffer *et al.*, 2009). I have shown that in some instances (e.g. whales on foraging grounds) we can expect to see shifts in vital rate state (i.e. rates varying around different means) in relation to boat exposure. It is therefore conceivable that rapid shifts in population dynamics can be caused by whale-watching disturbance exposure. Indeed, there is some indication in one population that this is precisely the case. An increase in exposure to whale-watching in the Doubtful Sound bottlenose dolphin population resulted in a marked stepwise decrease in calving success that influenced the population's growth rate leading to its current IUCN status of 'critically endangered' (Currey *et al.*, 2009a, 2009b, 2011). This chapter raises a number of hypotheses to help us guide management:

Hypothesis 1: Baleen whales have adapted to variability in foraging success on foraging grounds. Hence, whale-watching disturbances on foraging grounds may be less likely to affect the vital rates of individuals than disturbances on nursing grounds. However, if such effects were to exist on foraging grounds, we would expect to observe a step-change in vital rates that will be a function of the productivity of the foraging ground and also the heterogeneity in prey availability.

Hypothesis 2: The spatial ecology context in which disturbances occur will influence the likelihood that a given level of cumulative exposure will have population-level consequences.

Hypothesis 3: A population exposed to redder/brown environmental noise will have fewer opportunities to compensate for disturbances and therefore disturbances of these populations will be more likely to lead to population-level consequences.

Hypothesis 4: Socio-ecological feedback mechanisms forcing a synchronization of the variance in boat interactions (on a spatial and/or temporal scale) with the environmentally induced variance in the animal's activities will increase the propensity for population-level consequences.

Hypothesis 5: The choice of lactating females to maintain school membership benefits may push them to follow suboptimal, expensive, activity budgets leading them to reabsorb energy they were to invest in lactation to maintain their homeostasis.

Hypothesis 6: Marine environments tend to be redder than terrestrial ones and therefore I hypothesize that behavioural disturbances may be more likely to lead to population-level consequences in marine species than in their terrestrial counterparts.

These hypotheses are available to the scientific community to test in a concerted effort which may ultimately allow us to confidently move the management of whale-watching into a predictive paradigm. It is critical that managers make use of this information to inform managerial decisions under the precautionary principle. It may, for example, be advisable to minimize whale-watching development if a population already faces many ecological challenges (Figure 16.4). Predictive management may ensure that less (or different)

whale-watching in the short term affords more (sustainable) whale-watching in the long term.

Acknowledgements

I would like to thank Fredrik Christiansen and John Harwood for many fruitful discussions that have led to the development of some of the ideas I present here. This work was supported in part thanks to grant HR09011 from the Scottish Funding Council to the Marine Alliance for Science and Technology for Scotland.

REFERENCES

Anwar, S.M., Jeanneret, C.A., Parrott, L. & Marceau, D.J. (2007). Conceptualization and implementation of a multi-agent model to simulate whale-watching tours in the St. Lawrence Estuary in Quebec, Canada. *Environmental Modelling & Software* 22(12), 1775–1787.

Barros, N.B. & Wells, R.S. (1998). Prey and feeding patterns of resident bottlenose dolphins (*Tursiops truncatus*) in Sarasota Bay, Florida. *Journal of Mammalogy* 79(3), 1045–1059.

Beale, C.M. & Monaghan, P. (2004a). Behavioural responses to human disturbance: A matter of choice? *Animal Behaviour* 68, 1065–1069.

Beale, C.M. & Monaghan, P. (2004b). Human disturbance: People as predation-free predators? *Journal of Applied Ecology* 41(2), 335–343.

Bejder, L. (2005). Linking short and long-term effects of nature-based tourism on cetaceans. MS Thesis, Dalhousie University.

Bejder, L., Samuels, A., Whitehead, H., *et al.* (2006). Decline in relative abundance of bottlenose dolphins exposed to long-term disturbance. *Conservation Biology* 20(6), 1791–1798.

Bell, G. (1980). The costs of reproduction and their consequences. *American Naturalist* 116(1), 45–76.

Benoit-Bird, K.J. & Au, W.L. (2003). Prey dynamics affect foraging by a pelagic predator over a range of spatial and temporal scales. *Behavioral Ecology and Sociobiology* 53, 364–373.

Buckstaff, K.C. (2004). Effects of watercraft noise on the acoustic behavior of bottlenose dolphins, *Tursiops truncatus*, in Sarasota Bay, Florida. *Marine Mammal Science* 20(4), 709–725.

Connor, R.C., Read, A.J. & Wrangham, R.W. (2000). Male reproductive strategies and social bonds. In R.C. Connor, J. Mann, P.L. Tyack & H. Whitehead (Eds), *Cetacean Societies*. Chicago, IL: University of Chicago Press, pp. 247–269.

Corkeron, P.J. (2004). Whalewatching, iconography, and marine conservation. *Conservation Biology* 18(3), 847–849.

Corkeron, P.J. & Connor, R.C. (1999). Why do baleen whales migrate? *Marine Mammal Science* 15(4), 1228–1245.

Craig, A.S. & Herman, L.M. (1997). Sex differences in site fidelity and migration of humpback whales (*Megaptera novaeangliae*) to the Hawaiian Islands. *Canadian Journal of Zoology* 75, 1923–1933.

Currey, R.J.C., Dawson, S.M., Slooten, E., *et al.* (2009a). Survival rates for a declining population of bottlenose dolphins in Doubtful Sound, New Zealand: An information theoretic approach to assessing the role of human impacts. *Aquatic Conservation-Marine and Freshwater Ecosystems* 19(6), 658–670.

Currey, R.J.C., Dawson, S.M. & Slooten, E. (2009b). An approach for regional threat assessment under IUCN Red List criteria that is robust to uncertainty: The Fiordland bottlenose dolphins are critically endangered. *Biological Conservation* 142(8), 1570–1579.

Currey, R.J.C., Dawson, S.M. & Slooten, E. (2011), *Tursiops truncatus* (Fiordland subpopulation). In: IUCN 2011. IUCN Red List of Threatened Species. Version 2011.2. Retrieved from http://www.iucnredlist.org (accessed 18 June 2012).

Ellenberg, U., Setiawan, A.N., Cree, A., Houston, D.M. & Seddon, P.J. (2007). Elevated hormonal stress response and reduced reproductive output in Yellow-eyed penguins exposed to unregulated tourism. *General and Comparative Endocrinology* 152(1), 54–63.

Folke, C., Carpenter, S.R., Walker, B., *et al.* (2004). Regime shifts, resilience, and biodiversity in ecosystem management. *Annual Review of Ecology, Evolution, and Systematics* 35, 557–581.

Ford, J.K.B., Ellis, G.M., Olesiuk, P. & Balcomb, K.C. (2009). Linking killer whale survival and prey abundance. *Biology Letters* 6(1), 139–142.

Frid, A. & Dill, L.M. (2002). Human-caused disturbance stimuli as a form of predation risk. *Conservation Ecology* 6(1), 11.

Gannon, D.P., Berens McCabe, E.J., Camilleri, S.A., *et al.* (2009). Effects of *Karenia brevis* harmful algal blooms on nearshore fish communities in southwest Florida. *Marine Ecology Progress Series* 378, 171–186.

Hastie, G.D., Wilson, B., Wilson, L.J., Parsons, K.M. & Thompson, P.M. (2004). Functional mechanisms underlying cetacean distribution patterns: Hotspots for bottlenose dolphins are linked to foraging. *Marine Biology* 144(2), 397–403.

Hauser, D.D.W., Logsdon, M.G., Holmes, E.E., VanBlaricom, G.R. & Osborne, R.W. (2007). Summer distribution patterns of southern resident killer whales *Orcinus orca*: Core areas and spatial segregation of social groups. *Marine Ecology Progress Series* 351, 301–310.

Heithaus, M.R. & Dill, L.M. (2002). Food availability and tiger shark predation risk influence bottlenose dolphin habitat use. *Ecology* 83(2), 480–491.

Heithaus, M.R. & Dill, L.M. (2006). Does tiger shark predation risk influence foraging habitat use by bottlenose dolphins at mutiple spatial scales? *Oikos* 114, 257–264.

Higham, J.E.S. & Lusseau, D. (2007). Urgent need for empirical research into whaling and whale watching. *Conservation Biology* 21(2), 554–558.

Lima, S.L. & Zollner, P.A. (1996). Towards a behavioral ecology of ecological landscapes. *Trends in Ecology & Evolution* 11(3), 131–135.

Lusseau, D. (2003). Male and female bottlenose dolphins *Tursiops* sp. have different strategies to avoid interactions with tour boats in Doubtful Sound, New Zealand. *Marine Ecology Progress Series* 257, 267–274.

Lusseau, D. (2004). The hidden cost of tourism: Detecting long-term effects of tourism using behavioral information. *Ecology and Society* 9(1), 2.

Lusseau, D. (2005a). The residency pattern of bottlenose dolphins (*Tursiops* spp.) in Milford Sound, New Zealand, is related to boat traffic. *Marine Ecology Progress Series* 295, 265–272.

Lusseau, D. (2005b). The state of the scenic cruise industry in Doubtful Sound in relation to a key natural resource: Bottlenose dolphins. In M. Hall & S. Boyd (Eds), *Nature-based Tourism in Peripheral Areas: Development or disaster?* London: Channelview Publications, pp. 246–262.

Lusseau, D. & Bejder, L. (2007). The long-term consequences of short-term responses to disturbance. *International Journal of Comparative Psychology* 20, 228–236.

Lusseau, D. & Higham, J.E.S. (2004). Managing the impacts of dolphin-based tourism through the definition of critical habitats: The case of bottlenose dolphins (*Tursiops* spp.) in Doubtful Sound, New Zealand. *Tourism Management* 25, 657–667.

Lusseau, S.M. & Wing, S.R. (2006). Importance of local production versus pelagic subsidies in the diet of an isolated population of bottlenose dolphins *Tursiops* sp. *Marine Ecology Progress Series* 321, 283–293.

Lusseau, D., Williams, R.J., Wilson, B., *et al.* (2004). Parallel influence of climate on the behaviour of Pacific killer whales and Atlantic bottlenose dolphins. *Ecology Letters* 7, 1068–1076.

Lusseau, D., Slooten, E. & Currey, R.J.C. (2006). Unsustainable dolphin-watching tourism in Fiordland, New Zealand. *Tourism in Marine Environments* 3(2), 173–178.

McClung, M.R., Seddon, P.J., Massaro, M. & Setiawan, A.N. (2004). Nature-based tourism impacts on yellow-eyed penguins *Megadyptes antipodes*. *Biological Conservation* 119(2), 279–285.

McHugh, K.A., Allen, J.B., Barleycorn, A.A. & Wells, R.S. (2011). Severe *Karenia brevis* red tides influence juvenile bottlenose dolphin (*Tursiops truncatus*) behavior in Sarasota Bay, Florida. *Marine Mammal Science* 27(3), 622–643.

McMahon, C.R., Burton, H.R. & Bester, M.N. (2000). Weaning mass and the future survival of juvenile southern elephant seals at Macquarie Island. *Antarctic Science* 12, 149–153.

Mehtälä, J. & Vuorisalo, T. (2007). Conservation policy and the EU Habitats Directive: Favourable conservation status as a measure of conservation success. *European Environment* 17(6), 363–375.

Müllner, A., Linsenmair, K.E. & Wikelski, M. (2004). Exposure to ecotourism reduces survival and affects stress response in hoatzin chicks. *Biological Conservation* 118(4), 549–558.

Nowacek, S.M., Wells, R.S. & Solow, A.R. (2001). Short-term effects of boat traffic on bottlenose dolphins, *Tursiops truncatus*, in Sarasota Bay, Florida. *Marine Mammal Science* 17(4), 673–688.

Petchey, O.L., Gonzalez, A. & Wilson, H.B. (1997). Effects on population persistence: The interaction between environmental noise colour, intraspecific competition, and space. *Proceedings of the Royal Society B* 264(1389), 1841–1847.

Powell, J.R. & Wells, R.S. (2011). Recreational fishing depredation and associated behaviors involving common bottlenose dolphins (*Tursiops truncatus*) in Sarasota Bay, Florida. *Marine Mammal Science* 27(1), 111–129.

Rosenzweig, M.L. (1981). A theory of habitat selection. *Ecology* 62(2), 327–335.

Scheffer, M., Bascompte, J., Brock, W.A., *et al.* (2009). Early-warning signals for critical transitions. *Nature* 461, 53–59.

Stearns, S.C. (1977). The evolution of life history traits: A critique of the theory and a review of the data. *Annual Review of Ecology and Systematics* 8, 145–171.

Thomas, P.O. & Taber, S.M. (1984). Mother–infant interaction and behavioral development in southern right whales, *Eubalaena australis*. *Behaviour* 88, 42–60.

Vasseur, D.A. & Yodzis, P. (2004). The color of environmental noise. *Ecology* 85(4), 1146–1152.

Whitehead, H. & Mann, J. (2000). Female reproductive strategies of cetaceans. In J. Mann, R.C. Connor, P.L. Tyack & H. Whitehead (Eds), *Cetacean Societies: Field studies of dolphins and whales*. Chicago, IL: Chicago University Press.

Whitehead, H. & Payne, R. (1981). New techniques for assessing populations of right whales without killing them. *FAO Fisheries Series (5) [Mammals in the Seas]* 3, 189–209.

Williams, R., Trites, A.W. & Bain, D.E. (2002). Behavioural responses of killer whales (*Orcinus orca*) to whale-watching boats: Opportunistic observations and experimental approaches. *Journal of Zoology* 256, 255–270.

Williams, R., Lusseau, D. & Hammond, P.S. (2006). Estimating relative energetic costs of human disturbance to killer whales (*Orcinus orca*). *Biological Conservation* 133(3), 301–311.

Wirsing, A.J., Heithaus, M.R., Frid, A. & Dill, L.M. (2008). Seascapes of fear: Evaluating sublethal predator effects experienced and generated by marine mammals. *Marine Mammal Science* 24(1), 1–15.

17

The use of area–time closures as a tool to manage cetacean-watch tourism

Julian Tyne, Neil Loneragan and Lars Bejder

Introduction

The world's oceans have been exploited for generations. In some cases, this has led to the removal of top predators from ecosystems, resulting in a cascading effect through trophic levels altering ecosystems and restructuring food webs (Pauly *et al.*, 2002; Myers & Worm, 2003). Cetaceans (whales, dolphins and porpoises) have also been targeted, mainly for their meat and oil, and some populations being driven close to extinction. Fortunately, attitudes towards cetaceans have changed over the past two decades, and rather than harvest them, it is now more desirable to observe them in their natural environment (Bearzi *et al.*, 2010). Today, cetaceans are icons for marine conservation efforts. The USA was the first country to introduce legislation to protect marine mammals through the Marine Mammal Protection Act 1972 (MMPA). The MMPA was designed to minimize the capture or 'take', harassment and disturbance of marine mammals, primarily from fishing operations as by-catch and from cetacean hunting. The MMPA defines the term 'take' as 'hunting, killing, capture and harassment of a marine mammal or the attempt thereof'. Since the declaration of the MMPA, other countries have adopted their own legislation, e.g. The Marine Mammals Protection Act 1978 in New Zealand and the Environment Protection and Biodiversity Conservation Act 1999 in Australia.

Although protection of cetaceans is supported enthusiastically in many countries, only 1.3% of the world's oceans are protected from anthropogenic threats (Hoyt, 2011), and cetacean populations are still vulnerable to a multitude of anthropogenic impacts. These threats fall into two broad categories: *direct* impacts, i.e. those that are readily observable; and *cumulative* impacts, i.e. those that are not readily observable and are likely to cause effects through repeated exposure. Direct impacts are those that cause the death of individuals immediately, such as whaling (Gales *et al.*, 2005), ship strikes (Panigada *et al.*, 2006) and bycatch (Mangel *et al.*, 2010). Although the deaths of individual cetaceans are readily detected, quantifying the effects of direct impacts on the viability of cetacean populations is challenging as it requires information on the population size and the connectivity of populations. Cumulative impacts include sources of disturbance that are likely to affect behaviour and/or physiology and, as a consequence, are more difficult to identify and quantify. Indirect effects include noise pollution (Nowacek *et al.*, 2007; Tyack, 2008), chemical pollution (Reijnders *et al.*, 2009), tourism (Lusseau & Higham, 2004; Bejder *et al.*, 2006a, 2006b; Lusseau *et al.*, 2006), coastal development (Jefferson *et al.*, 2009), prey exploitation (Bearzi *et al.*, 2006), oil and gas exploration (Harwood & Wilson, 2001), shipping (Clark *et al.*, 2009), aquaculture (Watson-Capps

Whale-watching: Sustainable Tourism and Ecological Management, eds J. Higham, L. Bejder and R. Williams. Published by Cambridge University Press. © Cambridge University Press 2014.

& Mann, 2005) and climate change (Alter *et al.*, 2010).

Ironically, cetacean-watch operations, which are often promoted as beneficial, can cause significant impacts on cetaceans if not managed appropriately. Specifically, dolphin-watching can cause biologically significant impacts on exposed communities by causing habitat displacement and reducing the reproductive success of individuals (Lusseau, 2005; Bejder *et al.*, 2006b). As such, the International Whaling Commission (IWC) noted that 'there is compelling evidence that the fitness of individual odontocetes [toothed whales] repeatedly exposed to whale watching vessel traffic can be compromised and that this can lead to population-level effects'. The Whale Watching subcommittee of the IWC has noted that cetacean populations targeted by tourism operations can be divided into four categories: (1) resident populations where breeding, nursing, and feeding occur in the same area; (2) cetaceans on their breeding grounds; (3) cetaceans on their feeding grounds; and (4) cetaceans on their migratory corridors (International Whaling Commission, 2006). Each category is likely to require different levels and types of protection; for example, potentially, it is more important to protect cetaceans on their breeding grounds than on their migratory corridor. In addition, cetacean-watch tourism operates in varying social, cultural, economic and political environments (Higham *et al.*, 2008). Therefore, management frameworks for cetacean-watch operations should be designed based on the overall context in which the activity takes place. This raises the question of which management approach is the most appropriate to protect populations against impact(s) from tourism operations.

In this chapter we discuss area–time closures as a management approach to mitigate the impacts of commercial cetacean-watch tourism. We begin by evaluating the benefits and potential impacts of the cetacean-watch industry; discuss the variety of legislation currently available and then evaluate its effectiveness. We then discuss the development of area–time closures as part of management frameworks to help mitigate threats to cetacean populations. Finally, we identify the important issues for consideration when implementing area–time closures.

Benefits of the cetacean-watch industry

Cetacean-watch tourism is a rapidly growing industry with the potential to contribute to economic growth, education, conservation and the collection of scientific data. In 2009, cetacean watch tourism was a US $2 billion global industry with approximately 13 million tourists paying to observe cetaceans in their natural environment employing 13,000 people (O'Connor *et al.*, 2009; Cisneros-Montemayor *et al.*, 2010). Some coastal communities are highly dependent on the income generated by local cetacean-watch operations. For example, in 1986 the local community of Kaikoura, New Zealand, resurrected their ailing economy by developing commercial whale-watch operations (Hoyt, 2007; see Chapter 22). Before whale-watching, approximately 3400 tourists visited Kaikoura annually. Seven years after the commencement of whale-watching, the number of annual visitors increased to 80,000. In 1998, the number of tourists had increased 10-fold to an estimated 873,000 annually (Hoyt, 2007).

Cetacean-watch operations provide a platform to educate and raise awareness of the biology and environment of cetacean populations, population threats and population conservation, and many include an educational and interpretive component (Lück, 2003). Properly developed education programmes can be effective in managing tourist interactions with free-ranging animals in their natural environment (Orams, 1997). Furthermore, cetacean-watch experiences can lead to behavioural changes in tourists by encouraging a more environmentally aware behaviour (Orams, 1997 Ballantyne *et al.*, 2010).

Commercial cetacean-watch operations also offer a platform of opportunity for scientific research by providing frequent and relatively inexpensive access to study animals (Bejder & Samuels, 2003). Research from these vessels may, however, restrict the sampling methods and the type of abundance, distribution and behavioural data that can be collected (Bejder & Samuels, 2003). However, commercial tour vessels were used to study commercial swim-with-dolphin operations in the Bay of Islands, New Zealand (Constantine, 2001), and in controlled-approach experiments to record killer whale behaviours when approached by cetacean-watch vessels in Johnstone Strait, British Columbia (Williams & Ashe, 2007). The presence of researchers on a cetacean-watch vessel can also provide tourists with up-to-date knowledge information on the population of interest.

Costs of the cetacean-watch industry

Cetacean-watch tourism repeatedly seeks out prolonged close encounters with specific communities of free-ranging cetaceans. The cumulative impacts on cetacean populations from repeated encounters have the potential to cause significant biological effects on these populations. However, it is challenging to ascertain whether observed changes in population parameters (e.g. abundance, fecundity, survival rates) are attributable to a tourism operation or whether they are due to natural variation.

The National Research Council (2005) developed the Population Consequences of Acoustic Disturbance (PCAD) model to help identify possible effects of human activity on cetaceans. The PCAD model incorporates five groups of variables: sound, behaviour change, life functions, vital rates and population effect. The conceptual model identifies sound characteristics that may cause a disturbance with a resulting behavioural change (e.g. the sound may cause a change in dive behaviour or movement). It seeks to link behavioural changes to potential alterations in life functions (e.g. feeding and breeding), which can cause changes to vital rates (e.g. survival and reproduction) which, in turn, may have population-level effects (e.g. population growth rate). Although this model was developed specifically for acoustic disturbance, the framework can be applied to evaluating any human-induced impact on cetaceans. For example, repeated disruption to resting dolphins by a tour-vessel could result in a behavioural change from resting to travel behaviour (Lusseau, 2003a), resulting in an altered behavioural budget (i.e. affecting a life function). In turn, this could reduce the amount of available energy for reproduction (i.e. a vital rate), which has the potential to reduce population growth rates (i.e. population effects) (Bejder et al., 2006b). As such, repeated behavioural disruptions, mediated through cetacean-watch vessel disturbance, may cause long-term biologically significant effects on populations.

Behavioural responses of cetaceans to vessels vary greatly, ranging from attraction to avoidance. For example, northern resident killer whales (Orcinus orca) on Canada's Pacific coast alter their swimming path from a convoluted pattern to a more direct path with an increase in approaching whale-watching vessels (Williams & Ashe, 2007). In the Bay of Islands and Milford Sound, New Zealand, resting behaviour of bottlenose dolphins decreased as the number of boats increased (Constantine et al., 2004; Lusseau et al., 2006). In Shark Bay, Western Australia, long-term exposure to dolphin-watch vessels caused declines in relative abundance of bottlenose dolphins in an area where boat-based tourism occurred (Bejder, 2005; Bejder et al., 2006b).

The noise from cetacean-watching vessels has the potential to impact populations as anthropogenic noise affects the quality of habitat (Tyack, 2008). Noise pollution has the potential to impact cetaceans as their auditory capabilities are a primary means of communication, foraging and sensing their marine environment. Anthropogenic noise can interfere with cetacean acoustic systems and impair their communication, diminishing their ability to detect natural sounds including sounds generated by conspecifics (Nowacek et al., 2007; Tyack, 2008). Acoustic interference, referred to as acoustic masking (Clark et al., 2009; Jensen et al., 2009), may render cetaceans vulnerable to predation, affect their navigation and communication, and have long-term biologically significant effects. Noise from small vessels masks acoustic communications in bottlenose dolphins (*Tursiops* sp.) and short-finned pilot whales (*Globicephala macroryhnchus*) (Jensen *et al.*, 2009). Furthermore, avoiding sonar frequencies disrupts the swim path, navigation and detection of shallow waters (Zimmer & Tyack, 2007). Currently, we lack long-term data to evaluate the effects of acoustic pollution on cetacean populations (NRC, 2005), which is a significant gap in our knowledge.

Some tour operators offer swim-with-cetacean activities (Samuels & Spradlin, 1995; Bejder *et al.*, 1999; Constantine, 2001; Courbis, 2007; Kessler & Harcourt, 2010). Methods used to place swim-with customers in the path of wild cetacean groups alters their long-term behaviour (Constantine, 2001); e.g. the magnitude of avoidance response of dolphins to swimmers in the Bay of Islands, New Zealand, increased over time and the tour operator's success with swim-with attempts decreased over a three-year period (Constantine, 2001). It has not been possible to evaluate whether these behavioural changes have had long-term biologically significant impacts on the population.

Strategies for managing the cetacean-watching industry

Due to the rapid growth of the industry and lack of scientific information, management agencies face significant challenges in developing appropriate management strategies to mitigate possible impacts (Higham *et al.*, 2008). Carlson (2009) reviewed the cetacean-watching regulations of 47 jurisdictions worldwide and documented a wide variety of management frameworks for mitigating effects of cetacean-watch operations (Table 17.1). These include: unmanaged and unregulated cetacean-watch operations (Beasley *et al.*, 2010; Mustika *et al.*, 2012a), codes of conduct (Allen *et al.*, 2007), guidelines (Christiansen *et al.*, 2010; Schaffar *et al.*,

Table 17.1 The use of permits/licensing, general legislation and guidelines for cetacean-watch tourism in 47 jurisdictions (adapted from Carlson, 2009).

Jurisdiction	Permit/ licensed legislation	General regulations for the protection of cetaceans	Guidelines
ACCOBAMS*			×
Antarctica			×
Argentina	×		×
Australia	×		×
Azores		×	×
Bahamas			×
Brazil		×	×
British Virgin Islands			×
Canada	×		×
Canary Islands	×		×
Chile	×		×
Colombia			×
Dominica	×		×
Dominican Republic			×
Ecuador	×		×
France			×
Galapagos			×
Guadeloupe			×
Hong Kong			×
Iceland			×
Indonesia			×
Ireland			×
Japan			×
Madagascar			×
Mauritius			×
Mexico	×		×
Mozambique			×
New Caledonia		×	×
Newfoundland and Labrador			×
New Zealand	×		×
Niue	×		×
Norway			×
Oman			×
Pacific Islands Region	×		×
Philippines			×

(*cont.*)

Table 17.1 (*cont.*)

Jurisdiction	Permit/ licensed legislation	General regulations for the protection of cetaceans	Guidelines
Puerto Rico	×		×
South Africa	×		×
St Lucia		×	×
Tanzania			×
Tonga			×
Turks and Caicos			×
United Kingdom		×	×
United States		×	×
Uruguay			×

* Agreement on the Conservation of Cetaceans of the Black Sea, Mediterranean Sea and contiguous Atlantic area.

2010), general legislation (Wiener *et al.*, 2009) and permitting strategies (Bejder *et al.*, 2006b; Lusseau *et al.*, 2006; Notarbartolo di Sciara *et al.*, 2008).

Guidelines/codes of conduct

Few countries implement legislation to protect cetaceans from the effects of human disturbance and fewer countries have legislation that addresses commercial cetacean-watch tourism specifically (Table 17.1). In an attempt to offer some protection, numerous self-imposed voluntary codes of conduct and guidelines have been developed for commercial cetacean-watch operations to mitigate potential impacts (Garrod & Fennell, 2004). However, these agreements lack legislative power and legally binding rules. Often, adherence to codes of conduct is based on ethical obligation and peer pressure, which are often ineffective in reducing impacts on cetaceans (Garrod & Fennell, 2004). In Hawai'i, commercial cetacean-watch operators have been observed flouting voluntary guidelines, by steering bow-riding spinner dolphins (*Stenella longirostris*) directly to clients in the water (Wiener *et al.*, 2009). In Zanzibar, Tanzania, guidelines for

dolphin-watching have been violated with increasing frequency as the numbers of cetacean-watch vessels increased, causing detrimental effects on a local population of bottlenose dolphins (Christiansen *et al.*, 2010).

In Port Stephens, New South Wales, Australia, a variety of legislative measures were adopted after voluntary codes of conduct failed to adequately reduce tourism impacts on the local dolphin population (see Allen *et al.*, 2007). Subsequently, the New South Wales government introduced an amendment to National Parks and Wildlife Regulations to include marine mammals, adopting all aspects of the national guidelines as part of the regulations. A marine protected area (MPA) was declared within Port Stephens. This MPA includes different zoning areas, and commercial operations wishing to undertake dolphin-watching tours in the MPA must obtain a licence from the management agency. These amendments provided a mechanism that allows most stipulations within the formerly voluntary code of conduct to be enforced (Allen *et al.*, 2007). Moreover, Allen *et al.* (2007) suggested that dolphins in the MPA could be further protected by the implementation of spatial and temporal dolphin-watching zones. Speed restrictions were introduced in the MPA, as a mitigation measure to minimize boat impacts. However, speed restriction zones were ineffective at minimizing impacts on the local dolphins, and a revision on zone location was recommended (Steckenreuter *et al.*, 2012). Continuously monitoring the performance of a mitigation strategy that attempts to minimize impacts is important. Management agencies must be able to react quickly and adapt existing management measures as information on the effectiveness of management becomes available, an option that would be unavailable without a legislative framework.

General legislative framework

Legislation is often developed to allow the organizers of activities that have the potential to be

NMFS Permit GA LOC 15409

Figure 17.1 Tour boats and swimmers interact with Hawaiian spinner dolphins in their resting bays off the Kona Coast of the Hawai'i Island. Image taken under permit number GA LOC 15409.

detrimental to cetaceans to be prosecuted. However, general legislative frameworks are not specific to the cetacean-watch industry. For example, the US Marine Mammal Protection Act (MMPA) was designed to minimize harassment and disturbance to marine mammals, primarily for takes from commercial fishing operations as by-catch and from cetacean hunting. However, the interpretation of 'harassment' in the MMPA is a grey area and it is not clear how activities of the cetacean-watch operators fall within this Act. Under the MMPA, harassment is defined as 'any act of pursuit, torment, or annoyance which (i) has the potential to injure a marine mammal or marine mammal stock in the wild; or (ii) has the potential to disturb a marine mammal or marine mammal stock in the wild by causing disruption of behavioral patterns, including, but not limited to, migration, breathing,

nursing, breeding, feeding, or sheltering.' This leads to confusion and difficulty in determining when cetacean-watch activities are deemed to be harassing a cetacean(s). Under these circumstances, voluntary codes of conduct and voluntary guidelines are often agreed upon and implemented to help mitigate impacts as well as the general legislative framework. For example, in addition to the general protection legislation and codes and guidelines, specific legislation is being considered to further limit impacts on the local spinner dolphin population on Hawai'i Island (NOAA, 2005).

Due to growing concerns about the potential impact of cetacean-watch operations on the Hawaiian spinner dolphin (Figure 17.1), the Fisheries Service Pacific Islands Regional Office of NOAA is developing a legislative framework to reduce the exposure of resting spinner dolphins to human activity in Hawaiian waters (see Chapter 19). Hawaiian spinner dolphins display a highly predictable diurnal behaviour. At night they venture off shore to feed on shrimp, squid and fish that migrate towards the surface from the mesopelagic zone (Beniot-Bird & Au, 2003). During the day they move into coastal areas to socialize and rest (Norris & Dohl, 1980; Norris *et al.*, 1994). This predictable behaviour and their daytime reliance on sheltered bays that are easily accessible by people (see Figure 17.1) render them more exposed and more susceptible to human disturbance compared with other dolphin species. Recent studies suggest that the resting periods for Hawaiian spinner dolphins may be interrupted or truncated by exposure to human activity, but the biological significance of these impacts requires further investigation (Danil *et al.*, 2005; Delfour, 2007; Courbis & Timmel, 2009). Furthermore, the population of spinner dolphins along the Kona Coast of Hawai'i Island, which is the target for large-scale cetacean-watch operations, is genetically distinct from all other spinner dolphin populations in the Hawaiian Archipelago (Andrews *et al.*, 2010). Consequently, this population may be one of the most vulnerable to anthropogenic disturbance (Figure 17.1).

Permitted/licensed legislation framework

Some management frameworks exist where a legislative system requires a permit/license to engage in cetacean-watch activities (Table 17.1). This provides management agencies with the opportunity to regulate the level of tourism exposure of a cetacean population by, for example, having the option to revoke a licence should the cetacean-watch industry be shown to have a detrimental effect. For example, an unprecedented decision was made to reduce the number of dolphin-watch operators from two to one within a 'tourism zone' in Shark Bay, Western Australia, in response to research findings showing a decline in relative abundance of bottlenose dolphins (Bejder, 2005; Bejder *et al.*, 2006b; Higham & Bejder, 2008).

Management agencies rarely have sufficient scientific basis to determine the number of permits to allocate at the onset of a local industry. Thus, a conservative allocation of initial licences is important because of the difficulties in revoking licences once they have been issued (Higham *et al.*, 2008). The appropriate number of licences to be allocated is site-specific as each cetacean population is exposed to tourism under differing circumstances. For example, some areas are more important to cetaceans than others (critical habitats). Also, the susceptibility of individual animals to impacts varies with age (Stalmaster & Newman, 1978; Constantine, 2001; Müllner *et al.*, 2004), sex (Williams *et al.*, 2002; Lusseau, 2003b), previous experience (Bejder *et al.*, 2006a, 2009) and reproductive condition (Culik & Wilson, 1995; Nellemann *et al.*, 2000; Parent & Weatherhead, 2000; Beale & Monaghan, 2004). Some argue that cetaceans on breeding grounds are potentially more susceptible to the effects of cetacean-watch operations than on their migration corridor. Thus, it is important to gain appropriate insight into the specific characteristics of a targeted population in order to appropriately determine the number of tourism permits that is both biologically and economically sustainable.

Legislation, however, does not guarantee operator compliance, particularly when the laws are not well-known or not enforced (Keane *et al.*, 2011). For example, in Port Phillip Bay, Victoria, Australia, cetacean-watch tours that violate permit conditions, including: approach type, swim time, time in proximity of dolphins and interaction with new-born calves, were documented frequently (Scarpaci *et al.*, 2003). Improvements in operator compliance with regulations requires operator education, tourist education, enforcement of the regulations or a combination of these measures (Scarpaci *et al.*, 2003). Thus, legislation needs to be explicit for the protection of cetacean populations from cetacean-watch activities and it needs to be enforced and supported by programmes to ensure public awareness of the programme, to ensure compliance.

Important considerations for implementing time-area management strategies

Effective, long-term management strategies to monitor cetacean-watch operations need to establish thresholds of human–cetacean interactions and respond adaptively to operation impacts and natural phenomena (Higham *et al.*, 2008). Spatial management, including the use of protected areas or closures to commercial operations or no-take/no-watch areas, at the appropriate scale, is an effective approach in protecting both terrestrial and marine ecosystems (Pauly *et al.*, 2002; Hoyt, 2011). Protected areas have been implemented as precautionary measures when managing marine ecosystems to reduce the risks of over-exploitation, especially when scientific knowledge about the ecosystem is lacking (Hoyt, 2011). As a precaution where scientific knowledge is limited, the spatial range of a protected area might be increased to account for uncertainty in the available information. Spatial management has been used to conserve biodiversity, protect fish and cetaceans, and delineate areas for specific use to mitigate anthropogenic threats, enhance productivity and provide public focus for marine conservation (Lauck *et al.*, 1998; Hooker & Gerber, 2004; Hoyt, 2011). Recently, spatial management to reduce gillnet mortalities has improved the survival probability of Hector's dolphins (*Cephalorhynchus hectori*), an endangered cetacean species endemic to New Zealand (Gormley *et al.*, 2012). Spatial management is also a major part of fisheries management to protect spawning aggregations, immature individuals and critical habitats. In addition to spatial closures, limiting access to cetacean-watching in time, i.e. temporal closures, can be introduced to prohibit access to cetaceans during specific times that are critical to animals/populations (Constantine *et al.*, 2004; Notarbartolo di Sciara *et al.*, 2008).

The IUCN (1994) definition of a protected area is '[a]n area of land and/or sea especially dedicated to the protection and maintenance of biological diversity, and of natural and associated cultural resources, and managed through legal or other effective means'. Six categories of protected area, based on the main management purpose and the primary objective of the protected area have been defined (Table 17.2). Several of these categories are significant for the conservation of cetaceans, particularly areas that protect the habitat as well as the target species, such as: Nature Conservation Reserves, which are established to maintain, conserve and restore species and habitats; and Resource Reserves that are designed to protect natural ecosystems and use natural resources sustainably (Table 17.2; Dudley, 2008).

Protected areas for cetaceans are growing in number worldwide (Hoyt, 2011; Notarbartolo di Sciara *et al.*, 2008; Williams *et al.*, 2009; Table 17.3). Currently, the greatest number of protected areas are found in Australia and New Zealand (75), the Wider Caribbean (65) and the South Atlantic (56), while the number of reserves is likely to almost double in the Mediterranean and Black Seas should proposed protected areas be approved (Table 17.3). International boundary agreements between countries have been established to protect cetaceans. For example, a 'sister-sanctuary' relationship has been established to protect the North Atlantic humpback whale, *Megaptera novaeangliae*, and is situated between the US Stellwagen Bank National Marine Sanctuary (SBNMS), located between Cape Ann and Cape Cod in the southwest of the Gulf of Maine in the north, and Santaurio de Mamiferos Marinos de la República Dominicana (SMMRD), 3000 miles to the south (Table 17.3). The SBNMS (2181 km^2) protects the feeding and nursery areas of this population, while the SMMRD (2500 km^2) protects its mating and calving areas (Ward & MacDonald, 2009).

Area–time management frameworks have been developed to intervene when unregulated and unmanaged cetacean-watch tourism has been identified as a potential threat to cetacean populations. For example, Samadai Reef, on the coast of the Red Sea, Egypt, is an important area for spinner dolphins (Notarbartolo di Sciara *et al.*, 2008). In 2000, unregulated swim-with-dolphin tours began and

Table 17.2 Protected area categories, descriptions and primary objectives as determined by the International Union for the Conservation of Nature (IUCN) (Dudley, 2008).

Category	Managed for	Description	Primary objective
Ia. Scientific reserve	Strict nature reserve	Strictly protected areas set aside to protect biodiversity and also possibly geological/ geomorphological features, where human visitation, use and impacts are strictly controlled and limited to ensure protection of the conservation values. Such protected areas can serve as indispensable reference areas for scientific research and monitoring	To conserve regionally, nationally or globally outstanding ecosystems, species (occurrences or aggregations) and/or geodiversity features: these attributes will have been formed mostly or entirely by non-human forces and will be degraded or destroyed when subjected to all but very light human impact
Ib. Scientific reserve	Wilderness area	Usually large unmodified or slightly modified areas, retaining their natural character and influence, without permanent or significant human habitation, which are protected and managed so as to preserve their natural condition	To protect the long-term ecological integrity of natural areas that are undisturbed by significant human activity, free of modern infrastructure and where natural forces and processes predominate, so that current and future generations have the opportunity to experience such areas
II. National park	Ecosystem conservation and protection	Are large natural or near-natural areas set aside to protect large-scale ecological processes, along with the complement of species and ecosystems characteristic of the area, which also provide a foundation for environmentally and culturally compatible spiritual, scientific, educational, recreational and visitor opportunities	To protect natural biodiversity along with its underlying ecological structure and supporting environmental processes, and to promote education and recreation
III. National monument/ National landmark	Conservation of natural features	Are set aside to protect a specific natural monument, which can be a landform, sea mount, submarine cavern, geological feature such as a cave or even a living feature such as an ancient grove. They are generally quite small protected areas and often have high visitor value	To protect specific outstanding natural features and their associated biodiversity and habitats
IV. Nature conservation reserve	Conservation through active management	Aim to protect particular species or habitats and management reflects this priority. Many category IV protected areas will need regular, active interventions to address the requirements of particular species or to maintain habitats, but this is not a requirement of the category	To maintain, conserve and restore species and habitats

Table 17.2 (*cont.*)

Category	Managed for	Description	Primary objective
V. Protected landscape	Landscape/ seascape conservation and recreation	Where the interaction of people and nature over time has produced an area of distinct character with significant ecological, biological, cultural and scenic value: and where safeguarding the integrity of this interaction is vital to protecting and sustaining the area and its associated nature conservation and other values	To protect and sustain important landscapes/seascapes and the associated nature conservation and other values created by interactions with humans through traditional management practices
VI. Resource reserve	Sustainable use of natural resources	Conserve ecosystems and habitats, together with associated cultural values and traditional natural resource management systems. They are generally large, with most of the area in a natural condition, where a proportion is under sustainable natural resource management and where low-level non-industrial use of natural resources compatible with nature conservation is seen as one of the main aims of the area	To protect natural ecosystems and use natural resources sustainably, when conservation and sustainable use can be mutually beneficial.

the number of tourists increased dramatically. More than 800 swimmers were reported to be interacting with spinner dolphins in the small 1.5 km^2 lagoon in a single day (Notarbartolo di Sciara *et al.*, 2008). Spinner dolphin sightings decreased and concerns were raised about the effects on the local dolphin population (Notarbartolo di Sciara *et al.*, 2008). Management authorities suspended all tourist visits until a suitable management plan was developed. In 2004, an area–time management regime was introduced that included zoning the area into four different use areas: no tourist zone; diving and snorkelling zone; boat mooring zone; and dive sites zone. Interactions with people were confined to four hours each day and the number of visitors was restricted to 100 divers and 100 snorkellers per day. Visitor entrance fees were introduced, visits were allowed only under the supervision of trained and certified guides, and monitoring and enforcement programmes were introduced (Notarbartolo di Sciara *et al.*, 2008).

A number of development steps are needed to successfully implement an area–time management plan for cetacean-watching, including consultation with commercial tour operators, social science communities and natural scientists (Higham *et al.*, 2008). Management agencies are responsible for establishing and coordinating the development of the legislative framework, which should proceed prior to commencing commercial operations (Higham *et al.*, 2008). The size, location and access restrictions to these areas where the cetacean population is most vulnerable to disturbance should be identified as part of the management framework. Regulations for the allocation/revocation of permits and legislation to

Table 17.3 The number of existing and proposed protected areas with cetacean habitat across 18 marine regions. Adapted from Hoyt (2011).

Marine region	MPA or PA (marine protected area or protected area for river dolphins on land)			High seas MPA (marine protected area outside national waters of EEZ)		National EEZ Sanctuary (no hunting zone within national waters or EEZ)		Total
	Existing	Proposed	Existing with proposed expansion	Existing	Proposed	Existing	Proposed	
1. Antarctica	4	0	0	3	1	0	0	8
2. Arctic	29	8	4	0	4	0	0	45
3. Mediterranean and Black Seas	45	38	11	1	17	0	0	112
4. North West Atlantic	9	2	1	0	1	1	0	14
5. North East Atlantic	14	27	8	0	1	2	0	52
6. Baltic	6	6	0	0	0	0	0	12
7. Wider Caribbean	65	4	2	0	0	4	1	76
8. West Africa	40	4	1	0	0	0	1	46
9. South Atlantic	56	7	3	0	1	3	0	70
10. Central Indian Ocean	17	7	8	0	0	1	0	33
11. Arabian Seas	20	4	1	0	0	0	0	25
12. East Africa	22	2	2	0	1	1	0	28
13. East Asian Seas	23	8	1	1	0	0	1	34
14. North and South Pacific	21	3	4	0	0	12	0	40
15. North East Pacific	20	3	5	0	0	1	0	29
16. North West Pacific	21	3	0	0	0	0	0	24
17. South East Pacific	30	5	2	0	1	4	0	42
18. Australia – New Zealand	75	7	0	0	0	2	0	84
Total	517	138	53	5	27	31	3	774

control tourism operations, including restrictions on engine noise, speed of approach, distance, time with dolphins, and a visitor interpretation programme. The likely response of tour operators to restrictions of access to cetaceans should also be considered, as this may lead to increased interactions in other areas, which may have unforeseen detrimental effects on the population. As an example, if an area–time management framework was introduced in resting bays used by Hawaiian spinner dolphins, human–dolphin interactions might increase outside the resting bays.

Understanding the cetacean population of interest

Critical habitats are areas where a species executes behaviours essential to the viability of the population, and include foraging, breeding, nursing, socializing or resting habitats (Hoyt, 2011). Repeated disturbance of cetaceans within critical habitats has been implicated as a factor in reducing the viability of the population (Bejder, 2005; Bejder *et al.*, 2006b; Lusseau *et al.*, 2006). Understanding the abiotic and biotic environment and behaviour and biology of the focal population, particularly identifying critical areas, and the time of their use, prior to developing a management plan is important (Lusseau, 2003a; Lusseau et al., 2009). Sampling programmes should be developed that incorporate methodologies to collect comprehensive data on the population, and its habitat, that can be used to estimate population size, population structure, reproductive rates and behavioural budgets (Lusseau, 2004). These data provide the basis for developing models to identify critical areas and habitat use and assessing the potential threats to the population.

The NOAA initiated research programme on the spinner dolphins and their interactions with cetacean-watching activities along the west coast of Hawai'i Island, which provides an example of an integrated programme (NOAA, 2005). A suite of modern visual and acoustic techniques are being used and a systematic photographic identification sampling regime has been developed to study spinner dolphin populations in the study area. Group focal follows are undertaken to observe behaviours and human interactions outside resting bays. Land based theodolite tracking of spinner dolphins, their behaviours and human interactions within resting bays are also undertaken (Tyne et al., 2011). Cetaceans can display sequences of behaviour, or transitions between behaviours, such as foraging behaviour followed by socializing behaviour, which may be followed by resting behaviour. When cetaceans are disturbed, however, the probability of transitioning from one behavioural state to another is altered (Lusseau, 2003a), resulting in a change in their overall behavioural budget. Calculating the probability of transition from one behavioural state to another in the absence of cetacean-watch tourism, may provide an indicator for assessing the impact of tourism operations: changes in the transition probability may be an early warning of a deleterious effect of cetacean-watching tourism (Lusseau, 2004). Furthermore, 'show stoppers', or limit reference points, such as a decline in reproductive success or declines in abundance, should be identified as immediate evidence of a significant

impact resulting in an immediate management intervention, such as a 50% reduction of tourism on bottlenose dolphins in Shark Bay (Bejder et al., 2006b).

Baseline data on cetacean populations, prior to the start of tourism operations, should be used to develop monitoring programmes and management plans (Higham et al., 2008) and establish population parameters prior to or during the onset and growth of tourism activities (Bejder & Samuels, 2003). Reference points can be developed from the baseline data to provide target reference points and limit reference points, at which management actions are initiated, i.e. when predetermined acceptable thresholds have been exceeded. Target and limit reference points are commonly used in fisheries to monitor and manage the health of fish stocks and their ecosystems, e.g. spawning biomass at Maximum Sustainable Yield, percent of Virgin Biomass, and Spawning Potential Ratio. Similarly, the Potential Biological Removal (PBR; Wade, 1998) assesses the allowable limits of mortality on cetacean populations from anthropogenic disturbance (Williams et al., 2009). The PBR is calculated as the product of a minimum population estimate (N_{min}), one half of the maximum theoretical net productivity rate (R_{max}) and recovery factor (Fr) (PBR = N_{min} × $0.5R_{max}$ × Fr; Wade, 1998).

Area–time management systems require an understanding of why and when specific habitats/areas are critical to a population. Critical habitats encompass areas of high animal density, and areas essential to the viability of the population, e.g. a nursing area where only mothers and calves are present. Quantifying the importance of an area to a cetacean population by assessing habitat preference, and the behaviours in these habitats, is an important tool in the development of area–time management systems (Higham & Lusseau, 2007). A high proportion of the population of northern resident killer whales (Orcinus orca) in Johnstone Strait, Canada, uses a small proportion of their habitat for a rare behaviour called beach-rubbing, where individuals rub their bodies on the smooth pebble beaches, a behaviour thought to remove parasites or have some social significance (Williams *et al.*, 2009). This renders the population vulnerable due to the high proportion of the population that uses this small area, coupled with the heavy human use of Johnstone Strait by large ships (Williams *et al.*, 2009).

Bottlenose dolphins in Fiordland, New Zealand, are particularly sensitive to boat interactions while resting, and to a lesser extent, while socializing (Lusseau, 2004). Using behavioural state observations, Lusseau and Higham (2004) identified critical areas for dolphin resting and socializing in Doubtful Sound. Subsequently, a voluntary code of conduct for tour operators was developed which included some elements of zoning (unregulated). Similarly, the commercial dolphin-swim/watch industry in the Bay of Islands, New Zealand, altered the behaviour of bottlenose dolphins (Constantine *et al.*, 2004). Resting behaviour decreased significantly as the number of commercial vessels increased. Constantine *et al.* (2004) suggested that the local legislation was ineffective in protecting the dolphin population and recommended measures to minimize tour-boat impacts by restricting the number of boat trips, trip durations and limiting dolphin exposure to tour boats. In late 2004, The Department of Conservation (DOC) implemented changes to the dolphin-watch and swim-with-dolphin operations, some based on the findings of Constantine *et al.* (2004), others of their own making (Constantine, 2010, University of Auckland, pers. comm.).

Identifying critical habitat should also consider both the abiotic and biotic characteristics of the habitat, and link these characteristics to the focal cetacean population's behaviour within the area. These characteristics may include prey abundance, characteristics of the bathymetry, and substratum, temperature, salinity, turbidity, tide and currents. For example, Hawaiian spinner dolphins prefer to rest in sheltered sandy bays during the day (Norris & Dohl, 1980; Thorne *et al.*, 2012) and, as a consequence, any changes to this habitat may have significant biological consequences for the population.

Oceanographic features have also been used to identify critical cetacean habitats. Johnston and Read (2007) highlighted an ecological link between a predictable oceanographic feature in time and space that attracts cetaceans to the Bay of Fundy, Canada. In the summer and during flood tides, the Grand Manan Island wake attracts foraging fin whales (*Balaenoptera physalus*), minke whales (*Balaenoptera acutorostrata*) (Johnston *et al.*, 2005a) and harbour porpoise (*Phocoena phocoena*) (Johnston *et al.*, 2005b). Oceanographic observations provided an understanding of the spatial and temporal variability in the physical forces controlling the island wake (Johnston & Read, 2007). Secondary flows in the wake aggregate prey to predictable locations where the cetaceans focus their foraging efforts. These ecological links between the foraging habitats and foraging behaviour over space and time are therefore important factors when considering the spatial boundaries of a marine protected area in this region. Johnston and Read (2007) recommended that a proposed protected area at the Grand Manan Island encompass the island wake. The predictable nature of this oceanographic feature also allows human activities to be controlled within the area during the tidal flows that generate the island wake when mega fauna are foraging in the wake.

Socioeconomic considerations

Cetacean-watch operations are important for the economy of many coastal communities (Hoyt, 2007). Access restrictions to protect cetacean populations from tourism operations may have implications for the economic status of these communities. Therefore, it is important to highlight the benefits of sustainable cetacean-watch management and the significance of area–time management systems to local businesses and the wider community. The aim of these operations should be to maximize the economic viability of the local cetacean-watch industry, while sustainably managing the target cetacean population.

Raising awareness among visitors on cetacean biology and conservation is an important component in the development of any long-term management framework for cetacean-watching.

Many commercial cetacean-watch operations provide educational information on the target cetacean population, its environment, the associated conservation efforts, and to some degree the plight of cetaceans worldwide (Orams, 1997; Higham & Carr, 2002; Christensen *et al.*, 2007). In New Zealand, tourism operations must provide an educational component of the cetacean-watch experience, a condition for obtaining a cetacean-watch permit (Carlson, 2009). However, the effectiveness of educational programmes in raising environmental and conservational awareness and ultimately changing human behaviour is still being debated (Orams, 1996; Higham & Carr, 2002; Ballantyne *et al.*, 2010). Carefully designed educational nature-based programmes that incorporate strategies to facilitate behavioural change can instil greater environmental and conservational awareness in tourists and lead to changes in their behaviour and attitudes to interactions with marine fauna (Orams, 1997; Higham & Carr, 2002; Ballantyne *et al.*, 2010). The continued collection of visitor data and their perceptions of cetacean-watch operations enable the effectiveness of education programmes to be assessed and adapted if necessary.

Management considerations

Reliable and detailed scientific data on the critical habitat of cetaceans can provide management agencies with the information needed to establish protected areas. The spatial scale of the protected area should be large enough to be biologically relevant and small enough that cetacean-watch operations can be effectively managed within its boundaries (Ashe *et al.*, 2010; Ross *et al.*, 2011). Furthermore, the time when the critical behaviour occurs should inform management agencies when to restrict human access to the protected area. Population estimates, reproductive rates and changes in behavioural budget (Lusseau, 2004; Bejder *et al.*, 2006b) should provide management agencies with the information necessary to establish quantifiable Limits of Acceptable Change (LAC) in the target

cetacean population (Higham *et al.*, 2008). Spatial and temporal scales of habitats and cetacean presence in the region, and the LAC criteria should then be used to establish clearly defined legislation for operators and enforcement within the protected area boundaries.

The preceding discussion has demonstrated that rules and regulations are only part of the solution to minimizing impacts on cetacean populations from cetacean-watch operations. For a successful management plan, rules and regulations must also be supplemented with educational and enforcement programmes (Keane *et al.*, 2008) to help ensure compliance with regulations. Moreover, clearly defined legislation must have significant authority, including that to revoke operator licences (Bejder *et al.*, 2006b; Higham & Bejder, 2008). Without enforcement or legislation, management plans may fail to meet their goals and, ultimately, fail to protect cetacean populations and the long-term viability of the cetacean-watch tour operations.

Conclusions

Despite the rapid increase in cetacean-watch tourism since the 1970s and its recent expansion into developing countries, it continues to be perceived as a benign activity. However, in 2006, the International Whaling Commission noted that 'there is compelling evidence that the fitness of individual odontocetes [toothed whales] repeatedly exposed to whale-watching vessel traffic can be compromised and that this can lead to population-level effects' (International Whaling Commission, 2006).

In the absence of management regulations, tour operators have been observed overcrowding and encroaching extremely close to cetacean groups (Mustika *et al.*, 2012a, 2012b). Moreover, there are growing concerns for critically endangered cetacean species exposed to unmanaged cetacean-watching (Beasley *et al.*, 2010). In some areas voluntary codes of conduct have been employed; however, they are regularly flouted and often ineffective in providing cetaceans with adequate protection (Allen *et al.*,

2007; Wiener *et al.*, 2009). Where management legislation is in place, operator compliance is not guaranteed, particularly when the laws are not communicated or adequately enforced (Keane *et al.*, 2008). Spatial management has been shown to be effective in protecting cetaceans (Gormley *et al.*, 2012). In this chapter we argue that area–time management systems should be considered an important tool to manage cetacean-watch tourism and ensure its long-term viability.

The challenge for scientists and managers is to develop studies of appropriate temporal and spatial scales that quantify the population dynamics of tourism-exposed cetacean populations. When achievable, knowledge from such studies can help tease apart possible effects of anthropogenic disturbance of exposed populations from natural variability. By obtaining robust estimates on population size, identifying critical habitat and baseline population parameters, coupled with information on behavioural changes in response to cetacean-watching operations, scientists will be able to identify when and where cetacean populations are most vulnerable. Management agencies can use this information as a basis to develop appropriate legislative management frameworks to protect cetacean populations from disturbance when they are most vulnerable. The efficacy of a management strategy must be continuously monitored and should be able to adapt quickly when necessary. A management framework must also provide management agencies with a sufficient mandate to change operator conditions when rules have been violated repeatedly. As such, we argue that a licensing system should be considered more widely, so that management agencies have the authority to change and/or revoke a licence should it be deemed necessary.

REFERENCES

Allen, S., Smith, H., Waples, K. & Harcourt, R. (2007). The voluntary code of conduct for dolphin watching in Port Stephens, Australia: Is self-regulation an effective management tool? *Journal of Cetacean Resource Management* 9, 159–166.

Alter, E.S., Simmonds, M.P. & Brandon, J.R. (2010). Forecasting the consequences of climate-driven shifts in human behavior on cetaceans. *Marine Policy* 34, 943–954.

Andrews, K.R., Karczmarski, L., Au, W.W.L., *et al.* (2010). Rolling stones and stable homes: Social structure, habitat diversity and population genetics of the Hawaiian spinner dolphin (*Stenella longirostris*). *Molecular Ecology* 19, 732–748.

Ashe, E., Noran, D.P. & Williams, R. (2010). Animal behaviour and marine protected areas: Incorporating behavioural data into the selection of marine protected areas for an endangered killer whale population. *Animal Conservation* 13, 196–203.

Ballantyne, R., Packer, J. & Falk, J. (2010). Visitors' learning for environmental sustainability: Testing short- and long-term impacts of wildlife tourism experiences using structural equation modelling. *Tourism Management*, doi: 10.1016/j.tourman.2010.11.003.

Beale, C.M. & Monaghan, P. (2004). Behavioural responses to human disturbance: A matter of choice? *Animal Behaviour* 68, 1065–1069.

Bearzi, G., Pierantonio, N., Bonizzoni, S., Notarbartolo Di Sciara, G. & Demma, M. (2010). Perception of a cetacean mass stranding in Italy: The emergence of compassion. *Aquatic Conservation: Marine and Freshwater Ecosystems* 20, 644–654.

Bearzi, G., Politi, E., Agazzi, S. & Azzellino, A. (2006). Prey depletion caused by overfishing and the decline of marine megafauna in eastern Ionian Sea coastal waters (central Mediterranean). *Biological Conservation* 127, 373–382.

Beasley, I., Bejder, L. & Marsh, H. (2010). *Dolphin-watching Tourism in the Meekong River, Cambodia: A case study of economic interests influencing conservation*. International Whaling Commissions 62nd Annual Meeting in Agadir, Morroco, 2010.

Bejder, L. (2005). Linking short and long-term effects of nature-based tourism on cetaceans. PhD, Dalhousie University, Canada.

Bejder, L. & Samuels, A. (2003). Evaluating the effects of nature-based tourism on cetaceans. In N.M. Gales, M. Hindell & R. Kirkwood (Eds), *Marine Mammals and Humans: Towards a sustainable balance*. Collingwood, Australia: CSIRO Publishing.

Bejder, L., Dawson, S.M. & Harraway, J.A. (1999). Responses by hector's dolphins to boats and swimmers in Porpoise

Bay, New Zealand. *Marine Mammal Science* 15, 738–750.

Bejder, L., Samuels, A., Whitehead, H. & Gales, N. (2006a). Interpreting short-term behavioural responses to disturbance within a longitudinal perspective. *Animal Behaviour* 72, 1149–1158.

Bejder, L., Samuels, A., Whitehead, H., *et al.* (2006b). Decline in relative abundance of bottlenose dolphins exposed to long-term disturbance. *Conservation Biology* 20, 1791–1798.

Bejder, L., Samuels, A., Whitehead, H., Finn, H. & Allen, S. (2009). Impact assessment research: Use and misuse of habituation, sensitisation and tolerance in describing wildlife responses to anthropogenic stimuli. *Marine Ecology Progress Series* 395, 177–185.

Beniot-Bird, K.J. & Au, W.W.L. (2003). Prey dynamics affect foraging by a pelagic predator (*Stenella longirostris*) over a range of spatial and temporal scales. *Behavioral Ecology and Sociobiology* 53, 364–373.

Carlson, C. (2009). A review of whale watch guidelines and regulations around the world, version 2009. Retrieved from http://dolphincare.org/code_files/areviewof whalewatchguidelinesandregulationsaroundtheworld version2009pg96.pdf.202009%20pg%2096.pdf.

Christensen, A., Rowe, S. & Needham, M.D. (2007). Value orientations, awareness of consequences, and participation in a whale watching education program in Oregon. *Human Dimensions of Wildlife* 12, 275–279.

Christiansen, F., Lusseau, D., Stensland, E. & Berggren, P. (2010). Effects of tourist boats on the behaviour of Indo-Pacific bottlenose dolphins off the south coast of Zanzibar. *Endangered Species Research* 11, 91–99.

Cisneros-Montemayor, A.M., Sumaila, U.R., Kaschner, K. & Pauly, D. (2010). The global potential for whale watching. *Marine Policy* 34, 1273–1278.

Clark, C., Ellison, W., Southall, B., *et al.* (2009). Acoustic masking in marine ecosystems: Intuitions, analysis, and implication. *Marine Ecology Progress Series* 395, 201–222.

Constantine, R. (2001). Increased avoidance of swimmers by wild bottlenose dolphins (*Tursiops truncatus*) due to long-term exposure to swim-with-dolphin tourism. *Marine Mammal Science* 17, 689–702.

Constantine, R., Brunton, D.H. & Dennis, T. (2004). Dolphin-watching tour boats change bottlenose dolphin (*Tursiops truncatus*) behaviour. *Biological Conservation* 117, 299–307.

Courbis, S. (2007). Effect of spinner dolphin presence on level of swimmer and vessel activity in Hawai'ian bays. *Tourism in Marine Environments* 4, 1–14.

Courbis, S. & Timmel, G. (2009). Effects of vessels and swimmers on behavior of Hawaiian spinner dolphins (*Stenella longirostris*) in Kealake'akua, Honaunau, and Kauhako bays, Hawai'i. *Marine Mammal Science* 25, 430–440.

Culik, B.M. & Wilson, R.P. (1995). Penguins disturbed by tourists. *Nature* 376, 301–302.

Danil, K., Maldini, D. & Marten, K. (2005). Patterns of use of Maku'a Beach, O'ahu, Hawai'i by spinner dolphins (*Stenella longirostris*) and potential effects of swimmers on their behavior. *Aquatic Mammals* 31, 403–412.

Delfour, F. (2007). Hawaiian spinner dolphins and the growing dolphin watching activity in Oahu. *Journal of the Marine Biological Association of the UK* 87, 109–112.

Dudley, N. (2008). *Guidelines for Applying Protected Area Management Categories.* Gland, Switzerland: IUCN.

Gales, N.J., Kasuya, T., Clapham, P.J. & Brownell, R.L. (2005). Japan's whaling plan under scrutiny. *Nature* 435, 883–884.

Garrod, B. & Fennell, D.A. (2004). An analysis of whale watching codes of conduct. *Annals of Tourism Research* 31, 334–352.

Gormley, A.M., Slooten, E., Dawson, S., *et al.* (2012). First evidence that marine protected areas can work for marine mammals. *Journal of Applied Ecology* 49, 474–480.

Harwood, J. & Wilson, B. (2001). The implications of developments on the Atlantic Frontier for marine mammals. *Continental Shelf Research* 21, 1073–1093.

Higham, J.E.S. & Bejder, L. (2008). Managing wildlife-based tourism: Edging slowly towards sustainability? *Current Issues in Tourism*, 11, 75–83.

Higham, J. & Carr, A. (2002). Ecotourism visitor experiences in Aotearoa/New Zealand: Challenging the environmental values of visitors in pursuit of pro-environmental behaviour. *Journal of Sustainable Tourism* 10, 277–294.

Higham, J. & Lusseau, D. (2007). Defining critical habitats: The spatio-ecological approach to managing tourism–wildlife interactions. In J.E.S. Higham (Ed.), *Critical Issues in Ecotourism.* Oxford: Butterworth-Heinemann.

Higham, J.E.S., Bejder, L. & Lusseau, D. (2008). An integrated and adaptive management model to address the long-term sustainability of tourist interactions with cetaceans. *Environmental Conservation* 35, 294–302.

Hooker, S.K. & Gerber, L.R. (2004). Marine reserves as a tool for ecosystem-based management: The potential importance of megafauna. *BioScience* 54, 27–39.

Hoyt, E. (2007). *A Blueprint for Dolphin and Whale Watching Development*. Washington, DC: Humane Society International.

Hoyt, E. (2011). *Marine Protected Areas for Whales, Dolphins, and Porpoises: A world handbook for cetacean habitat conservation and planning*. London: Earthscan.

IUCN. (1994). *Guidelines for Protected Area Management Categories*. CNPPA with assistance of WCMC, IUCN, Gland, Switzerland and Cambridge, UK. x + 261 pp.

IWC (2006). Report of the Scientific Committee. Cetacean Research and Management. Cambridge: IWC, pp. 1–160.

Jefferson, T.A., Hung, S.K. & Würsig, B. (2009). Protecting small cetaceans from coastal development: Impact assessment and mitigation experience in Hong Kong. *Marine Policy* 33, 305–311.

Jensen, F.H., Bejder, L., Wahlberg, M., Aguilar Soto, N., Johnson, M. & Madsen, P.T. (2009). Vessel noise effects on delphinid communication. *Marine Ecology Progress Series* 395, 161–175.

Johnston, D.W. & Read, A.J. (2007). Flow-field observations of a tidally driven island wake used by marine mammals in the Bay of Fundy, Canada. *Fisheries Oceanography* 16, 422–435.

Johnston, D.W., Thorne, L.H. & Read, A.J. (2005a). Fin whales *Balaenoptera physalus* and minke whales *Balaenoptera acutorostrata* exploit a tidally driven island wake ecosystem in the Bay of Fundy. *Marine Ecology Progress Series* 305, 287–295.

Johnston, D.W., Westgate, A.J. & Read, A.J. (2005b). Effects of fine-scale oceanographic features on the distribution and movements of harbour porpoises *Phocoena phocoena* in the Bay of Fundy. *Marine Ecology Progress Series* 295, 279–293.

Keane, A., Jones, J.P.G., Edwards-Jones, G. & Milner-Gulland, E.J. (2008). The sleeping policeman: Understanding issues of enforcement and compliance in conservation. *Animal Conservation* 11, 75–82.

Keane, A., Ramarolahy, A.A., Jones, J.P.G. & Milner-Gulland, E.J. (2011). Evidence for the effects of environmental engagement and education on knowledge of wildlife laws in Madagascar. *Conservation Letters* 4, 55–63.

Kessler, M. & Harcourt, R. (2010). Aligning tourist, industry and government expectations: A case study from the swim with whales industry in Tonga. *Marine Policy* 34, 1350–1356.

Lauck, T., Clark, C.W., Mangel, M. & Munro, G.R. (1998). Implementing the precautionary principle in fisheries management through marine reserves. *Ecological Applications* 8, 72–78.

Lück, M. (2003). Education on marine mammal tours as agent for conservation – But do tourists want to be educated? *Ocean & Coastal Management* 46, 943–956.

Lusseau, D. (2003a). The effects of tour boats on the behavior of bottlenose dolphins: Using Markov chains to model anthropogenic impacts. *Conservation Biology* 17, 1785–1793.

Lusseau, D. (2003b). Male and female bottlenose dolphins *Tursiops* spp. have different strategies to avoid interactions with tour boats in Doubtful Sound, New Zealand. *Marine Ecology Progress Series* 257, 267–274.

Lusseau, D. (2004). The hidden cost of tourism: Detecting long-term effects of tourism using behavioural information. *Ecology and Society* 9 (1), art. 2.

Lusseau, D. (2005). Residency patten of bottlenose dolphins *Tursiops* spp. in Milford Sound, New Zealand is related to boat traffic. *Marine Ecology Progress Series* 295, 265–272.

Lusseau, D. & Higham, J.E.S. (2004). Managing the impacts of dolphin-based tourism through the definition of critical habitats: The case of bottlenose dolphins (*Tursiops* spp.) in Doubtful Sound, New Zealand. *Tourism Management* 25, 657–667.

Lusseau, D., Slooten, L. & Currey, R.J.C. (2006). Unsustainable dolphin-watching tourism in Fiordland, New Zealand. *Tourism in Marine Environments* 3, 173–178.

Lusseau, D., Bain, E.D., Williams, R. & Smith, J.C. (2009). Vessel traffic disrupts the foraging behaviour of southern resident killer whales *Orcinus orca*. *Endangered Species Research* 6, 211–221.

Mangel, J.C., Alfaro-Shigueto, J., Van Waerebeek, K., *et al.* (2010). Small cetacean captures in Peruvian artisanal fisheries: High despite protective legislation. *Biological Conservation* 143, 136–143.

Müllner, A., Eduard Linsenmair, K. & Wikelski, M. (2004). Exposure to ecotourism reduces survival and affects stress response in hoatzin chicks (*Opisthocomus hoazin*). *Biological Conservation* 118, 549–558.

Mustika, P.L.K., Birtles, A., Everingham, Y. & Marsh, H. (2012a). The human dimensions of wildlife tourism in a developing country: Watching spinner dolphins at Lovina, Bali, Indonesia. *Journal of Sustainable Tourism*, doi: 10.1080/09669582.2012.692881.

Mustika, P.L.K., Birtles, A., Welters, R. & Marsh, H. (2012b). The economic influence of community-based dolphin watching on a local economy in a developing country: Implications for conservation. *Ecological Economics* 79, 11–20.

Myers, R.A. & Worm, B. (2003). Rapid worldwide depletion of predatory fish communities. *Nature* 423, 280–283.

Nellemann, C., Jordhøy, P., Støen, O.-G. & Strand, O. (2000). Cumulative impacts of tourist resorts on wild reindeer (*Rangifer tarandus tarandus*) during winter. *Arctic* 53, 9–17.

NOAA (2005). *Protecting Spinner Dolphins in the Main Hawaiian Islands From Human Activities that Cause 'Take,' as Defined in the Marine Mammal Protection Act and Its Implementing Regulations, or To Otherwise Adversely Affect the Dolphins.* In N.O.a.A. Administration (Ed.), 051110296–5296–01; I.D.102405A.

Norris, K.S. & Dohl, T.P. (1980). Behavior of the Hawaiian spinner dolphin, *Stenella longirostris*. *Fisheries Bulletin* 77, 821–849.

Norris, K.S., Wursig, B., Wells, S. & Wursig, M. (1994). *The Hawaiian Spinner Dolphin*. Berkeley, CA: University of California Press.

Notarbartolo Di Sciara, G., Hanafy, M.H., Fouda, M.M., Afifi, A. & Costa, M. (2008). Spinner dolphin (*Stenella longirostris*) resting habitat in Samadai Reef (Egypt, Red Sea) protected through tourism management. *Journal of the Marine Biological Association of the UK* 89, 211–216.

Nowacek, D.P., Thorne, L.H., Johnston, D.W. & Tyack, P.L. (2007). Responses of cetaceans to anthropogenic noise. *Mammal Review* 37, 81–115.

NRC (2005). *Marine Mammal Populations and Ocean Noise: Determining when noise causes biologically significant effects*. Washington, DC: National Academy Press.

O'Connor, S., Campbell, R., Cortez, H. & Knowles, T. (2009). *Whale Watching Worldwide: Tourism numbers, expenditures and expanding economic benefits, a special report from the International Fund for Animal Welfare*. Yarmouth, MA: IFAW.

Orams, M.B. (1996). Using interpretation to manage nature-based tourism. *Journal of Sustainable Tourism* 4, 81–94.

Orams, M.B. (1997). The effectivenes of environmental education: Can we turn tourists into 'Greenies'? *Progress in Tourism and Hospitality Research* 3, 295–306.

Panigada, S., Pesante, G., Zanardelli, M., Capoulade, F., Gannier, A. & Weinrich, M.T. (2006). Mediterranean fin whales at risk from fatal ship strikes. *Marine Pollution Bulletin* 52, 1287–1298.

Parent, C. & Weatherhead, P.J. (2000). Behavioral and life history responses of eastern massasauga rattlesnakes (*Sistrurus catenatus catenatus*) to human disturbance. *Oecologia* 125, 170–178.

Pauly, D., Christensen, V., Guenette, S., *et al.* (2002). Towards sustainability in world fisheries. *Nature* 418, 689–695.

Pollock, K.H., Nichols, J.D., Brownie, C. & Hines, J.E. (1990). Statistical inference for capture–recapture experiments. *Wildlife Monographs* 107, 1–97.

Reijnders, P.J.H., Aguilar, A. & Borrell, A. (2009). Pollution and marine mammals. In F.P. Perrin, B. Würsig & J.G.M. Thewissen (Eds), *Encyclopedia of Marine Mammals second edition*. San Diego, CA: Academic Press.

Ross, P.S., Barlow, J., Jefferson, T.A., *et al.* (2011). Ten guiding principles for the delineation of priority habitat for endangered small cetaceans. *Marine Policy* 35, 483–488.

Samuels, A. & Spradlin, T.R. (1995). Quantitative behavioural study of bottlenose dolphins in swim-with-dolphin programs in the United States. *Marine Mammal Science* 11, 520–544.

Scarpaci, C., Dayanthi, N. & Corkeron, P.J. (2003). Compliance with regulations by 'swim-with-dolphins' operations in Port Phillip Bay, Victoria, Australia. *Environmental Management* 31, 342–347.

Schaffar, A., Garrigue, C. & Constantine, R. (2010). Exposure of humpback whales to unregulated whalewatching activities in their main reproductive area in New Caledonia. *Journal of Cetacean Research and Management* 11, 147–152.

Stalmaster, M.V. & Newman, J.R. (1978). Behavioral responses of wintering bald eagles to human activity. *The Journal of Wildlife Management* 42, 506–513.

Steckenreuter, A., Harcourt, R. & Möller, L. (2012). Are Speed Restriction Zones an effective management tool for minimising impacts of boats on dolphins in an Australian marine park? *Marine Policy* 36, 258–264.

Thorne, L.H., Johnston, D.W., Urban, D.L., *et al.* (2012). Predictive modeling of spinner dolphin (*Stenella longirostris*) resting habitat in the main Hawaiian Islands. *PLoS ONE* 7(8): e43167. doi:10.1371/journal.pone.0043167.

Tyack, P.L. (2008). Implications for marine mammals of large-scale changes in the marine acoustic environment. *Journal of Mammalogy* 89, 549–558.

Tyne, J.A., Johnston, D.W., Pollock, K.H. & Bejder, L. (2011). Quantifying the efficacy of a spatio-temporal management intervention on human-dolphin interactions in Hawai'i. In 19th Biennial Conference on the Biology of Marine Mammals, 27 November–2 December 2011, Tampa, Florida.

Wade, P.R. (1998). Calculating limits to the allowable human-caused mortality of cetaceans and pinnipeds. *Marine Mammal Science* 14, 1–37.

Ward, N. & Macdonald, C. (2009). Sister sanctuary: Care of marine mammal protected areas beyond borders – An innovative management tool for transboundary species. In First International Conference on Marine Mammal Protected Areas, 2009 Maui, Hawaii.

Watson-Capps, J.J. & Mann, J. (2005). The effects of aquaculture on bottlenose dolphin (*Tursiops* sp.) ranging in Shark Bay, Western Australia. *Biological Conservation* 124, 519–526.

Wiener, C.S., Needham, M.D. & Wilkinson, P.F. (2009). Hawaii's real life marine park: Interpretation and impacts of commercial marine tourism in the Hawaiian Islands. *Current Issues in Tourism* 12, 489–504.

Williams, R. & Ashe, E. (2007). Killer whale evasive tactics vary with boat number. *Journal of Zoology* 272, 390–397.

Williams, R., Trites, A.W. & Bain, D.E. (2002). Behavioural responses of killer whales (*Orcinus orca*) to whale-watching boats: Opportunistic observations and experimental approaches. *Journal of Zoology* 256, 255–270.

Williams, R., Lusseau, D. & Hammond, P.S. (2009). The role of social aggregations and protected areas in killer whale conservation: The mixed blessing of critical habitat. *Biological Conservation* 142, 709–719.

Zimmer, W.M.X. & Tyack, P.L. (2007). Repetitive shallow dives pose decompression risk in deep-diving beaked whales. *Marine Mammal Science* 23, 888–925.

Part IV

Sustainable management: insights and issues

18

The socioeconomic, educational and legal aspects of whale-watching

A Scottish case study

E.C.M. Parsons

Introduction

Between 1995 and 2000 the International Fund for Animal Welfare (IFAW) held a number of workshops on various aspects of whale-watching. The first discussed the scientific aspects (IFAW, 1995) of the activity, but in 1997 there were three workshops on the more social and managerial aspects of the industry, including a workshop on the educational impacts of whale-watching (IFAW, 1997) which also included debate about ways to persuade whale-watching operators to use codes of conduct to minimize their impacts, and another workshop on the legal aspects of whale-watching (IFAW, 2000). The third was on the socioeconomic impacts of whale-watching (IFAW, 1999). A global review of all of these aspects of whale-watching would arguably be a book, or books, by themselves. Therefore, this current chapter builds on some of the ideas and suggestions raised in these workshops for a specific case study in Scotland, an area where there has been substantive and quantitative research into these areas. The last workshop in the series mentioned above (IFAW, 2001) discussed issues such as the conflict between whale-watching and whaling in certain locations, and the potential negative socioeconomic impacts, and this is also briefly discussed in this chapter.

Whale-watching in Scotland

The first official whale-watching business (Sealife Surveys) in Scotland, indeed for the entire UK, began in 1989 on the island of Mull. Hoyt (2001) estimated that there were at least 35 land- and boat-based whale-watching activities in Scotland by 1998. Hughes (2001) suggested that in the UK there was an overall growth in public interest to seeing cetaceans in the wild from the late 1980s onwards, and a move away from tourism involving captive cetaceans – indeed, all of the captive cetacean facilities in the UK had closed by the mid-1990s, whereas there was a positive growth in wild cetacean-watching opportunities, particularly in eastern Scotland, focused on the resident common bottlenose dolphin (*Tursiops truncatus*) population in the Moray Firth.

The two main regions for whale-watching in Scotland are the Moray Firth on the east coast, and the western coast. For the latter, whale-watching is particularly concentrated on the islands of Mull, Skye, Islay, Lewis, Barra and around the Small Isles (Rum, Eigg, Muck and Canna) and Arisaig, Mallaig, Ullapool and Gairloch on the mainland. The primary target species on the east coast is the common bottlenose dolphin (*Tursiops truncatus*), although northern minke whales (*Balaenoptera*

Whale-watching: Sustainable Tourism and Ecological Management, eds J. Higham, L. Bejder and R. Williams.
Published by Cambridge University Press. © Cambridge University Press 2014.

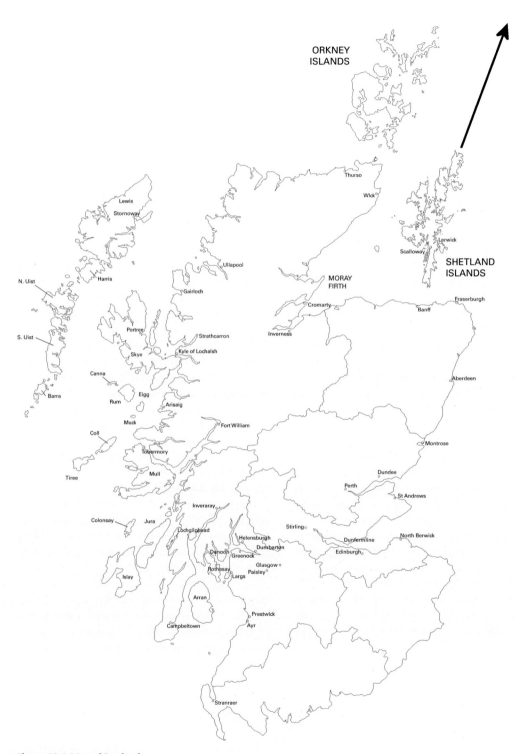

Figure 18.1 Map of Scotland.

acutorostrata) and harbour porpoises (*Phocoena phocoena*) are increasingly becoming a focus in some areas. On the west coast northern minke whales are the main target species, although depending on the area bottlenose dolphins, harbour porpoises or Risso's dolphins (*Grampus griseus*) can also be key target species. Non-cetacean marine megafauna may also be a component of whale-watching trips on the west coast, such as basking sharks (*Cetorhinus maximus*), harbour (*Phoca vitulina*) and grey seals (*Halichoerus grypus*), and even puffins (*Fratercula arctica*) and white-tailed sea eagles (*Haliaeetus albicilla*). Whale-watching activities in northern Scotland are also starting to develop both on the mainland (in areas such as Thurso) and the Orkney and Shetland Islands. For the mainland and Orkneys, harbour porpoises and white-beaked dolphins (*Lagenorhynchus albirostris*) may feature as a focus species, and in the Shetland Islands, both the ubiquitous harbour porpoise, and also killer whales (*Orcinus orca*), are sometimes key target species.

Economic value of whale-watching in Scotland

Tourism is the single largest industry in rural Scotland. The entire Scottish tourism trade grosses more than £2.5 billion (US $4 billion) annually. A study in 1996 on marine wildlife tourism noted that this particular sector of the tourism industry alone provided over 400 jobs, involving over 80 boat operators and was worth £9.3 million directly, but indirect income from marine wildlife tourism (i.e. associated income from accommodation, restaurants, etc., derived from marine wildlife tourists) was worth a total of £57 million and provided 2670 jobs (Masters *et al.*, 1998). An updated study was conducted in 2009 that estimated marine wildlife tourism had an economic impact of £15 million providing 633 additional full-time equivalent jobs, with coastal wildlife tourism having a net economic impact of £24 million providing 995 additional jobs (International Centre for Tourism and Hospitality Research, 2010).

Whale-watching is an important component of this marine wildlife industry, with whales and dolphins being the country's number one wildlife attraction. It has been noted that whales and dolphins are very effective at bringing tourists into an area: a survey determined that 23% of whale-watchers visited rural West Scotland specifically to go on whale-watching trips (Parsons *et al.*, 2003a). In 2000, a project funded by the UK Government estimated that the direct economic income (i.e. expenditure on excursion tickets) from cetacean tourism activities was estimated to be £1.77 million per annum (Parsons *et al.*, 2003a).The total economic value from tourists being brought to rural west Scotland solely due to the presence of whales represented an additional £5.1 million in tourism income for the region (Parsons *et al.*, 2003a). Moreover, tourists stayed in the region longer as the result of going on whale-watching trips, which adds a further £0.9 million of additional expenditure as a result of whale-watching (Parsons *et al.*, 2003a). Thus, the total gross income generated (directly and indirectly) by cetacean-related tourism in rural west Scotland was estimated at £7.8 million (~US $13 million; Parsons *et al.*, 2003a). In some rural areas it provides as much as 12% of local tourist income (Parsons *et al.*, 2003a), which as mentioned above is the major industry and employer in rural areas.

For the eastern coast of Scotland, Hoyt (2001) estimated that in 1998, land- and boat-based whale- and dolphin-watching in the Moray Firth generated £477,000 as direct expenditure and £2.34 million as total expenditure (when adding assumed indirect income from accommodation, food purchases, etc.). More recently, Davies *et al.* (2010) estimated that in 2009 the direct income from dolphin-watching in eastern Scotland was £10.4 million, of which at least £4 million (effectively translating into 202 full-time equivalent jobs) was solely due to the presence of a bottlenose dolphin population off eastern Scotland. Although Hoyt (2001) and Davies *et al.* (2010) use different methodologies to assess the economic impact of cetaceans, the data nonetheless suggest that there has been substantial growth in the economic value of the east coast dolphin-watching industry over a 10-year period. Hoyt (2001) also estimated that on the northern islands of Shetland, land-based whale-watching generated £109,000 in indirect expenditure.

When combining whale-watching expenditure for the whole of Scotland, a conservative estimate of the total income from Scottish cetacean-related tourism for the turn of the century (direct plus indirect expenditure) was a total of £10.7 million (~US $18 million; Parsons *et al.*, 2003a). Numerous new whale-watching businesses have been established in Scotland in the 14 years since the Hoyt (2001) evaluation, and in the 12 years since the Parsons *et al.* evaluation, and with the substantially higher economic value noted in 2009 for east coast whale-watching derived by Davies *et al.* (2010) this turn of the century figure would certainly be an under-estimate of the current economic value.

Woods-Ballard *et al.* (2003) investigated the nature of whale-watching operators in Scotland and the perceived sustainability of the industry. Most operators were found to be local people, supporting up to five full-time equivalent jobs (Woods-Ballard *et al.*, 2003). Many were ex-fishermen or farmers, and few had any formal training in wildlife tourism or business management (Woods-Ballard *et al.*, 2003).

In terms of growth of the industry, the east coast and western isles of Scotland were particularly perceived to be steadily increasing in tourist numbers (Woods-Ballard *et al.*, 2003). Many operators in Scotland considered the marine mammal numbers in their area to be increasing (Woods-Ballard *et al.*, 2003). Over one-half of the operators reported that they kept a record of cetacean sightings, many of which were already being used for research purposes (Woods-Ballard *et al.*, 2003) and in one case data collected from the whale-watching operator has been used in several major scientific publications in international journals on the ecology and behaviour of minke whales (e.g. Leaper *et al.*, 1997; Stockin *et al.*, 2001; Macleod *et al.*, 2004). Finally, most operators considered that their whale-watching businesses were economically

sustainable in the long term, and those operators that expressed concern over the future viability of their businesses were only a small percentage of the industry (Woods-Ballard *et al.*, 2003).

Whale-watching tourists in Scotland

Several studies have been conducted on the demographics of whale-watching tourists in western Scotland, and they were found generally to be middle-class, well-educated (63% educated to 21+) and middle-aged (35–55), although there is a notable proportion of younger participants (Parsons *et al.*, 2003b). In eastern Scotland, 'dolphin-enthusiast' tourists (for whom cetaceans were their main reason for their visit to the area) also tended to be middle-class (78%), although fewer 'dolphin enthusiasts' were likely to be middle-class when compared with other wildlife-watching tourists, or general tourists, in the region (Davies *et al.*, 2010). Whale-watchers were also more likely to be accompanied by children than general Scottish tourists in both western and eastern Scotland (Parsons *et al.*, 2003b; Davies *et al.*, 2010).

Most tourists taking whale-watching trips in western Scotland were domestic (British) tourists, a quarter of whom were Scottish (Parsons *et al.*, 2003b) and 70% were repeat visitors to the area (Parsons *et al.*, 2003b). For east Scotland, Davies *et al.* (2010) reported that 94% of 'dolphin-enthusiast' tourists were British, of which just over a quarter (28%) were Scottish. Of all whale-watchers in western Scotland, 62% were on their first whale-watching trip, and of those who had been whale-watching before, nearly half had done so in the UK (90% in Scotland; Parsons *et al.*, 2003b). In eastern Scotland, 85% of 'dolphin-enthusiast' tourists said that they would come back to the area again (versus 73% of 'wildlife-enthusiast' tourists and only 35% of general tourists) and 93% of 'dolphin-enthusiast' tourists, 88% of 'wildlife-enthusiast tourists' and 69% of general tourists said that they would try to visit new dolphin-watching locations, if they were made aware of these locations (Davies *et al.*, 2010).

In a survey of members of the general public conducted in Glasgow and Edinburgh it was noted that over half of those interviewed were aware of the possibilities to go on a whale-watching trip in Scotland; showing good awareness of whale-watching opportunities (Howard & Parsons, 2006a). Moreover, although only 7% had actually gone on a whale-watching trip, nearly 60% stated that they would like to go on a whale-watching trip, indicating a large, as yet untapped, domestic market for whale-watching (Howard & Parsons, 2006a), which may help the growth of the Scottish whale-watching industry.

Environmental awareness and whale-watching tourists in Scotland

It has also been found that tourists who went on marine wildlife-watching tours in western Scotland, particularly whale-watching, were very environmentally motivated and displayed great interest in animal welfare issues, e.g. over 90% of marine wildlife-watching tourists on the Isle of Mull were involved in environmental/wildlife-related activities; nearly 60% were members of environmental charities and an astonishing 18% stated that they actually engaged in voluntary work for environmental charities (Warburton et al., 2000). A survey conducted two years later discovered that the proportion of whale-watchers engaged in voluntary work had actually increased even further to 27% (Rawles & Parsons, 2004). Greater levels of environmental awareness and concern for animal welfare in whale-watchers were also apparent in other ways, e.g. nearly three-quarters of whale-watchers claimed that they only purchased cosmetic/hygiene products that had not been tested on animals and over 80% regularly recycled items (compared with only 18% of the general Scottish population; Rawles & Parsons, 2004). Also, just under half used energy-saving appliances or purchased organic or environmentally friendly products on a regular basis (Rawles & Parsons, 2004). The majority of whale-watching tourists had previously been aware of the occurrence of cetaceans in the waters of western

Scotland (Parsons et al., 2003b), which agrees with the data that whale-watchers are more interested in environmental and animal issues as noted above.

This high level of animal welfare and environmental concern amongst whale-watchers has several implications for whale-watching management in the area, including the possibility that environmental groups may be able to play an important role in educating whale-watching tourists about management issues, and also, through their awareness of animal welfare issues, tourists themselves may actually help to police whale-watching activities and ensure their sustainability (Rawles & Parsons, 2004).

The high levels of environmental motivation in whale-watching tourists are also apparent in other ways, for example when Scottish whale-watchers were questioned as to whether they would visit, or go whale-watching in a country that conducted whaling operations, they overwhelmingly stated that they would not: 79% of whale-watchers stated they would boycott visiting a country that conducted hunts for cetaceans, and a further 12.4% of whale-watchers stated that although they might visit a country that conducted whaling operations, they would not take a whale-watching trip in that country (i.e. 91.4% of whale-watchers would not go whale-watching in a country that hunted whales; Parsons & Rawles, 2003). These data refute the argument that whale-watching and whaling can peaceably coexist (which has been echoed in other studies, e.g. Parsons & Draheim, 2009), and suggests that whaling activities may very severely decrease whale-watching and wider tourism industry activities in those countries (see also Higham & Lusseau, 2008).

Education and whale-watching in Scotland

It has been proposed that whale-watching could have substantive benefits in terms of educating the public, especially about the biology and conservation status of cetaceans (IFAW, 1997), potentially even leading to more effective management of the impacts of tourism activities as a result

(Orams, 1994, 1995, 1996a, 1996b). To date, there has only been one, so far unpublished (and is thus described in this chapter in detail), study on education provision by operations that conducted whale-watching and marine wildlife tours. Bridgland (2002) obtained survey questionnaire data from 26 operators, of whom 50% stated that environmental education was a major feature of their trip. One operation had a small visitor centre with exhibits, and five operations said that they gave educational lectures or talks, all of which were less than 30 minutes in length (although one operation gave public lectures that could last more than an hour during the winter as well; Bridgland, 2002). Eighty per cent stated that the whale-watching boat skipper acted as a tour guide, providing information, but five operations also had onboard guides (four of which were paid; Bridgland, 2002). Eighty-eight percent included information about local history and culture, in addition to information about marine wildlife (Bridgland, 2002).

Three operators had degrees in biology or a 'relevant subject', and those operations with guides noted that they typically had, or were undergoing, university marine science degrees (Bridgland, 2002). Two-thirds had displays, books, leaflets or guides onboard the whale-watching vessels (primarily on marine wildlife, but information about historical and cultural topics were also provided by some operators; Bridgland, 2002). One operator had a bulletin board reporting recent cetacean sightings. Three operators further stated that they carried artefacts, such as feathers, teeth or baleen, to use as illustrative props for talks, etc. (Bridgland, 2002). Materials in languages other than English were limited: only 19% of operators had translated materials (with one operator having materials in seven languages) although comments were made that foreign tourists typically spoke English relatively fluently (Bridgland, 2002).

Fifty-four per cent of the operators stated that they were members, or linked with, a conservation charity (either personally or as a business) and 27% stated that they were involved with a research organization (the most frequently stated organization being the Hebridean Whale and Dolphin trust (http://www.hwdt.org), typically contributing information on sightings of cetacean species; Bridgland, 2002). Seventy-seven percent of operators responded that they had 'an educational role to play' although one operator rather negatively stated that 'this is a business, not an education service' (Bridgland, 2002). Bridgland (2002) asked whether a training course on education methods and content might help operators, but only a third of operators in the survey (35%) thought that such a course would necessarily help. However, almost a half (46%) stated that they would be willing to pay for such a course, although only 23% stated that they would want a certificate, or similar, if they took such a course (Bridgland, 2002).

In addition to boat-based commercial whale-watching operations in Scotland, there are several visitor centres run by cetacean conservation charities, such as the Hebridean Whale and Dolphin Trust (HWDT) visitor centre in Tobermory, Mull, and the Whale and Dolphin Conservation's (WDC) visitor centres in Spey Bay and North Kessock on the east coast, and nearby the Cetacean Research and Rescue Unit's (CRRU) centre in Gardenstown, Banff. There are also several community centres that highlight local wildlife and nature, including local cetaceans, such as the Arisaig Land, Sea and Islands Centre. All of these provide educational benefits and work closely with whale-watching companies in their local regions. Tourist information centres, particularly in the Highlands and Argyll regions, frequently have information and sell books about whales and dolphins, alongside promotional materials about whale-watching operations. Even the small aquarium belonging to the University (of London's) Marine Biological Station in Millport, on the Isle of Cumbrae (which is in the Firth of Clyde near Glasgow), has exhibits about whales and dolphins of Scotland, and the station itself runs an annual course on whales and dolphins utilizing Scottish whale-watching operators as an educational platform. Commercial visitor centres also provide educational information on whales and dolphins of Scotland, such as the Oban Sealife Sanctuary.

To date, there has only been one published study specifically investigating public knowledge of cetaceans in Scotland. In summer 2001, Scott and Parsons (2004) interviewed members of the public in southwest Scotland to determine their knowledge of the diversity and occurrence of cetaceans in the region. Most of those participating (46%) underestimated the number of cetacean species occurring in western Scottish waters with only 4.4% giving a correct answer. Knowledge was much better for members of the public in areas with thriving whale-watching industries and it was suggested that this perhaps played a role in increasing their knowledge (Scott & Parsons, 2004). Similarly members of the public from these locations were also more able to identify photographs of the four most common cetacean species (age, gender and level of environmental interest also had an effect on knowledge; Scott & Parsons, 2004). However, as yet, a full review on cetacean-related education in Scotland, its effectiveness and impacts on public attitudes and knowledge, and the role that whale-whaling and whale-related visitor centres play in this education, has not been conducted, but is perhaps merited.

Community benefits and whale-watching in Scotland

Whale-watching as a tourism activity is frequently, and wrongly, referred to as ecotourism. Ecotourism can be defined as:

environmentally responsible travel and visitation to relatively undisturbed natural areas, in order to enjoy and appreciate nature (and any accompanying cultural features – both past and present) that promotes conservation, has low negative visitor impact, and *provides for beneficially active socio-economic involvement of local populations*'(Ceballos-Lascuráin, 1996: 20, emphasis added).

This definition was officially adopted by the IUCN at the 1996 Congress (via Resolution CGR 1.67).

As can be seen in the definition, socioeconomic benefits of host communities are specifically included in the definition. The International

Whaling Commission's definition of 'whale ecotourism' also specifically mentions that the activity should 'Provide some benefits to the local host community within which the company operates' (Parsons *et al.*, 2006: 251). Do whale-watching activities in Scotland provide such socioeconomic benefits? Cetaceans definitely draw members of the public to Scotland, and without these animals hotels, restaurants and other businesses would not benefit from the additional revenue that whale-watchers bring (Parsons *et al.*, 2003a; Davies *et al.*, 2010). Whale-watchers certainly stay in locations longer as the result of whale-watching trips, and this also brings additional revenue (Parsons *et al.*, 2003a; Davies *et al.*, 2010). However, both of these sources of indirect values have been factored into the economic benefits that whale-watching brings, as noted above. Moreover, Warburton *et al.* (2001) noted that in one village where whale-watching occurred, of the 21 shops in the village 13 sold whale-related merchandise (from 0.2% to 70% with items ranging from £0.2 to £100 in price) and an estimated £15,300 would be generated for these local shops by the sale of this merchandise alone. Also, as Woods-Ballard *et al.* (2003) reported, and was noted above, many of the whale-watching companies are run by local Scottish residents, and often employ local people as guides, although the total extent of this host community employment has not been calculated.

Moreover, whale-watching companies may bring other local community benefits. For example, taking the example of just one company in Tobermory, on the Isle of Mull, the company sells locally produced snacks on board their boat and sells books, DVDs and photographs produced by local residents, in addition to a range of educational, recycled and environmentally friendly goods. The company also collects money for local charities and has also offered free trips to local school classes and small business owners at the end of the season. Several of the crew also volunteer for local community services, such as the lifeboat crew and the fire brigade. Arguably, this company, which has similar business practices to several other whale-watching companies in the country, provides substantive direct and indirect benefits to the host community.

Whale-watching codes of conduct and voluntary management schemes

In 1995, due to concerns over the possible impacts of whale-watching activities on dolphins in the Moray Firth, the Dolphin Space Program (DSP) was initiated (see http://www.dolphinspace.org). The DSP was set up with a multi-stakeholder steering group, which at the time was a novel approach for marine mammal management in Scotland, and this steering group includes members such as the regional police, coastguard, VisitScotland (the tourist board) and the WDC. The DSP developed an accreditation scheme, by which local whale-watching operators could join, if they committed to abiding by a whale-watching code of conduct produced by the DSP. Accredited members can also receive training, educational materials, and gain promotional benefits through being associated with the scheme. The DSP steering group meets at least once a year with accredited operators and arising issues and management problems are discussed.

There are as such no specific mandatory regulations with respect to whale-watching activities in Scotland. Despite this, in a 2000 survey it was found that nearly 90% of whale-watching operators stated that they voluntarily followed a whale-watching code of conduct or guidelines (Parsons & Woods-Ballard, 2003; Woods-Ballard *et al.*, 2003). The most frequently used whale-watching code or set of guidelines was one produced by a tour operators' association, and three codes of conduct produced by environmental non-governmental organizations were also used frequently (Parsons & Woods-Ballard, 2003). In a separate study, Bridgland (2002) noted that more than a half of marine tourism operators surveyed (58%) made their passengers aware that they used a code of conduct and of these 60% tried to relate their use of a code to the importance of whale-watching guideline use internationally.

However, only 27% of the operators surveyed by Parsons and Woods-Ballard (2003) were aware of specific whale-watching guidelines which had been produced by the UK government, and were available at the time. Moreover, none of the operators actually referred to these guidelines when watching cetaceans (Parsons & Woods-Ballard, 2003). This was interpreted as being evidence that although the majority of whale-watching tour operators had accepted codes of conduct, preferences were for local, or operator-produced, codes rather than governmentally appointed regulations, i.e. 'bottom-up' management (with scientific advice and guidance), rather than 'top-down' management by statutory bodies (Parsons & Woods-Ballard, 2003). It should be emphasized, however, that although whale-watching operators stated that they used a code of conduct, research has not been conducted to investigate how frequently these codes of conduct are fully complied with. Even despite sometimes being a legal requirement, compliance with whale-watching regulations and guidelines has been poor in many locations (e.g. Scarpaci *et al.*, 2003, 2004). Studies into whether guidelines are followed, partially or completely, factors that affect compliance, and the general efficacy of guidelines, should be a matter of priority.

Legal aspects of whale-watching in Scotland

Prior to 2004, coastal cetaceans in Scotland were protected under the UK Wildlife & Countryside Act of 1981. This Act made it illegal to intentionally kill, capture or injure cetaceans and, moreover, to intentionally disturb a cetacean (Parsons *et al.*, 2010a). This law extended to 12 nautical miles from the UK coastline. In 2004, the Scottish Parliament passed the Nature Conservation (Scotland) Act which superseded and slightly altered the provisions in the Wildlife and Countryside Act, effectively making it illegal to 'recklessly' (with the idea that reckless behaviour is easier to prove than 'deliberate' behaviour in a court of law) 'disturb or harass' cetaceans (although the meaning of harassment was not defined; Parsons *et al.*, 2010a). In addition, the Nature Conservation (Scotland) Act required that Scottish Natural Heritage (as the competent authority for nature conservation in Scotland) produce a 'Scottish Marine Wildlife Watching Code' (Nature Conservation (Scotland) Act; Part 3, Section 51). The code was to outline activities that might disturb marine wildlife, circumstances under which marine wildlife should be approached and ways to view marine wildlife 'with minimum disturbance' (Section 51 (2) (a),(b) & (c)).

There was a requirement for Scottish Natural Heritage to consult with persons 'appearing to them to have an interest in marine wildlife watching and other persons as it thinks fit' (Section 51 (5)), i.e. extensive consultation with stakeholders in the whale-watching industry, which was done via a series of workshops, meetings and subsequent consultation on drafts of the code, illustrating a high level of 'bottom-up' involvement with the development of the code, despite the act being a 'top-down' management action. There is no actual requirement in the act for Scottish Natural Heritage to enforce and police this code of conduct, but arguably the code would make enforcement of laws that prohibit disturbance of cetaceans easier as it can be taken that by adhering to behaviour in the guidelines you are not disturbing or harassing marine wildlife, and deviating from the proscribed guidelines could constitute reckless disturbance (Parsons *et al.*, 2010a).

As a member of the European Union, the UK (and thus Scotland) must also comply with the 1992 Council Directive on the Conservation of Natural Habitats and Wild Fauna and Flora (the Habitats Directive), which came into effect in the UK in 1994. Cetaceans are listed under Annex IV of the directive, which effectively requires member states to prohibit their deliberate capture and killing, as well as deliberate the disturbance of cetaceans. Unlike the Nature Conservation (Scotland) Act, which is limited to 12 nm from the coast, the Habitats Directive extends to 200 nm from the coastline, and so provides some protection for cetaceans while further ashore (Parsons *et al.*, 2010a). Under the

Habitats Directive, member nations are obliged to designate protected areas for species listed on Annex II (which for cetaceans includes harbour porpoise and common bottlenose dolphin), and as such a Special Area of Conservation (SACs) was designated in the Moray Firth, which provides some additional protective measures for animals in this protected area from whale-watching activities (Parsons *et al.*, 2010a).

Public attitudes to the impacts of whale-watching in Scotland

Although whale-watching activities are known to negatively impact cetaceans (Parsons, 2012), there have been few studies on the actual impacts of whale-watching activities on target cetaceans in Scotland. Studies on the impact of boat traffic, including whale-watching vessels, on bottlenose dolphins in the Moray Firth found that in 17 of the 22 cases where whale-watching boats approached dolphin groups, significantly fewer surfacings were recorded (Janik & Thompson, 1996). Another study noted increased synchronicity in dolphin surfacings in response to boat traffic (Hastie *et al.*, 2003), but the biological significance of whale-watching or other types of disturbance to cetaceans has not been assessed in the UK (Parsons *et al.*, 2010b).

In terms of public opinion on the impacts of whale-watching, the activity is not generally seen by the general public to be a major threat to the cetacean populations (Scott & Parsons, 2005; Howard & Parsons, 2006b; Zapponi & Robinson, 2007). Nearly a third of city-dwellers considered whale-watching to be only a minor problem, and nearly half considered it not to be a threat at all, with only 11% considering whale-watching activities to pose greater risk (Howard & Parsons, 2006b). Similar results were obtained in rural, coastal areas of western Scotland (Scott & Parsons, 2005), with less than 1% of the public considering whale-watching to pose a serious risk to cetaceans, a statistic that is not surprising considering the concentration of marine wildlife tourism in this locale. These results were also echoed for a slightly smaller survey of members of the general public in the Moray Firth (Zapponi & Robinson, 2007). There are, nonetheless, some areas of concern, noticeably an increase in numbers of companies offering high-speed whale-watching, i.e. boats capable of travelling to key whale-watching areas at speeds faster than 20 knots (Parsons & Gaillard, 2003).

Conclusion

In summary, for this case study in Scotland, whale-watching has a notable positive impact on the economy of coastal and rural areas (e.g. Parsons *et al.*, 2003a), and is considered to have potential for further growth (Howard & Parsons, 2006a). There appear to be educational benefits from whale-watching and cetacean-related tourism, although to date research into this area is largely limited to just one study (Bridgland, 2002). There are also some definable benefits to local communities that host whale-watching operations. Whale-watching tourists in Scotland appear to have a relatively high level of environmental awareness (Rawles & Parsons, 2004), which could be channelled to aid cetacean conservation, and certainly means that it is in tour operators' interests to ensure that their operations are as 'environmentally friendly' as possible to avoid alienating their clientele.

However, there has been a lack of specific research on the biological and population-level impacts of whale-watching activities on cetaceans in any region of Scotland, or indeed the UK as a whole (Parsons *et al.*, 2010b). There are laws within the UK that specifically prohibit the deliberate or reckless disturbance and harassment of cetaceans, and as such inappropriate behaviour by whale-watching trip operators that could negatively impact cetaceans is technically illegal. Moreover, although guidelines exist for whale-watching activities, including official governmentally produced guidelines, the efficacy of these guidelines, and their ability to reduce biological impacts on animals (such as stress; Parsons *et al.*, 2010b) or the degree to which whale-watching

operators comply with guidelines, have not been studied. Scotland has a valuable and maturing whale-watching industry, and for the majority of Scotland there has arguably been an impression that the industry as a whole is sustainable. At least, there have been no recent attempts to officially control or manage whale-watching activities by government agencies, or urgent calls to do so by environmental groups. Moreover, the general pubic do not consider whale-watching to be a threat to cetacean populations (Scott & Parsons, 2005; Howard & Parsons, 2006b; Zapponi & Robinson, 2007). However, there has not been a comprehensive investigation into whether this impression of sustainability is warranted, and such a study should be considered a priority. This is especially important when one considers the potential for growth of the whale-watching industry in the nation and the positive economic benefits that a healthy population of cetaceans can have; ensuring the long-term sustainability of this industry is important not just for cetaceans, but also for human communities in Scotland.

REFERENCES

Bridgland, R. (2002). A study of the extent of interpretation in marine wildlife tourism in western Scotland. MSc Thesis, School of Life Science, Napier University.

Ceballos-Lascuráin, H. (1996). *Tourism, Ecotourism, and Protected Areas.* Gland, Switzerland: IUCN.

Davies, B., Pita, C., Lusseau, D. & Hunter, C. (2010). *The Value of Tourism Expenditure Related to the East of Scotland Bottlenose Dolphin Population.* Aberdeen: Aberdeen Centre for Environmental Sustainability & Moray Firth Partnership.

Hastie, G.D., Wilson, B., Tuft, L. & Thompson, P.M. (2003). Bottlenose dolphins increase breathing synchrony in response to boat traffic. *Marine Mammal Science* 19, 74–84.

Higham, J.E.S. & Lusseau, D. (2008). Slaughtering the goose that lays the golden egg: Are whaling and whalewatching mutually exclusive? *Current Issues in Tourism* 11, 63–74.

Howard, C. & Parsons, E.C.M. (2006a). Public awareness of whalewatching opportunities in Scotland. *Tourism in Marine Environments* 2, 103–109.

Howard, C. & Parsons, E.C.M. (2006b). Attitudes of Scottish city inhabitants to cetacean conservation. *Biodiversity and Conservation* 15, 4335–4356.

Hoyt, E. (2001). *Whale Watching 2001: Worldwide tourism numbers, expenditures, and expanding socioeconomic benefits.* Crowborough: International Fund for Animal Welfare.

Hughes, P. (2001). Animals, values and tourism – Structural shifts in UK dolphin tourism provision. *Tourism Management* 22, 321–329.

IFAW. (1995). *Report of the Workshop on the Scientific Aspects of Whale Watching.* 30 March–4 April 1995, Monticastello di Vibio, Italy. Massachusetts: International Fund for Animal Welfare.

IFAW. (1997). *Report of the Workshop on the Educational Values of Whale Watching.* 8–11 May 1997, Province Town, Massachusetts, USA. Massachusetts: International Fund for Animal Welfare.

IFAW. (1999). *Report of the Workshop on the Socioeconomic Aspects of Whale Watching.* 8–12 December 1997, Kaikoura, New Zealand. Massachusetts: International Fund for Animal Welfare.

IFAW. (2000). *Report of the Workshop on the Legal Aspects of Whale Watching.* 17–20 November 1997, Punta Arenas, Chile. Massachusetts: International Fund for Animal Welfare.

IFAW. (2001). *Report of the Closing Workshop to Review Various Aspects of Whale Watching.* 6–10 February 2000, Tuscany, Italy. Massachusetts: International Fund for Animal Welfare.

International Centre for Tourism and Hospitality Research. (2010). *The Economic Impact of Wildlife Tourism in Scotland.* International Centre for Tourism and Hospitality Research, Bournemouth University. Retrieved from www.scotland.gov.uk/socialresearch (accessed 24 July 2012).

Janik, V.M. & Thompson, P.M. (1996). Changes in surfacing patterns of bottlenose dolphins in response to boat traffic. *Marine Mammal Science* 12, 597–602.

Leaper, R., Fairbairns, R., Gordon, J., Hiby, A., Lovell, P. & Papastavrou, V. (1997). Analysis of data collected from a whalewatching operation to assess relative abundance and distribution of the minke whale (*Balaenoptera acutorostrata*) around the Isle of Mull, Scotland. *Reports of the International Whaling Commission* 47, 505–511.

MacLeod, K., Fairbairns, R., Gill, A., *et al.* (2004). Seasonal distribution of minke whales (*Balaenoptera acutorostrata*) in relation to physiography and prey off the Isle of Mull, Scotland. *Marine Ecology Progress Series* 277, 263–274.

Masters, D., Nautilus Consultants & Carter, J. (1998). *Marine Wildlife Tourism: Developing a quality approach in the Highlands and Islands.* Inverness: Tourism and Environment Initiative and Scottish Natural Heritage.

Orams, M.B. (1994). Creating effective interpretation for managing interaction between tourists and wildlife. *Australian Journal of Environmental Education* 10, 2l–34.

Orams, M.B. (1995). Towards a more desirable form of ecotourism. *Tourism Management* 16, 3–8.

Orams, M.B. (1996a). A conceptual model of tourist–wildlife interaction: The case for education as a management strategy. *Australian Geographer* 27, 39–51.

Orams, M.B. (1996b). Using interpretation to manage nature-based tourism. *Journal of Sustainable Tourism* 4, 81–94.

Parsons, E.C.M. (2012). The negative impacts of whale-watching. *Journal of Marine Biology*, in press.

Parsons, E.C.M. & Draheim, M. (2009). A reason not to support whaling: A case study from the Dominican Republic. *Current Issues in Tourism* 12, 397–403.

Parsons, E.C.M. & Gaillard, T. (2003). Characteristics of high-speed whalewatching vessels in Scotland. Paper presented to the Scientific Committee at the 55th Meeting of the International Whaling Commission, 26 May–6 June 2003, Berlin, Germany. SC55/WW2.

Parsons, E.C.M. & Rawles, C. (2003). The resumption of whaling by Iceland and the potential negative impact in the Icelandic whalewatching market. *Current Issues in Tourism* 6, 444–448.

Parsons, E.C.M. & Woods-Ballard, A. (2003). Acceptance of voluntary whalewatching codes of conduct in West Scotland: The effectiveness of governmental versus industry-led guidelines. *Current Issues in Tourism* 6, 172–182.

Parsons, E.C.M., Warburton, C.A., Woods-Ballard, A., Hughes, A. & Johnston, P. (2003a). The value of conserving whales: The impacts of cetacean-related tourism on the economy of rural West Scotland. *Aquatic Conservation* 13, 397–415.

Parsons, E.C.M., Warburton, C.A., Woods-Ballard, A., *et al.* (2003b). Whalewatching tourists in West Scotland. *Journal of Ecotourism* 2, 93–113.

Parsons, E.C.M., Fortuna, C.M., Ritter, F., *et al.* (2006). Glossary of whalewatching terms. *Journal of Cetacean Research and Management* 8(Suppl.), 249–251.

Parsons, E.C.M., Clark, J. & Simmonds, M.P. (2010a). The conservation of British cetaceans: A review of the threats and protection afforded to whales, dolphins and porpoises in UK Waters, Part 2. *International Journal of Wildlife Law and Policy* 13, 99–175.

Parsons, E.C.M., Clark, J., Warham, J. & Simmonds, M.P. (2010b). The conservation of British cetaceans: A review of the threats and protection afforded to whales, dolphins and porpoises in UK waters, Part 1. *International Journal of Wildlife Law and Policy* 13, 1–62.

Rawles, C.J.G. & Parsons, E.C.M. (2004). Environmental motivation of whalewatching tourists in Scotland. *Tourism in Marine Environments* 1, 129–132.

Scarpaci, C., Nugegoda, D. & Corkeron, P.J. (2003). Compliance with regulations by 'swim-with-dolphins' operations in Port Philip Bay, Victoria, Australia. *Environmental Management* 31, 342–347.

Scarpaci, C., Nugegoda, D. & Corkeron, P.J. (2004). No detectable improvement in compliance to regulations by 'swim-with-dolphin' operators in Port Philip Bay, Victoria, Australia. *Tourism in Marine Environments* 1, 41–48.

Scott, N.J. & Parsons, E.C.M. (2004). A survey of public awareness of the occurrence and diversity of cetaceans in Southwest Scotland. *Journal of the Marine Biological Association of the United Kingdom* 84, 1101–1104.

Scott, N.J. & Parsons, E.C.M. (2005). A survey of public opinions in Southwest Scotland on cetacean conservation issues. *Aquatic Conservation* 15, 299–312.

Stockin, K., Fairbairns, R.S., Parsons, E.C.M. & Sims, D. (2001). The effects of diel and seasonal cycles on the dive duration of the minke whale (*Balaenoptera acutorostrata*) off the Isle of Mull, Scotland. *Journal of the Marine Biological Association of the United Kingdom* 81, 189–190.

Warburton, C.A., Parsons, E.C.M. & Goodwin, H. (2000). Whalewatching and marine wildlife tourism on the Isle of Mull, Scotland. Paper presented to the Scientific Committee at the 52nd Meeting of the International Whaling Commission, 11–28 June 2000, Australia. SC52/WW17.

Warburton, C.A., Parsons, E.C.M., Woods-Ballard, A., Hughes, A. & Johnston, P. (2001). *Whale-watching in West Scotland*. London: Department of the Environment, Food and Rural Affairs.

Woods-Ballard, A., Parsons, E.C.M., Hughes, A.J., Velander, K.A., Ladle, R.J. & Warburton, C.A. (2003). The sustainability of whalewatching in Scotland. *Journal of Sustainable Tourism* 11, 40–55.

Zapponi, L. & Robinson, K. (2007). Social attitudes to marine conservation in NE Scotland: Public perceptions and cetaceans in the Moray Firth. Poster presented at the 21st Annual Conference of the European Cetacean Society, 22–25 April 2007, San Sebastián, Spain. Retrieved from http://www.crru.org.uk/publications.asp.

Vigilance, resilience and failures of science and management

Spinner dolphins and tourism in Hawai'i

David W. Johnston

Introduction

This chapter provides a brief overview of the complex relationship between tourism operations in Hawai'i and spinner dolphins – informed by what is known about the specialized behaviour and habitat needs of these animals. The chapter begins by introducing readers to the biology and behavioural ecology of Hawaiian spinner dolphins and illustrates why chronic harassment from unregulated tourism operations jeopardizes their extraordinary lifestyle. A brief retrospective on what is known about the tourism industry in Hawai'i and spinner dolphins is presented, illustrating how both science and management have failed to study and protect this species over the past 15 years. A brief prospectus follows, and argues for precautionary management. Finally, a comparison with historical accounts of how human–bear relationships have been managed in the US over time reiterates that these issues have been faced previously in other systems, and that the management of dolphin–human interactions in Hawai'i, and indeed elsewhere, has not been informed by the lessons of history.

Spinner dolphins

Spinner dolphins are small odontocete cetaceans found globally in subtropical and tropical oceans

(Perrin, 1998). Spinners are widely distributed in the Pacific, where at least four subspecies are recognized (Perrin, 1998). One form – *Stenella longirostris longirostris* – often referred to as Gray's spinner dolphin or the Hawaiian spinner dolphin, is the most widely distributed (Perrin, 1998). Hawaiian spinners are easily identified visually by their distinctive long narrow beak, striking three-part grey colour pattern (dark grey dorsal cape, light grey lateral fields and white or light grey ventral surfaces) and behaviourally by their characteristic longitudinal aerial spins (Perrin, 1998). Hawaiian spinners use both pelagic and coastal habitats, and in many places in the Pacific Islands Region, spinner dolphins are found during daylight hours in shallow (< 50 m) and sheltered bays. In these locations spinner dolphins socialize, care for their young and exhibit a stereotypical behaviour referred to as resting. Resting spinner dolphins tend to remain in tight groups, surfacing synchronously but infrequently and spend longer periods of time at depth between surfacings. In many places they transition from energetic social interactions (leaps, spinning, etc.) into this resting behaviour in the early morning and then transition back into energetic social behaviour at or before midday. This behaviour has been documented in sheltered bays of the Big Island of Hawai'i (Karczmarski *et al.*, 2005), as well as in bays at Oahu, Maui, Lanai and Kauai. Similar patterns have been identified in Guam (Eldridge,

Whale-watching: Sustainable Tourism and Ecological Management, eds J. Higham, L. Bejder and R. Williams.
Published by Cambridge University Press. © Cambridge University Press 2014.

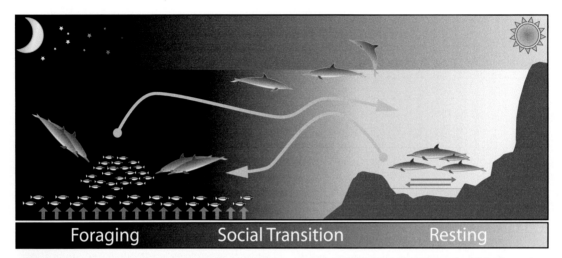

Foraging   Social Transition   Resting

Figure 19.1 A general overview of a day in the life of a Hawaiian spinner dolphin. *Note*: Available in poster form for download at the NOAA Pacific Islands Regional Office (PIRO) website http://www.fpir.noaa.gov/PRD/prd_swim_with_wild_dolphins.html.

1991), the Commonwealth of the Northern Mariana Islands (Trianni & Kessler, 2002), French Polynesia (Poole, 1995), and there are indications that spinner dolphins in American Samoa also rest in sheltered inshore regions during the day (Johnston *et al.*, 2008). Figure 19.1 provides a general schematic of a day in the life of a Hawaiian spinner dolphin.

There are several hypotheses addressing why spinner dolphins congregate in sheltered coastal regions during the day, all of which remain untested. The leading hypotheses suggest that spinners use coastal areas to limit predation risk, primarily from shark attacks but potentially from marine mammal predators such as killer whales (*Orcinus orca*), false killer whales (*Pseudorca crassidens*), pygmy killer whales (*Feresa attenuata*) and short-finned pilot whales (*Globicephala macrorhynchus*) (Norris *et al.*, 1994). Shallow, sheltered bays may simply limit encounter rates with pelagic predators during periods of reduced vigilance (while resting and socializing) and in these regions spinner dolphins may be better equipped to detect and avoid some predators by limiting stealth attacks from darker waters below. Indeed, spinner dolphins at most resting areas prefer open sandy bottoms (Norris *et al.*,

1994), which may highlight the dorsal profile of predators cruising below and facilitate lateral scanning for approaching dangers. Recent spatial modelling by Thorne *et al.* (2012) has confirmed many of the Norris *et al.* (1994) original hypotheses regarding predictors of resting habitats, but further research is required to fully understand why spinner dolphins frequent sheltered coastal regions during the day.

All forms of spinner dolphins appear to feed on small fishes (e.g. Myctophidae), squid (e.g. Omastrephidae) and shrimps (e.g. Sergestidae) obtained at depths between 100 and 300 m (Perrin, 1998; Benoit-Bird, 2004). Spinner dolphins do not generally feed in resting areas, but rather feed offshore on patches of prey found in near-surface waters during the night (Norris *et al.*, 1994). In pelagic systems, spinner dolphin prey species form deep horizontal patches during the day (often referred to as deep scattering layers) and these patches migrate into near-surface waters at night to exploit the higher productivity found in the euphotic zone (Perrin, 1998). In coastal regions like Hawai'i, these horizontal patches are part of the mesopelagic boundary layer or MBL (a phenomenon similar to the deep scattering layer seen in pelagic waters) that migrates both vertically and inshore during nighttime hours (Benoit-Bird & Au, 2003).

Spinner dolphins tend to leave coastal waters in the late afternoon and move offshore to meet the MBL and then track its progress inshore while foraging (Benoit-Bird & Au, 2003). As dawn approaches, spinner dolphins tend to track the MBL as it moves back offshore and into deeper waters, most likely until prey patches reach depths that make foraging dives inefficient for the dolphins (Benoit-Bird & Au, 2003). Prey patches within these scattering layers are not generally dense enough for individual spinner dolphins to forage on efficiently. To overcome this problem, spinner dolphins appear to exploit prey within the MBL in a coordinated fashion, herding patches of prey into tighter groups to further facilitate foraging (Benoit-Bird & Au, 2009). Indeed, it appears that dolphins are highly coordinated in these efforts, working in pairs within larger groups to herd and capture prey (Benoit-Bird & Au, 2009). Spinner dolphin prey items are generally small, and therefore each individual item is of low energetic value (Benoit-Bird, 2004). To satisfy the energetic requirements of maintenance, growth and reproduction, spinner dolphins must consume many small prey items during extended foraging trips.

Vigilance and resilience

Rest is a ubiquitous behaviour in animals, and especially crucial for animals that undertake lengthy and complex tasks (Cirelli & Tononi, 2008). Almost all animals exhibit periodic oscillations between various kinds of activity and a rest or sleep state (Cirelli and Tononi, 2008). During periods of activity, animals usually exhibit enhanced brain function, often referred to as vigilance. The concept of vigilance includes more than just behavioural alertness for predators (e.g. Lima & Dill, 1990). It encompasses any number of behavioural states exhibited by animals that require an enhanced ability to process information including social interaction, communication, foraging, feeding and navigation. Many

animals, including humans, exhibit decreases in performance of complex tasks during extended periods of vigilance (e.g. Mackworth, 1948), and this phenomenon is commonly referred to as a vigilance decrement (Dukas & Clark, 1995). The seminal study on vigilance decrements (Mackworth, 1948) was stimulated by the observed degradation of the ability of World War II radar-operators to detect objects on their screens over time. In higher vertebrates, vigilance decrement can manifest in a decreased ability to detect camouflaged predators or cryptic prey or in perhaps more abstract ways such as reduced decision-making capabilities. To recover from a vigilance decrement, animals must rest.

The processes governing vigilance decrement in animals and the transitions between activities requiring vigilance (e.g. foraging) and rest have been modelled quantitatively by Dukas and Clark (1995). In their model, vigilance decrements accrue to species as they undertake complex tasks as proportional rates scaled by species-specific constants. Their model focuses on assessing how vigilance decrement and recovery rates affect foraging times and predation risk, and makes useful predictions about how the length of foraging bouts and the costs incurred by switching from foraging to resting affect the length of time required for vigilance recovery. These relationships are conceptualized in Figure 19.2 for spinner dolphins and bottlenose dolphins (*Tursiops truncatus*). During a foraging bout t_f, an animal's vigilance level decreases until foraging ceases (v_1 to v_0), at which point vigilance decrement remains constant until rest (t_r) is initiated, and vigilance recovery can begin.

The length of time between transitioning from foraging to rest (τ) is dependent on the specific foraging tactics of the animal being modelled, and includes costs such as moving from foraging grounds to resting areas. The Dukas and Clark (1995) model predicts that the duration of foraging bouts is most greatly influenced by the rate at which animals recover from a vigilance decrement and the time delay incurred when transitioning to rest. This is conceptualized in Figure 19.3 for the same two species – the Hawaiian spinner dolphin and the

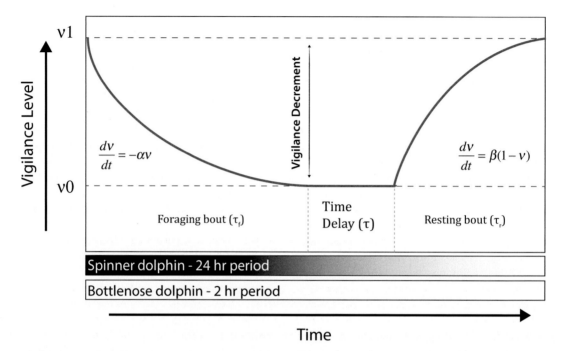

Figure 19.2 A model for changing vigilance levels in spinner dolphins (*Stenella longirostris longirostris*) and bottlenose dolphins (*Tursiops truncatus*) based on the Dukas and Clark (1995) model. The blue line represents vigilance levels across foraging and resting bouts. Note the differing time-scales for spinner dolphins and bottlenose dolphins.

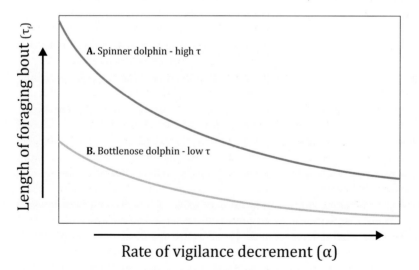

Figure 19.3 Optimal length of a foraging bout (t_f) as a function of the rate of vigilance decrement (α), for high (spinner dolphin) and low (bottlenose dolphin) costs of switching between foraging and resting (τ), modified from Dukas and Clark (1995).

bottlenose dolphin (with rate of vigilance recovery, β, held constant for both). In this plot, the length of a foraging bout, t_f, depends on the rate of vigilance decrement (α, which we assume is similar for both species) and on the time delay τ. If the costs of switching between foraging and resting are high (large τ), then longer foraging bouts become optimal. For spinner dolphins (Figure 19.3A), the costs of switching between foraging and resting are quite high, as animals must move some distance from foraging areas into sheltered bays to begin resting. It is further complicated if we consider the costs associated with social activities that occur between foraging and resting bouts, something not explicitly covered in the Dukas and Clark (1995) model. As such, longer foraging bouts would be optimal for spinner dolphins. Indeed, spinners may be especially adapted to compress foraging effort into one sustained foraging bout during the night when prey is most available (Figure 19.2). In doing so, they would accrue a significant vigilance decrement. In contrast, the bottlenose dolphin (Figure 19.3B) has a much more plastic behavioural schedule, with animals living proximate to food sources and alternating between periods of activity and rest more frequently (e.g. Bearzi *et al.*, 2009). In this case, the costs of switching from activity to rest are minor and optimal foraging bouts are shorter and more frequent. Indeed, the average foraging bout for a bottlenose dolphin is often measured in minutes, but can last as long as a few hours (Figure 19.2).

Hawaiian spinner dolphins exhibit a great degree of predictability in their behavioural patterns (Norris *et al.*, 1994), and this predictability increases their risk to human interactions. Recent studies of relationships among behavioural predictability and resilience in several odontocete cetaceans (using entropy estimates from Markov chain models of behavioral transitions: Lusseau *et al.*, 2009) suggest that behavioural systems that exhibit strong predictability in their transitional models (low entropy) were more strongly affected by perturbations than those that had high-entropy estimates and were less predictable. Further, the greatest factor

affecting the predictability of Markov chain models was bout duration. Longer bouts increased predictability and reduced resilience. Spinner dolphins engage in long resting bouts during daytime periods and long bouts of foraging at night. If the patterns revealed by Lusseau *et al.* (2009) translate to spinner dolphins, it would suggest that they exhibit low resilience in the face of behavioural perturbations such as harassment in resting bays.

Most recently, Wade *et al.* (2012) reviewed the available data on rates of recovery of depleted cetacean populations and concluded that odontocetes tend to exhibit less resilience to overexploitation than mysticetes, with some species exhibiting decreased birth rates following exploitation (including pelagic spinner dolphins). While it remains unclear exactly why this is so, they concluded that the strong social bonds exhibited by many odontocete species are clearly disrupted by anthropogenic removals and consistent harassment, and that the potential mechanisms for these observed reductions in birth rates include a deficit of adult females, a deficit of adult males and the disruption of mating systems (Wade *et al.*, 2012). Most relevant to the case of coastal spinner dolphins in Hawai'i is the fact that pelagic spinner dolphins in the Eastern Tropical Pacific do not appear to be recovering despite the fact that direct mortality during tuna sets has been dramatically reduced (e.g. Wade *et al.*, 2007; Gerrodette & Forcada, 2005). This suggests that indirect effects of continued chases, encirclements and entrapments of spinner dolphins – essentially intensive harassment – that occur during fishing operations constrain reproductive success (Wade *et al.*, 2012).

Spinner dolphins tend to partition their activities temporally, compressing certain behaviours into particular periods of the day to maximize foraging efficiency and avoid predation during periods of reduced vigilance. Because of this, spinner dolphins cannot compensate easily for disrupted resting periods during the day, as they must simply move on to the next foraging trip to exploit prey resources optimally. Furthermore, spinner dolphins must be efficient foragers to exploit prey during the night and this efficiency may be compromised if they cannot adjust for lost resting periods or if the social bonds that govern their complex cooperative foraging strategies are disrupted. The disruption of normal resting patterns poses significant and currently unquantified risks for both adult and young spinner dolphins (e.g. potentially reduced growth and reproduction and greater exposure to predation). Indeed, considering their compressed behavioural schedule, the low energetic value of prey items, the transient nature of their prey and greater susceptibility to perturbations (Benoit-Bird & Au, 2003; Benoit-Bird, 2004), spinner dolphins are likely more vulnerable to the adverse effects of human interactions than are many other species – such as bottlenose dolphins – that have more plasticity in their habits.

The effects of compromised rest periods on spinner dolphin physiology and energetics have not been formally investigated, but we do know that many animals exhibit marked negative responses when subjected to reduced/interrupted rest. In fact, in animals from insects to mammals, continual sleep deprivation leads to death (Cirelli & Tononi, 2008). It remains unclear how cetaceans in general may react to extended vigilance (Cirelli & Tononi, 2008). Regardless, the disruption of various 'normal' behaviour patterns proves problematic for many mammals, as their behaviour and physiology budgets can be greatly altered through stress (e.g. stampeding ungulate herds – reviewed in Stankowich, 2008), or they can be excluded or driven from important habitats (e.g. flushing marine birds from nests – see Ronconi & St. Clair, 2002) even if mediated by behavioural compensations such as habituation.

Retrospect

Decades of unregulated dolphin-based tourism in Hawai'i

Dolphin-based tourism is a growth industry in the Pacific Islands Region, with Hawai'i representing the most developed and mature (e.g. O'Connor *et al.*,

2009). In Hawai'i, spinner dolphins support the largest and most robust components of this industry, although other species (e.g. pilot whales) are also targeted. This industry comprises various types of operations ranging from beach-based kayak excursions to watching spinners at the surface through a variety of motorized boat trips to multi-day retreats featuring repeated boat-based snorkelling trips to swim with dolphins and physically interact with them.

In general, it is believed that the dolphin tourism industry in Hawai'i has grown significantly over the past 20 years, although there are few data available to actually quantify this trend. There are indications that the industry on the Waianae Coast of Oahu has grown from four operations to at least nine operations in the past eight years, although some of these may not be formally established commercial enterprises. Web searches for Hawaiian dolphin tourism advertisements suggest that the industry has expanded over the past decade (Stanton, 2005), although it is difficult to relate these results to real-world tourism effort. These indications are, however, consistent with expansion in marine mammal tourism seen in other areas of the Pacific Islands, where growth was as high as 49% in Guam during 1998–2008 (O'Connor *et al.*, 2009). Growth across Oceania, Pacific Islands and Antarctica was estimated at approximately 10% during the same period (O'Connor *et al.*, 2009).

At present, a simple Internet search for dolphin tours (searching the world wide web using keywords 'dolphin', 'swim' and the island name, e.g. 'Oahu') reveals that the greatest numbers of dolphin-based tourism operations are based on the Kona Coast of the Big Island of Hawai'i and on the Waianae coast of Oahu. On the Big Island, there are at least 14 tourism operations that advertise encounters with spinner dolphins along the Kona Coast, and there are also several periodic 'spiritual' retreat operations that include swimming with spinners as part of their package. On the island of Oahu, there are at least nine commercial operations focusing on dolphins along the Waianae Coast. This region has seen extensive growth over the past five years, coincident with the creation or refurbishment of large marinas. The four-islands area (Maui, Lanai, Molokai and Kahoʻolawe) also has at least nine dolphin tourism operations, focusing on La Perouse Bay, at Molokini and at bays on the nearby islands of Lanai. There are at least eight operators that include dolphin encounters as part of the experience during boat trips around the island of Kauai, primarily focusing on the Na Pali coast.

While several US management agencies at Federal and State levels are involved in marine mammal management issues (reviewed in Baur *et al.*, 1999), at present there is no centralized management of dolphin-based tourism in Hawaiʻi, and no legal permitting system that restricts entry into the industry. A brief overview of the main US legislation related to these issues is provided below.

The failure of science and management to protect spinner dolphins

Spinner dolphins have been studied in Hawaiʻi since the mid-1960s, and the most intensive studies were conducted by Ken Norris and his colleagues between 1974 and the mid-1990s. The results of these ground-breaking studies were published in the first comprehensive scientific natural history of a dolphin species in 1994 (Norris *et al.*, 1994). This book, entitled *The Hawaiian Spinner Dolphin*, leveraged an awesome array of scientific techniques (from radio-tracking to passive acoustic monitoring) to examine almost all aspects of spinner dolphin biology and ecology. When this book was published, Hawaiian spinner dolphins were one of the best-studied odontocetes in the world, surpassed only by bottlenose dolphins.

Since that time, many aspects of spinner dolphin research in Hawaiʻi have not kept pace with changes in the environment, coastal development and natural resource management. Several comprehensive studies of acoustics (e.g. see Lammers & Au, 2003) and foraging ecology (e.g. see Benoit-Bird & Au, 2003) have expanded on the foundations built by Norris and his colleagues, but almost

nothing new is known about the vital rates and social dynamics of Hawaiian spinner dolphins. At present, the only quantitative population estimate that exists for these animals (3351 individuals) is derived from a large-scale vessel-based survey in 2002 that did not sample dolphins in resting bays during the day (Barlow, 2006). Only very recently have genetic studies revealed that considerable population structure exists for spinner dolphins across the Hawaiian Archipelago (Andrews *et al.*, 2006, 2010), greatly increasing the complexity of much needed stock assessments for this species.

A limited amount of dated information on spinner dolphin abundance is available for the Kona Coast of the Big Island of Hawaiʻi, one newly proposed stock (Caretta *et al.*, 2010). Early photo-ID based estimates for the Kona coast of the Big Island during 1979–1981 (Norris *et al.*, 1994) indicated that there were approximately 960 dolphins using the region, but that the population was open, with animals moving into and out of the region. Subsequently, Östman (1994) provided an estimate of 2334 individuals in the same region during 1989–1992. More recently, Östman-Lind *et al.* (2004) suggested that between 855 and 1001 spinner dolphins were using these waters. These estimates should be viewed with caution, as they were not produced through a rigorous mark–recapture analysis, but rather by dividing the total number of identified individuals by the mean percent of individuals identified per school. These estimates have not been subjected to peer review, and there is no information on the variance of the estimates. Considering these caveats, it is difficult to make quantitative comparisons between time periods. Furthermore, almost nothing is known about the current vital rates of spinner dolphins in Hawaiʻi. These gaps make it almost impossible to assess how chronic harassment has affected spinner dolphins since the days of Norris and his colleagues, and they continue to hamper attempts to manage human interactions in resting bays. Local people have recently voiced concerns that the abundance of spinner dolphins in coastal waters is declining, and a number of these observations are captured in public testimony from federal public scoping meetings[1] about protecting dolphins held in 2006.

There is an increasing body of knowledge addressing the effects of tourism on dolphins. These studies span a range of species, from coastal common dolphins (*Delphinus delphis*) to killer whales and from this body of work a common theme is emerging. In cases where dolphin-based tourism is highly focused, the natural behaviours of focal animals are altered, often with deleterious effects (Lusseau & Bejder, 2008). For example, Bejder *et al.* (2006a, b) illustrated that bottlenose dolphins' relative abundance in Shark Bay, Australia declined significantly when tourism levels increased from one operation to two. Similarly, bottlenose dolphins in Doubtful Sound, New Zealand exhibit both short (avoidance, see Lusseau, 2006) and long-term negative effects (calf survival, see Currey *et al.*, 2009) from tourism interactions. In many cases it has been difficult to translate short-term reactions to population-level consequences. However, the number of studies that document short-term reactions with potential population-level effects is growing. Common dolphins in New Zealand have been negatively affected by tourism operations, primarily through interruptions of both foraging and resting bouts (Stockin *et al.*, 2008). Killer whale foraging behaviour has also been disrupted by intense whale-watching activity (Williams *et al.*, 2006, 2009) and efforts to model the energetic costs of interruptions in foraging activity indicate that substantial decreases in energy intake (up to 18%) may occur. Most recently, Visser *et al.* (2010) illustrated that the resting bouts of Risso's dolphins (*Grampus griseus*) can be interrupted by tourism activities.

At present, there are no studies that assess population-level effects of spinner dolphin harassment in resting areas, and most short-term studies have remained unpublished (e.g. Forrest, 2001), or are largely equivocal (e.g. Courbis & Timmel,

2009; Delfour, 2007), perhaps because of low power stemming from small sample sizes and suboptimal choice in response variables. Some studies have correlated human behaviour to dolphin presence (Courbis, 2007; Delfour, 2007), and others have reported changes in aerial activities and bay residency times associated with increased human interactions (Courbis & Timmel, 2009; Danil *et al.*, 2005). Almost all studies addressing the distribution and behaviour of spinner dolphins in Hawai'i express concern about increasing levels of tourism and potential effects on spinner dolphins (e.g. Delfour, 2007; Johnston, 2006; Lammers, 2004). However, no studies have directly and explicitly addressed the extent to which resting bouts are affected by various levels of human interactions, and there are no recent demographic studies available to assess population-level effects of this chronic harassment.

There are also no long-term studies of the growth of dolphin-based tourism in Hawai'i, despite the fact that many tourists choose Hawai'i as a vacation destination based on access to ocean-related forms of recreation and entertainment. At least one study has assessed the potential economic value of dolphin-based tourism in Hawai'i, and their results illustrate that young tourists place high value on opportunities to swim with dolphins in the wild (Hu *et al.*, 2009). Interestingly, this relationship decays as the age of tourist increases, indicating that older people value dolphin-based tourism that is less active (or perhaps less invasive). Furthermore, residents of Hawai'i tend to place lower value on swimming with dolphins than visitors (Hu *et al.*, 2009), perhaps because residents have already experienced dolphins in the wild (and indeed, large-scale dolphin-based tourism) while participating in other ocean-based recreational activities.

The US has adopted specific legislation to protect marine mammals. This legislation, called the Marine Mammal Protection Act (MMPA), was brought into force in 1972 (and subsequently amended many times – see Baur *et al.*, 1999) and has been effective at conserving species subjected to fisheries by-catch and greatly increased our

---

[1] The text of this public testimony is available at NOAA's Pacific Islands Regional Office (http://www.fpir.noaa.gov/PRD/prd_spinner_EIS.html), and makes for some compelling reading for those interested in the human dimensions of this situation.

understanding of many species (e.g. Read & Wade, 2000; Read *et al.*, 2006).

Under the MMPA, it is illegal for people to 'take' marine mammals without legal authorization (see section 101a). The concept of 'take' was originally defined under section 3(13) of the MMPA as any actions that 'harass, hunt, capture, kill or collect, or attempt to harass, hunt, capture, kill or collect' any marine mammal. Subsequently, this definition has been expanded to cover a range of more specific activities including, 'without limitation, any of the following: The collection of dead animals, or parts thereof; the restraint or detention of a marine mammal, no matter how temporary; tagging a marine mammal; the negligent or intentional operation of an aircraft or vessel, or the doing of any other negligent or intentional act which results in disturbing or molesting a marine mammal; and feeding or attempting to feed a marine mammal in the wild.' Furthermore, the term 'harassment' was defined in the 1994 amendments to the MMPA as 'any act of pursuit, torment, or annoyance which (i) has the potential to injure a marine mammal or marine mammal stock in the wild; or (ii) has the potential to disturb a marine mammal or marine mammal stock in the wild by causing disruption of behavioral patterns, including, but not limited to, migration, breathing, nursing, breeding, feeding, or sheltering.'

Much of this language seems pertinent to interactions with spinner dolphins in resting areas. Despite the creation of explicit legal rules protecting marine mammals from harassment in the US, there have been no attempts to apply these prohibitions on takes arising from consistent and numerous interactions between humans and dolphins in Hawai'i.

However, within the last nine years the governmental body that oversees the implementation of the MMPA – the US Marine Mammal Commission (MMC) – has repeatedly called for more science and stronger protective measures for spinner dolphins in Hawai'i. These recommendations illustrate the nature of the problem and call for local management authorities to take steps to reduce potentially harmful interactions within resting bays. For example, in a 2003 letter (Marine Mammal Commission 2003) to Vice Admiral Conrad C. Lautenbacher, Jr. (the US Department of Commerce Under Secretary for Oceans and Atmosphere at the time), the MMC stated 'commercial operators in Hawai'i are routinely offering the public opportunities to interact with dolphins in ways and at distances that result in unauthorized takings [see definition of takes above] and that are generally viewed as having adverse effects on the affected populations.' The letter states later: 'Despite the frequency of encounters between swimmers and dolphins, the predictability of when and where they will occur, and the clearly discernable disturbance that results, NOAA has taken little or no enforcement action to address the problem. At our meeting we heard from representatives of the National Marine Fisheries Service's Office of Protected Resources, the Office of Law Enforcement for the Service's Southwest Region, and the NOAA Office of the General Counsel. All of them agreed that, at least in some instances, the activities that are ongoing in Hawai'i constitute harassment [see definition of harassment under the MMPA above]. Yet, cases are not being brought, primarily because this issue is given low priority by the general counsel's office. As the senior enforcement attorney for the Southwest Region put it, "we tend to focus on cases that have a significant impact on whatever the resource is".'

Applying the precautionary approach to managing human–dolphin interactions in Hawai'i

In natural resource management, many systems are now focusing on the 'Precautionary Principle' for guidance, including the management of tourism (e.g. Fennell & Ebert, 2004). The Precautionary Principle addresses important issues of scientific consensus and the limits of science to understand, quantify and predict natural phenomena (see Ludwig *et al.*, 1993) – it urges us to take risk-averse and precautionary approaches to resource management in situations of scientific uncertainty (Stewart, 2002).

Marine mammals are notoriously difficult to count. These animals spend large proportions of their lives below the surface of the water and beyond the view of scientists wishing to enumerate their populations. To deal with these issues, scientists resort to sampling methods and statistical models that are subject to considerable biases stemming from variation in environmental conditions and animal behaviour. These methods tend to produce estimates with large levels of uncertainty, making it difficult to detect trends in the abundance of populations, especially if the populations are small and dispersed (Taylor & Gerrodette, 1993). For small odontocetes in US waters (like spinner dolphins), it is likely that 78% of precipitous declines in abundance of their populations would go undetected, despite significant efforts to enumerate them (Taylor *et al.*, 2007).

It has been argued publicly that because there is no definitive proof that spinner dolphins are depleted in Hawai'i and there is no need to take actions to protect them from anthropogenic insults. This argument raises pertinent questions about scientific uncertainty and where the burden of proof resides in contentious natural resource exploitation issues (Dayton, 1998). There are examples of what happens when sacrificing precaution because of uncertainty in managing natural resources. During the 1970s, commercial whaling was managed by the International Whaling Commission through the New Management Procedure, which provided a system to define stocks and set catch limits for large whales. This procedure relied heavily on science for management, but was hamstrung by the often-large levels of scientific uncertainty associated with defining management units and calculating the population sizes and trajectories required to set sustainable catch limits. These problems led to continued overexploitation and depletion of several whale stocks, finally culminating in the moratorium on whaling in 1986 and the listing of several species of large whales as endangered. More recently, Cadrin and Pastoors (2008) revealed that precautionary

management control rules are applied primarily to data-rich stock assessments in both the International Council for the Exploration of the Sea (ICES) and US marine fisheries management regimes, and that the precautionary approach is less frequently applied to management of fishery resources with the most uncertain stock assessments.

For spinner dolphins in Hawai'i, few estimates of abundance exist, they are not current, and comparisons of these data do not provide the statistical power required to detect any trend in dolphin numbers. At this point in time we simply cannot tell what the population trajectories are for spinner populations in Hawai'i – but should we wait for definitive scientific proof of their decline before limiting potentially harmful anthropogenic effects? In the case of managing human–dolphin interactions in Hawai'i, the available information clearly calls for the application of the precautionary principle – we can take risk-averse management actions now in the absence of a statistically demonstrated depletion of the population. A precedent in this area has already been set. In 2004, the Egyptian government protected a spinner dolphin resting area adjacent to Samadai Reef in the Red Sea, specifically to protect these animals from harassment during their resting period (Notarbartolo-Di-Sciara, 2008).

There are indications that the management of human interactions with spinner dolphins in Hawai'i is finally improving. In 2005, the NOAA Fisheries Service filed an announcement of proposed rule-making in the US Federal Register entitled 'Protecting spinner dolphins in the main Hawaiian Islands from human activities that cause "take," as defined in the Marine Mammal Protection Act and its implementing regulations, or to otherwise adversely affect the dolphins' (DOC, 2005). Subsequently, they have also announced an environmental impact assessment of various management options to reduce interactions with spinner dolphins, including the potential use of time/area closures (DOC, 2006). Most recently, NOAA has created the Dolphin Smart programme, a voluntary accreditation initiative that provides tourism operators with guidance on sustainable dolphin-watching practices and an opportunity for operations that adhere to these guidelines to eco-brand their tours as more dolphin-friendly.

Lessons from history: allowing spinner dolphins to remain wild

There are historical cases of human–wildlife interactions (in both terrestrial and marine systems) that can guide us in our assessments of spinner dolphin interactions with humans in the near-shore waters of Hawai'i (e.g. see Samuels *et al.*, 2000 and Bath & Enck, 2003). These examples can also help us make sound decisions regarding when and how humans can responsibly interact with these and other marine predators. In particular, the history of managing human–bear interactions in Yellowstone National Park provides a compelling analogy. There is an extremely important and valuable lesson to be taken from the history of human interactions with bears in Yellowstone. In the example below, we focus on the fact that animals do not always make the 'optimal' decision when interacting with humans and their habitats, and that these choices often have negative results for both animals and people. This means that real stewardship of wild animals requires that we provide ample opportunity for animals to *be* wild.

In the first century or so of Yellowstone National Park, garbage disposal was a large logistical problem. Current visitors have their food waste and trash exported from the park for disposal, but it was not always this way, as the park operated internal garbage dumps until 1970. However, from about 1890 until World War II, visitors to Yellowstone were entertained by nightly 'bear shows' at park dumps (Biel, 2006). At these shows, both black and grizzly bears would attend garbage heaps proximate to park hotels and camping areas to feed. In some cases humans constructed seating for these shows, including wooden bleachers to accommodate large groups of spectators (Biel, 2006). The bear shows were immensely popular. In fact, during the zenith of bear shows it could be difficult to find a place to

park your car. It was apparently quite popular with the bears as well; in 1920 an estimated 40 grizzly bears attended bear dumps in Yellowstone, and that number grew to over 250 individuals a decade later (Biel, 2006). People routinely provisioned bears by hand along roadsides, and at other camping locations within the park as well.

It was a recipe for disaster, for both people and bears. The closer bears came to people, the more habituated they became to human presence and the more injuries to bears and humans occurred. Some bears scared and injured people and, occasionally, bears even killed people. The first human mortality due to bear attack in Yellowstone occurred in 1942. Injuries were far more common. For example, between 29 and 38 people per year were injured while feeding bears during 1938–1942 (Biel, 2006). Bears suffered as well. Bears that were labelled repeat offenders were lethally removed, and lethal removals of bears peaked in Yellowstone during 1961, when 107 bears were shot (Biel, 2006).

The response by park management to human–bear interactions in Yellowstone was variable and often contradictory, and was to some extent a reflection of current values at the time. In the early days of Yellowstone, park management espoused what is often referred to as an aesthetic (as opposed to a utilitarian – see Lavigne *et al.*, 1999) conservation attitude toward the park and its resources (Biel, 2006). Bears were considered an important component of the Yellowstone landscape in terms of viewing pleasure. The park wanted what Alice W. Biel describes as 'the wild yet tame bear' (2006). As such, human–bear interactions were actually promoted. Later, as more ecologistic attitudes slowly permeated the National Park Service (NPS) management, interactions between bears and humans were officially discouraged, yet remained unofficially supported as the regulations prohibiting these interactions were rarely enforced (Biel, 2006). In fact, park management was initially unwilling to limit human–bear interactions as they feared that such measures would reduce park usage (Biel, 2006). Several logistical justifications were also used by park management to defend feeding of bears at roadsides and at campsites, including lack of enforcement capabilities and expenses of dealing with garbage (Biel, 2006). It wasn't until 1969 that the NPS finally produced a Bear Management Policy document that outlined specific objectives for education and enforcement, breaking from what Biel (2006: 96) described as the long-standing 'nudge nudge, wink wink' approach to reducing human provisioning of bears in Yellowstone. From then on, feeding bears was prohibited both officially and unofficially.

Let us compare the bear case with that of spinner dolphins in Hawai'i. As with bears, human activities alter the natural order for spinner dolphins to some extent, as they did not evolve within the current conditions created by humans. Clearly, we are not (yet) as worried about attacks on humans by spinner dolphins – and provisioning dolphins is not an issue in this case, but this does not minimize the importance of negative outcomes for dolphins stemming from interactions with humans. Spinner dolphins can act in two ways when encountered in resting bays. In some cases dolphins can be 'quietly shy' of intruders, edging away or in extreme cases vacating the resting bay entirely. In other situations spinners may immediately break from resting and approach vessels or swimmers (Norris *et al.*, 1994). In the latter case, close interactions are frequently justified by dolphin-swim participants and tourism proponents with arguments that dolphins choose to approach humans, and that they have no control over what occurs.[2] The implicit – and largely anthropomorphic – assumption is that dolphins prefer to interact with humans rather than socialize or recover their vigilance decrement; in effect, dolphins choose to forego crucial natural behaviours they need for social interaction and play with humans.

This anthropomorphic sentiment resonates with words penned by Horace M. Albright, the first NPS

Superintendent for Yellowstone during the zenith of bear shows in 1928: 'Bears are no longer wild animals to us. They have become personified. They are like people, and people want to treat them as such' (Biel, 2006). Albright further anthropomorphized bears, suggesting that male and female bears were poor parents because of their short (but natural) maternal and paternal investment periods: 'strange indeed that bears should prosper and increase in numbers under these harsh conditions of youth... no wonder they turn to [tourists] for kindness and candy' (Biel, 2006).

There are clear biological imperatives that govern a species response to invaders into their territory or habitat, including increased vigilance and reconnaissance of any potential threat (e.g. Cowlishaw, 1998). When viewed through a less anthropomorphic lens, it seems far more likely that dolphins approach humans that enter resting habitats for threat assessment purposes rather than for recreational purposes.

Even if the assumption that dolphins choose play with humans over rest is upheld, it remains unclear why (1) we expect animals to always make the correct choices in complex situations; and (2) why a bad choice on the dolphins' part would relieve humans of any responsibility if harm to dolphins stems from that choice. The existence of motor vehicle insurance in many developed nations illustrates that we do not even expect humans to always make the right choice, so why should we expect dolphins to?

Animals clearly make mistakes, and these mistakes can result in harm to them as individuals, and to their populations (e.g. Battin, 2004). One related example of this is the by-catch of odontocete cetaceans in longlines (Read, 2008). In Hawai'i, for example, false killer whales that frequently encounter longline gear learn to depredate either the bait or the catch (Forney *et al.*, 2011). In doing so, some animals make errors and become hooked or entangled in the gear (e.g. Baird & Gorgone, 2005), which results in various levels of injury or mortality. These harmful interactions are not ignored simply because the animal chose to interact with the

---

[2] See reactions by members of the public in response to the idea of an approach rule to limit interactions with spinner dolphins. Testimony in Public Scoping Meeting Transcripts associated with the Spinner Dolphin Human Interaction Environmental Impact Statement process available online at: http://www.fpir.noaa.gov/PRD/prd_spinner_EIS.html.

fishing gear. In the false killer whale/longline case, serious injuries or mortalities (DOC, 2012) are designated as 'takes' (as described above) under the MMPA and the US government is legally required to address these interactions if they are unsustainable and reduce them to a level that does not threaten the stock in question.

Furthermore, the MMPA currently prescribes a zero-rate mortality goal for marine mammal interactions with fisheries that do not transcend the unsustainable threshold, illustrating that negative interactions between humans and marine mammals are not dismissed as irrelevant even if the population-level effects are not immediately dire.

It is easy to see how this extends to spinner dolphins in Hawai'i. The fact that spinner dolphins are non-aggressive and choose to investigate visitors in resting areas should not disqualify them from protection from any harmful effects that may stem from those mistakes. In essence, the motivation of cetaceans is irrelevant when addressing the effects of humans on them, whether it is by-catch or harassment. In the case of spinner dolphins in Hawai'i, our actions alter their normal behavioural patterns in resting bays. In doing so, we reduce their ability to be wild.

There are growing parallels between early management of bear interactions in Yellowstone and current management of human–dolphin interactions in Hawai'i. For example, proponents of dolphin-based tourism often cite the aesthetic conservation value of viewing wild cetaceans, and emphasize the non-consumptive use of whales as beneficial (e.g. Barstow, 1986). This aesthetic attitude towards dolphins is similar to early attitudes towards human–bear interactions in Yellowstone. However, as in the case of bears in Yellowstone, unregulated interactions between humans and dolphins have become problematic. Anyone booking a trip to visit Kealakekua Bay or Honaunau on the Big Island of Hawai'i in the present day will be afforded a view of how many people attempt to interact with dolphins in these resting bays. Motorized vessels, kayaks, and swimmers/snorkellers will approach spinner dolphins while they are resting, and they will also be tracked when they exit resting bays by motorized vessels. The lackadaisical approach to enforcing early prohibitions on provisioning bears in Yellowstone is eerily similar to current management approaches for interactions between humans and spinner dolphins in Hawai'i. These interactions clearly change dolphin behaviour (e.g. Courbis & Timmel, 2009), and in some cases it causes animals to truncate resting periods entirely (Danil et al., 2005). This happens on a daily basis, across a number of resting bays when dolphins are present, and even NOAA legal counsel admit that harassment is occurring (MMC, 2003). Yet, not one prosecution has arisen from close encounters with spinner dolphins in resting bays in Hawai'i since harassment was clearly defined in the US MMPA by the 1994 amendments. As with bears in Yellowstone, it appears that official policy on dolphin interactions is often very different than on-the-ground implementation. Similarly, enforcement issues are now often cited as roadblocks to properly prosecuting people who purposefully harass spinner dolphins in Hawai'i (see MMC example above), deepening this parallel. Ironically, in 1992 a single person was assessed a fine of $1000 (two counts of harassment at $500.00 each) for swimming with spinner dolphins in Hawai'i (NOAA, 1992). This was two years before harassment was more clearly defined in the MMPA to better implement the law.

Conservation problems with cetaceans are often difficult to detect (e.g. Taylor et al., 2007), but once detected, they tend to follow a series of predictable steps as management agencies, scientists and other stakeholders ingest new information, adjust their views of the situation and proceed through responsive actions (Read, 2012). This process proceeds through structured scientific assessments of the problem to implementing mitigating measurements and monitoring programmes (Read, 2012). While there remains much science to do, we have detected a problem and it is possible to act now to begin mitigating the impacts of intensive tourism on spinner dolphins in Hawai'i.

The US MMPA provides considerable guidance on what constitutes unauthorized takes of marine mammals (see above) and includes specific definitions of what constitutes harassment of these animals (see above). Considering this guidance and definitions, many interactions between humans and spinner dolphins in resting bays are likely illegal. If so, they should be addressed through proper enforcement procedures. Any enforcement scheme should be accompanied by a strong outreach programme to educate local stakeholders, community members and tourists about the rules. It should be noted that education alone has not been effective in curtailing illegal interactions between humans and dolphins in other locations (e.g. McHugh *et al.*, 2011), so enforcement remains a key element to stimulate change. A compelling community-based monitoring programme also needs to be established in locations where human activities frequently overlap with spinner dolphin resting areas to keep concerned residents involved in the management and sustainability of their local dolphin population.

One promising approach to reducing interactions between humans and dolphins is to restrict human access to resting areas during periods when the dolphins are there. This approach, similar to time–area closures used in fisheries management to reduce by-catch of cetaceans (e.g. Murray *et al.*, 2000), would reduce overlap between humans and resting dolphins in both time and space. These areas and times could be explicitly described and communicated to people through a variety of media, including obvious markers in bays that delineate no-go areas. There are benefits to this type of system over others (e.g. approach rules). It eliminates any vagaries stemming from dolphins approaching humans in resting bays, and it provides clear guidelines for what humans involved in tourism activities can do. It also provides a clear path for monitoring and enforcement actions should people transcend these limitations. The NOAA is currently in the process of assessing the utility of time–area closures in four resting bays on the Kona coastline of the Island of Hawai'i as an adaptive management measure for human dolphin interactions (DOC, 2006).

Conclusions

Spinner dolphins are a highly visible and important component of the marine ecosystem in Hawai'i, and an important cultural and economic resource for the people who live there. Intensive dolphin-based tourism has clearly altered the behaviour and biology of coastal dolphins in other areas and considering the derived behavioural ecology of these animals and their specific habitat needs, the growth of unregulated dolphin-based tourism and associated harassment is an ongoing threat to Hawaiian spinner dolphins. Management authorities have largely ignored this issue, and only a minor amount of scientific effort has been directed towards assessing spinner dolphin abundance, demography and their responses to chronic harassment since the mid-1990s. For the past 15 years, both science and management have failed to address these issues and current resource management efforts for dolphins in Hawai'i have not tracked parallel processes in terrestrial systems. There is much to learn from how humans have dealt with terrestrial predators in parks and wilderness areas over time. From these lessons it can be concluded that there is an urgent need to carve out space and time for spinner dolphins in resting habitats, and restrict humans from interacting with them during rest periods. These actions will help allow spinner dolphins to retain some modicum of wildness, and in doing so it will help ensure the sustainability of their populations for future generations.

REFERENCES

Andrews, K.R., Karczmarski, L., Au, W.W.L., Rickards, S.H., Vanderlip, C.A. & Toonen, R.J. (2006). Patterns of genetic diversity of the Hawaiian spinner dolphin (*Stenella longirostris*). *Atoll Research Bulletin* 543, 65–73.

Andrews, K.R., Karczmarski, L., Au, W.W.L., *et al.* (2010). Rolling stones and stable homes: social structure, habitat diversity and population genetics of the Hawaiian spinner dolphin (*Stenella longirostris*). *Molecular Ecology* 19(4), 732–748.

Baird, R.W. & Gorgone, A.M. (2005). False killer whale dorsal fin disfigurements as a possible indicator of long-line fishery interactions in Hawaiian waters. *Pacific Science* 59(4), 593–601.

Barlow, J. (2006). Cetacean abundance in Hawaiian waters estimated from a summer/fall survey in 2002. *Marine Mammal Science* 22, 446–464.

Barstow, R. (1986). Non-consumptive utilization of whales. *Ambio* 15(3), 155–163.

Bath, A.J. & Enck, J.W. (2003). Wildlife–human interactions in National Parks in Canada and the USA. National Park Service and US Department of the Interior. Utah Regional Depository. Paper 424. Retrieved from http://digitalcommons.usu.edu/govdocs/424.

Battin, J. (2004). When good animals love bad habitats: ecological traps and the conservation of animal populations. *Conservation Biology* 18(6), 1482–1491.

Baur, D.C., Bean, M.J. & Gosliner, M.L. (1999). The laws governing marine mammal conservation in the United States. In J.R. Twiss Jr & R.R. Reeves (Eds), *Conservation and Management of Marine Mammals*. Washington, DC: Smithsonian Institute Press.

Bearzi, G., Fortuna, C.M. & Reeves, R.R. (2009). Ecology and conservation of common bottlenose dolphins. *Tursiops truncatus* in the Mediterranean Sea. *Mammal Review* 39(2), 92–123. doi:10.1111/j.1365-2907.2008.00133.x.

Bejder, L. & Samuels, A. (2003). Evaluating impacts of nature-based tourism on cetaceans. In N.J. Gales, M. Hindell & R. Kirkwood (Eds), *Marine Mammals: Fisheries, tourism and management issues*. Melbourne: CSIRO Publishing, pp. 229–256.

Bejder, L., Samuels, A., Whitehead, H., *et al.* (2006a). Decline in relative abundance of bottlenose dolphins exposed to long-term disturbance. *Conservation Biology* 20, 1791–1798.

Bejder, L., Samuels, A., Whitehead, H. & Gales, N.J. (2006b). Interpreting short-term behavioural responses to disturbance within a longitudinal perspective. *Animal Behaviour* 72, 1149–1158.

Benoit-Bird, K.J. (2004). Prey caloric value and predator energy needs: Foraging predictions for wild spinner dolphins. *Marine Biology* 145, 435–444.

Benoit-Bird, K.J. & Au, W.W.L. (2003). Prey dynamics affect foraging by a pelagic predator (*Stenella longirostris*) over a range of spatial and temporal scales. *Behavioral Ecology and Sociobiology* 53, 364–373.

Benoit-Bird, K.J. & Au, W.W.L. (2009). Cooperative prey herding by the pelagic dolphin, *Stenella longirostris*.

Journal of the Acoustical Society of America 125(1), 125–137.

Biel, A.W. (2006). *Do (Not) Feed the Bears: The Fitful History of Wildlife and Tourists in Yellowstone*. Lawrence, KS: University Press of Kansas.

Cadrin, S. & Pastoors, M. (2008). Precautionary harvest policies and the uncertainty paradox. *Fisheries Research* 94(3), 367–372.

Carretta, J.V., Forney, K.A., Oleson, E., *et al.* (2011). US Pacific marine mammal stock assessments: 2010. US Department of Commerce, NOAA. Technical Memorandum, NMFS-SWFSC-476.

Cirelli, C. & Tononi, G. (2008). Is sleep essential? *PLoS Biology* 6(8), 1605–1611. doi:10.1371/journal.pbio.0060216.

Cowlishaw, G. (1998). The role of vigilance in the survival and reproductive strategies of desert baboons. *Behaviour* 135(4), 431–452.

Courbis, S. (2007). Effect of spinner dolphin presence on level of swimmer and vessel activity in Hawai'ian bays. *Tourism in Marine Environments* 4(1), 1–14.

Courbis, S. & Timmel, G. (2009). Effects of vessels and swimmers on behavior of Hawaiian spinner dolphins (*Stenella longirostris*) in Kealake'akua, Honaunau, and Kauhako bays, Hawai'i. *Marine Mammal Science* 25, 430–440.

Currey, R.J.C., Dawson, S.M. & Slooten, E. (2009). An approach for regional threat assessment under IUCN Red List criteria that is robust to uncertainty: The Fiordland bottlenose dolphins are critically endangered. *Biological Conservation* 142(8), 1570–1579.

Currey, R.J.C., Dawson, S.M., Slooten, E., *et al.* (2009). Survival rates for a declining population of bottlenose dolphins in Doubtful Sound, New Zealand: An information theoretic approach to assessing the role of human impacts. *Aquatic Conservation-Marine and Freshwater Ecosystems* 19(6), 658–670.

Danil, K., Maldini, D. & Marten, K. (2005). Patterns of use of Maku'a Beach, O'ahu, Hawai'i, by spinner dolphins (*Stenella longirostris*) and potential effects of swimmers on their behavior. *Aquatic Mammals* 31(4), 403–412.

Dayton, P.K. (1998). Reversal of the burden of proof in fisheries management. *Science* 279, 821–822.

Delfour, F. (2007). Hawaiian spinner dolphins and the growing dolphin watching activity in Oahu. *Journal of the Marine Biological Association of the UK* 87(1), 109–112.

Department of Commerce. (2005). Protecting spinner dolphins in the main Hawaiian Islands from human activities that cause 'take,' as defined in the Marine Mammal

Protection Act and its implementing regulations, or to otherwise adversely affect the dolphins. *Federal Register* 70, 57923–57926.

Department of Commerce. (2006). Protection of marine mammals; notice of intent to prepare an environmental impact statement. *Federal Register* 71, 57923–57926.

Department of Commerce. (2012). *Process for Distinguishing Serious from Non-Serious Injury of Marine Mammals.* National Marine Fisheries Service Policy Directive PD 02–238. Department of Commerce, National Oceanic & Atmospheric Administration, National Marine Fisheries Service, Office of Protected Resources.

Dukas, R. & Clark, C.W. (1995). Sustained vigilance and animal performance. *Animal Behaviour* 49(5), 1259–1267.

Eldredge, L.G. (1991). Annotated checklist of the marine mammals of Micronesia. *Micronesica* 24, 217–230.

Fennell, D.A. & Ebert, K. (2004). Tourism and the precautionary principle. *Journal of Sustainable Tourism* 12, 461–479.

Forney, K.A., Kobayashi, D.R., Johnston, D.W., Marchetti, J.A. & Marsik, M.G. (2011). What's the catch? Patterns of cetacean bycatch and depredation in Hawaii-based pelagic longline fisheries. *Marine Ecology* doi:10.1111/j.1439-0485.2011.00454.x.

Forrest, A. (2001). The Hawaiian spinner dolphin *Stenella longirostris*: Effects of tourism. MSc thesis, Texas A&M University, Galveston, TX. 91 pp.

Gerrodette, T. & Forcada, J. (2005). Non-recovery of two spotted and spinner dolphin populations in the eastern tropical Pacific Ocean. *Marine Ecology Progress Series* 291, 1–21.

Hu, W., Boehle, K., Cox, L. & Pan, M. (2009). Economic values of dolphin excursions in Hawaii: a stated choice analysis. *Marine Resource Economics* 24(1), 61–76.

Johnston, D.W. (2006). A hard days night: Spinner dolphins need their rest too. *Ka Pili Kai* 28, 9–11.

Johnston, D.W., Robbins, J., Chapla, M.E., Mattila, D.K. & Andrews, K.R. (2008). Diversity, relative abundance and stock structure of odontocete cetaceans in the waters of American Samoa, 2003–2006. *Journal of Cetacean Research and Management* 10(1), 59–66.

Karczmarski, L., Wursig, B., Gailey, G., Larson, W. & Vanderlip, C. (2005). Spinner dolphins in a remote Hawaiian atoll: Social grouping and population structure. *Behavioral Ecology* 16(4), 675–685.

Lammers, M.O. (2004). Occurrence and behavior of Hawaiian spinner dolphins (*Stenella longirostris*) along Oahu's leeward and south shores. *Aquatic Mammals* 30(2), 237–250.

Lammers, M.O. & Au, Y. (2003). Directionality in the whistles of Hawaiian spinner dolphins (*Stenella longirostris*): a signal feature to cue direction of movement?. *Marine Mammal Science* 19(2), 249–264.

Lavigne, D.M., Scheffer, V.B. & Kellert, S.R. (1999). The evolution of North American attitudes toward marine mammals. In J.R. Twiss & R.R. Reeves (Eds), *Conservation and Management of Marine Mammals.* Washington, DC: Smithsonian Institute Press, pp. 10–47.

Lima, S.L. & Dill, L.M. (1990). Behavioral decisions made under the risk of predation: A review and prospectus. *Canadian Journal of Zoology* 68(4), 619–640.

Ludwig, D., Hilborn, R. & Walters, C. (1993). Uncertainty, resource exploitation, and conservation: Lessons from history. *Science* 260, 17–36.

Lusseau, D. (2006). The short-term behavioural reactions of bottlenose dolphins to interactions with boats in Doubtful Sound, New Zealand. *Marine Mammal Science* 22(4), 802–818.

Lusseau, D. & Bejder, L. (2008). The long-term consequences of short-term responses to disturbance: Experiences from whalewatching impact assessment. *International Journal of Comparative Psychology* 20, 228–236.

Lusseau, D., Williams, R., Bejder, L., *et al.* (2008). The resilience of animal behaviour to disturbance. Document SC/60/WW9. Santiago: Scientific Committee of the International Whaling Commission.

Lusseau, D., Bain, D.E., Williams, R. & Smith, J.C. (2009). Vessel traffic disrupts the foraging behavior of southern resident killer whales *Orcinus orca*. *Endangered Species Research* 6, 211–221.

Mackworth, N.H. (1948). The breakdown of vigilance during prolonged visual search. *Quarterly Journal of Experimental Psychology* 1, 6–21.

McHugh, K., Engleby, L., Horstman, S., *et al.* (2011). To beg or not to beg? Testing the effectiveness of enforcement and education activities aimed at reducing human interactions at a hotspot near Sarasota Bay, Florida. Abstract. 19th Biennial Conference on the Biology of Marine Mammals, 27 November–2 December, Tampa, FL.

MMC (2003). Comments on the increasing frequency with which marine mammals are being subjected to taking by harassment through directed human–marine mammal interactions. Letter to Vice Admiral Conrad C. Lautenbacher, Jr., Ph.D., USN (Ret.) Under Secretary for Oceans and Atmosphere Department of Commerce 14th and Constitution Avenue, N.W., Room 5128 Washington, DC 20230. Dated 6 May 2003.

Murray, K.T., Read, A.J., & Solow, A.R. (2000). The use of time/area closures to reduce bycatches of harbour porpoises: Lessons from the Gulf of Maine sink gillnet fishery. *Journal of Cetacean Research and Management 2*, 135–141.

NOAA. (1992). Joan P. Ocean, Case Nos. SW910586MM & SW910587MM (NOAA Sept. 17, 1992) (notice of violation & assessment).

Norris, K.S., Wursig, B., Wells, R.S. & Wursig, M (1994). *The Hawaiian Spinner Dolphin.* Berkeley, CA: University of California Press.

Notarbartolo Di Sciara, G., Hanafy, M.H., Fouda, M.M., Afifi, A. & Costa, M. (2008). Spinner dolphin (*Stenella longirostris*) resting habitat in Samadai Reef (Egypt, Red Sea) protected through tourism management. *Journal of the Marine Biological Association of the UK 89*, 211–216.

O'Connor, S., Campbell, R., Cortez, H. & Knowles, T. (Eds) (2009). *Whale Watching Worldwide: Tourism numbers, expenditures and expanding economic benefits.* Report to International Fund for Animal Welfare. Melbourne: Economists at Large. Available at: http://www.ifaw.org/international/resource-centre/whale-watching-worldwide.

Östman, J.S.O. (1994). Social organization and social behavior of Hawaiian spinner dolphins (Stenella longirostris). PhD dissertation, University of California, Santa Cruz.

Östman-Lind, J., Driscoll-Lind, A. & Rickards, S. (2004). Delphinid abundance, distribution and habitat use off the western coast of the island of Hawai'i. Southwest Fisheries Science Center Administrative Report LJ-04-02C. 28 pp. Available from SWFSC, PO Box 271, La Jolla, CA 92038.

Perrin, W.F. (1998). *Stenella longirostris. Mammalian Species* 599, 1–7.

Poole, M.M. (1995). Aspects of behavioral ecology of spinner dolphins (Stenella longirostris) in the nearshore waters of Mo'orea, French Polynesia. PhD Thesis, University of California, Santa Cruz.

Read, A.J. (2008). The looming crisis: Interactions between marine mammals and fisheries. *Journal of Mammalogy* 89, 541–548.

Read, A.J. (2012). Conservation biology. In I. Boyd, W.D. Bowen & S.J. Iverson (Eds), *Marine Mammal Ecology and Conservation.* Oxford: Oxford University Press, pp. 340–359.

Read, A.J. & Wade, P.R. (2000). Status of marine mammals in the United States. *Conservation Biology* 14, 929–940.

Read, A.J., Drinker, P. & Northridge, S. (2006). Bycatch of marine mammals in U.S. and global fisheries. *Conservation Biology* 20, 163–169.

Ronconi, R.A. & St. Clair, C.C. (2002). Management options to reduce boat disturbance on foraging black guillemots (*Cepphus grylle*) in the Bay of Fundy. *Biological Conservation* 108, 265–271.

Samuels, A., Bejder, L. & Heinrich, S. (2000). *A Review of the Literature Pertaining to Swimming with Wild Dolphins.* Report to the U.S. Marine Mammal Commission, Maryland.

Stankowich, T. (2008). Ungulate flight responses to human disturbance: a review and meta-analysis. *Biological Conservation* 141(9), 2159–2173. doi:10.1016/j.biocon.2008.06.026.

Stanton, K. (2005). Group wants limits on dolphin access: increased water traffic is disrupting the natural processes of spinner dolphins. Honolulu Star Bulletin, Honolulu US Census Bureau. Retrieved from: http://quickfacts.census.gov/qfd/states/15000.html (accessed 16 May, 2007).

Stewart, R.B. (2002). Environmental regulatory decision making under uncertainty. *Research in Law and Economics* 20, 76.

Stockin, K.A., Lusseau, D., Binedell, V., Wiseman, N. & Orams, M.B. (2008). Tourism affects the behavioural budget of the common dolphin *Delphinus* sp. in the Hauraki Gulf, New Zealand. *Marine Ecology Progress Series* 355, 287–295.

Taylor, B.L. & Gerrodette, T. (1993). The uses of statistical power in conservation biology: The vaquita and northern spotted owl. *Conservation Biology* 7(3), 489–500.

Taylor, B.L., Wade, P.R., De Master, D.P. & Barlow, J. (2000). Incorporating uncertainty into management models for marine mammals. *Conservation Biology* 14, 1243–1252.

Taylor, B.L., Martinez, M., Gerrodette, T., Barlow, J. & Hrovat, Y.N. (2007). Lessons from monitoring trends in abundance of marine mammals. *Marine Mammal Science* 23(1): 157–175.

Thorne, L.H., Johnston, D.W., Urban, D.L., *et al.* (2012). Predictive modeling of spinner dolphin (*Stenella longirostris*) resting habitat in the main Hawaiian Islands. *PLoS ONE* 7, e43167. doi:10.1371/journal.pone.0043167.

Trianni, M.S. & Kessler, C.C. (2002). Incidence and strandings genetic diversity of the Hawaiian spinner dolphin (*Stenella longirostris*), in Saipan Lagoon. *Micronesica* 34, 249–260.

Visser, F., Hartman, K.L., Rood, E.J.J., *et al.* (2010). Risso's dolphins alter daily resting pattern in response to whale watching at the Azores. *Marine Mammal Science* 27(2), 366–381. doi:10.1111/j.1748-7692.2010.00398.x.

Wade, P.R., Watters, G.M., Gerrodette, T. & Reilly, S.B. (2007). Depletion of spotted and spinner dolphins in the eastern tropical Pacific: Modeling hypotheses for their lack of recovery. *Marine Ecology Progress Series* 343, 1–14.

Wade, P., Reeves, R. & Mesnick, S. (2012). Social and behavioural factors in cetacean responses to overexploitation: Are odontocetes less 'resilient' than mysticetes? *Journal of Marine Biology* 2012, 1–15. doi:10.1155/2012/567276.

Williams, R., Lusseau, D. & Hammond, P. (2006). Estimating relative energetic costs of human disturbance to killer whales (*Orcinus orca*). *Biological Conservation* 133(3), 301–311.

Williams, R., Lusseau, D. & Hammond, P. (2009). The role of social aggregations and protected areas in killer whale conservation: The mixed blessing of critical habitat. *Biological Conservation* 142, 709–719.

20

Insights from agent-based modelling to simulate whale-watching tours

Influence of captains' strategy on whale exposure and excursion content

Clément Chion, Jacques-André Landry, Lael Parrott, Danielle Marceau, Philippe Lamontagne, Samuel Turgeon, Robert Michaud, Cristiane C. A. Martins, Nadia Ménard, Guy Cantin and Suzan Dionne

Introduction

Multi-agent models can bear several names depending on the field they were initially developed in (e.g. agent-based model in social science, individual-based model in ecology). Agent- and individual-based models (ABMs and IBMs) are becoming tools of choice to simulate complex social–ecological systems (Gimblett, 2002; Janssen & Ostrom, 2006; Monticino et al., 2007; Bennett & McGinnis, 2008). The recent development of dedicated programming platforms and libraries has also contributed to the expansion of multi-agent models coupled with geographic information systems (GIS) (Railsback et al., 2006). Such models have been applied in a wide variety of natural resource management contexts where heterogeneous actors interact, including rangeland management in arid zones (Gross et al., 2006), management of water use and access in river basins (Schlüter & Pahl-Wostl, 2007), control of irrigation channels (van Oel et al., 2010), agriculture management (Manson, 2005), and forest clearing for agriculture (Moreno et al., 2007). ABMs have also been used to support national parks and recreation areas' managers by simulating visitor movements to predict overcrowded areas along vehicular routes and hiking

trails (Itami et al., 2003), or along riverside rest areas and attraction sites for rafting trips on the Colorado River (Roberts et al., 2002).

ABMs of social–ecological systems where natural resource management is at stake are frequently used to explore outcomes of *what-if* scenarios of policy rules (Gimblett et al., 2002). Apart from testing policy rules, such models involving humans can also be used to explore the effects of alternative behaviours on the status of the natural resource. In this study, we developed a spatially explicit multi-agent model named 3MTSim (Marine Mammal and Maritime Traffic Simulator) to investigate whale-watching activities in the Saint-Lawrence Estuary and the Saguenay River, Québec, Canada (Parrott et al., 2011). Whale-watching activities in this area have increased dramatically since the 1990s (Dionne, 2001), raising concerns about the impact of intensive navigation on targeted whale populations, some of which were, and still are, of special concern (COSEWIC, 2005), threatened (Demers et al., 2011), or endangered (Beauchamp et al., 2009) under the Species at Risk Act (2002, Canada). Public pressure on governments led to the creation of the Saguenay–Saint-Lawrence Marine Park (referred to as marine park later) in 1998 (Guénette & Alder, 2007) whose limits are shown in

Whale-watching: Sustainable Tourism and Ecological Management, eds J. Higham, L. Bejder and R. Williams.
Published by Cambridge University Press. © Cambridge University Press 2014.

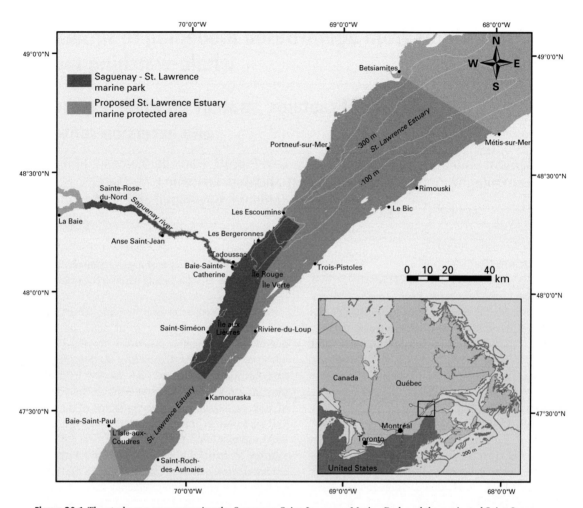

Figure 20.1 The study area encompassing the Saguenay–Saint-Lawrence Marine Park and the projected Saint-Lawrence Estuary marine protected area.

Figure 20.1. The implementation of regulations on marine activities followed in 2002 (Parks Canada, 2002), with law enforcement ensured by Parks Canada wardens. In addition to a series of rules regulating observation activities (e.g. maximum observation duration), the regulations also fixed a cap of 59 commercial permits for regular boats operating in the marine park (53 dedicated to whale-watching) (Parks Canada, 2002). Whale-watching activities in the marine park area rely on the relatively predictable presence of several whale species,

five of which represent 98.5% of the total number of observations (Michaud *et al.*, 2008). In 2007, we estimated that approximately 13,000 commercial excursions went to sea, 80% of which were dedicated to whale-watching within the marine park (Chion *et al.*, 2009). The projected Saint Lawrence Estuary Marine Protected Area (MPA), proposed by Fisheries and Oceans Canada, is expected to extend the protection of marine ecosystems beyond the marine park limits (Figure 20.1). As whale-watching activities are significantly less dense and abundant in the MPA than in the marine park, we decided to focus our study on excursions taking place in the marine park only.

3MTSim combines an ABM of navigation activities with an IBM of whale movements into a GIS-based representation of the geographic area. An asset of 3MTSim is that it allows the collection of exhaustive data of phenomena difficult or expensive to sample in the real system, such as the total amount of time each individual whale is exposed to observation boats. Major components of local marine activities are considered in 3MTSim with a special focus on whale-watching excursions. A great deal of effort was made to understand whale-watching captains' decision-making in order to reproduce realistically their behaviour in 3MTSim. Whale-watching excursions' data analysis and investigation of captains' decision-making processes through cognitive interviews revealed that they often favour *sure observations* of *potentially dramatic species* (Chion, 2011, chapters 3 and 4). Captains achieve *sure observations* mainly by exploiting the knowledge of current observations made by other whale-watching boats; a high level of cooperation at sea being the fundamental behavioural mechanism allowing the flow of information via the radio VHF communication channel. *Potentially dramatic species* are those well known for their spectacular displays (e.g. humpback whales' breaches or tail-slapping, fin whales hunting in large groups) or having notable characteristics (e.g. the blue whale is the largest animal ever on Earth; adult belugas are all white). Similarly, a vast collection of multi-platform observation data (enumerated later) was used to simulate the movements and distribution of whale species.

In this chapter, we investigate the effect of different captains' decision-making strategies on the exposure of targeted whales to observation vessels. Our investigation is aimed at demonstrating the feasibility of using an ABM for advisory purposes. After an overview of 3MTSim, we use the model to explore how alternative decision-making strategies, which could be suggested to whale-watching captains via a code of conduct or training sessions, might decrease whales' exposure to boats. We then discuss some lessons and insights that can be learned about the dynamics of whale-watching excursions using multi-agent modelling.

Overview of 3MTSim

3MTSim was developed as a decision-support tool for MPA managers. It integrates features dedicated to test the potential effects of alternative zoning and regulation plans (e.g. introducing speed limits, altering shipping routes, adding restricted access zones) on the patterns of traffic in and around the marine park and thus on the characteristics of whale–vessel encounters (e.g. rate, location). A description of the model and its functionalities is provided in Parrott *et al.* (2011).

The model combines a grid-based spatial environment (GIS) with an individual-based model of whale movements and an agent-based model of boats. During simulation runs, the movement of each individual whale and each boat is determined by algorithms and rules calibrated to reproduce observed patterns of behaviours (Grimm *et al.*, 2005). Simulations are run for short periods of time, based on realistic environmental conditions and known scenarios of whale abundances and patterns of habitat selection. The model time step currently used for simulations is one minute.

Spatial environment

The spatial environment of 3MTSim is represented by grid data (i.e. rasters) stored in an embedded GIS. The bathymetry is considered in the displacement and diving routines of whales, as well as for navigation. The state of the tide is modelled according to a simple daily cycle that selects the tide condition (flood, high, ebb and low tide) according to the date and time of day. While weather conditions are not explicitly modelled, visibility extent is represented by a single parameter for the whole area. This value remains constant for the duration of a simulation, mainly affecting whale-watching captains' ability to locate whales in their vicinity.

Whale individuals

The IBM of whale movements is described in detail in Lamontagne (2009). It includes the five most common species in the estuary: beluga (*Delphinapterus leucas*), minke (*Balaenoptera acutorostrata*), fin (*Balaenoptera physalus*), humpback (*Megaptera novaeangliae*) and blue whales (*Balaenoptera musculus*). Insufficient data were available on whales' food sources and on individuals' activity budgets, preventing any attempt to devise a behavioural model. Instead, whale movement patterns were extracted from:

- tracking VHF data: 80 tracks for more than 380 hours, for beluga (Lemieux Lefebvre, 2009), fin and blue whales (Giard & Michaud, 1997; Michaud & Giard, 1997, 1998); and
- land-based theodolite tracks of the four rorqual species: 140 focal follows with ~100 hours of tracking of individuals followed for more than 30 min (C.C.A. Martins, unpublished data).

Spatial distribution and aggregation patterns were derived from:

- sightings made from research vessels: ~550 baleen whales sightings from transect surveys (Group for Research and Education on Marine Mammals (GREMM), 2007); and
- sightings made from whale-watching vessels: 32,000 marine mammal sightings from more than 2100 sampled whale-watching excursions (Michaud *et al.*, 1997, 2008).

The model combines a simple diving routine with a displacement algorithm to determine each individual whale's depth, direction and speed at each model time step (i.e. 1 min). Diving and surface sequence durations are randomly selected from an empirically derived Weibull distribution computed for each species from land-based tracking data (Lamontagne, 2009). The diving routine uses a simple deterministic function to

calculate the amount of remaining oxygen as a function of the whale's depth and diving time, thus forcing the whale to surface regularly for breathing. Several displacement algorithms were implemented and tested, starting from a simple random walk and increasing in complexity to include residence indexes (Turchin, 1998) and social interaction between whales (Couzin *et al.*, 2005). The ability of each algorithm to successfully match the (often conflicting) patterns for each species was assessed. The MMNB algorithm (minimization of the mean normalized bias), a modification of the correlated random walk (Turchin, 1998), proved the most successful at reproducing the desired patterns, and is currently implemented in the model. For each species, MMNB randomly selects an individual's speed and move duration from the empirical distribution and then adjusts the turning angle to reduce the normalized mean difference between the real and simulated group size (animal density within a 2 km radius), turning angle and spatial distribution patterns (Lamontagne, 2009).

Whale-watching boat captain agents

Whale-watching excursions are challenging to model. Their dynamics, driven by captains' decisions, is highly dependent on several factors, such as whales' spatiotemporal distribution, species' abundance, and contextual factors (e.g. regulations, current observations made by concurrent companies, companies' guidelines and directions). These boat captains are goal-oriented and have to find a way to achieve their goal in a dynamic environment. Interviews with boat captains and marine park wardens conducted after excursions at sea, as well as VHF radio monitoring, revealed a number of attributes of their decision-making, that were included in the model. In particular, whale-watching boat captains: (1) take advantage of information on the most recent observations to explore space when no other information is available; (2) share information about whale locations; (3)

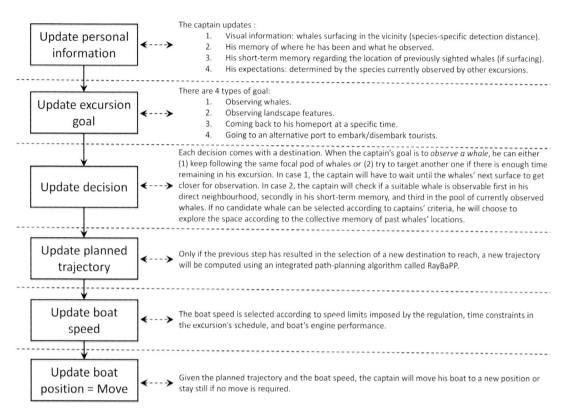

The captain updates :
1. Visual information: whales surfacing in the vicinity (species-specific detection distance).
2. His memory of where he has been and what he observed.
3. His short-term memory regarding the location of previously sighted whales (if surfacing).
4. His expectations: determined by the species currently observed by other excursions.

Update personal information

There are 4 types of goal:
1. Observing whales.
2. Observing landscape features.
3. Coming back to his homeport at a specific time.
4. Going to an alternative port to embark/disembark tourists.

Update excursion goal

Each decision comes with a destination. When the captain's goal is to *observe a whale*, he can either (1) keep following the same focal pod of whales or (2) try to target another one if there is enough time remaining in his excursion. In case 1, the captain will have to wait until the whales' next surface to get closer for observation. In case 2, the captain will check if a suitable whale is observable first in his direct neighbourhood, secondly in his short-term memory, and third in the pool of currently observed whales. If no candidate whale can be selected according to captains' criteria, he will choose to explore the space according to the collective memory of past whales' locations.

Update decision

Only if the previous step has resulted in the selection of a new destination to reach, a new trajectory will be computed using an integrated path-planning algorithm called RayBaPP.

Update planned trajectory

The boat speed is selected according to speed limits imposed by the regulation, time constraints in the excursion's schedule, and boat's engine performance.

Update boat speed

Given the planned trajectory and the boat speed, the captain will move his boat to a new position or stay still if no move is required.

Update boat position = Move

Figure 20.2 Sequence of actions (from top to bottom) that each captain agent goes through at each time step during the simulation.

give priority to more dramatic species such as the humpback whale; (4) try to adjust the content of their excursion according to that of their direct competitors; and (5) must respect navigational limits related to currents and bathymetry. In the model, at each time step the virtual captain agents follow a series of steps from information acquisition to movement execution (Figure 20.2).

A whale-watching captain's main objective is to observe whales during an excursion (although some also have subobjectives related to sightseeing, for example). In the model, excursions leave port according to planned schedules. Captains navigate using a path-planning algorithm to select the shortest path to their destination. Captains choose which whale to observe using a cognitive

heuristic decision-making module (Chion *et al.*, 2011) according to their preferences and constraints. The captains must make use of existing information (either from current data on whale locations if available, or retrieving from the memory of previous excursions' observations) to select where to navigate their excursion to observe whales. This type of decision is quick, based on limited information, with no optimal universal solution, and is repeated several times during an excursion. We assume, therefore, that the captains are operating in a context of bounded rationality, where they will select what appears to be the best choice given currently available information and the contextual setting both in time (e.g. what species they have already observed during the excursion will affect their choice of the next pod to target) and in space. The validation of the whale-watching vessel captain model is described in further detail in Chion *et al.* (2011).

Other whale-watching captains' cognitive and sensory capabilities are implemented within the model. Past observations are aggregated in a collective spatial memory according to a simple clustering algorithm that groups those past observations in clusters where the maximum pairwise distance does not exceed the visibility extent. This approach was chosen to represent the way captains aggregate past unique observations in broader regions where the action took place, rather than in precise locations where each given observation occurred. The distance of whale detection by a captain's visual module is species-dependent. For each species, the detection distance was calibrated using the knowledge of observers working at counting whales on the Saint-Lawrence during sea-based transects.

Methods

Rules considered by whale-watching captains to choose a whale to observe

Our investigation of whale-watching captains' decision-making revealed several notable characteristics and mechanisms which were subsequently implemented within the model (Figure 20.2). The following decision rules were elicited from field work mainly consisting of (1) seven semi-structured interviews (~10 hours) conducted with whale-watching captains after an excursion, (2) 15 hours of VHF radio monitoring, and (3) observations made during 30 excursions onboard all boats and ports in the marine park area. Extracted decision rules serve as the reference model for whale-watching captains' decision-making about which pod to target for observation. Within the sequence of actions detailed in Figure 20.2, these rules intervene at the step 'update decision', when the goal is 'observing whales' and a new pod of whales has to be targeted

among a set of candidate animals. Given a set of candidate whales, the captain agent jumps to the next rule until only one whale remains in the list. Rules are ranked as follows.

1. Captains try to find species that are not currently observed by any captain at sea. If such a species appears opportunistically in their surroundings, the captain will target it. This will give him/her an edge over the competition.
2. Captains favour species not already observed in their own excursion that have been observed in other excursions.
3. Captains prioritize the whales belonging to the overall top-ranking species in their decision (i.e. humpback, fin and blue whales). In fact, species' attractiveness is not the same for all species. When present in the area, data from sampled excursions show that humpback whales are responsible for the largest aggregations of boats followed by the fin, blue, minke and beluga whales (all pairwise differences statistically significant). Several characteristics have an impact on species' attractiveness, such as their potential spectacular displays, ease of observation (no fleeing behaviour), predictability of individual distribution, core habitat areas (e.g. proximity from departure ports), abundance, and species-specific regulations.
4. Captains prefer whales that are about to be lost from the pool of discovered ones (i.e. no boat observing them anymore). This is all the more true for individuals from species standing high in the preference ranking such as humpback whales.
5. The next criterion is the preference for whales with the lowest number of boats in their surroundings. Some captains, often those with more experience, give a higher priority to non-crowded sites.
6. Captains favour observations allowing subsequent observations in the area. This ability to anticipate and build an excursion in advance and adjust it as a function of upcoming information is expected to be more prevalent with experienced captains.

7. In case of a tie between candidate whales, captains will break the tie by choosing the closest whale.

We followed a naturalistic decision-making approach to investigate captains' decisions in action (Klein *et al.*, 1989; Klein, 2008) and modelled it following the bounded rationality framework (Simon, 1957; Gigerenzer & Selten, 2001). Being aware that all captains neither have the same experience nor the same values, using a single model to represent all captains' decision processes is a current limitation of the model. However, the validation process proved that this approach allowed the faithful reproduction of some key individual (total length, activity budget and contribution of species in observations) and collective (core areas of activity and boat aggregations) patterns of excursions (Chion *et al.*, 2011). This suggests that from the collective perspective, which is of particular interest, individual differences have a less critical influence on the global dynamics than individual similarities (e.g. overall preference ranking for given species) and shared collective mechanisms (e.g. cooperation via communication, prevalence of knowledge exploitation over space exploration).

The rules described above were implemented as cues within the *take-the-best* heuristic structure (Gigerenzer & Goldstein, 1996) that has proved to best reproduce excursion patterns (Chion *et al.*, 2011) among several cognitive heuristics taken from the bounded rationality literature (Gigerenzer & Selten, 2001). The prevalence of a non-compensatory heuristic such as *take-the-best* over compensatory ones (e.g. tallying) supports the fact that whales' characteristics do not have the same importance in captains' decisions.

Alternative decision-making strategies

To study how the decision-making process of captains can influence the dynamics of whale-watching excursions and ultimately affect the global dynamics of the system, we implemented two alternative decision-making strategies in 3MTSim

that virtual captains follow when deciding which whale to observe. Our objective was to foresee how such alternative behaviours could affect both whales' exposure and excursions' dynamics and content. This type of application could lead to a series of recommendations passed on to captains during seasonal training sessions. The simulations run with each alternative decision-making model (DMM) aimed at demonstrating the feasibility of such a utilization of ABM for advisory purposes.

We present hereafter the two simple alternative DMMs that were implemented and tested within 3MTSim. Rules contained within these two alternative models were implemented within the *take-the-best* heuristic structure. The alternatives were expected to mitigate whales' exposure without significantly affecting observation activities (e.g. time spent in observation).

Preference for less-crowded observation sites (DMM-1)

The idea of this DMM is to favour whales with fewer boats in observation. Taking into account this criterion in the process of selecting whales to observe is expected to decrease the aggregation of boats on observation sites, which is a goal pursued by the marine park managers regarding the management of whale-watching activities. Captains using this decision strategy will apply decision rules in the following order.

1. Captains try to find species that are not currently observed by any captain at sea. If such a species is visible in their surroundings, it will be targeted.
2. Captains favour species not already observed in their own excursion.
3. Captains will pick the observation site with the fewest boats on it, with a coordination mechanism allowing captains to account for others' intentions (i.e. captains heading to observe a whale but not currently observing it).
4. In case of a tie, captains will choose the closest site regardless of species.

No preference ranking of whale species (DMM-2)

This DMM gives the same weight (i.e. importance) to all species in captains' selection process of which pod of whales to target. The idea to test this DMM comes from an issue noticed repeatedly in the past, when some whales belonging to scarce species attract numerous boats in their vicinity, via a domino effect. Captains using this decision-making module will apply the following rules.

1. Captains try to find species that are not currently observed by any captain at sea. If such a species appears in their surroundings, it will be targeted.
2. Captains favour species not already observed in their own excursion.
3. Captains will not ground their decision based on a species preference ranking.
4. In case of a tie, captains will pick the closest whale between remaining candidates.

Design of experiment and simulation parameters

For the reference model and both alternative DMMs described above, we ran 10 replications of a one-week simulation. We fixed the number of runs to 10 by monitoring the inter-run variability.

The data from the first day of each simulation (transient state) were systematically discarded to keep only the model's steady state. The visibility parameter was set to 4 km for all simulations. The period of the year simulated is the peak tourist season (between mid-July and mid-August); this is the most critical time of the year in terms of the number of boats at sea. Excursion schedules and zodiac departures reach a peak at this time of the year in response to the maximum tourism demand.

All simulations were run with the same whale species' abundance and spatial distribution settings (Table 20.1). Except for belugas, abundances were selected to reflect the approximate proportion of each species compared to the others, as observed

Table 20.1 Whale species' setting used for simulations.

Species	Abundance	Years of spatial distribution data used
Minke	40	2007
Fin	20	2007
Blue	3	2007
Humpback	3	2007
Beluga	100	1994–2007

during recent seasons. For belugas, because they are often excluded from observation activities due to the minimum observation distance restriction (400 m), we lowered their number (from ~1000 in the real system to 100) to speed up simulations.

Output variables observed

In order to assess the impact of a given DMM strategy on the system, we observed several variables returned by the model. We distinguish variables characterizing the impact on whales' exposure from those impacting excursions' dynamics.

Variables characterizing whales' exposure to whale-watching boats

We chose four variables to characterize whale exposure to observation boats. We made the distinction between individuals, species and overall exposure.

- Exposure of individual whales.
- *Percentage of individual whales observed.* This variable provides insight on the proportion of individual whales that have been exposed to observation activities.
- *Duration of continuous sequences of observation.* This variable allows monitoring the duration of the continuous sequences of observation that animals are subject to. For instance, if two boats observe the same whale during 30 min successively (the first boat leaving when the

second arrives), the duration of the continuous observation sequence will be 60 min.

- Exposure of species.
- *Species' contribution to observation activities*. This variable tells us the contribution of each species to the budget of all whale-watching activities.
- Overall exposure of whales present in the area.
- *Time spent in observation activity*. This variable allows computing the total time whale-watching boats have been observing whales.

We are aware that some of these variables should be regarded cautiously as the level of knowledge introduced in the model may affect their accuracy. For instance, the percentage of individuals observed partially depends on the spatial location of individual whales; however, in reality some specific individuals may display some site fidelity, which is not fully known or modelled within 3MTSim.

Variables characterizing excursions' dynamics

Modifying a captain's strategy affects excursions' dynamics. We monitor changes by recording and analysing the following variables.

- *Success of the excursions*. This variable informs us about the percentage of excursions that made at least one observation during the outing.
- *Time spent in observation activity* (cf. description above).
- *Proportion of time boats are alone with the targeted pod*. This variable is an indicator of the quality of observations. Because the large number of boats at sea is one of the top-most sources of concern about whales' protection and the most negative element experienced by whale-watching tourists during their excursion (see Giroul *et al.*, 2000: 53–54), it can be reasonably inferred that decreasing boat concentrations would contribute to the enhancement of the visitors' experience.
- *Boat aggregations around observed pods of whales*. This is a critical variable for managers who wish to decrease boat aggregations at sea, known to modify whale behaviour (Michaud & Giard, 1997, 1998).

Table 20.2 Increase in the total number of individual whales observed each simulated day for both alternative DMM in comparison to the reference model.

DMM-1 (compared to reference)	DMM-2 (compared to reference)
+15%	+1.5%

More variables could be added to this analysis framework. However, the set of variables presented above are intended to give an insight into 3MTSim's capability to monitor effects induced by captains' changes of behaviour.

Results and discussion

We now present and discuss the simulation results for both alternative models (DMM-1 and DMM-2) and compare them to the reference model's outputs characterizing the current situation at sea.

Whale exposure

Simulations revealed that DMM-1 and DMM-2 both increase the total number of individual whales observed during a day compared to the current situation (modelled by the reference DMM). The strategy where captains favour observation sites with fewer boats leads to a 15% increase in the number of individuals observed, whereas the strategy where no preference ranking of species exists leads to a 1.5% increase in the number of observed individuals (Table 20.2). These relative increases are significant with both p-values < 0.01 (Wilcoxon rank sum test).

We compared the distributions of the duration of observation sequences produced by DMM-1 and DMM-2 (cf. Figure 20.3). Neither DMM-1 nor DMM-2 affected this metric significantly compared to the reference model (Wilcoxon rank sum test). This is a consistent result because the rule that controls the decision to leave the observation site remained the same for all tested models (a

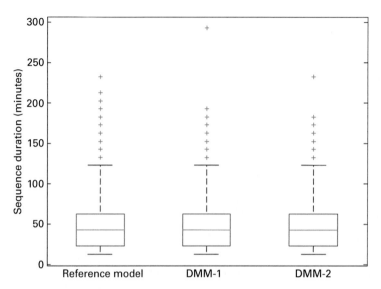

Figure 20.3 Boxplots of observation sequence durations for the three tested DMMs. No statistical difference was noticed between the distributions.

function of the number of the targeted whale's surfaces observed, the maximum time allowed by the marine park regulations for observing the same pod, the presence of other whales observable in the vicinity, and the remaining time in the excursion).

Several ways could be envisioned to reduce the duration of observation sequences in the real system: giving incentives to explore space to search for new whales instead of taking advantage of discovered whales; reducing the maximum authorized time in observation of the same pod (currently 60 min); or giving incentives to diversify activities at sea leading to more time spent discovering landscape features (e.g. lighthouses, sand dunes). Such strategies could be tested in the model to predict the effects on the duration of observation sequences.

Table 20.3 shows the repartition of observation effort on the four rorqual species. Both DMM-1 and DMM-2 led to a significant reduction of the proportion of time devoted to humpback whales observation. Attractiveness of this species is particularly high in the area for several reasons including its occasional spectacular behaviours, stability

Table 20.3 Contribution of each species to overall observation activities (%).

	Minke whale	Fin whale	Blue whale	Humpback whale
Reference model	32.8	31.1	4.0	32.1
DMM-1	44.3	35.9	3.9	15.9
DMM-2	34.4	37.7	8.2	19.7

of individual locations, and core habitat located in the vicinity of the busiest ports of excursion departure. In contrast, despite having the same abundance across simulation runs (3), blue whales always account for a smaller part of observations, especially because their home-range is located more downstream, farther from the most active homeports.

Excursion dynamics

The percentage of unsuccessful excursions (i.e. no whale observation) is similar for all tested models at approximately 3%. Again, this is consistent because

Table 20.4 Average and standard deviation of proportion of time spent in observation during excursions (%).

Reference model	DMM-1	DMM-2
49.4 ± 1.6	54.1 ± 0.8	53.0 ± 2.1

Table 20.5 Proportion of the total observation time an excursion is alone (1) or with another boat (2) observing a pod (%).

	Reference model	DMM-1	DMM-2
1. Alone with the pod	26.3	32.2	25.7
2. Two boats observing the same pod	22.8	27.7	23.6
Sum (1+2)	49.1	59.9	49.3

for all simulations captains favour (when possible) the exploitation of discovered whales rather than the more risky exploration of space. Consequently the success rate is not affected. Conversely, we found a slight change in the total amount of observation activities. Both DMM-1 and DMM-2 strategies lead to more observations than the reference model (Table 20.4).

The increase in time spent in observation is due to the fact that captains promote opportunistic observations for both DMM-1 and DMM-2. In the case of DMM-1, this is due to the fact that a whale surfacing opportunistically in the vicinity can be the best choice because there is no boat observing it. In the case of DMM-2, as the species is no longer a criterion for whale selection, whales surfacing in the vicinity of a boat will have more chance to be selected for observation. Observing close whales opportunistically reduces the travel time needed to reach a more distant site, thus explaining the increase in observation activities.

Reducing boat aggregations around pods of whales is positive both for whales and visitors' experience. Table 20.5 shows the proportion of time one or two boats are simultaneously observing the same

pod. As expected, using the DMM-1 strategy, captains significantly increase by ~11% the proportion of time they spend alone or with only another boat observing a pod when compared to the reference model. In contrast, DMM-2 does not significantly affect those variables. Let us point out that these figures take into account all excursions in a day, including early and late excursions where most observations occur alone as few boats are at sea at these times (compared to busier midday schedules).

Figure 20.4 shows the boxplots of the boat number distributions on observation sites. Only DMM-1 significantly reduces boat densities around whales, including median and maxima (Wilcoxon rank sum test).

Conclusion

Our goal was to provide insights on the use of multi-agent modelling to better understand the nature of interactions between whale-watching excursions and whales. 3MTSim is a spatially explicit multi-agent model representing whale movements and navigation activities within the Saint-Lawrence Estuary, Québec, Canada. This model was primarily developed to test alternative navigation-related management scenarios, including whale-watching activities. Because the whale-watching captains' decision process was modelled in detail, 3MTSim can also be used to predict outcomes from changes in captains' strategies to locate and observe whales, which was presented here.

Currently, whale-watching captains mostly ground their decisions of which whale to observe based on species, proximity, and competitor excursions' content. We demonstrated that prioritizing other criteria such as low aggregations of boats could help to decrease the overall density of boats in the vicinity of whales, without affecting important excursions' performance (e.g. time spent in observation). Additional decision strategies, not simulated, could also improve the situation at sea. For instance, captains could engage in more space *exploration* (as opposed to the currently widespread

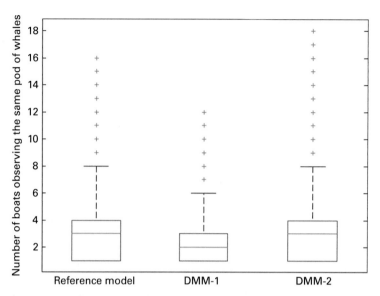

Figure 20.4 Boxplots representing the number of boats on observation sites for each tested DMM.

exploitation of discovered whales) or could systematically present some landscape or historical features (e.g. lighthouses, sand dunes) as part of their excursion instead of focusing exclusively on whale observation.

By simulating alternative decision strategies that could be followed by whale-watching captains, it is possible to devise a set of recommendations that marine park managers could communicate to captains during training sessions. As a decision-support tool, 3MTSim has the advantage of being able to illustrate the whole picture of the collective impact of navigation, including whale-watching activities, on whales. An appreciation of their collective impact on targeted whales was particularly absent from captains' discourses during interviews, 3MTSim could therefore help in raising further awareness about this issue. As new knowledge about the system's dynamics will become available, it will be possible to integrate it in 3MTSim. Expected model improvements include the implementation of noise emission by boats along with 3D propagation in the area, whales' reaction to the presence of boats, and captains' individual differences (e.g. values, preferences). This way, additional model output

variables could be accounted for to achieve a more complete impact analysis (e.g. spatial variations of boat–whale co-occurrences, whales' cumulative exposure to noise sources).

Acknowledgements

This project was made possible by a strategic project research grant from the Natural Sciences and Engineering Research Council of Canada (NSERC). Thanks to: Benoit Dubeau, Daniel Gosselin, Jeannie Giard, Véronique Lesage and Michel Moisan for contributing to discussions and data analysis. We are grateful to the following tour companies for their participation in the project: Croisières AML Inc., Croisière 2001 Inc., Groupe Dufour Inc., Croisières Charlevoix Inc., Les Croisières Essipit Inc., Croisières du Grand Héron, Les Écumeurs du Saint-Laurent, Société Duvetnor Ltée.

REFERENCES

Beauchamp, J., Bouchard, H., de Margerie, P., Otis, N. & Savaria, J.-Y. (2009). Recovery strategy for the blue whale

(Balaenoptera musculus), Northwest Atlantic population, in Canada [FINAL]. Ottawa: Fisheries and Oceans Canada, 62 pp.

Bennett, D.A. & McGinnis, D. (2008). Coupled and complex: Human–environment interaction in the Greater Yellowstone Ecosystem, USA. *Geoforum* 39, 833–845.

Chion, C. (2011). An agent-based model for the sustainable management of navigation activities in the Saint Lawrence estuary. PhD Thesis, École de technologie supérieure.

Chion, C., Turgeon, S., Michaud, R., Landry, J.-A. & Parrott, L. (2009). Portrait de la Navigation dans le Parc Marin du Saguenay–Saint-Laurent. Caractérisation des activités sans prélèvement de ressources entre le 1er mai et le 31 octobre 2007. Report presented to Parks Canada. Montréal: École de technologie supérieure and Université de Montréal, 86 pp.

Chion, C., Lamontagne, P., Turgeon, S., *et al.* (2011). Eliciting cognitive processes underlying patterns of human–wildlife interactions for agent-based modelling. *Ecological Modelling* 222(14), 2213–2226.

COSEWIC (2005). COSEWIC assessment and update status report on the fin whale *Balaenoptera physalus* in Canada. Ottawa: Committee on the Status of Endangered Wildlife in Canada, ix + 37 pp.

Couzin, I.D., Krause, J., Franks, N.R. & Levin, S.A. (2005). Effective leadership and decision-making in animal groups on the move. *Nature* 433(7025), 513–516.

Demers, A., Bouchard, H. & Beauchamp, J. (2011). Recovery strategy for the beluga (*Delphinapterus leucas*), St. Lawrence Estuary population, in Canada [PROPOSED]. Ottawa: Fisheries and Oceans Canada, 88 + X pp.

Dionne, S. (2001). Saguenay–St. Lawrence Marine Park Ecosystem Conservation Plan Summary. Quebec: Parks Canada.

Giard, J. & Michaud, R. (1997). L'observation des rorquals sous surveillance par la telemetrie VHF. *Le Naturaliste Canadien* 121, 25–29.

Gigerenzer, G. & Goldstein, D.G. (1996). Reasoning the fast and frugal way: Models of bounded rationality. *Psychological Review* 103(4), 650–669.

Gigerenzer, G. & Selten, R. (2001). *Bounded Rationality: The adaptive toolbox*. Cambridge, MA: The MIT Press.

Gimblett, H.R. (2002). *Integrating Geographic Information Systems and Agent-Based Modeling Techniques for Simulating Social and Ecological Processes*. Santa Fe, NM: Oxford University Press.

Gimblett, H.R., Richards, M.T. & Itami, B. (2002). Simulating wildland recreation use and conflicting spatial interactions using rule-driven intelligent agents. In *Integrating Geographic Information Systems and Agent-based Modeling Techniques for Simulating Social Ecological Processes*. Santa Fe, NM: Oxford University Press, pp. 211–243.

Giroul, C., Ouellet, G. & Soubrier, R. (2000). Étude des attentes de la clientèle des croisières aux baleines dans le secteur du parc marin du Saguenay–Saint-Laurent: Faits saillants. Presented to Parks Canada. Université du Québec à Trois-Rivières, 56 pp.

Grimm, V., Revilla, E., Berger, U., *et al.* (2005). Pattern-oriented modeling of agent-based complex systems: Lessons from ecology. *Science* 310, 987–991.

Gross, J., McAllister, R., Abel, N., Smith, D. & Maru, Y. (2006). Australian rangelands as complex adaptive systems: A conceptual model and preliminary results. *Environmental Modelling & Software* 21, 1264–1272.

Groupe de recherche et d'éducation sur les mammifères marins (GREMM) (2007). Recensement visuel et photographique systématique des grands rorquals dans le parc marin du Saguenay–Saint-Laurent – Projet pilote 2006. Final report presented to Parks Canada. Tadoussac (Qc), Canada.

Guénette, S. & Alder, J. (2007). Lessons from marine protected areas and integrated ocean management initiatives in Canada. *Coastal Management* 35, 51–78.

Itami, R., Raulings, R., MacLaren, G., *et al.* (2003). RBSim 2: Simulating the complex interactions between human movement and outdoor recreation environment. *Journal for Nature Conservation* 11, 278–286.

Janssen, M.A. & Ostrom, E. (2006). Empirically based, agent-based models. *Ecology and Society* 11(2), 37.

Klein, G.A. (2008). Naturalistic decision making. *Human Factors* 50(3), 456–460.

Klein, G.A., Calderwood, R. & MacGregor, D. (1989). Critical decision method for eliciting knowledge. *IEEE Transactions of Systems, Man and Cybernetics* 19(3), 462–472.

Lamontagne, P. (2009). Modélisation spatio-temporelle orientée par patrons avec une approche basée sur individus. Master in Engineering, École de technologie supérieure.

Lemieux Lefebvre, S. (2009). Déplacements et patrons de résidence chez la population de bélugas (*Delphinapterus leucas*) de l'estuaire du St-Laurent. Master, Université du Québec à Rimouski.

Manson, S.M. (2005). Agent-based modeling and genetic programming for modeling land change in the Southern Yucatan Peninsular Region of Mexico. *Agriculture Ecosystems & Environment* 111, 47–62.

Michaud, R. & Giard, J. (1997). Les rorquals communs et les activités d'observation en mer dans l'estuaire du Saint-Laurent entre 1994 et 1996: 1) Étude de l'utilisation du territoire et évaluation de l'exposition aux activités d'observation à l'aide de la télémétrie VHF. Joint project from the GREMM, ministère de l'Environnement et Faune du Québec, Fisheries and Oceans Canada, Heritage Canada, and Parks Canada. Group for Research and Education on Marine Mammals (GREMM), Tadoussac (Qc), Canada. 30 pp.

Michaud, R. and Giard, J. (1998). Les rorquals communs et les activités d'observation en mer dans l'estuaire maritime du Saint-Laurent entre 1994 et 1996: 2) Évaluation de l'impact des activités d'observation en mer sur le comportement des rorquals communs. Joint project from the GREMM, ministère de l'Environnement et Faune du Québec, Fisheries and Oceans Canada, Heritage Canada, and Parks Canada. Group for Research and Education on Marine Mammals (GREMM), Tadoussac (Qc), Canada. 22 pp.

Michaud, R., Bédard, C., Mingelbier, M. & Gilbert, M.-C. (1997). Whale watching activities at sea in the St. Lawrence marine estuary, 1985–1996: A study of spatial distribution of activities and factors favouring boat aggregation at whale watching sites. Final report submitted to Parks Canada. Le Groupe de Recherche et d'Éducation sur les Mammifères Marins, GREMM, Tadoussac, Québec, Canada.

Michaud, R., D'Arcy, M.-H., de la Chenelière, V. & Moisan, M. (2008). Les activités d'observation en mer (AOM) dans l'estuaire du Saint-Laurent: Zone de Protection Marine Estuaire du Saint-Laurent et Parc Marin du Saguenay–Saint-Laurent – Suivi annuel 2007. Final report submitted to Parks Canada. Tadoussac: GREMM, 71 pp.

Monticino, M., Acevedo, M., Callicott, B., Cogdill, T. & Lindquist, C. (2007). Coupled human and natural systems: A multi-agent-based approach. *Environmental Modelling & Software* 22, 656–663.

Moreno, N., Quintero, R., Ablan, M., *et al.* (2007). Biocomplexity of deforestation in the Caparo tropical forest reserve in Venezuela: An integrated multi-agent and cellular automata model. *Environmental Modelling & Software* 22, 664–673.

Parks Canada (2002). *Marine Activities in the Saguenay–St. Lawrence Marine Park Regulations: Saguenay–St. Lawrence Marine Park Act.* Canada: Parks Canada.

Parrott, L., Chion, C., Martins, C.C.A., *et al.* (2011). A decision support system to assist the sustainable management of navigation activities in the St. Lawrence River Estuary, Canada. *Environmental Modelling & Software* 26(12), 1403–1418.

Railsback, S.F., Lytinen, S.L. & Jackson, S.K. (2006). Agent-based simulation platforms: Review and development recommendations. *Simulation* 89(9), 609–623.

Roberts, C.A., Stallman, D. & Bieri, J.A. (2002). Modeling complex human–environment interactions: The Grand Canyon river trip simulator. *Ecological Modelling* 153, 181–196.

Schlüter, M. & Pahl-Wostl, C. (2007). Mechanisms of resilience in common-pool resource management systems: An agent-based model of water use in a river basin. *Ecology and Society* 12(2), art. 4.

Simon, H.A. (1957). *Models of Man: Social and rational.* New York, NY: Wiley.

Turchin, P. (1998). *Quantitative Analysis of Movement: Measuring and modeling population redistribution in animals and plants.* Sunderland, MA: Sinauer.

van Oel, P.R., Krol, M.S., Hoekstra, A.Y. & Taddei, R.R. (2010). Feedback mechanisms between water availability and water use in a semi-arid river basin: A spatially explicit multi-agent simulation approach. *Environmental Modelling & Software* 25, 433–443.

21

Cetacean-watching in developing countries

A case study from the Mekong River

Isabel Beasley, Lars Bejder and Helene Marsh

Introduction

Cetacean-watch tourism in developing countries remains poorly documented, and often poorly managed as a result of limited in-country capacity, ineffective governance, conflicting policy goals, and limited accountability. The dolphin-watch tourism that targets the population of Irrawaddy dolphins, *Orcaella brevirostris*, in the Mekong River in Cambodia and Laos is used as a case study to illustrate growing concerns associated with cetacean-watch tourism in developing countries. In the early 1990s, unregulated and unmanaged dolphin-watching tourism began in two of the most important habitats for this population, which now numbers less than 100 individuals. An Integrated Conservation Development Project, '*Dolphins for Development*', was initiated in Kampi Village, Cambodia in 2004. This Project included an attempt to manage the existing dolphin-watch tourism through: (1) promoting the sharing of tourism revenue to the local community; (2) encouraging effective management of the industry to minimize threats to the dolphins; (3) promoting visitor satisfaction; and (4) raising community and visitor awareness of the status of the dolphins and the need for conservation. Although the initial results were encouraging, subsequent government intervention has resulted in: (1) a large increase in the number of boats operating in prime dolphin habitat and an increase in the harassment of dolphins; (2) reduction in the benefit to local communities; and (3) little or no information being provided to national or international tourists. Although management agencies are implementing significant conservation measures to reduce the threat to the dolphins from gillnet fishing by subsistence fishers, few efforts are being directed towards management of the dolphin-watch tourism industry. The urgent need to develop dolphin-watching tourism management initiatives in Cambodia was highlighted at the 2010 International Whaling Commission meeting, where a 'no vessel-based' dolphin tourism policy was recommended. A precautionary approach to management is needed to address the problem of unsustainable cetacean-watching currently occurring in numerous developing countries. This approach should be informed by location-specific and comprehensive studies on both the ecology of the dolphin–tourism interactions and the social, economic, managerial and political influences on cetacean-watch operations.

Background

Whale- and dolphin-watching (hereafter referred to as cetacean-watching) is capable of delivering socioeconomic benefits to local communities, and

Whale-watching: Sustainable Tourism and Ecological Management, eds J. Higham, L. Bejder and R. Williams.
Published by Cambridge University Press. © Cambridge University Press 2014.

has been advocated as a tool for the conservation of cetaceans, particularly as a non-consumptive alternative to whaling (Orams, 2001; O'Connor *et al.*, 2009). As a result of its potential for delivering social, economic and environmental benefits, cetacean-watching is often referred to as a form of 'ecotourism' (Garrod & Wilson, 2003; Hoyt, 2005), although most operations do not conform to the strict definition of that term (Mustika *et al.*, 2012a,b). While some advocates of ecotourism emphasize its potential for promoting the well-being of both local peoples and the environment (e.g. Micronesia: Valentine, 1993), others caution that ecotourism is often merely used as a marketing tool (Thomlinson & Getz, 1996), with revenue and/or benefits rarely reaching local communities (Bookbinder *et al.*, 1998). Although the advantages for both operators and tourists involved in the cetacean-watch industry are clear, the potential contribution of such operations to the conservation of species and/or habitats remains debatable (Bejder & Samuels, 2003). There is also growing and widespread concern that cetacean-watching may have deleterious impacts on the cetaceans being targeted (Garrod & Fennell, 2004; Bejder *et al.*, 2006; Lusseu *et al.*, 2006; Neves, 2010). While many nations now have well-developed cetacean-watch industries with varying levels of management, little information is available on the burgeoning cetacean-watch industries in developing nations[1] that target some of the cetacean populations of greatest conservation concern.

The International Whaling Commission (IWC) has recently stated that '[t]here is compelling evidence that the fitness of individual odontocetes repeatedly exposed to whale-watching vessel traffic can be compromised and that this can lead to population level effects' (IWC, 2006: 47). Based on increasing concerns about the long-term effects of cetacean-watching on some populations, it has been recommended that an adaptive, precautionary approach is essential to managing tourism that targets small, closed, resident communities of cetaceans (Bejder *et al.*, 2006; Higham *et al.*, 2008). This precautionary approach is especially important for freshwater delphinids inhabiting riverine and estuarine habitats in developing countries, where numerous other anthropogenic threats (e.g. gillnet entanglement and habitat destruction) are already threatening their viability. Management of cetacean-watch activities in these situations requires special care, as endangered populations leave little margin for recovery from incorrect management decisions (Taylor & Gerrodette, 1993; Bejder *et al.*, 2006). The effects of tourism (no matter how well-managed) on the behavioural ecology and survival of small, highly endangered cetacean populations therefore needs to be carefully and critically evaluated.

O'Connor *et al.* (2009) provide a comprehensive account of the current status of cetacean-watching worldwide. Most tourism operations that target endangered cetaceans are located in Asia, including: Irrawaddy dolphins, *Orcaella brevirostris*, in the Mahakam River (Indonesia), the Ayeyarwady River (Myanmar) and Chilika Lagoon (India); Ganges River dolphins, *Platanista gangetica gangetica*, in the Koshi and Karnali Rivers (Nepal); and Indus River dolphins *Platanista gangetica minor* (Pakistan). In the Mekong River of Cambodia and southern Lao Peoples Democratic Republic (hereafter referred to as Laos), dolphin-watch tourism targets the Mekong River Irrawaddy dolphin subpopulation, which is classified as Critically Endangered by the World Conservation Union (IUCN) (Smith & Beasley, 2004).

In January 2001, a research and conservation project on Irrawaddy dolphins inhabiting the Mekong River (hereafter referred to as the Mekong dolphin population) was initiated. Research on the dolphin's population biology (abundance, distribution, mortality rates and causes) was the

---

[1] We acknowledge that there is much discussion regarding the definition of, and countries included within, 'developing nations'; however, in this chapter we follow the United Nations (2005) definition, where Japan, Hong Kong, Singapore and South Korea in Asia; Canada and the United States in northern America; Australia and New Zealand in Oceania; and Europe are considered 'developed' regions or areas.

primary focus of activities. Dedicated conservation activities began in 2003 in parallel with continuing research. Beasley (2007) found that the Mekong dolphin population is now restricted to approximately 12 deep-water pools (10–45 m in depth, ranging in size from 1 to 2 km^2) in a 190-km section of river between Kratie and the Laos/Cambodian border (Figure 21.1). Photo-identification studies indicated that dolphins exhibit high site fidelity to particular deep pool areas during the dry season (January to June), with some dolphins always identified in the same deep-water pool, while others travel between three and four neighbouring pool areas (Beasley *et al.*, 2009). During the wet season (July–December), water levels rise up to 30 m, and dolphins are distributed more widely throughout the Kratie to Laos/Cambodian border river section (Beasley, 2007). As of April 2007, the total Mekong dolphin population numbered no more than 100 individuals (Ryan *et al.*, 2011; Beasley *et al.*, 2013), with initial genetic analysis (nine samples) indicating very low genetic diversity within the population (Dove *et al.*, 2009). The mortality rate is high, with 46 carcasses recovered from 2003 to 2005 (54% of recovered carcasses were newborns; Gilbert & Beasley, 2006). Thus there are significant concerns for the future survival of this Critically Endangered population (Beasley, 2007). Accidental catch in gillnets and the dolphin-watch tourism industry (e.g. daily harassment by tour boats) are serious threats to this population (Beasley *et al.*, 2009).

Dolphin-watch tourism in the Mekong River is facilitated by the reliable occurrence of Irrawaddy dolphins in small deep-water pools throughout the year. There are currently two locations where tourists can view dolphins in the Mekong River: (1) Kampi Pool in Kratie Province, Cambodia; and (2) Chiteal Pool (known as Anlong Chiteal in Cambodia and Veun Nyang in Laos) on the Laos/Cambodian border (Figure 21.1; see Box 21.1 for the background to dolphin-watching tourism at each site). Although various problems associated with dolphin-watching tourism at Chiteal and Kampi Pools (such as operators chasing dolphins

in motorized boats) were evident by the early 2000s, the Cambodian and Laotian governments have never been actively involved in facilitating management to minimize boat disturbance on the dolphins.

In 2003, an Integrated Conservation Development Project at Kampi Village named '*Dolphins for Development*' (hereafter referred to as the 'Project') was initiated, which aimed to provide direct, tangible benefits to the local community in return for their cooperation with conservation activities to conserve the remaining dolphin population. This chapter discusses the outcomes of initial attempts to manage the existing dolphin-watch tourism industry at Kampi Pool as part of the Project, and provides a case study from which to discuss concerns associated with the management of cetacean-watch tourism in developing countries.

Materials and methods

Study area

The Kratie to Laos/Cambodia border river section (190 km) is an internationally important stretch of the Mekong River (Figure 21.1). Deep pools (10–90 m) in this river section are important fish-spawning sites and habitats for numerous animals and plants during the dry season. Much of this biodiversity is endangered or extinct elsewhere (Poulsen & Valbo-Jorgensen, 2001; Baird, 2006). This section of the river is also critical habitat for the remaining Mekong River dolphin population (Baird & Beasley, 2005).

The Project was undertaken at Kampi Village, 15 km north of Kratie Township, Kratie Province, Cambodia (Figure 21.1). Kampi Village is situated adjacent to Kampi Pool, which is the first major deep-pool habitat north of the Mekong Delta. This pool is one of the most important areas in the Mekong River for Irrawaddy dolphins (Beasley, 2007); 42 dolphins had been photo-identified (approximately 50% of the known total

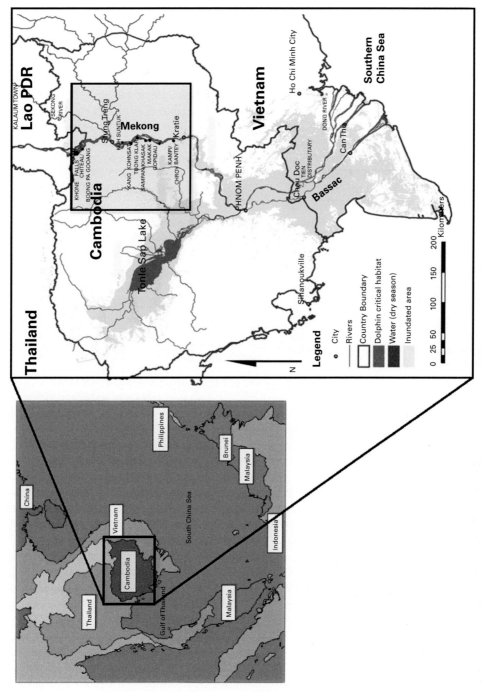

Figure 21.1 Map of Cambodia (left image) and the Kratie to Khone Falls river section of the Mekong Rover (right image: shown by the shaded box). Kampi Pool is located adjacent to Kratie Village (bottom right of shaded box), and Chiteal Pool is located near Khone Falls (top left of shaded box). Irrawaddy dolphins inhabiting the Mekong River now primarily occur in this 190 km river section from Kampi Pool north to the Laos/Cambodia border. Khone Falls on the Laos/Cambodian border prevents dolphin movement further north along the mainstream Mekong River.

Box 21.1

Dolphin-watch tourism at Kampi Pool

Kampi Pool (total area = 2 km^2) is located approximately 200 km north of Phnom Penh in Kratie Province, Cambodia (Figure 21.1). Dolphin-watch tourism at Kampi was originally land-based; tourists (and the local community) could freely view dolphins from the riverbank (30 m above the river during low water), overlooking the deep pool that dolphins consistently inhabit year-round. More formal tourism was initiated in 1997 by an international non-governmental organization (Community Aid Abroad) with a local committee of seven villagers from Kampi Village (approx. 135 families lived in Kampi Village as of 2004). From 1997 to 2000, viewing of dolphins was conducted sporadically from land, with no on-site management. Tourists were also able to view dolphins from small paddle-boats for a minimal fee (US $1). Only seven families were allowed to offer dolphin-watch tourism. In 2001, the seven villagers changed the small paddle-boats to larger 'stand-up' paddle-boats with motors and sunshades. These arrangements ensured tourist comfort and enabled dolphin viewing year-round (Beasley et al., 2009).

In 2002, the Kratie Tourism Department (a Cambodian government department) became formally responsible for dolphin-watch tourism at Kampi Pool, and cooperated with the seven families. No other families from Kampi Village were allowed to participate in the venture and the financial benefits (50% of revenue) were distributed only to the seven families, with Kratie Tourism Department receiving the remaining revenue. All other villagers were unable to participate in the tourism (apart from a few villagers being hired as casual boat drivers), and had also lost their rights to fish in the pool as a result of a Provincial Decree prohibiting fishing in Kampi Pool in the early 2000s. Conflict was rife in the village and the seven families became segregated from the other villagers. Local people were unaware that the sound from the boat motors and the boats' activities had the potential to interfere with the dolphins' daily activities. Additionally, villagers were unable to communicate with foreign tourists and no information (verbal or printed) was provided to the tourists regarding the dolphins, or their conservation status in the river. Thus, the situation was unmanaged and unregulated, and unable to contribute to dolphin conservation or management (Beasley et al., 2009).

Dolphin-watch tourism at Chiteal/Veun Nyang Pool

Dolphin-watch tourism was initiated at Chiteal Pool (locally known as Veun Nyang), total area = 1 km^2, in early 1993 by Hang Khone Village (located adjacent to Veun Nyang Pool). As described by Ian Baird (personal communication, 2010) who lived and worked in Hang Khone Village from 1993 to 1999:

'*In early 1993, Hang Khone village initiated a system that allowed all villagers in the community of 40+ households to participate in taking tourists to see dolphins, if they had motorized long-tailed boats to do so (non-motorized boats were rarely used). Hang Khone villagers worked with the Lao Community Fisheries and Dolphin Protection Project to create rules for ensuring that tourists would be safe, and that the income from tourism would be equitably distributed among the people in the community to provide incentives to assist with dolphin conservation efforts. Tourists sometimes went to view the dolphins from a large rock in the edge of the main pool where dolphins are found, but the use of the rock depended on various factors including whether the driver was paid extra to drive tourists into the pool, and season (e.g. in the rainy season the rock is totally underwater, so only boat-based observations could be conducted).*

Within a year or so of dolphin tourism beginning in Hang Khone, Veun Kham and Khone Tai villages (now just Khone Village) started to get involved on their own, without the involvement of the Lao Community Fisheries and Dolphin Protection Project, and without the rules or equitable rotational system established in Hang Khone. As a result of this increase in unregulated tourism, tension and bad feelings between villages developed.'

In 1994, Hang Khone village had 10 motor-boats and about 30 paddle-boats (Staccy & Hvenegaard, 2002). In 2000, larger motorized boats with shelters began to take tourists to view dolphins in the pool, expanding to at least 15 large motorized boats by the early 2000s. The change to large boats with shelters occurred because the Lao government provided a private company with a tourism concession that included the rights to manage all dolphin tourism activities in the region; a situation that caused considerable village resentment and excluded poorer people with smaller boats from the industry (after a few years the concession was revoked by the Lao government because the company had failed to fulfil all the conditions of their contract) (Ian Baird, personal communication, 2010). Cambodian nationals initiated small-scale tourism to opportunistically observe the dolphins at Chiteal/Veun Nyang Pool in the early 2000s, using fast speedboats from Stung Treng Township ($n = \geq 3$ boats). As early as 1994, the increasing number of boats taking tourists to view dolphins was raising concern as to the potential impact on dolphins in the pool (Baird & Mounsouphom, 1994). The first empirical confirmation that boat traffic was probably impacting the dolphins (at least in the short-term) was provided by Stacey (1996), who found that dolphins dived for longer durations when boats were present; dolphins surfaced closer to paddle- than motor boats; and fewer than expected occurrences of surface activities when large motor boats were within 100 m of dolphin groups (Stacey & Hvengaard, 2002).

population in the Mekong). Most of these animals have only been recorded in this pool (Beasley, 2007).

'Dolphins for Development' Project

In January 2004, a series of workshops were held in Kampi and the surrounding villages with the following aims: (1) update villagers on the findings of the dolphin research and conservation project from 2001 to 2003; (2) discuss the community's perceptions of threats to dolphins and the river system; and (3) discuss the community's perceptions regarding dolphin-watch tourism in Kampi Pool. Five major conclusions were drawn from the workshops: (1) Kampi villagers were dissatisfied with being prevented from fishing in Kampi Pool without adequate consultation (see Box 21.1); (2) fishing restrictions forced local people to fish illegally, an activity that most villagers did not want to conduct; (3) outsiders would often come to fish in the area unaware of the regulations; however, enforcement would often arrive too late to stop such activities; (4) the community supported the fishing regulations in principle, but required external assistance to develop alternative livelihoods if the ban was to be implemented effectively; and (5) the community were unhappy with the existing dolphin-watch tourism situation, and wanted the entire community to be involved in the industry.

Accordingly, the Project attempted to facilitate the conservation of dolphins and fisheries in Kampi Pool through: (1) encouraging rural development and livelihood diversification; (2) improving the management of the existing community-based dolphin-watch tourism; (3) developing environmental education and awareness raising material; and (4) strengthening stakeholder relationships. A major priority for all Project activities was cooperation with relevant stakeholders whenever possible, including; Kampi villagers, boat operators, the Kratie Fisheries and Tourism Offices, Provincial Governors, local NGOs, University of Phnom Penh students, and national and international tourists.

Project activities to assist the villagers manage the existing dolphin-watch tourism began in March 2004 with the objectives of: (1) promoting the sharing of the revenue within the local community; (2) encouraging management to minimize threats to dolphins; (3) promoting visitor satisfaction; and (4) raising awareness of the critical conservation status of the Mekong dolphin population.

Results

Results of the 'Dolphins for Development' Project: 2004

Improving the management of the dolphin-watch industry proved difficult. Established protocols were already in place that benefited few individuals within the community and caused significant conflict and resentment between villagers. Nonetheless, the Project's achievements (Table 21.1) were significant and included: (1) development and training of a Village Development Committee to manage funds and tourism; (2) production of a children's colouring book to sell to tourists to provide additional revenue to the Village Development Committee; (3) development of volunteer guidelines for boat operators; (4) initiation of a visitor recording system in cooperation with the Kratie Tourism Department; (5) a two-day local guiding course (including English lessons to potential guides) for local villagers interested in becoming involved in the industry; (6) infrastructure at the Kampi viewing site (Figure 21.2); and (7) a written agreement on revenue distribution between relevant stakeholders.

In December 2004, a written agreement was signed between the Kampi Village Development Committee, Kratie Fisheries and Tourism Offices, and Kratie Provincial Government to ensure that an entrance fee to the tourism site (US $2/international tourist, US $0.15/national tourist) would be introduced and shared between the Kampi community (40% for development activities), Kratie Tourism Office (30% to ensure maintenance of the tourism site) and Kratie Fisheries Office (30% for dolphin

Table 21.1 Summary of the objectives and achievements of the community-based tourism project component of the 'Dolphins for Development' Project, initiated at Kampi Village, Kratie Province.

Objective	Achievements
(1) Promoting community benefit from dolphin-watching tourism implemented prior to this projects' inception	– selling childrens' colouring books and T-shirts at a restaurant in Kratie Township, where all the profits were directed to community development activities – development and training of a **Village Development Committee (VDC)**, that was elected democratically in Kampi Village and responsible for facilitating the ICDP and management of funds obtained from tourism activities – a series of meetings with stakeholders involved, to secure an **agreement for the community to benefit financially** from the tourism through an entrance booth at the viewing site
(2) Encourage effective management of this industry to ensure it did not threaten the dolphin group inhabiting this area	– various meetings with boat owners and other stakeholders to **develop and finalize boat operating guidelines** – construction of signboards at the Kampi viewing site clearly explaining **regulations** for boat use and tourist behaviour – initiation of a **visitor recording system** at Kampi viewing site through **provision of a computer** to the Department of Tourism
(3) Promoting visitor satisfaction and awareness raising of dolphin conservation and status	– development of **educational materials** to raise national and international awareness of the dolphins and their habitat – a two-day **guide training course** to provide training for local guides from Kratie Township (including four individuals from Kampi village) – providing **English lessons** to individuals from Kampi Villages to facilitate communications with tourists – **infrastructure development** at the Kampi viewing site (e.g. toilets, car park, souvenir stalls, food stalls), to ensure its attractiveness for international and national visitors

conservation activities). All revenue from the hire of dolphin-watching boats (US $2–4/hour for each boat) continued to go to the Kratie Tourism Office and the boat owners.

Two matters were critical to the success of this agreement: (1) the capacity of the community to adequately manage the revenue raised through the dolphin-watch tourism industry; and (2) the accountability and transparency of all activities to pre-empt corruption. The capacity of the village to manage funds appropriately was enhanced by the democratic election of the Village Development Committee by the Kampi villagers. In addition, Committee members were provided with training in accounting and finance, and a local bank account was opened to facilitate financial transactions.

(a)   (b)

(c)

Figure 21.2 As part of the 'Dolphins for Development' Project, infrastructure at the Kampi viewing site was constructed in late 2004. Infrastructure included: (a) tourist information signs; (b) no swimming in Kampi Pool signs; and (c) an entrance booth, which was located at the entrance to the Kampi viewing site. (See colour plate section.)

In addition to the revenue agreement, boat operator guidelines were developed and focused on minimizing the use of boat engines in Kampi Pool during the dry season. An attempt to negotiate temporal closures (e.g. one day a week) where no tourist boats would enter Kampi Pool was not successful. However, it was formally agreed that a small pool (Chroy Banteay Pool = 500 m^2) adjacent to

Kampi Pool would not be visited by any tourist boats. Although Chroy Banteay was a small area less frequently used by dolphins, this arrangement provided an area where the dolphins could escape the tourist boats. 'Tourism Police' employed by the Kratie Government that were stationed at the Kampi viewing area to protect tourists from any harm facilitated the enforcement of these agreed guidelines.

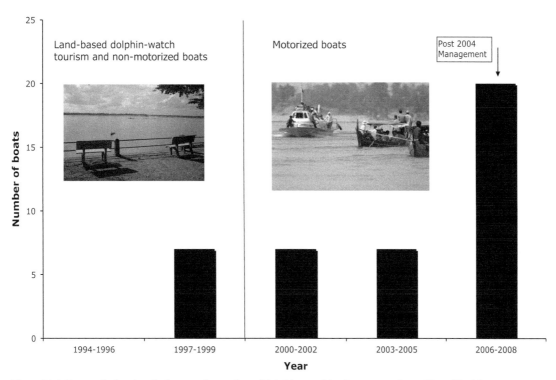

Figure 21.3 Bar graph showing the increase in number of dolphin-watching boats operating at Kampi Pool from 1997 to 2008 (non-motorized boats to the left of the vertical line and motorized boats to the right of the vertical line). Prior to 1997, only land-based observations were undertaken at the Kampi viewing site. Non-motorized boats were used in combination with land-based observations from 1997 to 2000 (left image). Post-2000, all dolphin-watch tourism boats were motorized with shelters for tourists (right image). (See colour plate section.)

Conflicting government policy goals: post-2004

In mid-2005, the Cambodian Government developed the 'Commission for Dolphin Conservation', with the mandate of managing and conserving Irrawaddy dolphins in the Mekong River. Since then, the Commission (recently renamed as the 'Commission for Dolphin Conservation and Development of Mekong River Dolphin Ecotourism Zone') has become responsible for both dolphin conservation and promotion of tourism development, a potential conflict of policy goals when dolphins are the major target of tourism efforts (Beasley *et al.*, 2009, 2010). A few months after the Commission was developed, the number of tourism boats operating at Kampi

Pool increased from 7 (2004) to 20, with the revenue from all new boats going directly to the Commission (Figure 21.3).

In January 2007, the newly formed Commission prohibited any NGO involvement in project activities at Kampi Village (including the continuing Project), and cancelled the previous community-based tourism agreement, despite Cambodia's national policy on poverty alleviation, where it is stated that 'social and economic growth should be equitable, and opportunities and benefits affordable and accessible to all, both geographically and between the rich and the poor' (see page i of IMF, 2006). Instead, the Commission allowed the community to operate 2 of the 20 tourist boats (Figure 21.4), and distribute the revenue gained

(a)   (b)

Figure 21.4 Dolphin-watch tourism has expanded quickly at Kampi Pool, and the majority of the local community now receives limited benefit from the dolphin-watch tourism. (a) During holiday periods there are many boats in the pool at one time. As a result of the new Commission for Dolphin Conservation, the number of boats at the dolphin-watch site increased from 7 in 2005 to 20 in 2007. Image (b) shows the small size of Kampi Pool, where almost all of the pool is visible in the image (2 km^2). (See colour plate section.)

from these two boats among the 128 families that did not own boats. To enter the site, a flat entrance fee of US $7 per person was charged, as of April 2007; US $1.50 was distributed to the private boat owners, the remaining proceeds went to the Commission. No other revenue from the entrance fee went back to the local community. Allowing more boats to operate in the pool significantly reduced the benefits to each boat owner, exacerbated village hostilities, and significantly increased the risk of dolphins being harassed by boats. Additionally, as of 2007, no awareness information was provided to tourists, and boat operators were encouraged to use their motors to approach dolphin groups. Tourists were also encouraged to swim with the dolphins, both at Kampi and Chiteal Pools.

Discussion

The Kampi Integrated Conservation and Development Project was developed to address poverty, increase the capacity of the local community, diversify local livelihoods to reduce the use of gillnets in Kampi Pool, and minimize tourism-boat

disturbance on dolphins using the Kampi area. Although many aspects of the Project were successful (Table 21.1), the management of the dolphin-watching tourism component ultimately failed as a result of government intervention that resulted in ineffectual governance and economic considerations influencing management.

As in most other countries, the governance of Cambodian natural resources operates at numerous levels, from the village, commune, district, province, to national government bodies. In addition to conflicting policy goals at a national level, local-level interests (i.e. village, commune and district) were major influences on the attempts to manage the dolphin-watch industry. The situation was also likely exacerbated by corruption, 'the unlawful use of public office for private gain'. Corruption is a serious problem in Cambodia. In 2009, the Corruption Perceptions Index (the perceived level of public-sector corruption) developed by Transparency International (2009) ranked Cambodia as 158 out of 180 nations (180 being most corrupt). Smith and Walpole (2005) point out that 'the impact of corruption on conservation is likely to manifest in two ways: (1) reduced effectiveness of conservation programs through a reduction in financial resources, law enforcement and political support; and (2) an incentive for over-exploitation of resources'. Management of the dolphin-watch industry in Cambodia was impeded by the managing agency having a direct economic interest in encouraging the continuation and expansion of the industry.

The dolphin-watch industry in both Kampi and Chiteal Pools grew quickly with the increasing numbers of international tourists and the local appreciation of the potential financial gains of tourism. Although there is potential for local Cambodian communities to gain significant financial and social benefit from dolphin-watch tourism, the current situation has resulted in little revenue accruing to the local community, and few restrictions on the number of boats in order to maximize profits for the management agency. Currently no data exist on: (1) tourist satisfaction with the experience and their likelihood of recommending it to others; (2) the effects of the 20+ tourist boats (as of April 2007) operating at Kampi Pool on the population biology of the resident dolphins; or (3) the economic influence of the industry at local, provincial, or national levels. However, it is likely that the cumulative effects of current dolphin-watch activities may have serious long-term impacts on the dolphins (Bejder et al., 2006), with resultant long-term reduction in the associated income from tourism.

Dolphin-watch tourism in the Mekong River was touted as a way to 'save' the species from extinction, when the Cambodian Ministry of Tourism announced development of the Mekong Discovery Trail in 2007 (www.mekongdiscoverytrail.com). Described as an 'ecotourism adventure', this route follows the current range of the Mekong dolphin along the Cambodian Mekong River from southern Laos south to Kratie, However, recent studies in other areas suggest that boat-based dolphin-watch tourism can cause biologically significant impacts (Bejder et al., 2006). At both the Kampi and Chiteal sites, it is currently possible to guarantee dolphin sightings from land during the dry season, as a

result of the dolphins' reliance on the small deep-water pools (although the Chiteal population of dolphins now numbers less than 10 individuals, and is critically at risk of local extinction). As a result of this sighting reliability, Cambodia may be one of the few places in the world where land-based dolphin-watch tourism could be a financially successful venture without impacting on the animals, thereby contributing significantly to dolphin conservation efforts through increased public awareness of the challenge of conserving dolphins in the Mekong River. Because few villagers currently receive significant income from hiring boats to watch dolphins, it should be possible to design economic incentives to offset their consequential loss of income from moving to land-based viewing of dolphins. Microeconomic tools are available to encourage the development of alternative livelihoods. Such tools include microfinance loans, which seek to eliminate poverty by providing fair, safe and ethical financial services for people who, because of their circumstances, are not able to access mainstream financial services. Microfinance institutions have had considerable success in alleviating poverty over the past two decades, particularly in Bangladesh (Davis & Kosla, 2007). This approach is being extended as a mechanism for conserving biodiversity by combining microfinance-lending approaches with a performance-based incentive structure for environmental stewardship (Mandel et al., 2009).

The Cambodian government is the largest potential short-term loser from a move to land-based tourism because it presently receives most of the income from the entrance fee and vessel hire. International pressure is being applied in an attempt to request the government to reduce the exposure of the Mekong River dolphins to vessel-based tourism. In 2010, a working paper outlining the status of dolphin-watch tourism in the Mekong River was presented at the annual Scientific Committee meeting of the International Whaling Commission (IWC) meeting in Morocco (Beasley et al., 2010). In response, the IWC Scientific Committee made the following statements:

The Committee expresses grave concern about the rapid and not fully explained decline of this riverine population. It commends the efforts by Cambodian government agencies and WWF-Cambodia to diagnose the cause(s) of the decline, and strongly recommends that every effort be made to stop and reverse it, e.g. by immediately eliminating entangling fishing gear in the pool areas used most intensively by the dolphins and by taking immediate steps to reduce the exposure of the dolphins to tour boat traffic …

The Committee reiterated its concern over the critically endangered Mekong River Irrawaddy dolphin population. In 2006, it had noted that there was compelling evidence that the fitness of individual odontocetes repeatedly exposed to tour vessel traffic can be compromised and that this can lead to population-level effects (IWC, 2006). It also stated that, in the absence of data, it should be assumed that such effects are possible until indicated otherwise – particularly for small, isolated and resident populations.

Accordingly, the Committee strongly recommends that the Cambodian government and relevant agencies make every effort to reduce the exposure of dolphins to vessel-based tourism in deep-water pools in the Mekong River (IWC, 2010).

In January 2012 a 'Mekong Irrawaddy Dolphin Conservation Workshop' was held in Cambodia to discuss future research and conservation efforts. Cambodian and international experts as well as government officials collaborated to produce 25 recommendations aimed at understanding and conserving the Mekong dolphin population.

The recommendations required resources and guidelines to facilitate and standardize studies concerning the causes of mortality, population dynamics, behaviour and ecology, and fisheries management. At the close of the workshop, the three agencies responsible for dolphin conservation in Cambodia, the Commission for Dolphin Conservation and Development of Mekong River Dolphin Ecotourism Zone, the Fisheries Administration, and the World Wide Fund for Nature, signed the 'Kratie Declaration on the Conservation of the Mekong River Irrawaddy Dolphins', committing to developing a strategy for implementing the recommendations.

Although the Declaration acknowledged 'the importance of Irrawaddy dolphins as a principal tourist attraction in North-Eastern Cambodia, and the importance of dolphins to the economies of Kratie and Stung Treng provinces', there is no mention in the recommendations of managing dolphin-watching tourism, despite the IWC's documented concerns.

Towards sustainable cetacean-watching in developing countries

Many of the issues associated with dolphin tourism in the Mekong are generic to cetacean-watch tourism in developing countries. The industry has typically been developed by subsistence fishers who opportunistically take advantage of the availability of local wildlife using their fishing boats to take visitors to watch dolphins. Entry to the industry is often flexible, allowing the fishers to switch between fishing and tourism depending on demand. The dolphin-watch industry is typically unregulated and by the time it comes to the attention of officials, it is impossible to stop, or modify on economic grounds and as a result of lack of capacity and political will. For example, at Lovina in north Bali, a fleet of 179 small, traditional vessels called *jukung* operate dolphin-watching focused on spinner dolphins (*Stenella longirostris*). A single school of dolphins may be surrounded by up to 80 boats (Mustika *et al.* (2012a,b)). At Chilika Lagoon in India, up to 600 converted fishing vessels participate in an industry based on a small (<150 animals), isolated and declining population of Irrawaddy dolphins, that is also subjected to gillnetting impacts (Dipani Sutaria and Coralie d'Lima, personal communication, 2010). The initial problems associated with such operations are generally governance and lack of capacity; however, as the industry develops and more tourists visit the site, the potential for conflicting policy goals increases.

Comprehensive studies that investigate ecological, social and political impacts are required to inform solutions to the problem of unsustainable dolphin watching in developing countries. As discussed by Mustika *et al.* (2012a), more research is required to understand the human dimensions of cetacean tourism, where research on the tourist experience (e.g. tourist satisfaction and opinions) coupled with research on the economic and managerial aspects of the industry are likely to be significantly cheaper than ecological research and more relevant to local stakeholders. An additional benefit of investigating human dimensions is the ability for conservation organizations to target national and international tourists to encourage self-regulation and responsible tourism, if management agencies fail to effectively minimize threats to endangered populations. An example for the Mekong dolphin population is to undertake national and international awareness campaigns to encourage tourists to view Mekong dolphins from land, rather than from a motorized boat.

Irrespective of whether the industry is new or established, management needs to be precautionary to pre-empt the potential long-term effects of cetacean-watching on the viability of the target species, regardless of the potential benefit of the industry to local communities. Dolphin-watching should not be encouraged in developing countries without appropriate management, prior regulation, and ongoing monitoring of compliance, conditions that may be impossible to achieve (Higham *et al.*, 2008). Both Corkeron (2004) and Bejder *et al.* (2006) stress that there are locations and situations where boat-based cetacean-watching tourism should not occur. Irrawaddy dolphin-watching in the Mekong River appears to clearly represent one such case.

When considering management options for new initiatives, in the absence of adequate information on tourism impacts, management deliberations must draw strong inference from the best documented sites, where long-term, individually specific information can be taken into account (Bejder *et al.*, 2006). Studies have shown that dolphin-watching tourism can have long-term

detrimental impacts on targeted populations. For example, in Shark Bay, Western Australia despite the industry being licensed and seemingly well-managed, bottlenose dolphins moved out of their preferred habitat in response to increased dolphin-watching tourism, reducing the calving success of most exposed females (Bejder *et al.*, 2006). Bejder *et al.* (2006) noted that while such a decline may not jeopardize the large, genetically diverse bottlenose dolphin population of Shark Bay, similar impacts would be dire for small, closed or isolated cetacean populations such as most riverine or lacustrine dolphin populations, including the Mekong dolphin population.

The requirement for adequate planning and precautionary management (see Higham *et al.*, 2008), is especially important in developing countries where cetacean-watching exists within challenging social, political, economic and environmental contexts. Most developing countries are evolving quickly in the face of burgeoning pressures from human over-population, excessive exploitation of resources, poverty, lack of basic services and corruption (Beasley, 2007). The positive and negative impacts of corruption on biodiversity conservation have recently been debated in the scientific literature (Kaufmann, 1997; Laurance, 2004; Ferraro, 2005; Katzner, 2005; Smith & Walpole, 2005; Walpole & Smith, 2005). Corruption is most prevalent in developing countries with low government salaries, weak regulatory institutions, high political patronage and almost non-existent accountability (Kaufmann, 1997; Laurance, 2004). As a result of such realities, ineffective governance and corruption are additional considerations accentuating the difficulty of managing cetacean-watch tourism in developing countries to ensure that operations are sustainable.

Conclusion

As pointed out by Neves (2010), the simplistic portrayal of cetacean-watching as diametrically opposite to whale-hunting obscures the existence of unsustainable cetacean-watching. Many of the issues described in this chapter are generic to cetacean-watch tourism in developing countries, where the industry is developed opportunistically, and entry is flexible allowing fishers to switch between fishing and tourism depending on demand. Such dolphin-watch industries are typically unregulated and by the time they come to the attention of officials, they are impossible to stop or modify on economic grounds and as a result of lack of capacity and political will.

When considering management options for new initiatives in the absence of adequate information on tourism impacts, management deliberations should draw strong inference from the best documented sites, where long-term, individually specific information can be taken into account. A precautionary approach to management is needed to address the problem of unsustainable dolphin-watching currently occurring in numerous developing countries. This approach should be informed by location-specific and comprehensive studies on both the ecology of the dolphin-tourism interactions and the social, economic, managerial and political influences on cetacean-watch operations.

Dolphin-watching tourism is not a benign industry. Direct deaths have been reported through boat collision, and the cumulative effects of long-term exposure on individuals and populations have been well-documented. There is great urgency to find solutions to the challenges of managing dolphin-watching tourism in developing countries, particularly at potential sites where dolphin-watching tourism is undeveloped or still in its infancy. Management and regulations are far easier to implement prior to development of industry rather than once established. There are numerous benefits to investigating the social dimensions of dolphin-watching tourism, as the dissatisfaction of tourists can be a powerful motive for improved management (Mustika et al., 2012a). Boat-based and 'swim-with-dolphin' tourism focused on critically endangered populations should be discouraged.

Acknowledgements

Funding for the 'Dolphins for Development' Integrated Conservation Development Project was provided by the British Embassy – Phnom Penh, and New Zealand Agency for International Development (NZAID) – Bangkok. Many thanks to Cambodian Rural Development team members who assisted to implement the Project, in particular Brendan Boucher, Or Channy, Sun Mao, Hean Pheap, Hang Vong, and Mao Savin. Thanks also to Yim Saksang and Lor Kim San, who assisted with all aspects of the Project. Thanks to the Cambodian Ministry of Forestry and Fisheries (MAFF) and all officials from the Cambodian Department of Fisheries for their full support of research and conservation activities throughout the duration of Project activities. Particular thanks to Excellency Chan Sarun (Minister of Agriculture, Forestry and Fisheries), Excellency Nao Thouk (Director, Department of Fisheries), Sam Kim Lun (Chief of Kratie Fisheries Office), Mao Chan Samon (Chief Stung Treng Fisheries Office), Sean Kin (Kratie Fisheries Office), and Kim Sokha (Department of Fisheries, Phnom Penh). Many thanks to Professor Ian Baird for providing essential background information on dolphin-watching tourism at Anlong Chiteal/Veun Nyang Pool. The lead author would like to acknowledge the significant support she received from the late Dr Peter Arnold, to initiate the Dolphins for Development Project.

REFERENCES

Baird, I.G. (2006). Strength in diversity: Fish sanctuaries and deep-water pools in Lao PDR. *Fisheries Management and Ecology* 13, 1–8.

Baird, I.G. & Beasley, I.L. (2005). Irrawaddy dolphin *Orcaella brevirostris* in the Cambodian Mekong River: An initial survey. *Oryx* 39, 301–310.

Baird, I.G. & Mounsouphom, B. (1994). Irrawaddy dolphins (*Orcaella brevirostris*) in southern Lao PDR and northeastern Cambodia. *Natural History Bulletin of the Siam Society* 42, 159–175.

Beasley, I.L. (2007). Conservation of the Irrawaddy dolphin, *Orcaella brevirostris* (Owen in Gray, 1866) in the Mekong River: Biological and social considerations influencing management. PhD Thesis submitted to the School of Earth and Environmental Science, James Cook University, Australia. 427 pp.

Beasley, I.L., Marsh, H., T.A. Jefferson. & P. Arnold. (2009). Conserving dolphins in the Mekong River: The complex challenge of competing interests. In I.C Campbell (Ed.), *The Mekong: Biophysical environment of an international river basin.* Sydney: Elsevier, pp. 363–387.

Beasley, I.L., Bejder, L. & Marsh, H. (2010). Dolphin-watching tourism in the Mekong River, Cambodia: A case study of economic interests influencing conservation. International Whaling Commission Working Paper SC/62/WW4. 9 pp.

Beasley, I.L., Pollock, K., Jefferson, T.A., *et al.* (2012). Likely future extirpation of another Asian river dolphin: The critically endangered population of the Irrawaddy dolphin in the Mekony River is small and declining. *Marine Mammal Science*, doi:10.1111/j.1748-7692.2012.00614.x.

Bejder, L. & Samuels, A. (2003). Evaluating the effects of nature-based tourism on cetaceans. In N. Gales, M. Hindell & R. Kirkwood (Eds), *Marine Mammals: Fisheries, tourism, management issues.* Melbourne: CSIRO Publishing, pp. 229–256.

Bejder, L., Samuels, A., Whitehead, H., *et al.* (2006). Relative abundance of bottlenose dolphins (*Tursiops* sp.) exposed to long-term anthropogenic disturbance. *Conservation Biology* 20, 1791–1798.

Bookbinder, M.P., Dinerstein, E., Rijal, A., Cauley, H. & Rajouria, A. (1998). Ecotourism's support of biodiversity conservation. *Conservation Biology* 12, 1399–1404.

Corkeron, P.J. (2004). Whale-watching, iconography and marine conservation. *Conservation Biology* 18(3), 847–849.

Davis, S. & Khosla, V. (2007). The architecture of audacity: Assessing the impact of the Microcredit Summit Campaign. *Innovations* Winter/Spring, 159–180.

Ferraro, P. (2005). Corruption and conservation: The need for empirical analysis. A response to Smith and Walpole. *Oryx* 39(3), 1–3.

Garrod, B. & Fennell, D.A. (2004). Analysis of whalewatching codes of conduct. *Annals of Tourism Research* 31(2), 334–352.

Garrod, B. & Wilson, J.C. (2003). *Marine Ecotourism: Issues and experiences.* Aspects of Tourism 7. Australia: Channel View Publications, 262 pp.

Gilbert, M. & Beasley, I.L. (2006). *Mekong River Irrawaddy Dolphin Stranding and Mortality Summary: January 2001–December 2005.* Phnom Penh: Wildlife Conservation Society – Cambodia program, 39 pp.

Higham, J.E.S., Bejder, L. & Lusseau, D. (2008). An integrated and adaptive management model to address the long term sustainability of tourist interactions with cetaceans. *Environmental Conservation* 35(4), 294–302.

Hoyt, E. (2005). Sustainable ecotourism on Atlantic Islands, with special reference to whale-watching, marine protected areas and sanctuaries for cetaceans. *Biology and Environment: Proceedings of the Royal Irish Academy* 105B(3), 141–154.

International Monetary Fund. (2006). Cambodia – Poverty Reduction Strategy Paper. IMF Country Report No. 6/266. Retrieved from http://www.imf.org/external/pubs/ft/scr/2006/cr06266.pdf

IWC (International Whaling Commission). (2006). Report of the Scientific Committee. *Journal of Cetacean Research and Management* 8(Suppl), 1–65.

IWC (International Whaling Commission). (2010). Report of the Scientific Committee. International Whaling Commission Meeting 62, Washington, DC.

Katzner, T.E. (2005). Corruption – A double-edged sword for conservation? A response to Smith and Walpole. *Oryx* 39(3), 1–3.

Kaufmann, D. (1997). Corruption: The facts. *Foreign Policy* 107, 114–131.

Laurance, W.F. (2004). The perils of payoff: Corruption as a threat to global biodiversity. *Trends in Ecology and Evolution* 19, 399–401.

Lusseau, D., Slooten, L. & Currey, R.J.C. (2006). Unsustainable dolphin-watching tourism in Fiordland, New Zealand. *Tourism in Marine Environments* 3(2), 173–178.

Mandel, J.T., Donlan, C.J., Wilcox, C., *et al.* (2009). Debt investment as a tool for value transfer in biodiversity conservation. *Conservation Letters* 2, 233–239.

Mustika, P.L.K., Birtles, A., Everingham, Y. & Marsh, H. (2012a). The human dimensions of wildlife tourism in a developing country: Watching spinner dolphins at Lovina, Bali, Indonesia. *Journal of Sustainable Tourism.* Published online, 1–23.

Mustika, P.L.K., Birtles, A., Welters, R. & Marsh, H. (2012b). The economic influence of community-based dolphin watching on a local economy in a developing country:

Implications for conservation. *Ecological Economics* 79, 11–20.

Neves, K. (2010). Cashing in on Cetourism: A critical ecological engagement with dominant E-NGO discourses on whaling, cetacean conservation, and whale watching. *Antipode* 42(3), 719–741.

O'Connor, S., Campbell, R., Cortez, H. & Knowles, T. (2009). *Whale Watching Worldwide: Tourism numbers, expenditure and expanding economic benefits*, a special report prepared by Economists at Large. Yarmouth MA: International Fund for Animal Welfare.

Orams, M.B. (2001). From whale hunting to whale watching in Tonga: A sustainable future? *Journal of Sustainable Tourism* 9(2), 128–146.

Poulsen, A.F. & Valbo-Jorgensen, J. (2001). Deep pools in the Mekong River. *Catch and Culture Magazine* 7(1), 10–11.

Ryan, G.E., Dove, V., Trujillo, F. & Doherty, P.F. (2011). Irrawaddy dolphin demography in the Mekong River: An application of mark–resight models. *Ecosphere* 2, 1–14.

Smith, B.D. & Beasley, I.L. (2004). *Orcaella brevirostris* (Mekong River subpopulation). IUCN 2004. 2004 IUCN Red List of Threatened Species, www.redlist.org. Downloaded 1 September 2010.

Smith, R.L. & Walpole, J. (2005). Should conservationists pay more attention to corruption. *Oryx* 39(3), 1–6.

Stacey, P.J. (1996). Natural history and conservation of Irrawaddy dolphins, *Orcaella brevirostris*, with special reference to the Mekong River, Laos PDR. MSc Thesis, University of Victoria. 374 pp.

Stacey, P.J. & Hvengaard, G.T. (2002). Habitat use and behaviour of Irrawaddy dolphins (*Orcaella brevirostris*) in the Mekong River of Laos. *Aquatic Mammals* 28(1), 1–13.

Taylor, B.L. & Gerodette, T. (1993). The uses of statistic power in conservation biology: The vaquita and spotted owl. *Conservation Biology* 7(3), 489–500.

Thomlinson, E. & Getz, D. (1996). The question of scale in ecotourism: Case study of two small ecotour operators in the Mundo Maya region of Central America. *Journal of Sustainable Tourism* 4(4), 183–200.

Transparency International. (2009). *Annual Report 2009*. Berlin: Transparency International.

United Nations. (2005). Standard Country or Area Codes for Statistical Use. Series M, No. 49, Rev. 4 (United Nations Publications, Sales No. M.98.XVII.9). Available in part at: http://unstats.un.org/unsd/methods/m49/m49regin.htm

Valentine, P.S. (1993). Ecotourism and nature conservation: A definition with some recent developments in Micronesia. *Tourism Management* 24, 107–116.

Walpole, M.J. & Smith, J. (2005). Focusing on corruption: A reply to Ferraro and Katzner. *Oryx* 39(3), 1–2.

22

Kaikoura (New Zealand)

The concurrence of Māori values, governance and economic need

David G. Simmons

Introduction

This chapter traces the history of human settlement in Kaikoura, focusing in particular on its origins as an important centre of Māori settlement (c. 950–1130 AD) and livelihood, subsequent European settlement (from the 1830s AD) to a world-class tourism destination based on whale-watching. Historical patterns of settlement, (indigenous) Māori–Pakeha (settler) relations, the early use of whales for commercial trade, subsequent periods of resource extraction and agricultural cycles of growth and decline establish the historical context of tourism development. Contemporary tourism development is, however, primarily based on desperate local economic conditions brought about by restructuring of the New Zealand economy in the mid-1980s.

The history of Kaikoura as a whale-watching destination must first be explored by examining its evolution within a larger set of Māori values. For Kaikoura these are centred around key concepts of:

mother earth (papatuanuku);

Mauri (the physical life force that imbues all living things) and

kaitiakitanga (guardianship of natural resources).

Although these are argued to be particularly poignant in Kaikoura where the (Ngai Tahu) Māori owned-and-operated Whalewatch Kaikoura has become the cornerstone tourist experience, elsewhere in New Zealand they have provided a unique

context for resource management.[1] Thus Māori values, alongside more recent resource exploitations, subsequent 'boom and bust' economic cycles and finally significant 'restructuring' of the late 1980s have paved the way for a cautious approach to contemporary tourism management based on scientific research and destination benchmarking and certification. The early adoption of the GreenGlobe21/earthcheck™ certification system, and Kaikoura's claim as the world's first independently certified tourism destination, have engendered sufficient momentum to see this approach to sustainability permeate all the District's activities and many of its productive sectors. This pathway, in turn, is argued to have arisen out of the unique concurrence of Māori values, tourism focused on marine mega-fauna, and a set of scientific partnerships that have been able to integrate and support a strong focus on governance and management for sustainability.

---

[1] Kaitiakitanga is a widely accepted concept in contemporary New Zealand (planning) legislation. For example, the Resource Management Act (1991) is directed at the sustainable management of environmental resources and records that those managing the act must take kaitiakitanga into account. The Act defines kaitiakitanga as the exercise of 'guardianship by the tangata whenua (indigenous Māori) of an area in accordance with Tikanga Māori in relation to natural and physical resources; and includes the ethic of stewardship'.

Whale-watching: Sustainable Tourism and Ecological Management, eds J. Higham, L. Bejder and R. Williams.
Published by Cambridge University Press. © Cambridge University Press 2014.

Location and history: waves of human settlement and exploitation

The Kaikoura District (South Island, New Zealand) is located on the east coast of the south Island of New Zealand (Figure 22.1). It spans an area from the Haumuri bluffs in the south to the Kekerengu valley in the north, covering 2048 km² of diverse landscape. The inland boundary of the district is the Clarence River and Inland Kaikoura Range, climbing 2885 m and snow-covered for much of the year. The spectacular coastline provides excellent sightseeing (by both private vehicle and train), fishing, sporting and recreation for locals and visitors alike. The contemporary township of Kaikoura is situated on a peninsula protruding from the rugged coastline approximately half-way along the major north–south national highway between the major south island city, Christchurch, and the ferry terminal at Picton which connects to the more populated North Island. Within the sea the Kaikoura Canyon, at approximately 1300 m deep, is remarkably close to shore (500 m; see NIWA, 2010), provides the natural habitat for a permanent community of sperm whales, and is also host to over 200 species of marine life, including dolphins and seals which are also key attractions for tourists.

At first glance the relationship between Māori and whales (Tohora) might seem to be an ambivalent one. Māori oral history records that, on the one hand, whales were traditionally regarded as a sign of plenty and a source of food. On the other, they had spiritual significance as taniwha (mythical sea creatures), such as those that accompanied the founding canoes to Aotearoa/New Zealand. They also acted as kaitiaki or guardians to particular individuals or tribes (such as the local Kati Kuri hapu (extended family)), and could be called upon in time of need. That is, they occupied a place in Māori mythology and history that was both physical and metaphysical. In short, there is a consistent deep belief in their role as 'taonga' or prized possessions of the Māori people. Māori history and culture is an integral part of Kaikoura, and there is evidence of Māori settlement in the area up to 1000 years ago. Māori legend tells that it was from the Kaikoura peninsula that Maui fished up the North Island from out of the ocean. The name itself, as is common with New Zealand place names, reflects key natural features, in this case the abundant seafood (Kai = food; Koura = crayfish/lobster) on which they founded their settlement.

While the history of Māori settlement in Kaikoura is contested (Anderson, 1991) as the first or 'one of the first' Māori settlements in all of New Zealand, there is no doubt that the region was favoured by early Māori (generally thought to have arrived between 950 and 1130 AD). Ancient Māori settlement was based around fishing and the hunting of seals, moa and other birds (Challis, 1991). With an abundance of seafood, Kaikoura remained an attractive location for Māori development and was the site of local conflicts in the sixteenth and mid-seventeenth centuries and again in a period from 1827. From the 1670s onwards the land is generally recognized as being of Ngai Tahu (the principal Iwi (local tribal group)) that occupied almost all of the South Island (McAloon *et al.*, 1998).

The region was inspected sporadically by European ships in the 1830s and 1840s and there are fragmentary reports of commercial whaling during the 1830s (Sherrard, 1966). The first definite shore-based whaling station was established early in 1843. This station was financed from Wellington and was run on the ground by Robert Fyffe and John Murray on the Kaikoura peninsula itself. Operations were successful, but the crew took fright after a nearby affray (in the adjoining northern Marlborough province) in June 1843 and the station was abandoned. Word of Kaikoura's whaling potential had spread, however, and by the end of 1844 there were stations at Waipapa, Kaikoura, South Bay and the Amuri Bluff within the Kaikoura District. Throughout this period whaling was commonly undertaken by young seafaring Māori who are reported to have revelled in both the activity and trade. Whaling continued on a small scale until 1920, although by 1870 there were only two stations in use and in 1905 only one, at South Bay on the

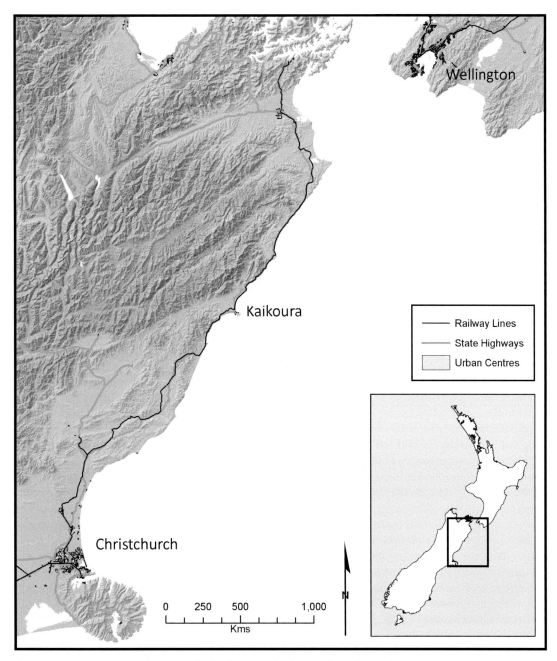

Figure 22.1 Location Map: Kaikoura, South Island, New Zealand. (See colour plate section.)

peninsula. By the end of the whaling period only a couple of mammals a year were caught (Cyclopedia, 1906: 446) as local stock was close to exhausted and whaling had become locally uneconomic. Kaikoura Māori worked with the whalers until shore-based whaling ended in 1922, although commercial whaling in New Zealand waters continued until the 1960s, with the last whale harpooning by a New Zealand vessel occurring off the coast of Kaikoura on 21 December 1964 (Hutching, 1990).

The period from 1847 to 1859 remains as a contested and, in retrospect, uncomfortable period of land acquisition by the Crown acting on behalf of early European settlers/runholders. Many of these agreements have remained contested to the present, and are slowly being redressed by the Government's Waitangi Tribunal. Notwithstanding, through one mechanism or another much of the Kaikoura lands had been acquired by European farming interests and local Māori had secured only £300 and some 2250 ha of reserves by the end of this period – much of it 'of the most useless and worthless description' as Mackay himself said (Sherrard, 1966; Evison, 1993: 382–384). Even following earlier designations Ngai Tahu's reserves were considerably reduced after 1900, when the Crown compulsorily acquired them for scenic or railway purposes. Over time, half of the area reserved was lost, much of it to the Crown for these purposes. With only a number of small reserves, local Māori were forced to rely on seasonal farm work and on fishing, and whatever crops and cattle they had (Sherrard, 1966). Fishing was a major source of subsistence and petty cash: in 1870 all but two of the fishers in Kaikoura were Māori. Barracouta – reasonably sized fish and easily caught – were the main species with hapuku, trumpeter, tarakihi, moki, rock cod and kahawai comprising most of the catch, but the local market varied according to the availability of meat (AJHR, 1870: D-9, p. 5).

During the 1850s various authorities reserved some 6000 ha around the Kaikoura Peninsula, and in so doing set the foundation for the development of the present township. Town and small farm sections, mostly of 15–20 ha, were surveyed out between 1861 and 1864, and a few whalers had already been given grants for 4-ha blocks. Few small farmers took up these swampy sections, for not only was the task of draining such land time-consuming and hard, but there was little prospect of a market for their produce, and over time much was picked up by existing runholders. Small farming spread steadily in the 1850s, and by 1865 there were 12 runs on what would later be the Kaikoura County.

Life was hard and isolated. Aside from coastal trading, access remained the next key development in the opening up of the District. The route now known as the Inland Kaikoura road, linking Kaikoura with North Canterbury via Rotherham, was built between 1882 and 1888. This was the main route for a considerable time, for although the road via the coastal route (now SHW1) was opened in the early 1890s the major rivers were not bridged until 1914 (Sherrard, 1966). Railway development had been under consideration since the 1870s. There was considerable argument in the 1870s and 1880s over the best route by which to connect Christchurch, Kaikoura, Picton and the North Island beyond. The eastern coastal route (roughly, that which was eventually chosen) was very rough and steep in places, but gave access to the good land at Cheviot to the south. The alternative, which would have followed the Inland Kaikoura road, was easier but gave access to much less good land (AJHR 1876: E-1; 1883: D-2). The global depression and First World War intervened and it was not until the end of 1945 that the rail link was completed (Sherrard, 1966). The population had increased with rail works after 1935, but fell slightly after completion in 1945 (Ministry of Works, 1962).

In 1962, a government survey (Ministry of Works, 1962) reviewed present resources and future potential for the broader Marlborough region (of which Kaikoura District was still a part). By this time, the town labour force was concentrated in transport and servicing occupations. Public-sector departments were major employers, particularly the Ministry of Works and Railways Department. The railway repair and maintenance shops accounted for many of the 80 railway employees. Road transport company depots also accounted for a number of workers in service stations and vehicle repairs. The dairy factory was small; stock and station, bank agencies and private shops accounted for a few. Fishing and lime extraction were in decline. Over two-thirds of the district population was concentrated in the town and the surrounding small plain.

In the wider District five distinct farming zones were described: dairy farming was concentrated on the fertile crescent around the Peninsula; sheep were raised to the south of that area, and arable farming was carried on to the north. Sheep and beef cattle were raised along the coastal strip north of the Clarence River, and wool growing predominated further inland (Marlborough Regional Development Council, 1976). Fishing was not a major industry. By 1960 it was apparent that the easy crayfish grounds had been over-fished in the years since 1945, and by 1975 the same was true of other species as well (Marlborough Regional Development Council, 1976). Despite Kaikoura's status as the second largest fishing port in the northern South Island after Nelson, with a catch valued at more than half a million dollars, fishing was generally small-scale. In 1975 there were 97 registered fishing vessels, and 149 fishers. Many of these were part-time operators; 38 boats and 59 fishers were classed as 'active', that is, making more than NZ$4000 a year from fishing. Whaling was but a historical curiosity throughout this period.

Tourism – early beginnings

Although, at the turn of the century, the town had been 'recognised as being one of the most picturesque and healthy spots in New Zealand it (was) often referred to as the Scarborough of the colony' (Cyclopedia, 1906: 446), little development had occurred. In a Marlborough Regional Council report (1976) the potential for tourism was referred to only briefly. Kaikoura was basically a 'tea and pee' stop for motorists and rail passengers, although there had been some development of baches[2] at South Bay in the 1950s. There was, reportedly,

considerable reluctance in the town to see further tourist development (Marlborough Regional Council, 1976). What little tourism occurred was largely a spin-off from road traffic to and from the Picton (North Island) ferry link. In Kaikoura, there had been one 40-bed motel before the Cook Strait ferries started running in 1962. In 1975, there were 304 motel beds and another 500 hotel and camping ground beds. It was observed, with some understatement, that there was 'concentration of capital investment [in tourism] in the main centres of Blenheim, Picton and Christchurch and not, as could be expected, in the areas of great scenic attraction' (Marlborough Regional Council, 1976: 127). In other words, accommodation, rather than tourist attractions as such, were the emphasis.

Like most small rural regions, Kaikoura was hard hit by the recession of the 1970s and the restructuring of the New Zealand economy which followed after 1984. Farm incomes dropped throughout the country after 1984, which had a significant effect on towns such as Kaikoura. Public-sector employment declined considerably, with 170 jobs lost in a town of 3000 (Brett, 1992).

The 1980s were also notable for a greatly increased emphasis on tourism as a generator of overseas funds, and for the increasing visibility of Māori claims under the Treaty of Waitangi. These two factors intersected in Kaikoura. In 1985, confronted with a wave of redundancies and a falling population, a group of Kaikoura people established a promotion association and tourist centre. Originally the emphasis was on the District's scenery and walkways, but once the whale-watching ventures started, visitor numbers soared (*Press*, 4 September 1996: 25).

Whale-watching: contested beginnings, turbulent initiation

It has been noted that by the 1980s Kaikoura's economy and employment depended largely on farming and fishing industries and on government organizations such as the Railways, Public Works and

[2] A local name for holiday 'cribs' – taking its name from bachelor's residences/cribs.

Post Office. With the closure of many of these organizations, Kaikoura witnessed a growing trend in unemployment – in particular, with a worrying 90% Māori youth unemployment. For Kaikoura Māori it was time to search for alternative ways of employment creation. Naturally many local discussions ensued, the majority of which have not been formally reported. From a long set of generally environmentally based ideas, whale-watching emerged as the favoured venture (Poharama *et al.*, 1998). Again, while the outcome is now clear, the early history of development of whale-watching as a commercial tourism venture remains contested. At about the same time as local Māori were considering whale-watching options, a similar venture was being researched by an alternative partnership – but whether they were working for (Keene, 1992), with (Ansley, 1991), or in competition to local Māori interests remains unclear (APEC, 1997). For a short period in March 1988, nine months before Kaikoura Tours started up, an alternative operation 'Nature Watch Charters Limited' had been established out of scientific interest by two Americans (Todd and Sutherland). For some, controversy still lingers over whose 'moemoea' or dream gave rise to the initial venture, but whether that of Todd and Sutherland or the Māori people of Kaikoura, the latter consistently maintain that the original idea was theirs (Poharama *et al.*, 1998).

Twenty years on, it is commonly accepted by local Māori that 'Kaikoura Tours', now known as 'Whale Watch', was initiated by the five Māori families who attended the initial meeting (Poharama *et al.*, 1998). Unable to obtain a loan from the banks, the families mortgaged their homes and cars as collateral for a loan which they obtained from the Māori venture-capital agency, Mana Enterprises, a branch of the then national government's Department of Māori Affairs. The Māori Whale Watch enterprise took two years to plan and develop. Beginning in 1985, monthly meetings were held to discuss finance, boats, training needs, legal aspects and other requirements. These meetings were held prior to initiating a feasibility study. The Marlborough Development Board gave a business development grant of NZ$5000 for an initial study into the feasibility of whale-watching in Kaikoura waters (Brett, 1992), and additional support in unemployment relief and training programmes. The Department of Conservation (who enact the Marine Mammals Protection Act) were kept informed that a feasibility study was being undertaken on behalf of local iwi (the largest social unit in Māori culture, most commonly translated as 'tribe') and subsequently granted a concession[3] to begin operating a whale-watching venture.

According to Bill Solomon, the Upoko Runanga ('head') of Takahanga Marae:

The operation was originally set up as an employment creation venture to help discourage young Māori locals from leaving the district, and to provide a source of funds for the development of Takahanga Marae. (Scott, 1991)

Turbulent waters

Formally established in 1987 as Kaikoura Tours Limited, Whale Watch demonstrated for local iwi a sense of sovereignty in that they are a people whose customs come from a marine environment, and it is from this environment that they derive their kawa or 'ways of doing things'. Whale Watch was officially opened in July 1989 by the Honourable Koro Wetere, the then Minister of Māori Affairs, who described it as 'a positive example of Māori people helping themselves by providing employment and conserving New Zealand's unique natural and cultural heritage' (*The Press*, 1989).

When Whale Watch began, it had three business operations: the Whale Watch enterprise, the Whale Watch Craft Shop, and the Whale Watch Restaurant. Not all locals were behind the ventures, and in its first year of operation the only businesses to support it in terms of bringing it to the attention of visitors were two local accommodation providers (Poharama *et al.*, 1998). After many years of being a

[3] 'Concessions' (to permit commercial activities under specified environmental management protocols) are issued by the Department of Conservation, in this case they refer to the provisions of the Marine Mammals Act (1978).

'government' town, racism escalated in Kaikoura in those early years, to the point where non-Māori who supported the venture were reported to be receiving abusive telephone calls, and petitions were circulated to try and stop Whale Watch from berthing their boats in South Bay. In November 1990 all eight outboard motors on the boats were sabotaged by having the gearbox plugs removed – which added NZ$25,000 worth of damage to the NZ$14,000 stolen two weeks earlier, when thieves broke into the Whale Watch office (Brett, 1992). In February 1991 the company's 44-seater shuttle bus was gutted by fire in suspicious circumstances, after which '[their] insurance company refused to insure them' (Ansley, 1991). The police were unable to solve these reported crimes and so Whale Watch was obliged to set up its own nightly patrol roster to check boats, offices and other assets. This state of affairs did not continue indefinitely.

As the business grew both in size and complexity, financial support was provided by the Ngai Tahu Trust Board (the corporate body of the major South Island iwi), which now holds a 47% shareholding in Whale Watch while the Kaikoura Runanga (tribal council) holds the remaining 53%. The acquisition by Whale Watch of the government's tourism concession was seen as a conservation 'monopoly' which 'polarised the community' (Brett, 1992: 60; APEC, 1997: 35), especially in the early days when Todd and Sutherland's Nature Watch tours were still in operation.

Since the establishment of Whale Watch, Kaikoura has witnessed a rapid increase in tourism. The economy and employment opportunities in Kaikoura are now largely dependent on the tourism industry, and Whale Watch is the biggest tourist attraction Kaikoura has to offer. Today it records an enviable set of accomplishments. From its humble beginnings, it has turned into a multi-million dollar business that is tribally owned and operated. To local South Island iwi, tourism is a young but growing enterprise, and many see Whale Watch as the beginning of their participation in the tourism industry in time comparable in extent to North Island iwi (Te Arawa), which had already held a long involvement in tourism centred on the

volcanic plateau around Rotorua (Te Awekotuku, 1981; Tahana et al., 2000). Reporting some 14 years ago, Poharama et al. (1998) present survey data indicating that 70% of Māori respondents had been involved in tourism and that 60% had other household members currently employed in the tourism sector (on average, two persons per household). Thus one can conclude that the early aspirations of local Māori and their leaders were well fulfilled.

Today the company structure of Whale Watch Kaikoura Limited is made up of capital and directors from Te Runanga O Kaikoura Trust Board (the corporate body of the local hapu, Ngati Kuri), and Te Runanga O Ngai Tahu (the corporate body of the major South Island iwi). Described in the literature as 'a greenfields ecotourism venture' (Te Karaka, 1997: 11; Te Māori, 1997: 4), Whale Watch portrays a mix of indigenous history, culture, heritage and natural environment. According to APEC (1997: 34), it is 'run by indigenous people with a heritage and a view of the future based on strong principles'. Those principles are that everything it does must be culturally acceptable, economically viable and environmentally sensitive (Stone, quoted in APEC, 1997: 40). This is reflected in the path the company has followed since its inception. However, the path has not been easy or straight and now the presentation of Māori culture is more muted as the scale of operation makes it more difficult to practise the principles that emphasize culture.

Broader aspects of tourism development and consolidation

Whales were not the only species of marine megafauna that were attractive to tourists – and the well-known seal breeding grounds close to the township (a separate swimming-with-seals enterprise was established in 1987) along with dolphin-watching and -swimming (established 1989) also soon attracted visitor interest. A significant point of consolidation appears to be centred around Kaikoura being chosen for the first in four extensive national tourism planning case studies in 1996 (Simmons & Fairweather, 2005). A research team from Lincoln University undertook benchmarking studies of visitor flows and effects, economic and environmental impacts, community and Māori engagement and reported their integration and summary report under the title 'Towards a Tourism Plan for Kaikoura' (Simmons & Fairweather, 1998). Key findings from that time indicate that, in the 12 years since the forming of Whale Watch (in 1986), tourism had grown rapidly from small beginnings, to 873,000 visits (356,000 overnight) per year. Current growth in visitor numbers was estimated at that time as 14% per annum.

Three key visitor groups were identified and quantified (Fairweather *et al.*, 1998):

- *Short stop visitors* (those staying less than two hours), for whom Kaikoura primarily fills the role of a 'convenient break' (79.7% of first choice of attraction to Kaikoura). The core of this group comprises domestic (New Zealand) visitors (75%) engaging in more extensive trips. While average expenditure per person per visit is relatively low (estimated at NZ$2.40 per visitor) some activities are undertaken, including 16% who visit the Kaikoura Information Tourism Incorporated (KITI) visitor centre. Short stop visitors represented a large group – 44% of all visitation or an estimated 380,000 visitors annually.
- *Day visitors* (those staying more than two hours, but not overnight) were numerically the smallest of the three visitor groups. Visitor numbers were estimated at 137,000 annually. For these visitors, Kaikoura is a specific destination, as evidenced by their high interest in whale-watching (49%) and visiting the seal colony (44%). Engagement in commercial activities and supporting industries lifts daily average per person expenditure for this group to NZ$47.50, the highest of all three groups.
- *Overnight visitors* (those staying one or more nights) were mainly international visitors who make up seven of eight overnight visitors. Visitor numbers were estimated at 356,000 annually. Average length of stay was reported as 1.8 days.

Within this pattern, domestic visitors tend to stay for shorter periods. Lower-cost forms of accommodation (backpackers and motor camps) were used mostly, while commercial activities (whale-watching and swimming/viewing dolphins especially) provided the key focus for commercial activity. Informal activities (visiting the seal colony (63%) and the visitor centre (77.4%)) were important to their overall experience in Kaikoura. These activities were paced throughout their visit to generate an average daily per person expenditure of NZ$45.73.

Total direct spending by visitors was estimated to be NZ$28 million per year with variable estimates of flow on effects ranging from NZ$21 to NZ$36 million (Butcher *et al.*, 1998). Approximately 330 persons (Full-Time Equivalent, FTE) are employed directly in tourism. While almost 700 people in the district worked in businesses that are wholly or partly tourism based, this figure is adjusted to reflect the part-time and seasonal nature of the work, and the fact that many businesses sell only part of their turnover to visitors.

Every job in tourism was reported to lead, on average, to a further 0.21 jobs elsewhere in the district economy. Overall the flow-on effects were reported as quite small, reflecting both the very limited business support infrastructure in Kaikoura and also the very low demand for external inputs in some businesses (such as homestay accommodation). The flow-on employment effects mean that in total approximately 400 FTE jobs are generated in the district by tourism. This excludes any jobs in social services (such as teaching) that might be lost if tourism (and hence employment) declined, and people emigrated from the district. Total employment in the Kaikoura district is believed to be around 1400 Full-Time Equivalent jobs (FTEs). In summary, employment figures indicated that some 30% of the economy either directly or indirectly arises from activities associated with the tourism sector. Importantly, in terms of the history described above in the 10 years since the establishment of Whale Watch, tourism development appeared to recapture the 156 FTE jobs lost during the

Table 22.1 Employment (full-time equivalent) by sector: 1996–2006.

Industry	1996	2001	2006[4]	% change 96/06
Accommodation, cafes, restaurants	165	222	288	75
Cultural and recreational services	42	52.5	79.5	89
Retail trade	157.5	156	178.5	13
Transport and storage	43.5	67.5	75	72
Property and business services	28.5	40.5	61.5	116
Construction	70.5	69	97.5	38
Agriculture, forestry and fishing	315	286.5	264	−16
Mining	6	4.5	3	−50
Manufacturing	123	115.5	114	−7
Electricity, gas and water supply	6	0	3	−50
Wholesale trade	16.5	22.5	45	173
Communication services	3	10.5	6	100
Finance and insurance	27	16.5	15	−44
Government administration and defence	24	24	24	0
Education	85.5	66	85.5	0
Health and community services	42	64.5	70.5	68
Personal and other services	45	40.5	52.5	17
Not elsewhere included	45	16.5	25.5	−43
Total	1245	1275	1488	20

Notes:

1. Measured as full-time plus half of part-time, as at census date (March of the various years). The number of persons employed at census date is between seasonal maximum and minimum. Consequently the census figures are likely to represent a reasonable annual average.

2. Due to different data collection and assessment processes these data sets are not exactly comparable. They are included here to support the discussions on longer-term trends in the business activity base of the Kaikoura community. (*Source*: Butcher *et al.*, 1998). Key tourism-related sectors are listed first.

significant economic restructuring of the mid-1980s when FTE units dropped from 1266 to 1110 from the 1986 to 1991 census, and then subsequently rose to 1386 for the 1996 census (Butcher *et al.*, 1998: 2; Table 22.1).

In fact, it was not only the outstanding success of tourism but the rate and direction of change that emerged as a central concern, especially among long-term residents, many of whom were not linked directly to the tourism sector as investors or employees (Horn *et al.*, 1998). For example, at that time

4 Retrieved from: http://www.stats.govt.nz/Census/ 2006CensusHomePage/Tables/AboutAPlace (accessed 2 September 2010).

many residents mistakenly believed that tourism was forcing up property taxes ('rates') by imposing large demands on infrastructure, particularly water and sewerage treatment. Importantly, the economic study (Butcher *et al.*, 1998) was able to estimate that tourism was directly responsible for about 10% of peak demand for water and 25% of peak demand for sewerage treatment.

In summarizing their study Simmons and Fairweather (1998) argued that at that time Kaikoura was at a crossroads in its development: it has successfully managed its involvement stage and now there are signs that the development stage is occurring. 'If tourism is to grow and retain both its personal character and broad distribution of benefits then specific attention to planning is required to manage future growth' (Simmons & Fairweather, 1998: 16). It was clear that if tourism were to continue to serve the key needs of locals (jobs, income), and if the town is to retain the essential elements of a 'small coastal community' which attract visitors and residents alike, tourism planning could not be left to chance.

This central government-funded study is still regarded as a seminal point in the management of tourism in Kaikoura as it led directly to the development of a tourism plan (to sit alongside other official planning requirements such as the District Plan required by the national Resource Management Act). In a more pragmatic sense, the strong local desire to establish tourism on a path to sustainability led to the early engagement with the Green-Globe/earthcheck's environmental benchmarking and certification programme for tourism businesses and destinations. Again led by the scientific community (Lincoln University and LandcareResearch, a New Zealand Crown Research Institute), Kaikoura undertook the necessary studies to become the world's first global local government to secure the Certified GreenGlobe21 (now 'earthcheck') gold destination status. Since this time the requirements of the earthcheck environmental standard have become a focal point for all District resource management activity. Resource management activities have reached out from the tourism sector to the large agricultural sectors particularly on the issue of waterways management (a weakness highlighted in the first assessment) and to embody all aspects of the District's management and operations (Kaikoura District Council, 2008).

Tourism: whales and community life – the pathway to sustainability

A key objective of this chapter has been to review the ongoing evolution of Kaikoura as a community and whale-watching tourism destination in the 12 years since the major scientific studies described

above. The issues raised in the late 1990s have, to some variable extent, been able to be addressed. At the centre of this has been the acceptance of the GreenGlobe21/earthcheck™ reporting and certification system as a major operational platform across all council activities (see e.g. Kaikoura District Council, 2011a), and the formal recognition of the tourism as both a major source of economic activity and agent of change that requires regular community input and district-wide management.

Official documents developed by the District Council indicate that there is now an official tourism strategy that stands alongside the District Plan. A tourism and economic development committee provide regular reports on visitor volumes and impact (Kaikoura District Council (nd), 2011b). Ongoing progress has been made in waste and energy management, the enhancement and management of amenities, and broader community engagement and development. Planning tools, such as specific tourism areas and precincts, have been notified in the Long Term Council Community Plan (the Local Government's (District's) official planning tool) which seeks to integrate land, asset and budgetary plans.

While accurate visitor numbers are always challenging for small destinations, based on visitation patterns in New Zealand which to October 2010 had recorded a cumulative growth of 8% in the last decade (and the rapid growth in visitation to Kaikoura in the late 1990s) it would be reasonable to assume that visitor numbers have grown from the 873,000 in 1998 to in excess of one million today. Such an estimate is supported by official statistics that demonstrate ongoing growth in sectors readily identified with tourist activities (cultural and recreation services, and accommodation cafes and restaurants, in particular) and flow-on effects to tourism support sectors (property and business services, and wholesale trade). Continuing employment declines in agriculture and some public services, first reported by Butcher *et al.* (1998, table 1: 2) for the decade of economy-wide 'restructuring' prior to 1996, are also evident, although at an abated rate (Table 22.1). Notwithstanding this overall trend, while it appears that tourism has generated a significant increase in employment, it has been brought about by growth in the volume of tourism, rather than by changes in the structure of the industry.

Since the early establishment of Kaikoura's environmental benchmarking and certification programme, alternative environmental certification programmes, at least in a 'light' form, have been migrated to the national assessment framework via the national government-sponsored Qualmark Green programme (see http://www.qualmark.co.nz/responsibletourism.php). These too are reinforced by the very successful '100% pure brand' promoted by TourismNZ (the national tourism promotion agency). Notwithstanding these up-link initiatives, pragmatic issues do remain on the ground.

Major challenges remain in how to manage a small (secondary) destination in the midst of bullish national tourism marketing efforts. Without clear links to national strategy, smaller destinations are to an important sense volume and price takers. The national profile of Kaikoura and its international standing and awards do, however, allow it to have visibility far above its size as a tourist destination. At present, it is estimated that tourism might be in excess of 30% of the local economy (as measured by employment metrics). Allen *et al.* (1993) have noted that this is at the upper limits of 'thresholds' where tourism, because of its physical presence, size and market risk, becomes increasingly difficult for a community to manage. In the bio-physical sense there is increasing concern (and a robust set of scientific measures) about the capacity of whales to sustain increasing visitation (not just in terms of visit frequency and duration, but boat sizes, distance and the like[5]), which is also manifest in the ongoing search for scientific evidence to inform limits on the number of 'swimmers'

alongside dolphins, and with seals. Such concerns have already led Whale Watch Kaikoura to undertake significant investment into onshore visitor experience opportunities to both ensure ongoing growth in visitation and as a buffer against the naturally variable weather.

Notwithstanding the remarkable success of Kaikoura over the past two decades, key questions remain concerning the ongoing growth and management of tourism in the face of increasingly known capacity issues and environmental management. However, tourism destinations exist in a much broader, global distribution system (Simmons & Becken, 2004). This system in itself is energy-intensive and long-haul destinations, such as those in New Zealand, are particularly exposed to these risks (Becken, 2008). A decade is but one in a long journey to sustainability.

Māori values, governance and science – a unique leadership mix

In this discussion, I seek to distil the unique features of governance and resource management that have placed Kaikoura at the forefront of sustainable tourism destination management. There can be little doubt that Kaikoura is seen as a highly successful tourism destination, based as it is on its iconic whale-watching attraction. Indeed, some might argue that the 'whale-watching' profile of Kaikoura is a brand that has become as big as the town itself.

When one sets aside the relatively short time exposure of tourism development in Kaikoura and examines the broader aspects of sustainable tourism development, there are a series of key features of tourism development that stand out as unique. First among these is the nature of tourism attractions in Kaikoura. While many destinations have a broad collection of often diffuse attractions, the three key attractions of Kaikoura draw deeply on its geography and marine environment. These are: whales, dolphins and seals. While each of these are common property resources, available

---

[5] See, for example, The New Zealand Whale and Dolphin Trust and their publications: http://www.whaledolphintrust.org.nz/publications.html, or (NZ) Department of Conservation (1999).

to be engaged by the passing public, each has, in turn, been brought into well-managed, successful commercial operations. Furthermore, the 'boom and bust' scenarios that pervade the history of human settlement at Kaikoura (whales, seals, fisheries, forestry, etc.) have laid a cautious grounding for the approach to marine-based tourism, which is well reinforced in legislative and planning processes.

The second key factor relates to local leadership, Māori history and ideology, and subsequent legislative and planning processes. Māori values have both underpinned the daily operation of Whale Watch and informed the general planning legislation in New Zealand. As an example, New Zealand's primary planning legislation, which has the overarching goal of ensuring the 'sustainable management of resources', the Resource Management Act (1991), which in defining 'matters of national importance' (S6) specifically (S6 e)) requires attention to 'the relationship of Māori and their culture and traditions with their ancestral lands, water, sites, waahi tapu (sacred sites), and other taonga (treasures)' and that those managing natural resources need to take kaitiakitanga into account. Thus, attention to Māori values is increasingly evident over time in many aspects of New Zealand life, from national policy statements to regional and district plans, and is readily evident in the planning approach and outcomes in significant Māori settlements such as Kaikoura.

A third key feature, and perhaps the glue that has brought Kaikoura as a destination to the global forefront of sustainable management, has been the unique set of science partnerships that have formed around and supported destination planning and management. This strong tri-partite (local marine operators, District Council, research providers) 'leadership' as the cornerstone of all tourism visitation and activity in Kaikoura has led to a united voice to represent what is often a disparate group of private-sector interests. There can be little doubt that the engagement of Kaikoura as the first in a series of four in-depth national tourism planning case studies brought a scientific

framework and data set and analytical rigour that remains the envy of many. That this too was then able to be directly linked to the emerging benchmarking and certification programmes for sustainable tourism (GreenGlobe21/earthcheck™) and become the first territorial authority globally to achieve tourism certification, and then be rapidly promoted to finalists within the Global Sustainable Stewardship awards and international conferences (some hosted within Kaikoura) provided a confidence that this was a path forward. Backed by resolute local management and community support, the earthcheck environmental standards today pervade all council planning and reporting metrics (Kaikoura District Council, 2008, 2011a, 2011b), and demonstrate the power of the tourism sector to reach out across other sectors (such as dairy, fishing and transport) to engage communities in a common vision and set of actions.

Conclusion

Many studies of tourism take a 'point in time' view of sustainability. A unique feature of this Kaikoura case study lies in its ability to demonstrate strong benchmarking of environmental performance and continuous improvement for more than a decade. An evidence base from marine ecology and environmental benchmarking (with external verification) provides evidence to suggest that the Kaikoura community has developed a governance model and set of partnerships to potentially avoid the 'boom and bust' resource cycles that mark their past and which are outlined in the opening sections of this chapter. It is a moot point whether the congruence of forces around the development and enactment of a sustainability platform for tourism are able to be socialized across all sectors of the economy and continue to unite the community into the future, but the public-sector governance model demonstrated in Kaikoura provides hope that this is the case. For the present, the focus on whales as 'the' key attraction, their unique place in Māori mythology and history, their recognition as unique mega-fauna and recent re-'evaluation' as tourist attractions have provided a strong foundation for consideration of the myriad issues that compel a broader consideration of sustainable development, and evidence that tourism can be an advocate for a wider sustainability platform. Notwithstanding these hopes, one is drawn inevitably back to the conclusion that it is the wider set of Māori values, (focused as it is on whales in Kaikoura), strong partnership with legislative governance models and the underpinning of science to inform policy and practice that has made Kaikoura a unique exemplar of a robust pathway to sustainable development.

REFERENCES

AJHR. (1870). Appendices to the Journals of the (New Zeland) House of Representatives. Various years including 1870, D-9, p. 5.

AJHR. (1876). Appendices to the Journals of the (New Zeland) House of Representatives, p. E-1.

AJHR. (1883). Appendices to the Journals of the (New Zeland) House of Representative, p. D-2.

Allen, L.R., Hafer, H.R., Long, P.T. & Perdue, R.R. (1993). Rural residents' attitudes toward recreation and tourism development. *Journal of Travel Research* 31(4), 27–33.

Anderson, A. (1991). The chronology of colonisation in New Zealand. *Antiquity* 65, 767–795.

Ansley, B. (1991). In the wake of the whales. *Listener and TV Times*, 6 May 1991.

APEC. (1997). *Tourism and Best Practice in APEC Member Economies*. Singapore: APEC Secretariat.

Becken. S. (2008). Developing indicators for managing tourism in the face of peak oil. *Tourism Management* 29(4), 695–705.

Brett, C. (1992). The Great Whale Hunt. *North & South*, August 1992.

Butcher, G., Fairweather, J.R. & Simmons, D.G. (1998). *An Input–Output Model of the Kaikoura Tourism Economy*. Report No. 8. Canterbury, New Zealand: Tourism Research Education Centre (TREC), Lincoln University.

Challis, A. (1991). The Nelson–Marlborough region: An archaeological synthesis. *New Zealand Journal of Archaeology* 13, 101–142.

Cyclopedia Company. (1906). *The Cyclopedia of New Zealand: Volume 5, Nelson, Marlborough and Westland*. Wellington: The Cyclopedia Company.

Department of Conservation. (1996). *Conservation Management Strategy: Nelson/Marlborough Conservancy*. Wellington: Department of Conservation.

Evison, H.C. (1993). *Te Wai Pounamu, The Greenstone Island: A history of the southern Maori during the European colonisation of New Zealand*. Wellington: Aoraki Press.

Fairweather, J.R., Horn, C.M. & Simmons, D.G. (1998). *Estimation of Visitors to Kaikoura*. Report No. 2. Tourism Research Education Centre (TREC), Lincoln University, Canterbury, New Zealand.

Horn, C.M. & Simmons, D.G. (2000). Community adaptation to tourism: Comparisons between Rotorua and Kaikoura. *Tourism Management* 23(2), 133–143.

Horn, C.M., Simmons, D.G. & Fairweather, J.R., (1998). *Kaikoura Residents' Experience of, and Adaptations to Tourism*, Report No. 6. Canterbury, New Zealand: Tourism Research Education Centre (TREC), Lincoln University.

Hutching, G. (1990). Kaikoura, nature tourism town. *Forest & Bird* 21(4).

Kaikoura District Council. (2008). Kaikoura District Plan. Retrieved from: http://www.kaikoura.govt.nz/council_documents/district_plan/) and http://www.kaikoura.govt.nz/green_globe/ (accessed 20 September 2010).

Kaikoura District Council (2011a). Annual Plan 2011/12. Retrieved from: http://www.kaikoura.govt.nz/docs/Council%20Documents/Annual%20Plans%20and%20Reports/annual_plan_20112012.pdf (accessed 12 September 2012).

Kaikoura District Council (nd) (2011b). *Annual Report 2011*. Kaikoura: Kaikoura District Council.

Keene, H. (1992). Whales a blessing for Maoris at Kaikoura. *The Press*, 19 May 1992.

Marlborough Regional Development Council. (1976). *Marlborough: A survey of present resources and future potential*. Marlborough: Marlborough Regional Development Council.

McAloon, J., Simmons, D.G. & Fairweather, J.R. (1998). *Kaikoura: Historical background*. Report No. 1. Canterbury, New Zealand: Tourism Research Education Centre (TREC), Lincoln University.

Ministry of Works Town and Country Planning Branch. (1962). *Marlborough Region: A brief survey of present resources and future potential*. Wellington: Ministry of Works.

NIWA. (2010). Retrieved from: http://www.niwa.co.nz/our-science/aquatic-biodiversity-and-biosecurity/research-projects/all/kaikoura-canyon

Poharama, A., Henley, M., Smith, A., Fairweather, J.R. & Simmons, D.G. (1998). *The Impact of Tourism on the Māori Community in Kaikoura*. Report No. 7. Canterbury, New Zealand: Tourism Research Education Centre (TREC), Lincoln University.

Press, 'The'. (1989). Wetere opens Kaikoura marae-based venture. *Christchurch Press* (J. Dunbar), 17 July.

Press, 'The'. (1996). 4 September, p. 25.

Resource Management Act. (1991). Retrieved from (NZ) Ministry for the Environment: http://www.mfe.govt.nz/rma/index.html/ (accessed 23 July 2012).

Scott, H. (1991). Whale watching at Kaikoura – The birth and development of a unique tourism venture. Thesis, Central Institute of Technology, Heretaunga.

Sherrard, J.M. (1966). *Kaikoura – A history of the district*. Kaikoura: Kaikoura County Council.

Simmons, D.G. & Fairweather, J.R. (1998). *Towards a Tourism Plan for Kaikoura*. Report No. 10. Canterbury, New Zealand: Tourism Research Education Centre (TREC), Lincoln University.

Simmons, D.G. & Fairweather, J.R. (Eds). (2005). *'Understanding the Tourism Host–Guest Encounter in New Zealand: Foundations for adaptive planning and management*. Christchurch, New Zealand: EOS Ecology.

Simmons, D.G. & Becken, S. (2004). The cost of getting there: Impacts of travel to ecotourism destinations. In R. Buckley (Ed.), *Environmental Impacts of Ecotourism*. Wallingford: CABI, pp. 15–23.

Statistics New Zealand. (2006). NZ Census of Population and Dwellings, 2006. Retrieved from: http://www.stats.govt.nz/Census/2006CensusHomePage/Tables/AboutAPlace/ (accessed 2 September 2010).

Tahana, N., Te O Kahurangi Grant, K., Simmons, D.G. & Fairweather, J.R. (2000). *Tourism and Māori Development in Rotorua*. Canterbury, New Zealand: Tourism Research Education Centre (TREC), Lincoln University.

Te Awekotuku, N. (1981). Sociocultural impact of tourism on the Te Arawa people of Rotorua, New Zealand. PhD dissertation, New Zealand, University of Waikato.

Te Karaka (the Ngai Tahu Magazine) (1997). Koanga (Spring), 'The Ngai Tahu – Giants of the eco-tourism business'. *Te Karaka* 6, 8–12.

Te Māori (maori News weekly) (1997). p. 4.

The New Zealand Whale and Dolphin Trust. (http://www.whaledolphintrust.org.nz/)

23

Management of dusky dolphin tourism at Kaikoura, New Zealand

David Lundquist

Introduction

The desire to view and interact with animals in their natural environment has driven strong growth of cetacean-based tourism since its inception in California in the 1950s. After years of steady growth in the US, the economic potential of the cetacean-watching tourism industry was recognized worldwide (see Chapter 1). In the early years of the industry, cetacean-based tourism developed with little or no management, regulation or oversight (Hoyt, 2008). Legislation to protect cetaceans was often designed to limit the effects of whaling and harpoons, not tourism boats and cameras. Perhaps as a consequence, little consideration was given to the effects of tourism on the target species. It was perceived to be less harmful than the alternative, and therefore additional protection was not needed (O'Connor *et al.*, 2009).

Growth of cetacean tourism has led to increasing concerns about the effects of repeated close encounters with tour boats on the health of the target animals and the sustainability of the industry (Bejder & Lusseau, 2008; Parsons, 2012). Researchers have described a host of short-term reactions by cetaceans to the presence of boats, including changes in behaviour, movement, respiration, dive characteristics, habitat use, communication and group cohesion (e.g. Au & Green, 2000; Barr & Slooten, 1999; Bejder *et al.* 2006a, 2006b;

Buckstaff, 2004; Constantine, 2001; Dans *et al.*, 2008; Erbe, 2002; Gordon *et al.*, 1992; Jahoda *et al.*, 2003; Lusseau, 2006; Nowacek *et al.*, 2001; Scarpaci *et al.*, 2000; Stensland & Berggren, 2007; Williams *et al.*, 2002).

In most cases, it is not clear whether these short-term reactions carry long-term consequences for the target species. Building the requisite data set to analyse impacts requires foresight into the relevant biological indicators to collect, financial support and manpower over an extended period of time to collect data, as well as a reasonable amount of luck in the choice of location and timing of the study versus growth and development of tourism. Not surprisingly, there are relatively few places where this confluence of events has occurred. Bejder *et al.* (2006a, 2006b), Lusseau *et al.* (2006) and Constantine *et al.* (2004) demonstrated long-term effects on bottlenose dolphin (*Tursiops* spp.) populations targeted by tour boats. Probable impacts seen in these populations included habitat displacement (Bejder *et al.*, 2006a, 2006b; Lusseau *et al.*, 2006), horizontal avoidance and declining population size (Lusseau *et al.*, 2006), and sensitization to tour-boat approaches (Constantine *et al.*, 2004).

Management of cetacean tourism is complex, with different ecological, social, political, economic and cultural contexts in each place where it occurs. Stakeholders must be identified and each perspective included in the overall management plan.

Whale-watching: Sustainable Tourism and Ecological Management, eds J. Higham, L. Bejder and R. Williams.
Published by Cambridge University Press. © Cambridge University Press 2014.

Conservation parameters must be established and agreed upon by all stakeholders, with regular checkpoints built in to gauge the effect of the activity. These are complicated time-consuming procedures, and implementing them after tourism has been established is difficult. Unfortunately, due to the historical context of the growth and development of cetacean tourism, this is the most common situation.

Higham *et al.* (2009) proposed a generalized management framework for cetacean tourism which attempts to address the complicated contexts in which this activity occurs (Figure 23.1). The model integrates a minimum of four key stakeholders (commercial tourism operators, natural scientists, social scientists and planning/management agencies) in an adaptive framework designed to promote sustainability. Each stakeholder is assigned specific responsibilities before tourism begins and at regular checkpoints after tourism is established.

In the pre-tourism phase, management and planning agencies are responsible for establishing the legislative groundwork to manage cetacean tourism. Social scientists evaluate the support for tourism within the community and share this information with management agencies and potential operators. Natural scientists collect baseline ecological data on the target species, determine appropriate monitoring criteria, and find a suitable control site outside the commercial operating area. Management agencies then determine Limits of Acceptable Change (LAC; Stankey *et al.*, 1985) criteria for the target population, set up operating guidelines and issue permits.

Once tourism begins, researchers collect data at the tourism and control sites, and provide a monitoring report to the management agency. Social scientists evaluate visitor perceptions and satisfaction and provide feedback to management agencies and tour operators. The management agency reviews permits, makes changes based on scientific evidence and amends the management scheme if necessary. Operators modify their tours according to the new permit conditions and any visitor suggestions. A new tourism cycle then begins, with

researchers evaluating effects under new operating conditions and providing a report to the management agency.

The management model proposed by Higham *et al.* (2009) is intentionally idealized and general (Figure 23.1). In order to evaluate the effectiveness of this framework critically, it is helpful to compare it to a real-life situation. Dusky dolphin tourism at Kaikoura, New Zealand provides a good example, as multiple rounds of research and management action have been taken in this location since the activity began in 1989. This chapter will use the dusky dolphin tourism industry in Kaikoura as a case study to examine the following questions.

1) Can the proposed model be effectively applied to real-world (i.e. non-ideal) tourism development?

2) Is management of dolphin-based tourism in this location falling short of best practices to ensure sustainability?

3) If the sustainability of tourism is uncertain, what suggestions does the model offer for further research and management action?

Case study: applying the Higham *et al.* (2009) model to swimming with dusky dolphins near Kaikoura, New Zealand

Tourism targeting cetaceans near Kaikoura began in 1988, with the inception of a whale-watching industry focusing on local sperm whales. The industry grew quickly, and in the summer of 1989–1990 a pair of local former fishermen looking for a similar opportunity began taking tourists on fishing, diving and wildlife-viewing tours (Buurman, 2010). Tours quickly focused on swimming with dusky dolphins, and the business expanded rapidly. Initially, tours were seasonal, but by 1995 operations were year-round and more than 11,000 tourists went swimming with dolphins (Dennis Buurman, Encounter Kaikoura, pers. comm.). By the early 2000s, dolphin-swimming numbers levelled off, with 23–27,000 customers each year (Buurman, pers. comm.).

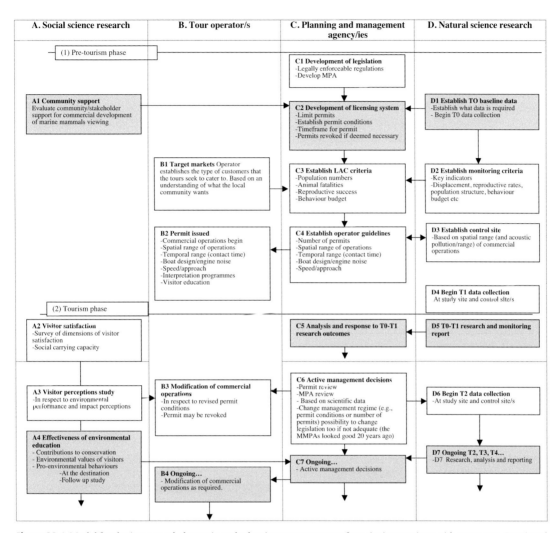

Figure 23.1 Model for the integrated, dynamic and adaptive management of tourist interactions with cetaceans. Reprinted from Higham *et al.* (2009) with permission of the authors.

Management context of Kaikoura tourism

At the macro-level (global; Higham *et al.*, 2009), management of this activity is influenced by international policy-setting organizations such as the International Union for Conservation of Nature and Natural Resources (IUCN) and the International Whaling Commission (IWC). Policy decisions and public communications by such

organizations are widely distributed and have the potential to influence tourist behaviour. A large percentage of tourism dollars spent in New Zealand are from overseas visitors (42%; New Zealand Ministry of Economic Development[1]), and the

---

[1] http://www.tourismresearch.govt.nz/Documents/Key% 20Statistics/KeyTourismStatisticsJune2011.pdf (accessed 2 July 2011).

'clean, green' image of New Zealand is a vital part of this industry, so conservation management is influenced by declarations by international policy organizations which might jeopardize this image.

At the national level, the primary piece of legislation relevant to conservation of dusky dolphins is the New Zealand Marine Mammals Protection Act (MMPA, 1978). The MMPA led to the establishment of the Marine Mammals Protection Regulations (MMPR) in 1988, which were amended in 1990 (SR 1990/287), 1992 (SR 1992/322) and 2008 (SR 2008/255). The MMPR defines rules for issuing, revoking, suspending, restricting and amending commercial tourism permits focusing on marine mammals. It also provides guidelines for behaviour of vessels (boat or aircraft) in the presence of marine mammals: how close and from which relative direction a vessel may approach, how fast the vessel can move, and how many vessels may be within 300 m of a group of marine mammals at the same time. These regulations establish the general operation of tourism vessels targeting dusky dolphins in Kaikoura and are enforced by the New Zealand Department of Conservation (DOC). Altering operational conditions attached to an individual permit is the primary method used by DOC to manage this activity. The MMPA is not the only legislation that affects conservation of dusky dolphins, however. Fisheries-related by-catch is governed by the Ministry of Fisheries under the Fisheries Act (SR 1996/88). The Resource Management Act (SR 1991/69) is administered by the Ministry for the Environment, generally via regional and district councils. Traditional cultural relationships relative to management of the natural world are granted to New Zealand Māori under the Treaty of Waitangi, and consultation with Māori is a common part of natural resource management.

At the local level, dusky dolphin tourism is managed by the DOC Nelson/Marlborough Conservancy. Researchers associated with national (NZ) and international universities have provided recommendations relative to management of tourism

since 1997. Additional local stakeholders include tour operators, recreational and commercial boating and fishing organizations, the Ngāti Kuri sub-tribe of Ngāi Tahu, and other concerned local individuals. Te Korowai o Te Tai o Marokura (Kaikoura Coastal Guardians) is a group which was created with the intent of bringing local stakeholders together to map out a long-term strategy for management of natural resources in the Kaikoura area, and includes representatives from most of the aforementioned groups. Each of these stakeholders has an interest in management of dusky dolphin tourism, and should be consulted during the decision-making process.

Pre-tourism phase

Some of the steps outlined by Higham *et al.* (2009) for the pre-tourism phase of management fit well with what has occurred in Kaikoura. The New Zealand MMPA was passed in 1978, well before the inception of dusky dolphin tourism in Kaikoura (C1, Figure 23.1). The biology of dusky dolphins in other parts of the world was first studied in the early 1970s (Würsig & Würsig, 1980), but research into the behaviour, movement patterns and life history of dusky dolphins near Kaikoura did not begin until 1984 (D1, Figure 23.1; Cipriano, 1992). Cipriano (1992) collected data which were not specifically designed to establish a baseline for tourism management, but nevertheless provide an important snapshot of dusky dolphin behaviour in the absence of tour vessels (D1, Figure 23.1). The MMPR were established by DOC in 1988, also prior to the advent of tourism (C2, Figure 23.1). Despite this, when dusky dolphin tourism first began in 1989, DOC did not require a permit, perhaps because it was not clear if this form of tourism was commercially viable (Buurman, 2010). It was not until 1990 that the first permit to swim with dusky dolphins was issued (B2, Figure 23.1; Buurman, 2010). In combination with the MMPR, these permits established the geographic area where the operation could occur, defined how many swimmers could enter the water simultaneously, required an educational component to each tour, and regulated many aspects of vessel operation in the presence of dolphins, including approach speeds, approach direction and maximum number of vessels within 300 m of a dolphin pod (B2, C4, Figure 23.1).

Some steps critical to sustainable management of dusky dolphin tourism as outlined by Higham *et al.* (2009) did not occur prior to the activity beginning. Community support for this form of tourism (A1, Figure 23.1) was not studied scientifically. It is likely, however, that there was substantial support for any business which could successfully revive a depressed local economy, and no local opposition to the activity was evident (Simmons & Fairweather, 1998; see also Chapter 22). The operators did not target specific types of customers based on a pre-defined community desire (B1, Figure 23.1), but Buurman (2010) described how the business began as a general combination of fishing, diving and nature tours, and quickly focused on swimming with dolphins as the primary activity. Essentially, interest in a particular type of tour defined the target market during the first season of operation (B1, Figure 23.1).

The most important missing pieces were a failure to establish monitoring criteria (D2, Figure 23.1) and the resulting lack of LAC criteria (C3, Figure 23.1). There was no research at the time specifically designed to determine the effects of tourism on dolphins, so failing to establish these criteria is not surprising. However, the combination of missing these two steps is quite significant, as they form the basis for making management decisions based on quantifiable criteria. The last missing step is the establishment of a non-tourism control site for comparison of effects (D3, Figure 23.1). In this case, the behavioural ecology of dusky dolphins in the Kaikoura Canyon system is different than at other parts of New Zealand (Würsig & Würsig, 2010), so there is no other site suitable for use as a control. However, with appropriate foresight, it might have been possible to establish a control site within the tourism operating area to use for comparison.

Tourism phase – Cycle 1

The first research effort dedicated to examining the effects of tourism (D4, Figure 23.1) occurred between 1993 and 1995 (Barr, 1997). Additional research on the effects of vessels (D4, Figure 23.1) was conducted by Brown (2000) and Yin (1999). Barr and Slooten (1999) provided a report to DOC (D5, Figure 23.1) detailing the number and types of vessels approaching dolphins, changes to dolphin group spacing and aerial behaviour, and violations of the MMPR (1992). They found that dolphins rested most often at midday, aerial behaviour increased after late morning when vessels were present, and groups became tighter (individuals were closer together) in mid to late afternoon (Barr & Slooten, 1999). They found that dolphins were accompanied by boats 72% of the time during daylight hours in summer, and most often these were commercial tour boats (84% of all boats; Barr & Slooten 1999). Additionally, they reported that more than three vessels were within 300 m of the same pod of dolphins 8.2% of the time, and that 7.4% of boat approaches violated the guidelines provided in the MMPR (1992). Yin (1999) found that small groups changed direction of travel more often when boats were present. Brown (2000) found changes in dolphin behaviour in the presence of tour boats, and suggested that some southward habitat shift may have occurred as a response to vessels coming from the Kaikoura peninsula at the northern end of the observation area.

As a result of the studies by Barr (1997), Barr and Slooten (1999), Yin (1999) and Brown (2000), DOC took a number of management actions (C5, C6, Figure 23.1). A 10-year moratorium on new permits was instituted in 1999 (Childerhouse & Baxter, 2010), effectively limiting the growth of the industry. Additionally, commercial operators agreed to institute a voluntary (i.e. outside the legally binding permit process) midday rest period from December to

March, wherein they would refrain from approaching dolphin pods between the hours of 11:30 a.m. and 1:30 p.m. (Childerhouse & Baxter, 2010). No specific action was taken with regard to violation of regulations regarding the number of vessels or method of approaching dolphins. Observation of the rest period by commercial operators began in 1999 (B3, Figure 23.1; Brown, 2000).

A series of social science studies of tourism in Kaikoura were conducted by Lincoln University researchers during this period (A3, Figure 23.1; summarized by Simmons & Fairweather, 1998). These studies described the number and type of visitors to Kaikoura, the economic impact of tourism, responses of local citizens to tourism development, and the impact of tourism on the Māori community (Simmons & Fairweather, 1998). Perceptions of dusky dolphin tourism were not specifically analysed, but marine mammals were identified as a key feature that visitors valued (Simmons & Fairweather, 1998). Locals and visitors valued Kaikoura for its small-town community and disliked overt signs of development and crowding (Simmons & Fairweather, 1998). These studies provide important context for the social environment in which dusky dolphin tourism was occurring at that time.

Tourism phase – Cycle 2

In the early 2000s, the number of tourists swimming with dusky dolphins near Kaikoura stabilized at 23–27,000 per annum (Dennis Buurman, Dolphin Encounter, pers. comm.). In 2000, Lück (2003) conducted the first study of visitor satisfaction specific to dusky dolphin tourism (A2, Figure 23.1). Tourists were generally independent travellers, young, educated, members of various environmental organizations and on their first dolphin-oriented tour (Lück, 2003). While most were satisfied with the tour, thought the staff did a good job and would recommend the tour to friends, they did not think they had learned much about dolphins or marine life (A4, Figure 23.1; Lück, 2003). Education is an important component of the concept of ecotourism, and is a

legislative requirement of any commercial cetacean tourism enterprise in New Zealand (MMPR 1992). Ideally, Lück's (2003) results should have instigated a review of the educational portion of the tours to determine if operators were meeting their legal requirement under the MMPR (1992), but no apparent action was taken by managers based on these findings.

Further research into dusky dolphin tourism near Kaikoura continued during this period. Dahood (2009) analysed a set of sighting data collected by the swim-with-dolphin tour operator from 1995 to 2006. The 95% kernel home-range (KHR) calculated by Dahood indicated a northward shift from 1995–2000 to 2001–2006, but the 50% KHR – the core area used – changed very little in size or location, and was very similar to the pre-tourism and current core areas described by Lundquist (2012). This suggests that dusky dolphins have not shifted their usage of the Kaikoura area in a significant manner as tourism has developed. In 2005, Duprey et al. (2008; D6, Figure 23.1) analysed the effectiveness of the voluntary midday rest period begun in 1999 and found that commercial vessels continued to approach dolphin groups during the rest period, although at a reduced rate compared to non-rest periods. These visits were attributable solely to the dolphin-viewing companies, as the dolphin-swimming company strictly observed the rest period (Duprey et al., 2008). This conclusion was confirmed by Markowitz et al. (2009; Ds6, Figure 23.1) in a study commissioned by DOC near the end of the 10-year moratorium on new permits. Similar non-compliance with voluntary regulations has been reported elsewhere (Allen et al., 2007), and is likely not an effective management strategy. Markowitz (2004) and Duprey (2007) both noted that calves were present in more than 70% of large dolphin groups targeted by tourism vessels. Swimming with juvenile dolphins is forbidden under the MMPR (1992), although 'juvenile' is not defined within the MMPR (1992) and it is unclear whether the prohibition on swimming with calves is legally enforceable.

Markowitz et al. (2009) described a number of different short-term reactions of dusky dolphins in the presence of tourism vessels, including reduced resting and socializing, slower swimming speeds, more frequent changes in bearing of travel and increased leaping. The magnitude of behavioural change was correlated positively with the number of vessels and number of vessel approaches (Markowitz *et al.*, 2009). Markowitz *et al.* (2009) recommended that DOC (D7, Figure 23.1) maintain the current level of permits (i.e. renew existing permits but not add additional ones), make the midday rest period mandatory and extend it from October to March (to provide additional protection during the October–December calving season), provide better education and enforcement of the three-vessel rule, raise the number of swimmers allowed per vessel, limit the number of times per tour each vessel could approach dolphin groups to put swimmers in the water, and a number of other operational and training suggestions.

Following submission of the Markowitz *et al.* (2009) report, DOC advertised the results and solicited feedback from operators, independent scientists and the general public (C7, Figure 23.1; Andrew Baxter, DOC, pers. comm.). Meetings were held to present results to local Māori, operators, and other interested parties. Upon expiration of the 10-year moratorium on additional permits, a subsequent five-year moratorium on additional motorized boat-based permits was issued through 2014 (C8, Figure 23.1; A. Baxter, pers. comm.). The number of swimmers allowed per vessel was increased from 13 to 16, each vessel was limited to five swim approaches per trip, the midday rest period was made mandatory from November to February (and still voluntary in March), and all operators were required to adhere to a Code of Conduct to be developed in conjunction with DOC (A. Baxter, pers. comm.). These changes to permit conditions were instituted in 2010 (B4, Figure 23.1; A. Baxter, pers. comm.). The possibility of additional permits for non-motorized or airborne tour permits was left open, and a local kayaking tour company subsequently received a permit to view dusky dolphins in a limited geographic area around the Kaikoura Peninsula (C8,

Figure 23.1; A. Baxter, pers. comm.). Lundquist *et al.* (2013) found that dolphin groups reorient more in the presence of aircraft, so any expansion of this industry should be considered carefully. Dusky dolphins often seek shallow, near-shore waters for shelter when predators are present (Cipriano, 1992; Srinivasan, 2009), and nursery groups are often found close to shore (Würsig *et al.*, 1997). Permitting kayak tours to approach dolphins in these same areas might have a greater effect than expected for non-motorized craft, as the animals being targeted may be vulnerable individuals.

Tourism phase – Cycle 3

The current phase of tourism began after permit conditions were changed in 2010 and will presumably conclude with the end of the current moratorium in 2014. In addition to the short-term effects described in the Markowitz *et al.* (2009) report, Lundquist (2012) described long-term changes in the response of dusky dolphins to vessels. The magnitude of responses have grown over time: slower speeds, less resting, and more milling (Lundquist, 2012). Changes in the activity budget of dusky dolphins in the presence of vessels were described by Lundquist *et al.* (2012), and may be used to help determine LAC criteria. Additional research will be needed to determine whether changes instituted in 2010 have reduced the effects described by Markowitz *et al.* (2009) and Lundquist (2012). Te Korowai (http://www.teamkorowai.org.nz) have made multiple proposals in their long-term strategy document which may affect future management decisions. These include designation of the Kaikoura area as a UNESCO World Heritage Site, and creating a marine mammal sanctuary encompassing the entire tourism operating area plus a buffer zone of several miles outside the tourism area. The marine mammal sanctuary has been proposed to protect local wildlife from the effects of seismic exploration, but exact details have not been set forth, so it is unknown how this might affect the long-term sustainability of dusky dolphin tourism.

Evaluating the sustainability of tourism in Kaikoura

The largest gaps identified in this case study are that tourism began prior to: (1) permits being issued and (2) monitoring criteria being established to define LACs. Unfortunately, both of these gaps are common occurrences in animal-based tourism (Higginbottom, 2004). This makes retroactive application of regulations more difficult, and forces wildlife managers into the position of having to prove that significant negative effects are occurring, rather than forcing tourism companies to prove the opposite. While permits were issued within a year of tourism beginning, LAC criteria are still undefined and this shortcoming brings into question the sustainability of the industry. LAC criteria are generally used to manage stocks of marine mammals which are subject to 'takes', either directly or via incidental by-catch. Direct take and by-catch are not expected with tourism (although boat strikes may occur), but research has shown that population-level effects can still occur (Bejder *et al.*, 2006a, 2006b; Lusseau *et al.*, 2006). Long-term impacts of non-consumptive wildlife tourism have been demonstrated for other non-marine mammal species in New Zealand (Higham, 1998), even in the absence of behavioural responses that may indicate an effect is occurring. Not only is the resource degraded over time, but the tourism experience is also correspondingly degraded (Higham, 1998).

Dusky dolphins in New Zealand are a large and mobile population, and therefore it is difficult to detect a change in population size, but the LAC concept can be applied nonetheless. Other variables (activity budgets, habitat use, local abundance, population structure) may be feasibly evaluated and used to set LAC, as long as studies are done to provide a basis of support for the limits which are chosen. Changes to these variables can be evaluated and a threshold for acceptable change set. These measurements can be used by managers as a quantitative way of determining when changes to permit

conditions (or numbers of permits) must occur. In the absence of LAC criteria, managers must rely on a more subjective qualitative decision-making process, which leaves open the possibility of the management decisions being affected by personal, political, economic or social factors which ignore the ecological consequences for the target species and hence the sustainability of the activity.

In order to support monitoring of LAC criteria, a mechanism needs to be in place to manage and fund long-term data collection. In Kaikoura, most of the results which informed management decisions were collected by university students, with partial or no funding from DOC. While this was a low-cost way of providing support for management decisions, it resulted in multiple sets of data which were collected using different methods, over different timescales, and owned by different people. The result was a disjunct set of data which answered similar questions in different ways, rather than a cohesive set of data collected to answer specific questions in a consistent manner. A more appropriate strategy to monitor an industry now worth several millions of dollars locally (Simmons & Fairweather, 1998) would be to establish a research programme, funded by the industry but administered by DOC, that is consistent over time and integrated with the overall management scheme.

In the case of Kaikoura, management action has been taken at 10-year intervals. Community support, visitor satisfaction, environmental effects and ecological changes can change in a much shorter time span than this, particularly if tourism activity is growing as quickly as it did in the first decade in Kaikoura. It is also difficult to obtain the buy-in of all stakeholders on changes to permit conditions after 10 years, as there is a strong tendency to accept that practices which have been followed for a decade must be correct, even if research shows otherwise. The model proposed by Higham *et al.* (2009) suggests determining the timeframe for permit review in the pre-tourism stage (C2), but no firm guideline has been set for this in New Zealand. Permits and permit conditions could potentially be reviewed at any time, but in practice action is only taken at the end of each moratorium period. The current moratorium of five years ends in 2014, which is a reasonable timeframe for taking action to ensure sustainability. Results of long-term comparisons described by Lundquist (2012) will be particularly relevant to the next permit review. Because social perceptions and community desires can change dramatically over the years, studies to update the results presented by Simmons and Fairweather (1998) and Lück (2003) are highly desirable prior to the next round of management decisions.

Other management issues at Kaikoura seem to be ignored, overlooked or rely upon voluntary action by operators. Voluntary measures were not an effective way of regulating operations in Kaikoura, as compliance with the midday rest period varied depending on the operator (Duprey *et al.*, 2008). Swimming with juvenile dolphins (Markowitz, 2004; Duprey, 2007) and cutting off the path of dolphin groups (Markowitz *et al.*, 2009) are activities specifically forbidden in the MMPR (1992), but both have been reported without any apparent enforcement action being taken by DOC. There are no explicit rules for differentiating an adult from a 'juvenile' (or a juvenile from a calf), so the regulation is apparently unenforceable as written. A more proactive approach to enforcement is warranted, as these regulations are designed based on known classes of vulnerable animals and activities likely to cause disturbance. Tours are required to have sufficient educational value to participants (MMPR 1992), but 'sufficient' is a vague, undefined term. The educational value of tour presentations is largely unknown, beyond the report by Lück (2003) indicating that tourists did not feel they had learned enough.

Recommended future research and management action

Analysis of this case study provides valuable information to managers about the responses of dusky dolphins to tourism vessels and how those responses have changed over time. Compared to the model proposed by Higham *et al.* (2009), it is clear that much more work is needed to ensure that this industry is sustainable. The historical methodology of behavioural observation appears to have reached its limit of effectiveness in determining population-level effects, and management decisions ultimately rely upon the precautionary principle (Foster *et al.*, 2000) rather than explicit measurements of biological impact. Studies are needed to link the physiological effects of vessel traffic on dusky dolphins with more readily observed behavioural responses. Such studies will provide managers with the tools to more accurately assess the biological significance of responses observed via the current methodology. Potential studies include:

A **Energetics**. Behavioural responses to vessel traffic have the potential to affect the energetic requirements of the target animals, so it would be useful to quantify the cost of the observed responses. Techniques most commonly used in wildlife ecology to calculate the field metabolic rate of free-ranging animals include doubly labelled water (DLW; Speakman, 1997) and heart rate monitoring (Butler *et al.*, 2004). Using the DLW method may be of limited effectiveness in endothermic aquatic mammals such as dusky dolphins (Butler *et al.*, 2004), but heart rate monitoring has been successful with pinnipeds (Cooke *et al.*, 2004). Historically, this method has required implantation of a collection device, which necessitates capture, anaesthetization, and post-anaesthetic restraint, all of which are difficult with cetaceans. Devices which may be attached externally to monitor heart rate have been developed (Cooke *et al.*, 2004), which are more realistic to use with dolphins. The use of controlled vessel interactions during the study would permit calculation of energetic costs associated with the presence of vessels, similar to what has been done for killer whales (*Orcinus orca*) by Williams *et al.* (2006). Combining such data with estimations of daily energy expenditure for dusky dolphins (Cipriano, 1992; Kastelein *et al.*, 2000; Srinivasan, 2009) might allow for useful approximation of the energetic difference between behavioural budgets such as those reported by Lundquist *et al.* (2012). This would allow managers to establish threshold LAC values relative to behavioural budget, which could be evaluated with ongoing monitoring to ensure dolphins are not detrimentally affected.

B **Communication**. Vessel noise has the potential to mask the vocal communication (Richardson *et al.*, 1995) of dusky dolphins, particularly noise generated by tour vessels which approach closer than 50 m (Jensen *et al.*, 2009). There is the possibility that some of the behavioural responses described here are a result of such masking. This theory could be investigated by attaching acoustic tags to individuals, perhaps similar to those used by Burgess *et al.* (1998) with northern elephant seals. Their tags recorded ambient noise, vessel noise, respirations and possibly vocalizations, in addition to other information (Burgess *et al.* 1998). Similarly, Jensen *et al.* (2009) used DTAGS (Johnson & Tyack, 2003) attached to short-finned pilot whales to record background and vessel noise. The ability to record the vocalizations of the tagged animal, conspecifics and vessel noise during controlled vessel interactions would allow analysis of a number of effects. Most important would be determining whether vocalizing changes in the presence of a vessel, the received sound level at which changes occur, and if this changes when multiple vessels are present. The technology exists to conduct this research in Kaikoura, and evaluating the type and intensity of noise dusky dolphins are exposed to would provide valuable information for wildlife managers. Threshold values for source noise levels could be developed and mitigation criteria established by managers.

C **Hormonal responses**. Chronic increases in glucocorticosteroid levels have been linked to reduced individual fitness (Fowler, 1999; Romero & Wikelski, 2001; Sapolsky *et al.*, 2000). Rolland (2012) used faecal samples to demonstrate that stress hormone levels decreased in North Atlantic right whales (*Eubalaena glacialis*) when underwater noise levels decreased by 6 dB following the events of 11 September 2001. Researchers investigating the effects of tourism on penguins (Ellenberg *et al.*, 2007) took blood samples of animals during controlled interactions in tourism and non-tourism areas to analyse glucocorticosteroid levels in response to the interaction. A similar analysis could be done with dusky dolphins by sampling individuals within the tourism area, and just outside it. This would provide much-needed information about physiological responses to controlled vessel approaches.

D **Population size and local abundance**. Line-transect aerial surveys are widely used to estimate population sizes for marine mammals (Buckland *et al.*, 1993). Survey methods such as those employed by Slooten *et al.* (2006) could be replicated for dusky dolphins to estimate the population size around New Zealand. Using unmanned aerial vehicles equipped with high-definition still or video cameras to conduct surveys is an emerging methodology that may be useful in the future (Koski *et al.*, 2009; Hodgson *et al.*, 2010). The use of multi-spectral (infrared and visible) cameras for semi-automated detection of marine mammals is also developing rapidly, and such sensors can be mounted on aircraft (manned or unmanned) or on land (Schoonmaker *et al.*, 2008a, 2008b; Podobna *et al.*, 2009; Graber, 2011). Technological advances should continue to reduce the cost and increase the effectiveness of survey techniques. In addition to estimating the size of the population of dusky dolphins in New Zealand, surveys could be conducted to determine local abundance in the Kaikoura area. Regular surveys could be useful in establishing whether area avoidance or habitat displacement is occurring in response to vessel traffic. Unfortunately, the dusky dolphin population in New Zealand is not well-suited to abundance estimation. The population is large, highly mobile, poorly marked, and local abundance varies seasonally. These characteristics mean it is unlikely that existing abundance techniques will be able to detect a trend useful for management purposes.

Management of dusky dolphin tourism at Kaikoura generally follows the form of the model proposed by Higham *et al.* (2009), and it appears the model is flexible enough to accommodate the differences found in other locations. This case study shows the value in comparing real-life management to an idealized model. The comparison identified a number of actions which would improve management and sustainability of dolphin-oriented tourism in Kaikoura, and the model provided reasonable guidance for making improvements. While legislation, regulation and permit conditions are in place, certain aspects need to be defined, enforced or clarified. Based on the results presented here, the following recommendations are suggested to improve management of this activity:

1. **Define limits of acceptable change (LAC).** This is a critical aspect of management which has been overlooked in Kaikoura. LAC is generally tied to population-level effects, and while these are difficult to detect in a large, mobile population of animals like dusky dolphins off Kaikoura, the projects suggested above would provide quantitative data which could be used to set LAC criteria. These could then be evaluated and compared over time to determine whether the activity is likely to be ecologically sustainable or not.

2. **Define a consistent time period for evaluation of changes relative to LAC.** A 5-year cycle for evaluation is appropriate, as this promotes active responses to change in a dynamic system. It also allows managers to learn and apply best practices for cetacean tourism management demonstrated in other parts of the world. LAC-related data should be collected every year, and it is possible that collection of certain data may be automated. Ongoing work could be funded via a combination of DOC funding and user levies (or donations) obtained directly from cetacean tourists and included in the cost of the tour,

a model which is used in other parts of New Zealand.

3. **Clarify rules on swimming with juvenile dolphins.** This is a forbidden activity under the MMPR (1992), but occurs regularly when vessels interact with large pods. The rule is apparently unenforced because a clear definition of 'juvenile' has not been established. Allowing an activity which is not permitted under the regulations is poor practice, and either the regulations should be updated to indicate situations where this is acceptable or extended to create explicit rules for evaluating whether an animal is a juvenile/calf or an adult.

4. **Enhance monitoring and enforcement.** Current monitoring of cetacean tourism in Kaikoura is primarily conducted using a secret shopper method, whereby DOC representatives ride aboard tours without being identified as official observers (Childerhouse & Baxter, 2010). Shore-based monitoring would be a simple, cost-effective way of enhancing this monitoring, as many of the regulations and permit conditions can be observed: number of vessels within 300 m, approach method of vessels, observation of the rest period, number of times swimmers enter the water, and presence of juveniles in the pod. Rather than observing a single tour from a single boat, multiple tours and boats can be observed with the same amount of effort.

5. **Actively learn from the public.** Updated information is needed about user perceptions, community desires, education of tourists and other aspects of dolphin tourism. The only social science studies performed in Kaikoura occurred more than a decade ago and it is reasonable to expect that things have changed since then. The wishes of the local community should be reflected in management decisions, and the best way of understanding this is through research surveys. Educational programmes should be improved if tourists do not feel they are learning, as this is a requirement based on the MMPR (1992).

6. **Reserve the right to reduce tour boat traffic near dolphins**. Commercial tour boats spend more time with dolphins than any other type of vessel, and the number of vessels present within 300 m is linked to the magnitude of the effects (Lundquist, 2012; Markowitz *et al.*, 2009). As tourism has grown, so has the magnitude of responses of dolphins to vessels (Lundquist, 2012). Changes to permit conditions instituted in 2010 (increased number of swimmers, limited number of swim drops) may help reduce the number and intensity of tour boat interactions, but this must be actively monitored and tracked to determine if vessel traffic decreases and dolphin responses lessen. If not, DOC must strongly consider reducing the number of permits or reducing the number of commercial vessels allowed within 300 m of a single pod.

Conclusions

There is a growing body of evidence that tourism activity can affect the long-term well-being of cetacean populations, and the model proposed by Higham *et al.* (2009) provides a strong map for sustainable planning and management. Comparing tourism targeting dusky dolphins near Kaikoura to this model has highlighted a number of issues of critical importance, and suggested changes which should improve the management of this activity. The recommendations above will serve to mitigate the effects of vessel-based tourism, better inform wildlife managers, represent the interests of a larger group of stakeholders, more clearly elucidate activities which are allowed or forbidden, and thereby result in a greater chance of sustainable tourism in Kaikoura. Differences exist between sites and species, but similar comparison to this management model should be conducted in other places where cetacean tourism occurs, resulting in low-cost, high-value management recommendations. Wildlife managers responsible for protecting species targeted by tourism in other locations should seek to use such a model to ensure that the public is able to experience the natural world, communities benefit economically, and animals are protected for future generations to enjoy.

REFERENCES

Allen, S., Smith, H., Waples, K. & Harcourt, R. (2007). The voluntary code of conduct for dolphin watching in Port Stephens, Australia: Is self-regulation and effective management tool? *Journal of Cetacean Research and Management* 9, 159–166.

Au, W.W.L. & Green. M. (2000). Acoustic interaction of humpback whales and whale-watching boats. *Marine Environmental Research* 49, 469–481.

Barr, K. (1997). The impacts of marine tourism on the behaviour and movement patterns of dusky dolphins (*Lagenorhynchus obscurus*) at Kaikoura, New Zealand. Thesis, University of Otago, Dunedin, New Zealand. 97 pp.

Barr, K. & Slooten. E. (1999). *Effects of Tourism on Dusky Dolphins at Kaikoura*. Conservation Advisory Science Notes No. 229. Wellington, New Zealand: Department of Conservation.

Bejder, L. & Lusseau, D. (2008). Valuable lessons from studies evaluating impacts of cetacean-watch tourism. *Bioacoustics* 17, 158–161.

Bejder, L., Samuels, A., Whitehead, H. & Gales, N. (2006a). Interpreting short-term behavioural responses to disturbance within a longitudinal perspective. *Animal Behaviour* 72, 1149–1158.

Bejder, L., Samuels, A., Whitehead, H., *et al.* (2006b). Decline in relative abundance of bottlenose dolphins exposed to long-term disturbance. *Conservation Biology* 20, 1791–1798.

Brown, N.C. (2000). The dusky dolphin, *Lagenorhynchus obscurus*, of Kaikoura, New Zealand: A long-term comparison of behaviour and habitat use. Thesis, University of Auckland, Auckland, New Zealand. 153 pp.

Buckland, S.T., Breiwick, J.M., Cattanach, K.L. & Laake, J.L. (1993). Estimated population size of the California gray whale. *Marine Mammal Science* 9, 235–249.

Buckstaff, K.C. (2004). Effects of watercraft noise on the acoustic behavior of bottlenose dolphins, *Tursiops truncatus*, in Sarasota Bay, Florida. *Marine Mammal Science* 20(4), 709–725.

Burgess, W.C., Tyack, P.L., Le Boeuf, B.J. & Costa, D. (1998). A programmable acoustic recording tag and first results from free-ranging northern elephant seals. *Deep-Sea Research II* 45, 1327–1351.

Butler, P.J., Green, J.A., Boyd, I.L. & Speakman, J.R. (2004). Measuring metabolic rate in the field: The pros and cons of the doubly labelled water and heart rate methods. *Functional Ecology* 18, 168–183.

Buurman, D. (2010). Dolphin swimming and watching: One tourism operator's perspective. In B. Würsig & M. Würsig (Eds), *The Dusky Dolphin: Master acrobat off different shores*. London: Academic Press, pp. 277–289.

Childerhouse, S. & Baxter, A. (2010). Human interactions with dusky dolphins: A management perspective. In B. Würsig & M. Würsig (Eds), *The Dusky Dolphin: Master acrobat off different shores*. London: Academic Press, pp. 245–275.

Cipriano, F.W. (1992). Behavior and occurrence patterns, feeding ecology, and life history of dusky dolphins (*Lagenorhynchus obscurus*) off Kaikoura, New Zealand. Dissertation, University of Arizona, Tucson, USA. 216 pp.

Constantine, R. (2001). Increased avoidance of swimmers by wild bottlenose dolphins (*Tursiops truncatus*) due to long-term exposure to swim-with-dolphin tourism. *Marine Mammal Science* 17, 689–702.

Constantine, R., Brunton, D.H. & Dennis, T. (2004). Dolphin-watching tour boats change bottlenose dolphin (*Tursiops truncatus*) behaviour. *Biological Conservation* 117, 299–307.

Cooke, S.J., Hinck, S.G., Wikelski, M., *et al.* (2004). Biotelemetry: A mechanistic approach to ecology. *Trends in Ecology and Evolution* 19(6), 334–343.

Dahood, A.D. (2009). Dusky dolphin (*Lagenorhynchus obscurus*) occurrence and movement patterns near Kaikoura, New Zealand. MSc Thesis, Texas A&M University. 89 pp.

Dans, S.L., Crespo, E.A., Pedraza, S.N., DeGrati, M. & Garaffo, G.V. (2008). Dusky dolphin and tourist interaction: Effect on diurnal feeding behavior. *Marine Ecology Progress Series* 369, 287–296.

Duprey, N.M. (2007). Dusky dolphin (*Lagenorhynchus obscurus*) behaviour and human interactions: Implications for tourism and aquaculture. MSc thesis, Texas A&M University. College Station, Texas, USA.

Duprey, N.M.T., Weir, J.S. & Würsig, B. (2008). Effectiveness of a voluntary code of conduct in reducing vessel traffic around dolphins. *Ocean and Coastal Management* 51, 632–637.

Ellenberg, U., Setiawan, A.N., Cree, A., Houston, D.M. & Seddon, P.J. (2007). Elevated hormonal stress response and reduced reproductive output in Yellow-eyed penguins exposed to unregulated tourism. *General and Comparative Endocrinology* 152, 54–63.

Erbe, C. (2002). Underwater noise of whale-watching boats and potential effects on killer whales (*Orcinus orca*), based on an acoustic impact model. *Marine Mammal Science* 18(2), 394–418.

Foster, K.R., Vecchia, P. & Repacholi, M.H. (2000). Science and the Precautionary Principle. *Science* 288, 979–981.

Fowler, G.S. (1999). Behavioral and hormonal responses of Magellanic penguins (*Spheniscus magellanicus*) to tourism and nest site visitation. *Biological Conservation* 90, 143–149.

Gordon, J., Leaper, R., Hartley, F.G. & Chappell, O. (1992). *Effects of Whale-watching on the Surface and Underwater Acoustic Behaviour Off Kaikoura, New Zealand*. Science and Research Series 52. Wellington, New Zealand: Department of Conservation.

Graber, J. (2011). Land-based infrared imagery for marine mammal detection. MS Thesis, University of Washington, USA. 95 pp.

Higginbottom, K. (2004). *Wildlife Tourism: Impacts, management, and planning*. Altona, Australia: Common Ground Publishing.

Higham, J.E.S. (1998). Tourists and albatrosses: The dynamics of tourism at the Northern Royal Albatross Colony, Taiaroa Head, New Zealand. *Tourism Management* 19(6), 521–531.

Higham, J.E.S., Bejder, L. & Lusseau, D. (2009). An integrated and adaptive management model to address the long-term sustainability of tourist interactions with cetaceans. *Environmental Conservation* 35(4), 294–302.

Hodgson, A.J., Noad, M., Marsh, H., Lanyon, J. & Kniest, E. (2010). Using unmanned aerial vehicles for surveys of marine mammals in Australia: Test of concept. Final report to the Australian Marine Mammal Centre, 76 pp.

Hoyt, E. (2008). Whale watching. In W.F. Perrin, B. Würsig & J.G.M. Thewissen (Eds), *Encyclopedia of Marine Mammals* (2nd edn). San Diego, CA: Academic Press, pp. 1223–1227.

Jahoda, M., Lafortuna, C.L., Biassoni, N., *et al.* (2003). Mediterranean fin whale's (*Balaenoptera physalus*) response to small vessels and biopsy sampling assessed through passive tracking and timing of respiration. *Marine Mammal Science* 19(1), 96–110.

Jensen, F.H., Bejder, L., Wahlberg, M., Aguilar Soto, N., Johnson, M. & Madsen, P.T. (2009). Vessel noise effects on delphinid communication. *Marine Ecology Progress Series* 395, 161–175.

Johnson, M.P. & Tyack, P.L. (2003). A digital acoustic recording tag for measuring the response of wild marine mammals to sound. *IEEE Journal of Oceanic Engineering* 28, 3–12.

Kastelein, R.A., van der Elst, C.A., Tennant, H.K. & Wlepkema, P.R. (2000). Food consumption and growth of a female dusky dolphin (*Lagenorhynchus obscurus*). *Zoo Biology* 19, 131–142.

Koski, W.R., Allen, T., Ireland, D., *et al.* (2009). Evaluation of an Unmanned Airborne System for monitoring marine mammals. *Aquatic Mammals* 35(3), 347–357.

Lück, M. (2003). Environmentalism and on-tour experiences of tourists on wildlife watch tours in New Zealand: A study of visitors watching and/or swimming with dolphins. PhD Thesis, University of Otago, Dunedin, New Zealand. 249 pp.

Lundquist, D. (2012). Behaviour and movement patterns of dusky dolphins (*Lagenorhynchus obscurus*) off Kaikoura, New Zealand: Effects of tourism. Thesis, University of Otago, New Zealand. 142 pp.

Lundquist, D., Gemmell, N. & Würsig, B. (2012). Behavioural responses of dusky dolphin (*Lagenorhynchus obscurus*) groups to tour vessels off Kaikoura, New Zealand. *PLoS ONE* 7(7): e41969. doi:10.1371/journal.pone.0041969.

Lundquist, D., Gemmell, N., Würsig, B. & Markowitz, T. (2013). Dusky dolphin movement patterns: Short-term effects of tourism. *New Zealand Journal of Freshwater and Marine Research* 47(4), 430–449.

Lusseau, D. (2006). The short-term behavioural reactions of bottlenose dolphins to interactions with boats in Doubtful Sound, New Zealand. *Marine Mammal Science* 22(4), 802–818.

Lusseau, D., Slooten, L. & Currey, R.J.C. (2006). Unsustainable dolphin-watching tourism in Fiordland, New Zealand. *Tourism in Marine Environments* 3(2), 173–178.

Markowitz, T. (2004). Social organization of the New Zealand dusky dolphin. Dissertation, Texas A&M University, Galveston, Texas, USA. 278 pp.

Markowitz, T.M., DuFresne, S. & Würsig, B. (2009). Tourism effects on dusky dolphins at Kaikoura, New Zealand. Report submitted to New Zealand Department of Conservation, Wellington. 93 pp.

Nowacek, S., Wells, R.S. & Solow, A.R. (2001). Short-term effects of boat traffic on bottlenose dolphins, *Tursiops truncatus*, in Sarasota Bay, Florida. *Marine Mammal Science* 17(4), 673–688.

O'Connor, S., Campbell, R., Cortez, H. & Knowles, T. (2009). *Whale Watching Worldwide: Tourism numbers, expenditures and expanding economic benefit.* A special report prepared by Economists at Large. Yarmouth, MA: International Fund for Animal Welfare.

Parsons, E.C.M. (2012). The negative impacts of whale-watching. *Journal of Marine Biology.* http://dx.doi.org/10.1155/2012/807294.

Podobna, Y., Schoonmaker, J., Boucher, C. & Oakley, D. (2009). Optical detection of marine mammals. *Proceedings of SPIE* 7317, 73170J. http://dx.doi.org/10.1117/12.818359.

Richardson, W.J., Greene, C.R., Malme, C.I. & Thompson, D.H. (1995). *Marine Mammals and Noise.* San Diego, CA: Academic Press.

Rolland, R.M., Parks, S.E., Hunt, K.E., *et al.* (2012). Evidence that ship noise increases stress in right whales. *Proceedings of the Royal Society B* 279, 2363–2368.

Romero, L.M. & Wikelski, M. (2001). Corticosterone levels predict survival probabilities of Galapagos marine iguanas during El Nino events. *Proceedings of the National Academy of Sciences of the United States of America* 98, 7366–7370.

Sapolsky, R., Romero, L. & Munck, A. (2000). How do glucocorticoids influence stress responses? Integrating permissive, suppressive, stimulatory and preparative actions. *Endocrine Reviews* 21, 55–89.

Scarpaci, C., Bigger, S.W., Corkeron, P.J. & Nugegoda, D. (2000). Bottlenose dolphins (*Tursiops truncatus*) increase whistling in the presence of 'swim-with-dolphin' operations. *Journal of Cetacean Research and Management* 2, 183–185.

Schoonmaker, J., Dirbas, J., Podobna, Y., Wells, T., Boucher, C. & Oakley, D. (2008a). Multispectral observations of marine mammals. *Proceedings of SPIE* 7113, 711311. http://dx.doi.org/10.1117/12.800024

Schoonmaker, J., Wells, T., Gilbert, G., Podobna, Y., Petrosyuk, I. & Dirbas, J. (2008b). Spectral detection and monitoring of marine mammals. *Proceedings of SPIE* 6946, 694606. http://dx.doi.org/10.1117/12.777740

Simmons, D.G. & Fairweather, J.R. (1998). *Towards a Tourism Plan for Kaikoura.* Tourism Research and Education Center Report No. 10. Christchurch, New Zealand: Lincoln University. 47 pp.

Slooten, E., Rayment, W. & Dawson, S. (2006). Offshore distribution of Hector's dolphins at Banks Peninsula, New Zealand: Is the Banks Peninsula Marine Mammal sanctuary large enough? *New Zealand Journal of Marine and Freshwater Research* 40(2), 333–343.

Speakman, J.R. (1997). *Doubly Labelled Water: Theory and practice*. London: Chapman & Hall.

Srinivasan, M. (2009). Predator influences on behavioral ecology of dusky dolphins. PhD dissertation, Texas A&M University, College Station, TX, USA. 161 pp.

Stankey, G.H., Cole, D.N., Lucas, R.C., Petersen, M.E. & Frissell, S.S. (1985). *The Limits of Acceptable Change (LAC) System for Wilderness Planning*. General Technical Report INT-176. Ogden, UT: US Department of Agriculture, Forest Service, 37 pp.

Stensland, E. & Berggren, P. (2007). Behavioural changes in female Indo-Pacific bottlenose dolphins in response to boat-based tourism. *Marine Ecology Progress Series* 332, 225–234.

Williams, R., Trites, A.W. & Bain, D.E. (2002). Behavioural responses of killer whales (*Orcinus orca*) to whale-watching boats: Opportunistic observations and experimental approaches. *Journal of Zoology, London* 256, 255–270.

Williams, R., Lusseau, D. & Hammond, P.S. (2006). Estimating relative energetic costs of human disturbance to killer whales (*Orcinus orca*). *Biological Conservation* 133, 301–311.

Würsig, B. & Würsig, M. (1980). Behavior and ecology of the dusky dolphin, *Lagenorhynchus obscurus*, in the South Atlantic. *US Fishery Bulletin* 77(4), 871–890.

Würsig, B. & Würsig, M. (2010). *The Dusky Dolphin: Master acrobat off different shores*. London: Academic Press. 441 pp.

Würsig, B., Cipriano, F., Slooten, E., Constantine, R., Barr, K. & Yin, S. (1997). Dusky dolphins (*Lagenorhynchus obscurus*) off New Zealand: Status of present knowledge. *Report of the International Whaling Commission* 47, 715–722.

Yin, S. (1999). Movement patterns, behaviors, and whistling sounds of dolphin groups off Kaikoura, New Zealand. Thesis, Texas A&M University, Galveston, USA. 107 pp.

24

Save the whales Part II

A new science advocacy communication framework

Wiebke Finkler

If a solution is to speak to a people and not end up as the private answer of a sect, it needs to find roots in their life, language and thought

(Campbell, 1974: 444)

Introduction

The context for this chapter is the uncontrolled explosive growth of global whale-watching and the failure of widespread sustainable practices. Economic imperatives and environmental marketing have driven the global growth of whale-watching (Neves, 2010). In this chapter I argue that the widespread failure of sustainability is, in part, due to ineffective public communication and poor uptake of science related to impact assessments of whale-watching. The production and traditional scientific dissemination of research alone, while influencing the discourse and development of whale-watching in some contexts, have not been enough to contribute to the long-overdue evolution towards sustainable whale-watching. This chapter calls for the need to raise and address a range of new questions relating to the efficacy of science and the urgent need to improve science communication. New questions need to be asked to address the failure of collaboration between whale-watch stakeholders in order to catch up with the realities of global unsustainable whale-watching

practices (Corkeron, 2004; Neves-Graca, 2004; Higham & Bejder, 2008). The current work proposes a new science advocacy communication framework for the whale-watching setting drawing on elements of science communication, media studies, marketing and tourism management. This audiovisual Science Communication Commercial, or SciCommercial for short, presents a potential management tool for the whale-watching industry by advocating sustainable practices to stakeholders, increasing awareness about impacts and managing visitors' expectations.

The dilemma for the whale-watching industry

In 1982, the International Whaling Commission (IWC) adopted a moratorium on commercial whaling that went into effect in 1986. The major reason for the moratorium was scientific uncertainty about whale stocks (Morishita, 2006). Since then, two groups, one in favour of sustainable whaling and one completely opposed to any killing of whales, continue to argue at the IWC and other political arenas (Cisneros-Montemayor *et al.*, 2010). The fundamental arguments against whaling have changed from being an ecological argument (they are endangered) to a moral and ethical argument (whales are special) (Kalland, 1993). Most analysts seem

Whale-watching: Sustainable Tourism and Ecological Management, eds J. Higham, L. Bejder and R. Williams.
Published by Cambridge University Press. © Cambridge University Press 2014.

to agree that the debate on whaling has reached a stalemate (Rethmann, 2008), leading some to characterize the IWC organization as hamstrung (Bailey, 2012). Overshadowed by this debate, yet connected, is the growth of the global whale-watching industry.

Since the collapse of the commercial whaling industry, whale-watching has developed into a multi-billion dollar global business (Hoyt 2001; O'Connor *et al.*, 2009) widely regarded as the antithesis to whaling (Neves, 2010). 'The Whale' has become a human passion (Chen, 2011) arousing controversies more intense than most other instances of animal politics (Blok, 2007). People want to observe cetaceans (whales, dolphins and porpoises) in the wild and the demand for these up-close and personal experiences fuels an ever-expanding industry worldwide (Hoyt, 2001). There is no doubt about the socioeconomic benefits: in 2008, 13 million people participated in whale-watching in 119 countries generating $US 2.1 billion (O'Connor *et al.*, 2009). However, a significant body of scientific research outlining the short-term and long-term consequences of whale-watching on whales and dolphins has accumulated over the last three decades (Constantine *et al.*, 2004; Bejder *et al.*, 2006a; Lemon *et al.*, 2006; Williams *et al.*, 2006, 2009; Schaffar *et al.*, 2009; Lachmuth *et al.*, 2011; see also Chapters 13–17).

The most commonly cited impact on whales and dolphins is disturbance due to repeated close approaches of boats as well as associated underwater noise pollution (Orams, 2000; Garrod & Fennell, 2004; Weilgart, 2007; Tyack, 2008; Jensen *et al.*, 2009; Chen, 2011; Wright *et al.*, 2011). Whale-watching can result in reduced reproduction rates and ultimately affect overall population fitness (Bejder *et al.*, 2006a). While debate exists about the methodology of impact research assessment (Bejder & Samuels, 2003; Bejder *et al.*, 2006b), one issue has clearly crystallized: whale-watching is not a benign activity, but instead has the potential for serious long-term impacts on populations of whales and dolphins (Garrod & Fennell, 2004; Bejder *et al.*, 2006a; IWC, 2007; Wright *et al.*, 2011; see also Chapter 2).

Typically, environmental non-governmental organizations (NGOs) would question the exponential growth of whale-watching (Corkeron, 2004), yet many NGOs actively promote an increase in commercial operations as a benign activity and potent tool for whale conservation and, as such, a way to save a species or population from direct exploitation such as whaling (Neves, 2010; Bailey, 2012). Marketing by NGOs has played a key role in the growth of whale-watching and contributed to the dilemma that whale-watching, which is now known to cause significant impacts (Bejder *et al.*, 2006a), continues to be equated with whale conservation (Neves, 2010). This contradicts an overwhelming body of science and diminishes the ability to differentiate between good and bad whale-watching practice. Historically, NGOs played a key role in saving whales from extinction through hunting practices. In regard to bad whale-watch practices, however, NGOs appear stuck in a time warp (Corkeron, 2004).

The critical challenge of sustainability

The rapid growth of whale-watching constitutes a challenge to the sustainable management of the industry and demands a more stringent approach to ensure its long-term sustainability (Gordon *et al.*, 1992; Higham *et al.*, 2009; Kessler & Harcourt, 2010; Schaffar *et al.*, 2009). Key stakeholders involved in whale-watching include scientists (social and natural sciences), commercial operators, planning agencies, environmental NGOs and the public (Higham *et al.*, 2009; see also Chapter 1). The latter comprises whale-watch tourists, host communities and the wider public as such. It is crucial to increase communication and collaborations among the different stakeholders (Garrod & Fennell, 2004; Neves-Graca, 2004; Higham *et al.*, 2009). Management of whale-watching around the globe ranges from government regulations to guidelines and voluntary codes of conduct to no management at all. Guidelines typically include restrictions on the number of boats in close proximity, boat speed limits, minimum approach distance, boat approach patterns and interaction patterns with whales and dolphins (Chen, 2011). Overall management is inconsistent between different states and across international waters (Blewitt, 2008), with considerable variation among whale-watching codes of conduct (Garrod & Fennell, 2004).

At present, no workable mechanism exists for the development and implementation of whale-watch regulations. Self-regulatory measures are insufficient (Wiley *et al.*, 2008; Schaffar *et al.*, 2009) and enforcement of regulations almost non-existent. Of all guidelines, 62% are currently voluntary in nature (Garrod & Fennell, 2004). The whale-watching industry continues to grow while its management, which has been described as highly fragmented, patchy and even haphazard, has largely failed (Gjerdalen & Williams, 2000; Garrod & Fennell, 2004). If whale-watching is to live up to its lofty aspirations of conservation-oriented ecotourism instead of mass tourism, it is crucial to improve compliance, develop precautionary management strategies that are adaptive to research findings and urgently address multi-stakeholder involvement (Berrow, 2003; Garrod & Fennell, 2004; Allen *et al.*, 2007; Higham *et al.*, 2009). Operators represent a considerable human capital as cumulative source of knowledge through practice (Gjerdalen & Williams, 2000; Neves-Graca, 2004), yet only 6.9% of whale-watching codes of conduct have been developed by the industry itself (Garrod & Fennell, 2004). More appropriate business models have to allow greater community involvement in whale-watching opportunities and regulatory development (Neves, 2010).

It is timely to raise and investigate other questions such as 'what are the underlying causes of unsustainable whale-watching?'; 'why are scientific concerns largely ignored?' and 'what actions can be taken?' One core problem lies in the most fundamental ingredient in any relationship: communication. In the case of whale-watching, in particular, the communication and underlying language used by NGOs and scientists have left the general public largely uninformed. NGOs play a key role in the political arena of the IWC as well as the public sphere (Bailey, 2012). Yet there is a chasm in the communication of whale-watching issues to the public. A significant body of scientific work has been accumulated yet with a few exceptions there has been a complete failure on a global level in terms of translating research into management outcomes or public awareness.

NGOs strongly influence the dominant conservation discourse and how the public perceives environmental problems (Betsill & Corell, 2001; Neves, 2010). Anti-whaling protests became one of the defining markers of environmental organizations (Rethmann, 2008) and an essential part of their narrative. Some of the major NGOs have profited greatly in the form of public contributions (Skodvin & Andresen, 2003) as images of butchered whales became commodities in the highly competitive world of NGO fundraising and whales as flagship species became part of their branding (Neves, 2010). NGOs continue to raise public awareness about whaling, lobby state decision-making, coordinate boycotts, participate in international negotiations, and help monitor and implement international agreements (Betsill & Corell, 2001). Yet NGOs fail to do the same for whale-watching, instead using it as an advocacy tool against whaling. NGOs 'get away' with this due to widespread public naivety and lack of awareness about the impacts of whale-watching. The majority of people cannot recognize impacts such as noise effects on foraging behaviour, making them invisible to most consumers and hard to identify for untrained eyes (Neves, 2010). Current NGO discourses and aggressive marketing of whale-watching exacerbates this and inhibits critical assessment of the tourism industry (Neves, 2010).

Management of whale-watching is as much about managing people's attitudes and expectations as it is about the science of populations of whales and dolphins (Orams, 2000; Lunney *et al.*, 2008; Kessler & Harcourt, 2010). After all, it is not the whales that need to be managed, but the humans that target them (Forestell & Kaufman, 1994). Education has been advocated as an effective management strategy to encourage pro-conservation knowledge, attitudes and behaviour change (Zeppel, 2008; see also Chapters 10 and 11). Managers and operators appear to assume that by simply providing information a 'magical transformation' into environmentally aware and responsible tourists will occur (Maiteny, 2002; Orams & Taylor, 2005). These claims, however, are largely untested and there is little research on the management effectiveness of whale-watch interpretation (Lück, 2003; Powell & Ham, 2008).

Competition between operators can increase pressure to find a competitive edge and if operators respond to this by engaging in less-desirable practices such as going closer to wildlife it can contribute to negative impacts (Moscardo & Saltzer, 2004). Unrealistic visitor expectations of the whale-watching experience can increase pressure on tour operators to provide ideal experiences leading to a tendency to ignore existing guidelines (Kessler & Harcourt, 2010). There is a significant correlation between visitor satisfaction and proximity to whales and subsequent breaches of regulations and guidelines to satisfy visitor expectations of close encounters (Scarpaci *et al.*, 2004; Valentine *et al.*, 2004). The business perspective consequently dominates animal welfare.

Tourism advertising, commercials and mass media play an important part in how people form expectations. Too often, people arrive with unrealistic expectations to watch whales perform the most dramatic behaviours at close range (Sironi *et al.*, 2009). A survey of swim-with-dolphin participants showed that half of the swimmers that came within 5 m stated that it was still too distant (O'Neill *et al.*, 2004). Consequently, calls for educational programmes designed to target operators, whale-watchers and recreational users are increasing (Allen *et al.*, 2007). The obvious challenge lies in managing visitors' desire for close interaction with wildlife.

Refashioning cetaceans

It is time to evolve and 'refashion' the iconic status of whales and dolphins (Corkeron, 2004;

see also Chapter 4). Whale-watching as a global mass (eco-)tourism industry adds potentially significant levels of stress to populations of whales and dolphins already under pressure due to other anthropogenic factors, e.g. pollution, overfishing, whaling (Corkeron, 2004). Further, the impacts of whale-watching are likely to be cumulative and long-term rather than catastrophic, and thus risk being unnoticed for decades.

Conservation and consumerism have become closely entwined (Neves, 2010). There is a real danger of revisiting the tragedy of the commons, i.e. in the absence of restrictions people act on the basis of self-interest resulting inevitably in the over-exploitation of the resource (Neves-Graca, 2004). The various stakeholders involved in whale-watching are either ignorant of or ignoring scientific research findings about the dangers of current global whale-watching practices: most NGOs as well as whale-watching operators continue their comfortable journey on the anti-whaling bandwagon. This context points to a failure of science communication, which invites questions to be asked about the efficacy of science.

The efficacy of science: implementation, communication and uptake

Scientists increasingly express frustration that their management advice remains ignored and whale-watching instead is determined more by political or industry influences (Hughes, 2001; Lusseau *et al.*, 2006; Cater & Cater, 2007; Higham & Bejder, 2008). Clearly, communication is a two-way process between communicator and audience, and it is crucial that the stakeholders involved (i.e. the audience) are receptive to scientific information. Part of the problem, however, is due to the language of science. All too often scientists preach to the converted (i.e. the scientific community), publishing their work in academic journals with a view to impact factors, citation counts and career advancement, and disseminating their work to their academic peers at scholarly conferences. Consequently, the primary communication format of scientific publication all too often is limited to an isolated academic audience instead of the public domain and science remains the language of a chosen few, largely inaccessible and incomprehensible to non-scientists. This inhibits understanding between stakeholders (Neves-Graca, 2004) and urgently requires a translator.

There is a tendency among the public and many academics to view science as an entirely objective practice in pursuit of revealing the truth about reality: a reality that is assumed to exist independently of us, regardless of what we might think or say about it (Heazle, 2004). Scientists are uniquely qualified to participate in public policy and not just publish findings in scholarly papers. They have a responsibility to bring scientific findings to the attention of all stakeholders especially when the public funds their research (Scott et al., 2007). Unfortunately, all too often expressing opinion and advocating personal policy preferences is considered inappropriate as many scientists see their role as unimpeachable providers of policy-neutral information (Lackey, 2007). However, science is not value-neutral and cannot operate independently of the larger political environment in which it exists (Heazle, 2004). It is pointless to try and explain competing scientific interpretations of reality solely on the basis of which one is more or less true. Instead, we should investigate to what extent governments and NGOs use empirical scientific methods and why (Heazle, 2004). Science deals with uncertainty and doubt. Yet, it is not scientific uncertainty itself that influences policy-making but how we choose to use it, which in turn is ultimately determined by political choices, i.e. the extent to which it fulfils a recognized need or utility (Heazle, 2004). The IWC is a great example in terms of the whaling debate, e.g. 'all whales are endangered' or 'only some whales are endangered', as well as related utility of whale-watching for NGO anti-whaling politics as described earlier in this chapter.

Modern science tends to communicate via academic publication and conducts public outreach through education. In comparison, NGOs tend to use advocacy as their vehicle for public communication. However, advocacy does not exist as a clear concept with a fixed meaning, but occurs in various shades (Jickling, 2003). Environmental advocacy is defined as symbolic discourse (i.e. legal, educational, expository, artistic, public and/or interpersonal communication) aimed at supporting conservation and the preservation of finite resources (Bryant, 1990). Some authors point out that anyone who tries to mould opinions and policies about how to treat the Earth could be considered an environmental advocate (Cantrill, 1993). This tension between science and advocacy is very relevant for conservation (Scott et al., 2007). By leaning away from advocacy scientists risk sending the message that actions should be avoided and participation in controversial issues as well as adoption of a position are unimportant. 'Advocates, consultants and researchers must take a stand if ours is not merely an exercise in academic sophistry' (Cantrill, 1993: 89).

Increasingly, scientists are recognizing the need to communicate their work to the general public. Still, the academic discourse tends to be primarily on the safe side, i.e. dominated by calls for more future research. Aristotle believed that perfect knowledge is not achievable concerning practical affairs or ethics: we must use our best judgement. As Ophuls (2011) points out 'in fact, there may be no such thing as an airtight scientific case. Science is a work in progress that occasionally undergoes revolutions and that sometimes discovers the extraordinary' (p. 27). The scientific quest for certainty neglects the crucial communication of current knowledge to the wider public. It is not surprising that the IWC's five-year strategic plan for whale-watching (IWC, 2011) lists more research as its first objective while education is listed last, somewhere within the wider management objective. Key issues that emerge include the epistemic responsibility of scientists and the role of conservation and advocacy in whale-watching interpretation (Russell & Hodson, 2002). Clearly, sustainable management requires new tools to assist decisions on marketing of wildlife-based activities (Tremblay, 2001).

The importance of language or the 'how' instead of the 'what'

The environmental movement successfully contributed to stronger public concern over whales. Activities as well as mobilization campaigns were the single most important factor (Skodvin & Andresen, 2003). The instruments and language employed by the environmental movement are fundamentally different to those of the scientific community. As outlined previously, one key difference is that the scientific community largely believes that it gains legitimacy from the provision of objectivity and that any active mobilization can jeopardize scientific credibility (Skodvin & Andresen, 2003). In contrast, one important means of public mobilization by the environmental movement included a strong focus on qualities of whales such as intelligence, social characters and beauty (Skodvin & Andresen, 2003; Bailey, 2009). The discussion about whaling shifted from a debate on the sustainable management of the whale resource to a debate on ethics and morality of whaling more generally. When issues turn into a debate over values, scientific input will have limited effect (Skodvin & Andresen, 2003). Emotionally charged 'Save the Whale' slogans clearly mobilize the public more effectively than scientific publications on sustainable management.

'The Origin' was probably the last major scientific work that a popular audience could both read and at least in outline, begin to understand. (Campbell, 1974: 442).

Charles Darwin, one of the critical founders of modern science, used normative language many scientists now are reluctant to use, such as taking heart, knowing despair, or being courageous. 'Darwin does not describe nature with the heartless objectivity of a scientist, but with the detached yet sympathetic engagement of a poet. Darwin gives nature personality' (Campbell, 1974: 447).

Over the past 30 years, the cognitive and brain sciences have shown us that reason is not conscious, unemotional, logical, abstract or universal. Instead, real reason is mostly (98%) unconscious, requires emotion, narratives and logic of frames (Lakoff, 2010). People think in terms of typically unconscious structures called frames, which have direct connections to emotional regions of the brain (Lakoff, 2010). Hearing a word can activate its frame and can be chosen to activate desired frames. Facts to be communicated must be framed properly in order to make sense in terms of their system of frames. In short, one cannot avoid framing. While solid science is crucial to underpin arguments, many people trained in public policy and science often believe that if you just give people the facts they will reason to the right conclusion (Lakoff, 2010). The success of any social movement, however, does not depend on the objective facts of a given case, but instead on how the issue is argued or framed (Bailey, 2009). This is what effective communicators do. Any issue can be framed in multiple ways: the key to success is finding the right frames.

Mass media coverage is a key contributor in shaping and affecting science and policy discourse as well as public understanding and action (Boykoff & Roberts, 2007). The media has been accused of lacking adequate scientific input (Greenberg *et al.*, 1989) and exaggerating scientific claims for the sake of the story, leading some to argue that it is futile for science to invest much hope in the enlightenment of the media (Weingart *et al.*, 2000). This outlook appears somewhat simplistic and irresponsible. Clearly, apart from understanding the right crisis, it is also crucial to understand what to do about it and to provide numbers and facts in ways so their overall significance is apparent (Lakoff, 2010). It is time for scientists to actively 're-frame' their frustrations, move outside their academic science camp and get involved in 'pro-active communication'. Ultimately the most profound enemies of

progress are not scientific sceptics or industry oppo-
nents but general misunderstanding and apathy
(Carpenter, 2001).

SciCommercials: a new science advocacy communication framework

Social movements have been effective due to a
simple basic framing. Frames are communicated
via language, visual imagery and experience of the
natural world (Lakoff, 2010). While whale-watching
clearly provides strong visual imagery and experiences, the language, framing and marketing of
whale-watch science urgently has to evolve. Science
Communication is an emerging discipline that aims
to communicate science (both natural and social) to
the layperson. It has the potential to provide practical solutions and to develop and underpin a campaign for action in order to link science with applied
management. Science Communication plays a crucial role in communicating complex scientific issues
to the wider public via the use of different social
and multimedia platforms, including film-making
(Holliman *et al.*, 2009).

One key ingredient for science communication is
storytelling and when it comes to communicating
factual material through film, the traditional format
is the documentary (Leon, 2007). The documentary
represents a long format for communication using
film (Lee-Wright, 2010), and whales have been the
subject of various documentaries of the past. But
when trying to change attitudes, one potential and
largely unexploited avenue is to use commercials to
communicate scientific facts and change attitudes:
effectively a short form of film-making. The challenging goal of a commercial is to communicate a lot
of information in less than one minute. TV commercials must be entertaining and eye-opening while
getting the main message across and convincing
people that this message is directly relevant to their
own lives (Roberts, 2004; Wiedemann, 2009). People
should feel a part of the issue raised and responsible for solving the problem, or at least wanting to
provide support to anyone who is trying to solve the

problem. If people know more about the environment and are led to believe that they can make a difference specific advocacy campaigns can be effective (Cantrill, 1993).

We live in a visual culture and some NGOs
have long realized the importance of strong marketing and inventive television advertising campaigns. Clearly, images of factory-like whaling ships,
butchered whales and war-like harpoons have
greater visual shock value than the viewfinder
of a camera and whale-watching boat. NGOs
should consider developing less-simplistic messages for consumers of whale-watching (Neves,
2010), while scientists needs to employ less-
complicated communication strategies to advocate
research relevant to sustainable whale-watch management and increase public awareness. It is time
to evolve science communication strategies and
explore the realm where education turns to advocacy: a new whale-watch science communication
commercial framework, or short SciCommercial
(see Figures 24.1 and 24.2).

The proposed SciCommercial is an audiovisual marketing and communication tool that can
be applied to various whale-watch stakeholders
including operators, public, NGOs and resource
managers (see Figure 24.1). Before initiating a particular advocacy campaign including SciCommercials, it is crucial to identify which stakeholders have
the greatest investment, relevant self-interests of
those stakeholders, and each stakeholder's beliefs
relevant to the environment (Cantrill, 1993). The
SciCommercial can advocate for the sustainable
development of whale-watch management frameworks by resource managers, realistic expectations
of whale encounters by the public including whale-
watchers, and responsible business practices by
operators. It may also influence NGOs to evolve
beyond the whale-hunting/watching binary (outlined in the early part of this chapter) towards
responsible advocacy support. Effective whale-
watch communication should provide opportunities to consider similarities rather than differences
to other stakeholders, focus on their motivational
handicaps and provide directions for easily adopted

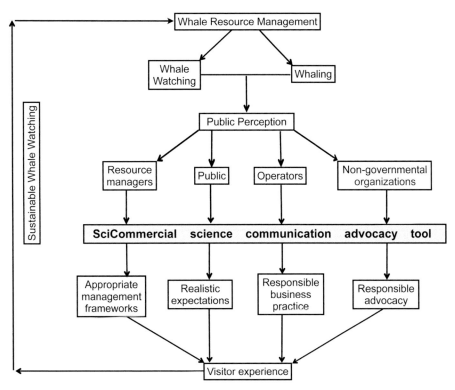

Figure 24.1 Whale-watch stakeholders and potential effects of SciCommercial as science communication advocacy tool for sustainable whale-watch management.

actions supporting the advocated policy (Cantrill, 1993).

The SciCommercial framework merges elements of documentary film-making such as storytelling with techniques used in TV commercials to frame and communicate scientific information on sustainable whale-watching practice (see Figure 24.2). Lakoff (2010) pointed out that there is limited possibility for changing communication frames. Any new language must make sense in terms of already existing systems of frames, work emotionally and be introduced in communication systems that allow for sufficient spread over population, repetition and trust in messenger. Considering prominent whale conservation discourse outlined earlier, the challenge lies in how best to frame (see Figure 24.2) or 're-frame' sustainable whale-watching to the public. One potential narrative would be to use existing anti-whaling language already embedded in the general public's mind and evolve it in terms of communication strategy and content to address the current whale-watching dilemma.

The audiovisual SciCommercial will advocate sustainable whale-watching by blending elements of documentary film-making such as storytelling/ narrative and scientific content with communication strategies and techniques employed by television advertising. Binet and Field (2007) outline key communication strategy models commonly defined by how advertising campaigns work including emotional involvement, fame, information and persuasion. The broad conclusion from the data is that communication models that use emotional appeal (emotional involvement and fame)

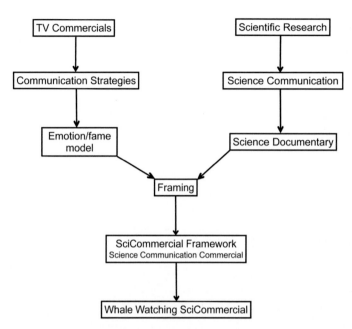

Figure 24.2 Whale-watching SciCommercial framework drawing on film-making elements of television commercials and science documentaries.

are more effective and profitable than rationally based models (information and persuasion). They are better at generating awareness and commitment as well as creating authority (fame) for brands. Fame is not the same as awareness: it is a perception of authority in the category rather than the state of knowledge (Binet & Field, 2007). The data argue strongly for a more widespread adoption of emotion and fame as an influence model and the SciCommercial will utilize this model in order to become more emotionally engaging and relevant to the public (see Figure 24.2).

In the commercial world, campaign objectives are specified in terms of hard results (i.e. sales) rather than intermediate ones (beliefs, attitudes and awareness) and the aim of marketing is almost always to change people's behaviour, not just what they think and feel (Binet & Field, 2007). Apart from economic benefits, hard results in terms of sustainable whale-watching relate to behaviour change. In order to make informed choices, whale-watchers, for example, have to know what to think and

consequently which operator to choose based on sustainable whale-watching practice (the aim of advocacy) not how to think (the aim of education). Sustainable whale-watching has to include hard results for whale conservation, i.e. action for sustainable behaviour change 'within' the industry.

Future empirical research and conclusion

The proposed SciCommercial framework is broadly applicable to any whale-watch stakeholder to influence communication, best practice and marketing of whale-watching (as outlined in the previous section; see Figure 24.1). One apparent chasm outlined in this chapter is the lack of public awareness about scientific concerns relating to sustainable whale-watching practices. Scientists have raised the issue of unrealistic visitor expectations and its negative domino effect on operational practice (Moscardo & Saltzer, 2004; Kessler & Harcourt, 2010). The public (including tourists) needs to know what the whale-watch regulations are and should be given realistic expectations of their encounters (Kessler & Harcourt, 2010). Thus one obvious avenue for future empirical SciCommercial research lies in addressing and managing visitors' expectations.

For whale-watching to become sustainable the scientific evidence suggests that operators need to take a more conservative approach (literally) by operating at a greater distance from the whales so as to lessen the impacts of their presence and accompanying factors such as noise pollution. However, in order to remain competitive with other operators, they must alter the perceptions and demands of the customers by increasing public awareness of the conflict between close-proximity viewing of whales and sustainable practice. The solution then requires that the public's attitude to whale-watching is changed and that they become aware of the need to keep their distance from the whales. Research identified boat behaviour as one of the main impacts of whale-watching (Orams, 2000; Garrod & Fennell, 2004; Weilgart, 2007; Tyack, 2008; Jensen *et al.*, 2009; Chen, 2011; Wright *et al.*, 2011). The whale-watch SciCommercial aims to get the consumer to 'buy into' the concept that successful whale-watching must be experienced from a safe distance for the whales in order to reduce impacts of boats and associated underwater noise.

In sum, scientists can no longer afford to simply assume that information alone can stimulate change towards sustainable whale-watching. Change is more likely if rooted in and driven by significant and meaningful experiences and if a person's heart is in it (Maiteny, 2002). Experience, as a central element of whale-watching, requires sound and emotional science advocacy support including new media of communication and re-framing of scientific language as proposed in the SciCommercial framework. If shown to be successful, such SciCommercials would present an important advocacy management tool for the whale-watching industry as well as a new approach to marketing practices of whale-watching on a micro- and potentially macro-management level.

REFERENCES

Allen, S., Smith, H., Waples, K. & Harcourt, R. (2007). The voluntary code of conduct for dolphin watching in Port Stephens, Australia: Is self-regulation an effective management tool? *Journal of Cetacean Research and Management* 9(2), 159–166.

Bailey, J.L. (2009). Norway, the United States, and commercial whaling. *The Journal of Environment & Development* 18(1), 79–102.

Bailey, J.L. (2012). Whale watching, the Buenos Aires Group and the politics of the International Whaling Commission. *Marine Policy* 36(2), 489–494.

Bejder, L. & Samuels, A. (2003). Evaluating the effects of nature-based tourism on cetaceans. In N. Gales, M. Hindell & R. Kirkwood (Eds), *Marine Mammals: Fisheries, tourism and management issues.* Collingwood: CSIRO Publishing, pp. 229–256.

Bejder, L., Samuels, A., Whitehead, H., *et al.* (2006a). Decline in relative abundance of bottlenose dolphins exposed to long-term disturbance. *Conservation Biology* 20(6), 1791–1798.

Bejder, L., Samuels, A., Whitehead, H. & Gales, N. (2006b). Interpreting short-term behavioural responses to disturbance within a longitudinal perspective. *Animal Behaviour* 72(5), 1149–1158.

Berrow, S.D. (2003). An assessment of the framework, legislation and monitoring required to develop genuinely sustainable whalewatching. In B. Garrod & J.C. Wilson (Eds), *Marine Ecotourism: Issues and experiences.* Wallingford: CABI, pp. 66–78.

Betsill, M.M. & Corell, E. (2001). NGO influence in international environmental negotiations: A framework for analysis. *Global Environmental Politics* 1(4), 65–85.

Binet, L. & Field, P. (2007). *Marketing in the Era of Accountability*. World Advertising Research Centre. Oxfordshire: Alden Press.

Blewitt, M. (2008). Dolphin–human interactions in Australian waters. *Royal Zoological Society of New South Wales Forum Proceedings – Too close for comfort*, pp. 197–210.

Blok, A. (2007). Actor-networking ceta-sociality, or, what is sociological about contemporary whales? *Distinktion: Scandinavian Journal of Social Theory* 8(2), 65–89.

Boykoff, M.T. & Roberts, J.T. (2007). Media coverage of climate change: Current trends, strengths, weaknesses. In *Human Development Report*. New York, NY: United Nations Development Programme.

Bryant, B.I. (1990). *Environmental Advocacy: Concepts, issues, and dilemmas*, Ann Arbor, MI: Caddo Gap Press.

Campbell, J.A. (1974). Charles Darwin and the crisis of ecology: A rhetorical perspective. *Quarterly Journal of Speech* 60(4), 442–449.

Cantrill, J.G. (1993). Communication and our environment: Categorizing research in environmental advocacy. *Journal of Applied Communication Research* 21(1), 66–95.

Carpenter, C. (2001). Businesses, green groups and the media: The role of non-governmental organizations in the climate change debate. *International Affairs* 77(2), 313–328.

Cater, C. & Cater, E. (2007). *Marine Ecotourism: Between the devil and the deep blue sea.* Wallingford: CABI Publishing.

Chen, C. (2011). From catching to watching: Moving towards quality assurance of whale/dolphin watching tourism in Taiwan. *Marine Policy* 35(1), 10–17.

Cisneros-Montemayor, A., Sumaila, U., Kaschner, K. & Pauly, D. (2010). The global potential for whale watching. *Marine Policy* 34(6), 1273–1278.

Constantine, R., Brunton, D.H. & Dennis, T. (2004). Dolphin-watching tour boats change bottlenose dolphin (*Tursiops truncatus*) behaviour. *Biological Conservation* 117(3), 299–307.

Corkeron, P.J. (2004). Whale watching, iconography, and marine conservation. *Conservation Biology* 18(3), 847–849.

Forestell, P.H. & Kaufman, G.D. (1994). *Resource Managers And Field Researchers: Allies or adversaries?* Translated by Great Barrier Reef Marine Park Authority, 17.

Garrod, B. & Fennell, D.A. (2004). An analysis of whale-watching codes of conduct. *Annals of Tourism Research* 31(2), 334–352.

Gjerdalen, G. & Williams, P. (2000). An evaluation of the utility of a whale watching code of conduct. *Tourism Recreation Research* 25(2), 27–37.

Gordon, J. & New Zealand Department of Conservation. (1992). *Effects of Whale-Watching Vessels on the Surface and Underwater Acoustic Behaviour of Sperm Whales off Kaikoura, New Zealand.* Science & research series. Wellington: Department of Conservation.

Greenberg, M.R., Sandman, P.M., Sachsman, D.B. & Salomone, K.L. (1989). Network television news coverage of environmental risks. *Environment: Science and Policy for Sustainable Development* 31(2), 16–44.

Heazle, M. (2004). Scientific uncertainty and the International Whaling Commission: An alternative perspective on the use of science in policy making. *Marine Policy* 28(5), 361–374.

Higham, J.E.S. & Bejder, L. (2008). Managing wildlife-based tourism: Edging slowly towards sustainability? *Current Issues in Tourism* 11(1), 75.

Higham, J.E.S., Bejder, L. & Lusseau, D. (2009). An integrated and adaptive management model to address the long-term sustainability of tourist interactions with cetaceans. *Environmental Conservation* 35(4), 294–302.

Holliman, R.E.W., Scanlon, E., Smidt, S. & Thomas, J. (2009). *Investigating Science Communication in the Information Age: Implications for public engagement and popular media.* Oxford: Oxford University Press.

Hoyt, E. (2001). Whale Watching 2001. Unpublished Report, IFAW and UNEP.

Hughes, P. (2001). Animals, values and tourism – Structural shifts in UK dolphin tourism provision. *Tourism Management* 22(4), 321–329.

IWC (2007). *Report of the Scientific Committee.* International Whaling Commission. SC/59/REP1 presented to the IWC Scientific Committee, 98 pp.

IWC (2011). Five year strategic plan for whalewatching 2011–2016, translated by IWC.

Jensen, F., Bejder, L., Wahlberg, M., Soto, N., Johnson, M. & Madsen, P. (2009). Vessel noise effects on delphinid communication. *Marine Ecology Progress Series* 395, 161–175.

Jickling, B. (2003). Environmental education and environmental advocacy: Revisited. *The Journal of Environmental Education* 34(2), 20–27.

Kalland, A. (1993). Management by totemization: Whale symbolism and the anti-whaling campaign. *Arctic* 46(2), 124.

Kessler, M. & Harcourt, R. (2010). Aligning tourist, industry and government expectations: A case study from the swim with whales industry in Tonga. *Marine Policy* 34(6), 1350–1356.

Lachmuth, C.L., Barrett-Lennard, L.G., Steyn, D. & Milsom, W.K. (2011). Estimation of southern resident killer whale exposure to exhaust emissions from whale-watching vessels and potential adverse health effects and toxicity thresholds. *Marine Pollution Bulletin* 62(4), 792–805.

Lackey, R.T. (2007). Science, scientists, and policy advocacy. *Conservation Biology* 21(1), 12–17.

Lakoff, G. (2010). Why it matters how we frame the environment. *Environmental Communication* 4(1), 70–81.

Lee-Wright, P. (2010). *The Documentary Handbook.* New York, NY: Routledge.

Lemon, M., Lynch, T.P., Cato, D.H. & Harcourt, R.G. (2006). Response of travelling bottlenose dolphins (*Tursiops aduncus*) to experimental approaches by a powerboat in Jervis Bay, New South Wales, Australia. *Biological Conservation* 127(4), 363–372.

Leon, B. (2007). *Science on Television: The narrative of scientific documentary.* Lutton: Pantaneto Press.

Lück, M. (2003). Education on marine mammal tours as agent for conservation – But do tourists want to be educated? *Ocean & Coastal Management* 46(9–10), 943–956.

Lunney, D., Munn, A. & Meikle, W. (2008). Contentious issues in human–wildlife encounters: Seeking solutions in a changing social context. *Royal Zoological Society of New South Wales Forum Proceedings – Too Close for Comfort*, pp. 285–292.

Lusseau, D., Slooten, L. & Currey, R.J.C. (2006). Unsustainable dolphin-watching tourism in Fiordland, New Zealand. *Tourism in Marine Environments* 3(2), 173–178.

Maiteny, P.T. (2002). Mind in the Gap: Summary of research exploring 'inner' influences on pro-sustainability learning and behaviour. *Environmental Education Research* 8(3), 299–306.

Morishita, J. (2006). Multiple analysis of the whaling issue: Understanding the dispute by a matrix. *Marine Policy* 30(6), 802–808.

Moscardo, G. & Saltzer, R. (2004). Understanding wildlife tourism markets. In K. Higginbottom (Ed.), *Wildlife Tourism: Impacts, management and planning.* Altona: Common Ground Publishing, pp. 167–185.

Neves, K. (2010). Cashing in on cetourism: A critical ecological engagement with dominant E-NGO discourses on whaling, cetacean conservation, and whale watching. *Antipode* 42(3), 719–741.

Neves-Graca, K. (2004). Revisiting the tragedy of the commons: Ecological dilemmas of whale watching in the Azores. *Human Organization* 63(3), 289–300.

O'Connor, S., Campbell, R., Cortez, H. & Knowles, T. (2009). *Whale Watching Worldwide: Tourism numbers, expenditures and expanding economic benefits.* Yarmouth, MA: International Fund for Animal Welfare.

O'Neill, F., Barnard, S. & Lee, D. (2004). *Best Practice and Interpretation in Tourist–Wildlife Encounters: A wild dolphin swim tour example.* Gold Coast: CRC for Sustainable Tourism.

Ophuls, W. (2011). *Plato's Revenge. Politics in the age of ecology.* Cambridge, MA: Massachusetts Institute of Technology.

Orams, M. (2000). Tourists getting close to whales, is it what whale watching is all about? *Tourism Management* 21(5), 561–569.

Orams, M.B. & Taylor, A. (2005). Making ecotourism work: An assessment of the value of an environmental education programme on a marine mammal tour in New Zealand. In C. Ryan, S.J. Page & M. Aicken (Eds), *Taking Tourism to the Limits: Issues, concepts and managerial perspectives – Advances in tourism research.* Oxford: Elsevier Science, pp. 83–98.

Powell, R.B. & Ham, S.H. (2008). Can ecotourism interpretation really lead to pro-conservation knowledge, attitudes and behaviour? Evidence from the Galapagos Islands. *Journal of Sustainable Tourism* 16(4), 467–489.

Rethmann, P. (2008). Fantasies at the International Whaling Commission: Management, sustainability and conservation. In L.W. Pauly & W.D. Coleman (Eds), *Global Ordering: Institutions and autonomy in a changing world.* Vancouver: University of British Columbia Press.

Roberts, K. (2004). *Lovemarks: The future beyond brands.* Auckland: Reed.

Russell, C.L. & Hodson, D. (2002). Whalewatching as critical science education? *Canadian Journal of Math, Science & Technology Education* 2(4), 485–504.

Scarpaci, C., Nugegoda, D. & Corkeron, P.J. (2004). No detectable improvement in compliance to regulations by swim-with-dolphin operators in Port Phillip Bay, Victoria, Australia. *Tourism in Marine Environments* 1(1), 41–48.

Schaffar, A., Madon, B., Garrigue, C. & Constantine, R. (2009). *Avoidance of whale watching boats by humpback whales in their main breeding ground in New Caledonia.* Paper SC/34/WW6 presented to the International Whaling Commission Scientific Committee.

Scott, J.M., Rachlow, J.L., Lackey, R.T., *et al.* (2007). Policy advocacy in science: Prevalence, perspectives, and implications for conservation biologists. *Conservation Biology* 21(1), 29–35.

Sironi, M., Leske, N., Rivera, S., Taboada, D. & Schteinbarg, R. (2009). *New regulations for sustainable whalewatching at Península Valdés, Argentina.* Paper SC/61/WW10 presented to the International Whaling Commission Scientific Committee, Portugal.

Skodvin, T. & Andresen, S. (2003). Nonstate influence in the International Whaling Commission, 1970–1990. *Global Environmental Politics* 3(4), 61–86.

Tremblay, P. (2001). Wildlife tourism consumption: Consumptive or non-consumptive? *International Journal of Tourism Research* 3(1), 81–86.

Tyack, P.L. (2008). Implications for marine mammals of large-scale changes in the marine acoustic environment. *Journal of Mammalogy* 89(3), 549–558.

Valentine, P.S., Birtles, A., Curnock, M., Arnold, P. & Dunstan, A. (2004). Getting closer to whales – Passenger expectations and experiences, and the management of swim with dwarf minke whale interactions in the Great Barrier Reef. *Tourism Management* 25(6), 647–655.

Weilgart, L.S.W.L.S. (2007). The impacts of anthropogenic ocean noise on cetaceans and implications for management. *Canadian Journal of Zoology* 85(11), 1091–1116.

Weingart, P., Engels, A. & Pansegrau, P. (2000). Risks of communication: Discourses on climate change in science, politics, and the mass media. *Public Understanding of Science* 9(3), 261–283.

Wiedemann, J. (2009). *Advertising Now. TV commercials.* London: Taschen.

Wiley, D.N., Moller, J.C., Pace, R.M. & Carlson, C. (2008). Effectiveness of voluntary conservation agreements: Case study of endangered whales and commercial whale watching. *Conservation Biology* 22(2), 450–457.

Williams, R., Lusseau, D. & Hammond, P.S. (2006). Estimating relative energetic costs of human disturbance to killer whales (*Orcinus orca*). *Biological Conservation* 133(3), 301–311.

Williams, R., Bain, D.E., Smith, J.C. & Lusseau, D. (2009). Effects of vessels on behaviour patterns of individual southern resident killer whales *Orcinus orca. Endangered Species Research* 6(3), 199–209.

Wright, A.J., Deak, T. & Parsons, E. (2011). Size matters: Management of stress responses and chronic stress in beaked whales and other marine mammals may require larger exclusion zones. *Marine Pollution Bulletin* 63(1–4), 5–9.

Zeppel, H. (2008). Education and conservation benefits of marine wildlife tours: Developing free-choice learning experiences. *Journal of Environmental Education* 39(3), 3–17.

25

Time to rethink

Fostering the nascent 'sustainability paradigm'

James Higham, Lars Bejder and Rob Williams

This book addresses human interactions with cetaceans in the wild, at the heart of which lies the considerable challenge of sustainable management. The urgency of this subject arises from spectacular growth in demand and widespread evidence of unsustainable management of tourist interactions with cetaceans. The foregoing chapters serve to highlight the complex interplay of the macro- (global), meso- (national/regional) and micro-level (local/site-specific) policy, planning and management settings. The dynamic nature of these contexts combined with the urgent need for integrated and adaptive management approaches are most evident (Higham *et al.*, 2009). Given the current failing of the long-term sustainable management of many whale-watching activities, a new whale-watching paradigm is clearly required (Lusseau *et al.*, 2013). Any such paradigm must be informed by change at three spatial scales of analysis (Higham *et al.*, 2009).

While attending to the management of tourist interactions with cetaceans, it is necessary to understand the policy, planning and management inputs which take place at the macro- (global), meso- (national) and micro (local–regional) levels at specific sites in different parts of the world (Figure 25.1). The *macro-level* context is characterized by growing concerns for declining levels of global biodiversity and the increasing instability of complex ecosystems (Tilman, 1999; Worm *et al.*, 2009). In the special case of the large whales, we are not talking about recent human activities degrading an otherwise pristine system. On the contrary, we are talking about recent human activities that may be hindering the recovery of many populations from the near-extinction caused by over-harvest during twentieth-century industrial whaling. Species and population changes in the marine environment have been influenced by human activities that result in such phenomena as depleted fisheries, global climate change and marine pollution (Gössling & Hall, 2006). Various inter-governmental panels, non-government organizations, international environmental groups and media continue to discuss and debate concerns surrounding the issues of global environmental change, biodiversity conservation and habitat protection (Table 25.1). In 2006, for example, the International Whaling Commission (IWC) actively addressed impact issues associated with whale-watching (see Chapter 6), responding to empirical research and engaging in dialogue with IWC member countries where concerns for sustainable tourist interactions with cetaceans exist (Higham *et al.*, 2009). Global governance agencies and organizations such as the IWC, the Intergovernmental Panel on Climate Change (IPCC), the International Union for the Conservation of Nature (IUCN; e.g. the Red List) and prominent environmental non-government organizations (eNGOs), must strive to exert the effective influence

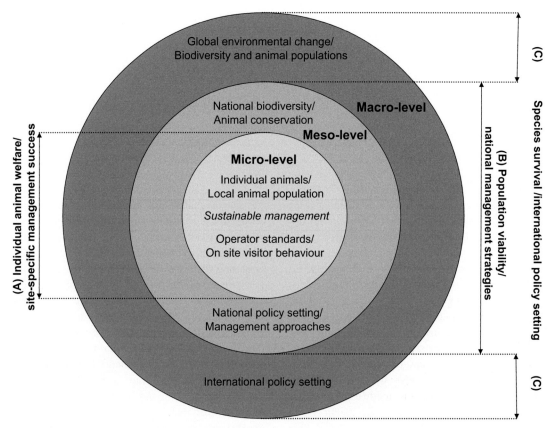

Figure 25.1 Macro-, meso- and micro-level policy and planning contexts (*source*: Higham *et al.*, 2009: 296).

of the macro-level upon the meso- and micro-levels of policy, planning and management. The challenges of doing so are considerable (Neves, 2010).

Within the global or macro-level setting, the *meso-level* context includes the establishment of national policy, planning and management priorities relating to tourism development and the conservation of marine mammals. Since the 1970s, the development of Marine Mammal Protection Regulations (MMPRs) pre-dates the growth of large-scale commercial whale-watching (Lusseau, 2003). These MMPRs were designed to address particular management needs that demanded immediate action; namely, the unsustainable removals of dolphins through by-catch in tuna fisheries. It is fair to say that the MMPRs have not yet kept pace with

the science emerging on the cumulative effects of sublethal removals (see Chapter 2). Subsequent growth in whale-watching has been so swift that planning and management agencies have been poorly prepared or otherwise unable to adequately meet the challenge of sustainable management. After all, the benign image of whale-watching hardly looks like an activity as demanding of management intervention as the sight of dolphins drowning in seine nets chasing schools of tuna with catcher boats and helicopters.

In developing countries, recent and rapid growth in whale-watching activities represents a major challenge because advanced public policy mechanisms are generally not in place (see Chapter 21). Furthermore, in different parts of the

Table 25.1 Macro-, meso- and micro-level factors influencing the management of human interactions with cetaceans (adapted from Higham *et al.* 2009: 295).

Dynamic factors	Global environmental change Environment/ecology	Socioeconomic/ Geopolitical
Macro-level factors	Global biodiversity Species population levels Global climate change Pollution of the oceans – Noise pollution – Chemical pollution – Oil and gas exploration Marine transportation – Vessel strikes	International policy setting – UNESCO world heritage – The World Conservation Union (IUCN) – International Whaling Commission (IWC) – International eNGOs – Environmental groups Inter-governmental agreements – Oceans Policies – Reduction of carbon emissions International media Public interest/demand
Meso-level factors	National biodiversity Rare/endangered status Resource use conflicts Animal population conservation – Animal mortality/morbidity – By-catch – Pollution	Government/Policy setting – Environment policies – Conservation policies – Economic policies National identity Indigenous/First Nations peoples National environmental and conservation lobby groups School education programmes
Micro-level factors	Individual animal welfare – Disease – By-catch Local population survival – Reproductive success – Population biology Local ecology – Species fluctuations – Food chain stability – Predator/prey relations	Management approach Tour operators – Business models Visitor demand – Visitor satisfaction – Visitor education programmes Private/recreational use Local community interests – School education programmes – Local environmental stewardship Research community – Social science community – Natural science community – Researcher impacts

world, whale-watching exists in contrasting social, cultural, economic and political contexts. These contrasts extend to the historical relationships with cetaceans, which despite a groundswell of disdain towards whale-hunting may represent deeply entrenched expressions of political autonomy and cultural identity (Levine & Levine, 1987; Sawada & Minami, 1997). In some such cases, traditional whale-hunting may form part of the eco-tourist experience (Hinch, 1998). Whale-watchers in Norway and Iceland may (unsuspectingly or otherwise) witness the killing and flensing of whales

(Aftenposten, 2006). Responses to such visions are socially and culturally constructed, some of which may elicit sentiments of revulsion or interpretations of a long-standing and intriguing expression of local coastal culture (Ris, 1993; Smested, 1997).

The meso-level context varies greatly between countries, regions and local communities where concerns vary markedly, including but not limited to national biodiversity, conservation, animal ethics and resource use, government priorities (relating to economic and social development), employment creation, environmental protection and conservation, as well as cultural and environmental values. Nonetheless, the dominant meso-level context is generally one of non-existent, inadequate or dated policy settings for the management of marine tourism (Corkeron, 2004; Lusseau, 2003). Policy and planning responses aimed at sustainable long-term tourism management have been hindered by political priority given to social and economic development through tourism (Higham & Lusseau, 2007; Lusseau *et al.*, 2013), at the expense of sustainable management of natural resources. Meanwhile, the mortality or morbidity rates in focal animal populations may change due to extreme weather events, climate change, changing predator/prey relationships, the availability and accessibility of food sources, pollution and bacterial disease (Shelton & McKinley, 2007; Lusseau & Higham, 2004). However, they are rarely, if ever, accommodated in existing management approaches which lack the adaptability to respond to immediate, short- or medium-term changes in the mortality or morbidity of local or regional populations of wild animals (Higham *et al.*, 2009).

Finally, the *micro-level* context focuses on site-specific issues where whale-watch management needs are particularly pressing (e.g. Bejder *et al.*, 1999, 2006; Constantine, 2001; Corkeron, 2004; Lusseau, 2003; Williams *et al.*, 2002, 2006). In recent years, researchers and management agencies alike have been consumed with trying to understand the causes and consequences of human impacts upon cetaceans in the wild. Meanwhile, the whale-watch industry has grown in a manner that is almost entirely unchecked and largely unmanaged (Hoyt, 2011; Corkeron, 2004). There are important exceptions to the rule. In Kaikoura (*Aotearoa*/New Zealand), whale-watching is emblematic of major economic transition, community revival and cultural renaissance (see Chapter 22). Similarly, at Cape Ann Whale Watch (CAWW) in Gloucester, MA (USA), 20 years of collaboration between tourism operators and scientific researchers have informed all aspects of sustainable human interactions with cetaceans (see Chapter 10). In Canada, commercial whale-watchers that focus on killer whales were instrumental in identifying and protecting gravel beaches that the whales use for beach-rubbing; without pressure from whale-watching, the Robson Bight (Michael Bigg) Ecological Reserve would not exist as a voluntary, no-entry Marine Protected Area (Ashe *et al.*, 2010; Williams *et al.*, 2006, 2009). In Washington State (USA), the commercial whale-watching fleet responded quickly to science that showed that particular styles of whale-watching elicited strong evasive reactions from killer whales, and abandoned that style of whale-watching long before any laws required the industry to do so (Williams *et al.*, 2002). Understanding the immediate, medium- and long-term consequences of tourist interactions with cetaceans, both at the individual and population levels, has been the highest order of priority. These interests have been well served by the naturalistic/positivist research tradition which has dominated the field to date. This volume, however, highlights the parallel need for the sustainable whale-watch paradigm to be informed by critical analysts engaging a subjectivist epistemology and applying the research methodologies of the social sciences (e.g. sociology, social anthropology, political science). For example, action research, which engages social scientists with those who play a part in the sustainable interplay of people with cetaceans at specific sites, is required to lead change and inform the drive towards sustainability at the micro-level. Such research must see cetaceans themselves as stakeholders in, not merely the subjects of, sustainable whale-watching.

Rethinking (sustainable) whale-watching

Global governance

Rethinking (sustainable) whale-watching to inform a new paradigm must occur within all parts of the system outlined in Chapter 1 (see Figure 1.1). Various critical elements in the whale-watching system (e.g. policy settings, planning approaches, management strategies, business strategies) must be informed by rigorous ecological science. In the first instance, global politics bears considerable influence over the prospects for sustainable whale-watching. It has been stated ad infinitum that whales are highly political (see Chapter 5). Neves (2010: 728) observes that the dominant conservation discourse situates whale-watching in distinction to whale-hunting, which greatly hinders 'the emergence of a more critical view of this activity in places where practices are more closely dictated by short-term profit goals than by environmental concerns' (Neves, 2010: 728). She argues that in the current late capitalist society, cetaceans have been transformed from providers of material commodities to providers of 'services' through commercial marine tourism development. The influence of commercial (business) and regional economic development interests 'silences alternative voices' (Neves, 2010), which may question the ecological effects of whale-watching. This elevates the need for ecological science to be conducted in different environmental, socio-cultural and developmental contexts (Higham & Lusseau, 2007), and furthermore, for eNGOs to actively and decisively intervene in cases of ecologically unsustainable practice. Yet many eNGOs steadfastly hold the unitary and uncritical belief or position that whale-watching is more desirable than whale-hunting (see Chapters 1 and 4). Such a position negates a critical analysis of whale-watching, leading to paralysis of intervention in cases of clearly unsustainable whale-watching practices. Neves (2010: 737) succinctly states that important 'distinctions will remain invisible so long as whale watching is presented and approached in homogenized form'.

While this remains the starting position in discussion of the merits of whale-watching, critical debate will be inhibited and sustainability interests constrained.

Biological science

While set within the broad context of global environmental change, conservation of the world's oceans and the loss of marine biodiversity, the focus of biological science has been at the local, site-specific level. Fortunately, with the rapid development of commercial whale-watching has come the concerted scholarly effort aimed at understanding the impacts of tourist interactions with cetaceans in the wild, and the management initiatives that are required to mitigate those impacts (Bejder et al., 2006). The potential impact of whale-watching has been studied for over two decades (Baker & Herman, 1989; Corkeron, 2004), during which time a wide variety of short-term effects has been detected on many species (e.g. Au & Green, 2000; Bejder et al., 1999; Hastie et al., 2003; Lusseau, 2006; Nowacek et al., 2001; Van Parijs & Corkeron, 2001; Williams et al., 2002). These short-term effects include changes in vocalization patterns, respiration patterns, variation in path directedness and other short-term behavioural alterations resulting from apparent horizontal and vertical avoidance tactics (Frid & Dill, 2002). However, it has been difficult to move from the description of short-term changes, which sometimes appear contradictory, to a comprehensive understanding of the biological relevance of these effects (Corkeron, 2004; Bejder et al., 2006).

Set within a positivist research paradigm and guided by a realist ontology, these studies are framed within immutable natural laws that allow for the articulation of findings that can be generalized (Guba, 1990). However, in this field, generalizations and transferability cannot be confused. Wildlife disturbance studies have shown that interpreting behavioural responses outside of the biological and ecological context in which they are studied may be uninformative at best and, worse still, misleading (Beale & Monaghan, 2004; Gill *et al.*, 2001; Bejder *et al.*, 2009). The body of research also highlights the difficulties inherent in extrapolating findings across short-, medium- and long-term temporal scales. In recent years, some short-term avoidance tactics have been shown to have long-term consequences for individuals and their populations (Bejder *et al.*, 2006; Foote *et al.*, 2004; Lusseau, 2005; Lusseau *et al.*, 2006; Williams *et al.*, 2006). These studies have taken a multi-contextual approach to elucidate the mechanisms linking short-term avoidance tactics to long-term impacts. Comparisons between control and impact sites with long-term life-history data have revealed how whale-watching disturbance has short-term effects on the lives of cetaceans that can lead to long-term consequences for the viability and fitness of individuals and their populations.

Some studies also show that populations respond non-linearly to impacts (see Chapter 19). There are thresholds which, if exceeded, may lead to greater impacts on the viability of populations exposed to whale-watching. In the traditions of the Limits of Acceptable Change (LAC) management planning framework, understanding these thresholds and the triggers that determine progression from one to the next are critical to sustainability (Duffus & Dearden, 1990). Urgent gaps and challenges clearly remain: how best to advance the ecological science required to inform sustainable management; which financial models are available to fund the necessary science; which business models are most appropriate to this unique operational context; and how can management agencies best receive coherently and act decisively upon rigorous science (Higham & Bejder, 2008)? Furthermore, what is the best approach to empower local communities and, indeed, the most centrally located stakeholders of all: the cetaceans themselves, who are largely removed from stakeholder discourses and analyses beyond the field of animal ethics. Where do whales and dolphins now stand among the manifold stakeholders who influence, or are influenced by, the late capitalist model which privileges the unrestrained sale and consumption of whale-viewing 'services' (Neves, 2010)?

The production and consumption of viewing experiences

Knight (2009) presents an important problematization of the wildlife viewing experience. In doing so, he accommodates multiple stakeholder perspectives, including those who produce the experience (the business), those who consume it (the visitor), and those who are 'consumed' (the focal animals). He notes that wildlife viewing refers 'not to distant, fleeting glimpses of retreating animals, but to close-up, protracted observation of animals who stand their ground' (Knight, 2009: 171). With proximity as a key determinant of satisfaction (Muloin, 1998), proximity variables have been shown to include eye contact, touching distance, physical contact, and/or holding (Servais, 2005; Gilders, 1995; Kertscher, 2000, cited by Knight, 2009). Thus, the production of wildlife viewing experiences on the part of business operators and the mediation of experiences by guides and interpreters will inevitably centre on proximity variables as determinants of visitor satisfaction. The marketing of such experiences gives rise to unrealistic and, in some cases, physically unobtainable expectations. The presentation of images that depict impossible viewing opportunities places the promise of immediate and prolonged proximity to wildlife to the forefront of visitor expectations.

The role of eNGOs in the production of cetacean viewing experiences must also be critiqued. Advocating for whale-watching in such a way that exerts pressure to locate visitors in close proximity to whales perpetuates the capitalist production of cetacean viewing (Neves, 2010). The use of powerful imagery in online media has been central to this imperative (see Chapter 24). Images of visitors in close and apparently benign interaction with whales on the surface or the water (in some instances set alongside images of whales being flensed) perpetuate the homogenized form of whale-watching which Neves (2010) has thoroughly and most usefully critiqued. The result at many whale-watching sites is not unlike other popular wildlife viewing locations in which a continual sequence of visitors arrive to and depart from the immediate proximity of focal animals (Muloin, 1998; Spong & Symonds, 2003; Bearzi, 2003). Influenced by product marketing, powerful imagery and the resulting expectations, the consumption of wildlife viewing experiences is assumed to involve close, intimate and prolonged interactions with wild animals. Under this scenario, sustainability is inevitably called into question.

These discourses raise important and timely questions relating to the production and consumption of whales. It has been demonstrated that the late capitalist production of whale 'services', as described by Neves (2010), gives rise to serious concerns for sustainability. In this respect, it may be argued that unsustainable whale-watching has little to differentiate it from whale-hunting. Such arguments may then extend to the right of small and often remote, poor and underdeveloped communities to use all resources at their disposal, including whales, in any way to sustain themselves (Bailey, 2012). Under this set of circumstances, asserting the right to hunt whales may also represent a manifestation of autonomy, resistance to neo-imperialism and/or the survival and renaissance of indigenous cultural practices. 'Such framing can be very powerful' (Bailey, 2012: 490). The counter-argument is that whale-watching may offer an avenue of sustainable, long-term economic development that offers a means to develop and sustain poor, remote/peripheral communities (Hall & Boyd, 2003). According to Bailey (2012: 490), this 'may provide a moral counterweight to the powerful arguments used by whalers in defense of their activities'. The key is that whale-watching *must be sustainable*. This requires that the production and consumption of whale experiences are comprehensively and critically inspected and reframed so as to move from the late capitalist paradigm onto a sustainability pathway.

The science–policy nexus

The degree to which 20 years of science has informed sustainable whale-watch policy and planning has been limited to date. This is not due to the lack of rigorous science that is available to inform a sustainability paradigm. Other factors have intervened: the very rapid development of whale-watching from humble origins; the swift development of commercial business models; the absence of policy frameworks for sustainable development; the under-resourcing of management agencies; and the failure of science communication. In terms of inputs and outputs, the complexity of human–animal interactions is also a significant factor. Duffus and Dearden (1990) contend that this complexity arises from the interplay of three key components of the wildlife experience: site users, focal wildlife species (both individual animals and local animal populations) and the wider ecology of the viewing site. Duffus and Dearden (1990) also served initial warning of the dynamic temporal dimensions of wildlife tourism, insofar as tourist interactions with wildlife animals vary within diurnal, seasonal and life course timeframes (see Chapter 19) (Higham & Shelton, 2011). Science that addresses the complex spatiotemporal dimensions of animal ecology must likewise inform sustainable whale-watching. Evidently, until the cost of whale-watching can be simply understood (i.e. this number are effectively killed by sublethal disturbance), it will remain difficult for research managers to evaluate sustainability.

The current shortcomings of whale-watch management in many parts of the world arise from the fundamental failure of policy and planning agencies to develop and implement adequate legislation. The politics of affording adequate protection to critical marine habitats are fundamental to this failure. The designation of marine protected areas (MPAs) is central to protecting wild animal populations, and access to MPAs for wildlife viewing must be addressed in policy and planning arrangements. The science is available to inform the appropriate designation of MPAs based on 'identifying critical habitat for threatened populations, and ensuring meaningful protection' (Williams *et al.*, 2009: 709). The science highlights that '[a] recurring challenge for cetacean MPAs is the need to identify areas that are large enough to be biologically meaningful while being small enough to allow effective management of human activities within those boundaries'

(Ashe *et al.*, 2010: 196). Ashe *et al.* (2010: 196) also note that MPAs 'present an attractive option to mitigate impacts of anthropogenic activities, but they run the risk of tokenism if placed arbitrarily'. These failures are demonstrated most acutely in instances where scientific information relating to all human activities that impact the mortality and morbidity of cetaceans is available but not acted upon (Bearzi, 2003). In such cases, it is critical to afford animal populations maximum protection where responsiveness to anthropogenic disturbance is greatest. These areas may be spatially and temporally unstable (Lusseau & Higham, 2004), highlighting the need for spatiotemporal protection measures to be monitored over time.

Following Duffus and Dearden (1990), some species of wild animals and some individual animals demonstrate a well-established response of inquisitiveness to humans or signs of human activity. Other animals are intensely private, remain concealed from onlookers and flee readily from human interaction. However, a behavioural formulation of human–wildlife interactions must accept the imperfect relationship between observable animal behaviour and internal state (Higham & Shelton, 2011). The apparent tolerance that some wild animals display towards human approach and observation cannot be interpreted as absence of negative impact (Ellenberg *et al.*, 2006; Bejder *et al.*, 2009). This is particularly applicable to animals engaged in critical behaviours, such as female cetaceans with calves (Bejder *et al.*, 2006). The science that is required to inform these understandings is now widely available. The widespread failure to engage with and act upon such science contributes to the continuing inadequacy to move onto a sustainability pathway (see Chapter 24).

Management: enacting a suite of interventions

Given the significant threats facing the viability of global cetacean populations (see Chapter 2), of which whale-watching is but one, governments have come somewhat belatedly to accept the urgency of management intervention (Gjerdalen & Williams 2000; Garrod & Fennell, 2004). Approaches to whale-watch management, according to Allen *et al.* (2007: 161), 'ranges from government regulation, to a variety of guidelines and codes of conduct, to no management whatsoever' (see Chapter 17). Such variability has been described as 'haphazard' and 'highly fragmented' (Garrod & Fennell, 2004). The debate has circulated as to the merits of different management approaches and interventions. It is now evident that the sustainability paradigm must be informed by a suite of management interventions.

The relationship between proximity to wild animals and visitor satisfaction has been documented (Knight, 2009). A competitive advantage may be achieved by commercial businesses who provide their passengers with high and prolonged proximity to animals. This point is central to the widespread failure of sustainability. Codes of conduct and voluntary guidelines have been implemented at many whale-watch sites. They offer the advantage of swift development and implementation in cases where the legislative context is inadequate and the pace of change is rapid (Garrod & Fennell, 2004). However, Garrod and Fennell's (2004: 349) analysis of whale-watch codes of conduct generated results that were not 'especially encouraging, either from the perspective of the further development of the sector (and those holding a stake in its growing socioeconomic benefits) or from the viewpoint of cetacean conservation.' Allen *et al.* (2007) report on the adoption of a voluntary code of conduct to oversee dolphin-watch operations in Port Stephens, New South Wales (Australia) in 1996. Their analysis reported high operator compliance for four variables: number of dolphin watching boats per dolphin school; time spent by individual operators with dolphins; method of approach to dolphins; and frequency of cruises conducted per day. However, a number of operators failed on various counts, including the ability to discriminate between groups of dolphins with and without calves and boat-handling around dolphins. This study confirmed that voluntary codes and guidelines do not work in isolation or without monitoring of compliance.

The sustainability pathway requires more extensive protection and management measures (Garrod & Fennell, 2004). Such measures must include compliance checks, monitoring and enforcement, all of which are critical aspects of all management regimes. In many cases, voluntary codes fail to achieve the necessary levels of compliance (Allen *et al.*, 2007). In cases where legally binding management regimes are in place, non-compliance is related to an absence of law enforcement (Jett & Thapa, 2010). The designation of MPAs is a critical first step towards effective protection measures for cetaceans (Hooker & Gerber, 2004) and must reflect the biology of target species (Wilson *et al.*, 2004; Ashe *et al.*, 2010). The effectiveness of MPA protection measures requires careful planning in response to the manifold stressors that contribute to cetacean morbidity and mortality (see Chapter 2), effectively communication and appropriate enforcement (Lusseau & Higham, 2004).

Business models and community integration

Terms such as 'green economy', 'triple bottom line' and 'ecopreneurship' are no longer based upon shallow rhetoric. These terms describe business ethics founded on more than financial performance. The philosophy of the green economy is to achieve low-carbon and socially inclusive sustainable growth, while ecopreneurs place contributions to environment and society ahead of profit margins. Traditional business models centred on profit margins and returns to shareholders are now being openly questioned. These emerging business philosophies recognize the new reality, that business must demonstrate responsibility for negative outputs. Fundamentally, this involves a clear commitment to mitigation of negative externalities. Social media have empowered public scrutiny of business performance, and this will continue to be the case in in the future.

In reference to commercial whale-watching interests, Neves (2010: 721) recognizes that 'business models are a prime factor in shaping how whale watching is designed and practiced in specific contexts'. Her analysis interrogates the 'implicit assumption' that 'whale watching companies have the right to profit from putting humans in contact with the animals that provide these services' (Neves, 2010: 733). This reflects the antiquated assumption of the right to kill whales without restraint for commercial profit. Efforts to position whale-watching as contributing to marine conservation may be seen as a means by which to cement this assumption. Employing a comparative analysis of two business cases in the Canary Islands and the Azores, Neves (2010) demonstrates that some business models are more attuned to social and ecological externalities and outputs than others. The case from the Canary Islands is characterized by business models based on high volume, mass consumption, cost competitiveness and the high/fast throughput of visitors. In contrast, the Azorean case is carefully regulated to guard against the rapid development of high volume visitor operations. The tourism systems contrast in terms of accessibility and connectivity, volume of visitor arrivals and course of tourism development, all of which underpin the development of the respective whale-watch business models (Neves, 2010). While these systems are contrasted in terms of sustainable business operations, it remains questionable how the local community benefits in both instances (Neves, 2010).

Business models are critical to a nascent sustainable whale-watching paradigm (Neves 2010). This is so not only in terms of sound ecological practice, but also in terms of meaningful integration with wider constituents. Such constituents may include indigenous peoples, scientific communities, policy-makers, resource managers, environmental groups, school children and other resident groups, all of which may seek to contribute to the stewardship of local resources. The flow of economic and non-economic benefits to local constituents is critical to the sustainability pathway, as opposed to the 'implicit assumption' of the right to profit from whale 'services' (Neves,

2010). In this respect, alternative rather than traditional business models may be more appropriate. Not-for-profit trusts offer one such alternative, as they are typically community-owned, overseen by an elected board, and committed to reinvestment in local conservation projects. Indigenous business models present another alternative. Whale Watch Kaikoura (*Aotearoa*/New Zealand) offers one example of a powerful and successful indigenous business model, one that is firmly centred upon long-term ecological sustainability and intergenerational wealth creation. Consequently, business success is measured not only in financial terms, but equally in terms of the cultural renaissance, wider community well-being, and protection of *taonga* (treasures), which in this case refers to the whales (see Chapter 22).

Tourists and environments

Appropriate business models should also be designed with the types of tourists that businesses seek to serve, the qualities of experience that they are committed to providing, and the relationship between the visitor experience and marine conservation. The failure to communicate the science addressing the impacts of visitors upon cetaceans is a critical challenge to be addressed (see Chapter 24). Negative ecological outputs are the most studied aspect of the whale-watch phenomenon; yet visitors still want to get closer to whales and dolphins (see Chapter 9), and eNGOs continue to do little to counter these misconceptions. Unregulated whale-watching occurs in places such as Greenland, where high-volume/high-speed encounters with repeatedly targeted individual animals (Boye *et al.*, 2010) represent an exploitation against which tourists must be empowered to respond. Given the importance of the tourism–conservation nexus, it is remarkable how few academic studies inform this aspect of whale-watching.

Tourists, knowingly or otherwise, are central components in the drive for proximity to animals. They are empowered to discern between sustainable and unsustainable practices. With the support of eNGOs, policy-makers, resource managers and local communities, businesses must find effective avenues to inform visitors of the potential adverse impacts of human–wildlife interactions, not only for the sake of the focal animals, but also from the business and scientific standpoint with the aims of: (a) reducing avoidance, flight or retreat responses, (b) mitigating adverse impacts on animals that may otherwise discontinue critical, site-specific behaviours (e.g. by instigating site abandonment); and (c) promoting sustainable visitor experiences. Targeting visitors who seek to be engaged rather than entertained is central to this approach. Delivering demonstrably sustainable experiences becomes the competitive advantage.

Increasingly engaged tourists seek to be assured that their mere presence at wildlife viewing sites (and their associated behaviours) is not to the detriment of the animals that they seek to experience, either in terms of the welfare of individual animals or wider population fitness (Muloin, 1998; Higham & Lusseau, 2004). They also seek to be educated (Lück, 2003; see also Chapters 10 and 11). Empowering visitors as environmental observers is critical to the sustainability paradigm. However, sentiments of stewardship must extend beyond the on-site experience if visitors are to move from short-term experiences to long-term marine conservation. The sustained engagement of visitors as environmental ambassadors represents a significant challenge that whale-watch businesses are yet to adequately address (see Chapter 10).

Conclusion: the urgency of meaningful protection

This book addresses the pressing need for whale-watching to move onto a sustainability pathway. In order to do so, the most urgent priority is to afford cetaceans meaningful protection from causes of significant anthropogenic impact, of which commercial tourism is only one. A foremost challenge is to develop widespread and effective management that monitors tourist activities, establishes appropriate thresholds of those activities, and responds to the impacts of tourist activities, thereby becoming

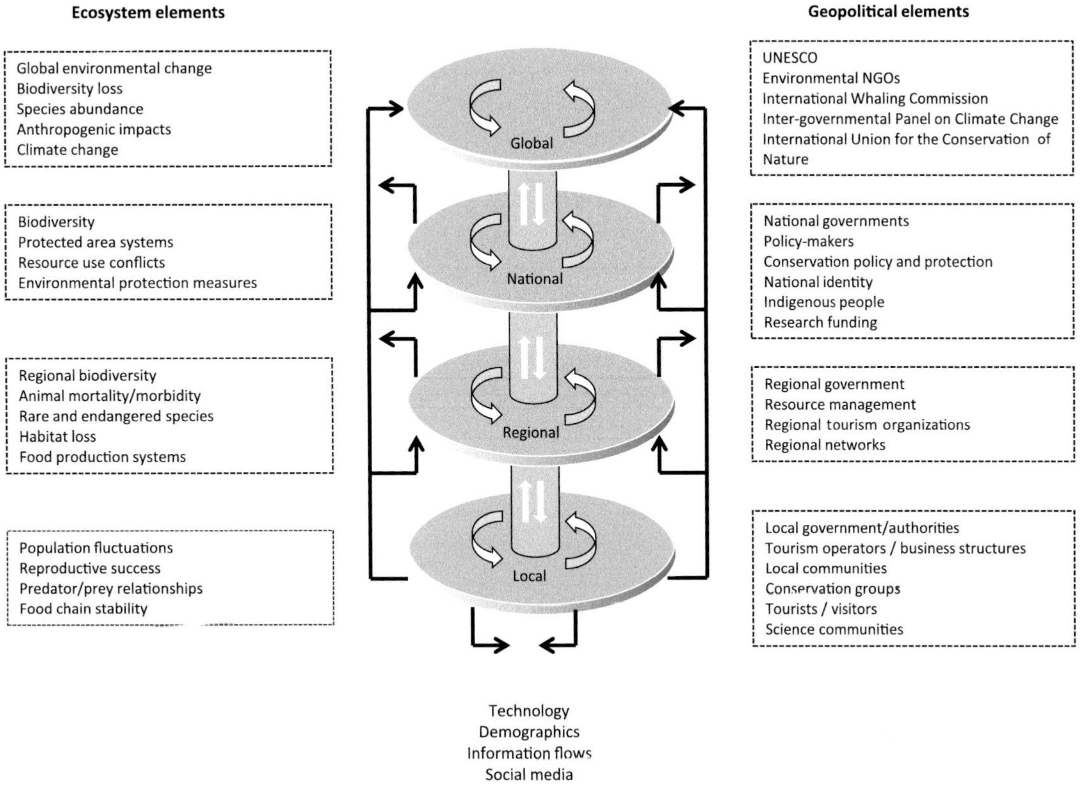

Ecosystem elements

Global environmental change
Biodiversity loss
Species abundance
Anthropogenic impacts
Climate change

Biodiversity
Protected area systems
Resource use conflicts
Environmental protection measures

Regional biodiversity
Animal mortality/morbidity
Rare and endangered species
Habitat loss
Food production systems

Population fluctuations
Reproductive success
Predator/prey relationships
Food chain stability

Global

National

Regional

Local

Geopolitical elements

UNESCO
Environmental NGOs
International Whaling Commission
Inter-governmental Panel on Climate Change
International Union for the Conservation of
Nature

National governments
Policy-makers
Conservation policy and protection
National identity
Indigenous people
Research funding

Regional government
Resource management
Regional tourism organizations
Regional networks

Local government/authorities
Tourism operators / business structures
Local communities
Conservation groups
Tourists / visitors
Science communities

Technology
Demographics
Information flows
Social media

Figure 25.2 Cetaceans, tourism and the global–local nexus (*source*: adapted from Milne & Ateljevic, 2004).

actively adaptive to change over time (Higham *et al.*, 2009). This requires leadership and a strong commitment to sustainability, which involves a range of stakeholders interacting at various cross-cutting levels (Figure 25.2). While such interactions have hitherto been uncommon, it is critical that they are broadly and diligently developed. There exists a critical role for science, specifically in terms of the use of biodiversity data and spatiotemporal ecology studies to inform systematic conservation planning and the designation of a global–local network of MPAs (Hoffman *et al.*, 2008; Hoyt, 2011). While significant progress has been made in the development of terrestrial protected areas (such as National Parks and World Heritage Areas), protection of the marine environment remains woefully neglected and shamefully inadequate. Protection measures must extend to area–time closures (see

Chapter 19) to provide animals sufficient relief from anthropogenic pressures. Until such time, it can be argued that the sustainability paradigm has been deliberately ignored or resisted, so as to perpetuate the development interests of the neo-liberal economic agenda.

Meanwhile, new challenges are rapidly emerging. In many parts of the world, whale-watching remains largely or entirely unregulated and/or poorly or ineffectively managed. Mustika *et al.* (2012) observe that more than half (54.7%) of all nations with whale-watch industries are now developing world countries. While the International Whaling Commission has called for codes of practice to be developed and implemented in countries that participate in cetacean watching (see Chapter 6), the limited regulatory settings that exist in most developing world contexts, combined with limited capacity to deliver effective compliance (Mustika *et al.*, 2012), represent a significant and pressing concern (see Chapter 21).

This book addresses a number of paradigms that define the relationships that exist between humans and cetaceans. The two prevailing (competing) paradigms may be termed 'extractive' (hunting) and 'appreciative' (watching). Terms such as 'exploitative' and 'consumptive' must be used advisedly. Caution is required given that critical analysts have come to see that these paradigms share more traits in common than those that differentiate them (Lemelin, 2006; Knight, 2009). These paradigms are not polar opposites and in some cases actually coexist. Reference is made to the *Capitalist transformation paradigm* (Neves, 2010), which argues a transition from physical extraction to the selling of 'services' (visual experiences), both of which must be recognized as exploitative and consumptive in different ways and to varying degrees. This book, ultimately, makes the case for a new pathway, which we refer to as the *sustainability paradigm*.

Central to the *sustainability paradigm* is the recognition of intrinsic worth: a philosophical position that argues that to some degree, cetaceans should be afforded protection for their own sake, and left to their own devices, free for significant periods of their lives from the insatiable human demands. The sustainability paradigm also requires spatial and temporal measures to limit and manage interaction and intrusion. In order for sustainability outcomes to be achieved, this book calls for an inclusive, multi-stakeholder approach to achieving in full measure the emerging sustainability. Until such time, there will remain a sense of uncertainty as to which paradigm will prevail. We hope that the reader takes up the challenge to actively inform the sustainability discourse and contribute to resolving this uncertainty.

REFERENCES

Aftenposten.no (2006). *Whale shot in front of tourists*. 5 July 2006. Retrieved from: http://www.aftenposten.no/ english/local/article1376980.ece (accessed 27 November 2009).

Allen, S., Smith, H., Waples, K. & Harcourt, R. (2007). The voluntary code of conduct for dolphin watching in Port Stephens, Australia: Is self-regulation an effective management tool? *Journal of Cetacean Research & Management* 9, 159–166.

Ashe, E., Noren, D.P. & Williams, R. (2010). Animal behaviour and marine protected areas: Incorporating behavioural data into the selection of marine protected areas for an endangered killer whale population. *Animal Conservation* 13, 196–203.

Au, W.W.L. & Green, M. (2000). Acoustic interaction of humpback whales and whale-watching boats. *Marine Environmental Research* 49, 469–481.

Bailey, J.L. (2012). Whale-watching, the Buenos Aires Group and the politics of the International Whaling Commission. *Marine Policy* 36, 489–494.

Baker, C.S. & Herman, L.M. (1989). *Behavioral responses of summering humpback whales to vessel traffic: Experimental and opportunistic observations*. Technical report NPS-NR-TRS89-01. Anchorage, AK: National Park Service, Alaska Regional Office.

Beale, C.M. & Monaghan, P. (2004). Behavioural responses to human disturbance: A matter of choice? *Animal Behaviour* 68, 1065–1069.

Bearzi, G. (2003). At home with the dolphins. In T. Frohoff & B. Peterson (Eds), *Between Species: Celebrating the dolphin–human bond*. San Francisco, CA: Sierra Club Books, pp. 104–109.

Bejder, L., Dawson, S.M. & Harraway, J.A. (1999). Responses by Hector's dolphins to boats and swimmers in Porpoise Bay, New Zealand. *Marine Mammal Science* 15, 738–750.

Bejder, L., Samuels, A., Whitehead, H., *et al.* (2006). Decline in relative abundance of bottlenose dolphins exposed to long-term disturbance. *Conservation Biology* 20, 1791–1798.

Bejder, L., Samuels, A., Whitehead, H., Finn, H. & Allen, S. (2009). Impact assessment research: Use and misuse of habituation, sensitisation and tolerance to describe wildlife responses to anthropogenic stimuli. *Marine Ecology Progress Series* 395, 177–185.

Boye, T.K., Simon, M. & Madsen, P.T. (2010). Habitat use of humpback whales in Godthaabsfjord, West Greenland, with implications for commercial exploitation. *Journal of the Marine Biological Association of the United Kingdom* 90(8), 1529–1538.

Constantine, R. (2001). Increased avoidance of swimmers by wild bottlenose dolphins (*Tursiops truncatus*) due to long-term exposure to swim-with-dolphin tourism. *Marine Mammal Science* 17(4), 689–702.

Corkeron, P.J. (2004). Whalewatching, iconography, and marine conservation. *Conservation Biology* 18, 847–849.

Duffus, D.A. & Dearden, P. (1990). Non-consumptive wildlife-oriented recreation: A conceptual framework. *Biological Conservation* 53(3), 213–231.

Ellenberg, U., Mattern, T., Seddon, P.J. & Jorquera, G.L. (2006). Physiological and reproductive consequences of human disturbance in Humboldt penguins: The need for species-specific visitor management. *Biological Conservation* 133(1), 95–106.

Foote, A.D., Osborne, R.W. & Hoelzel, A.R. (2004). Whale-call response to masking boat noise. *Nature* 428, 910.

Frid, A. & Dill, L. (2002). Human-caused disturbance stimuli as a form of predation risk. *Conservation Ecology* 6, 11–26.

Garrod, B. & Fennell, D.A. (2004). An analysis of whale-watching codes of conduct. *Annals of Tourism Research* 31(2), 334–352.

Gilders, M.A. (1995). *Reflections of a Whale-Watcher*. Bloomington, IN: Indiana University Press.

Gill, J.A., Norris, K. & Sutherland, W.J. (2001). Why behavioural responses may not reflect the population consequences of human disturbance. *Biological Conservation* 97, 265–268.

Gjerdalen, G. & Williams, P. (2000). An evaluation of the utility of a whale watching code of conduct. *Tourism Recreation Research* 25(2), 27–37.

Gössling, S. & Hall, C.M. (Eds). (2006). *Tourism and Global Environmental Change. Ecological, social, economic and political interrelationships*. London: Routledge.

Guba, E.G. (1990). The alternative paradigm dialog. In E.G. Guba (Ed.), *The paradigm dialog*. London: Sage, pp. 17–27.

Hall, C.M. & Boyd, S. (2003). *Ecotourism in Peripheral Areas*. Clevedon, UK: Channel View Publications.

Hastie, G.D., Wilson, B., Tufft, L.H. & Thompson P.M. (2003). Bottlenose dolphins increase breathing synchrony in response to boat traffic. *Marine Mammal Science* 19, 74–84.

Higham, J.E.S. & Bejder, L. (2008). Managing wildlife-based tourism: Edging slowly towards sustainability? *Current Issues in Tourism* 11(1), 63–74.

Higham, J.E.S. & Lusseau, D. (2004). Ecological impacts and management of tourist engagements with cetaceans. In R. Buckley (Ed.), *Environmental Impacts of Ecotourism*. Wallingford: CAB International.

Higham, J.E.S. & Lusseau, D. (2007). Urgent need for empirical research into whaling and whale-watching. *Conservation Biology* 21(2), 554–558.

Higham, J.E.S. & Lusseau, D. (2008). Slaughtering the goose that lays the golden egg: Are whaling and whale-watching mutually exclusive? *Current Issues in Tourism* 11(1), 63–74.

Higham, J.E.S. & Shelton, E. (2011). Tourism and wildlife habituation: Reduced population fitness or cessation of impact? *Tourism Management* 32(4), 1290–1298.

Higham, J.E.S., Bejder, L. & Lusseau, D. (2009). An integrated and adaptive management model to address the long-term sustainability of tourist interactions with cetaceans. *Environmental Conservation* 35, 294–302.

Hinch, T.D. (1998). Ecotourists and indigenous hosts: Diverging views on their relationship with nature. *Current Issues in Tourism* 1(1), 120–124.

Hoffman, M., Brooks, T.M., de Fonseca, G.A.B., *et al.* (2008). Conservation planning and the IUCN Red List. *Endangered Species Research* 6, 113–125.

Hooker, S.K. & Gerber, L.R. (2004). Marine reserves as a tool for ecosystem-based management: The potential importance of megafauna. *Bioscience* 54, 27–39.

Hoyt, E. (2011). *Marine Protected Areas for Whales, Dolphins and Porpoises* (2nd edn). London: Earthscan.

Jett, J.S. & Thapa, B. (2010). Manatee zone compliance among boaters in Florida. *Coastal Management* 38, 165–185.

Knight, J. (2009). Making wildlife viewable: Habituation and attraction. *Society and Animals* 17, 167–184.

Lemelin, R.H. (2006). The gawk, the glance, and the gaze: Ocular consumption and polar bear tourism in Churchill, Manitoba, Canada. *Current Issues in Tourism* 9(6), 516–534.

Levine, H.B. & Levine, M.W. (1987). *Steward Island: Anthropological perspectives on a New Zealand fishing community*. Wellington: Victoria University of Wellington.

Lück, M. (2003). Environmentalism and on tour experiences on wildlife watch tours in New Zealand: A study of visitors watching and/or swimming with dolphins. PhD Thesis, University of Otago, Dunedin, New Zealand.

Lusseau, D. (2003). Male and female bottlenose dolphins *Tursiops* sp. have different strategies to avoid interactions with tour boats in Doubtful Sound, New Zealand. *Marine Ecology Progress Series* 257, 267–274.

Lusseau, D. (2005). The residency pattern of bottlenose dolphins (*Tursiops* spp.) in Milford Sound, New Zealand is related to boat traffic. *Marine Ecology Progress Series* 295, 265–272.

Lusseau, D. (2006). The short-term behavioral reactions of bottlenose dolphins to interactions with boats in Doubtful Sound, New Zealand. *Marine Mammal Science* 22, 802–818.

Lusseau, D. & Higham, J.E.S. (2004). Managing the impacts of dolphin-based tourism through the definition of critical habitats: The case of bottlenose dolphins (*Tursiops* spp.) in Doubtful Sound, New Zealand. *Tourism Management* 25(5), 657–667.

Lusseau, D., Slooten, E. & Currey, R.J.C. (2006). Unsustainable dolphin-watching tourism in fiordland, New Zealand. *Tourism in Marine Environments* 3, 173–178.

Lusseau, D., Bejder, L., Corkeron, P., Allen, S. & Higham, J.E.S. (2013). Learning from past mistakes: A new paradigm for managing whale-watching. *Conservation Letters* (under review).

Milne, S. & Ateljevic, I. (2004). Tourism economic development and the global–local nexus. In S. Williams (Ed.), *Tourism: Critical concepts in the social sciences*. New York, NY: Routledge, pp. 81–103.

Muloin, S. (1998). Wildlife tourism: The psychological benefits of whale-watching. *Pacific Tourism Review* 2, 199–213.

Mustika, P.L.K., Birtles, A., Everingham, Y. & Marsh, H. (2012). The human dimensions of wildlife tourism in a developing country: Watching spinner dolphins at Lovina, Bali, Indonesia. *Journal of Sustainable Tourism* 20(2), 1–23.

Neves, K. (2010). Cashing in on cetourism: A critical ecological engagement with dominant E-NGO discourses on whaling, cetacean conservation, and whale watching. *Antipode* 42(3), 719–741.

Nowacek, S.M., Wells, R.S. & Solow, A.R. (2001). Short-term effects of boat traffic on bottlenose dolphins, *Tursiops truncatus*, in Sarasota Bay, Florida. *Marine Mammal Science* 17, 673–688.

Ris, M. (1993). Conflicting cultural values: Whale tourism in Northern Norway. *Arctic* 46(2), 156–163.

Sawada, H. & Minami, H. (1997). Peer group play and co-childrearing in Japan: A historical ethnography of a fishing community. *Journal of Applied Developmental Psychology* 18(4), 513–526.

Servais, V. (2005). Enchanting dolphins: An analysis of human–dolphin encounters. In J. Knight (Ed.), *Animals in Person: Cultural perspectives on human–animal intimacies*. Oxford: Berg, pp. 21–229.

Shelton, E.J. & McKinlay, B. (2007). Shooting fish in a barrel: Tourists as easy targets. In J.E.S. Higham & M. Lück (Eds), *Marine Wildlife and Tourism Management: Insights from the natural and social sciences*. Wallingford: CABI, pp. 219–231.

Spong, P. & Symonds, H. (2003). The ocean's chalk circle. In T. Frohoff & B. Peterson (Eds.), *Between Species: Celebrating the dolphin–human bond*. San Francisco, CA: Sierra Club Books, pp. 311–321.

Tilman, D. (1999). The ecological consequences of changes in biodiversity: A search for general principles. *Ecology* 80(5), 1455–1474.

Van Parijs, S.M. & Corkeron, P.J. (2001). Boat traffic affects the acoustic behaviour of Pacific humpback dolphins, *Sousa chinensis*. *Journal of the Marine Association UK* 81, 533–538.

Williams, R., Trites, A.W. & Bain, D.E. (2002). Behavioural responses of killer whales (*Orcinus orca*) to whale-watching boats: Opportunistic observations and experimental approaches. *Journal of Zoology* 256, 255–270.

Williams, R., Lusseau, D. & Hammond, P. (2006). Estimating relative energetic costs of human disturbance to killer whales (*Orcinus orca*). *Biological Conservation* 133(3), 301–311.

Williams, R., Lusseau, D. & Hammond, P. (2009). The role of social aggregations and protected areas in killer whale conservation: The mixed blessing of critical habitat. *Biological Conservation* 142, 709–719.

Wilson, B., Reid, R.J., Grellier, K. & Thompson, P. (2004). Considering the temporal when managing the spatial: A population range expansion impacts protected areas-based management for bottlenose dolphins. *Animal Conservation* 7, 331–338.

Worm, B., Hilborn, R., Baum, J.K., *et al.* (2009). Rebuilding global fisheries. *Science* 325, 578–585.

Index

Printed in the United States
by Baker & Taylor Publisher Services